Essentials of Engineering Hydraulics

Macmillan International College Edition

Titles of related interest:

J A Fox: *An Introduction to Engineering Fluid Mechanics*

B Henderson-Sellers: *Reservoirs*

N T Kottegoda: *Stochastic Water Resources Technology*

Essentials of Engineering Hydraulics

JONAS M. K. DAKE
B.Sc (Eng.) (London); M.Sc.Tech. (Man.); Sc.D. (M.I.T.)

ANSTI

First edition 1972
Reprinted with corrections 1974
Second edition 1983

Published by
THE MACMILLAN PRESS LTD
London and Basingstoke
Companies and representatives throughout the world.

In association with;
African Network of Scientific and Technological Institutions

P.O. Box 30592
Nairobi
Kenya

ISBN 333 34334 4 (cased)
333 34335 2 (paperback)

Typeset by MULTIPLEX techniques ltd

Printed in Hong Kong

Contents

4 Flow in non-erodible open channels

5 Experimental fluid mechanics

6 Water pumps and turbines

PART TWO *SPECIALIZED TOPICS IN CIVIL ENGINEERING*

7 Flow in erodible open channels

8 Physical hydrology and water storage

9 Groundwater and seepage

10 Sea waves and coastal engineering

11 Fundamental economics of water resources development

Problems

Appendix: Notes on Flow Measurement

Foreword to the First Edition

by

J. R. D. Francis, B.Sc. (Eng.), M.Sc., M.I.C.E., F.R.Met.S.
Professor of Fluid Mechanics and Hydraulic Engineering,
Imperial College of Science and Technology, London

It is a pleasure to have the opportunity of commending this book. The author, a friend and former student of mine, has attempted to bring out the principles of physics which are likely to be of future importance to hydraulic engineering science, with particular reference to water resources problems. With the greater importance and complexity of water resource exploitation likely to occur in the future, our analysis and design of engineering problems in this field must become more exact, and there are several parts of Dr. Dake's book which introduce new ideas. In the past half-century, the science of fluid mechanics has been largely dominated by the demands of aeronautical engineering; in the future it is not too much to believe that the efficient supply, distribution, drainage and re-use of the world's water supply for the benefit of an increasing population will present the most urgent of problems to the engineer.

I feel particularly honoured, too, in that this book must be among the first technical texts to come from a young and flourishing university, and is, I think, the first in hydraulic engineering to come from Africa. Over many years, academics in Britain and elsewhere have attempted, with varying success, to help the establishment of degree courses at Kumasi, and to produce skilled technological manpower. That a book of this standard should now come forward is a source of pleasure to all those who have helped, and an indication of future success.

J. R. D. FRANCIS
1972

Preface to the Second Edition

The Second Edition of *Essentials of Engineering Hydraulics* has retained the primary objectives and structure of the original book. However, the rational metric system of units (Système International d'Unités) has been adopted generally although a few examples and approaches have retained the imperial units.

The scope of the book has been increased by inclusion of section 1.8, '*Fluids in Static Equilibrium*' and sub-sections 8.7.3 and 9.3.5 '*Routeing of Floods in River Channels*' and '*The Transient State of the Well Problem*', respectively. There has been general updating.

A guide to the solution of the tutorial problems at the end of the book is available for restricted distribution to lecturers upon official request to the publisher.

Jonas M. K. Dake
Nairobi 1982

Preface to the First Edition

Teaching of engineering poses a challenge which, although also relevant to the developed countries, carries with it enormous pressures in the developing countries. The immediate need for technical personnel for rapid development and the desire to design curricula and training methods to suit particular local needs provide strong incentives which could, without proper control, compromise engineering science and its teaching in the developing world.

The generally accepted role of an engineering institution is the provision of the scientific foundation on which the engineering profession rests. It is also recognized that the student's scientific background must be both basic and environmental. In other words, engineering syllabuses must be such that, while not compromising on basic engineering science and standards, they reflect sufficient background preparation for the appropriate level of local development.

This text has been written to provide in one volume an adequate coverage of the basic principles of fluid flow and summaries of specialized topics in hydraulic engineering, using mainly examples from African and other developing countries.

A survey of fluid mechanics and hydraulics syllabuses in British universities reveals that the courses are fairly uniform up to second year level but vary widely in the final year. This book is well suited to these courses. Students in those universities which emphasize civil engineering fluid mechanics will also find this book useful throughout the whole or considerable part of their courses of study.

Essentials of Engineering Hydraulics can be divided into two parts. Part I, Elementary Fluid Mechanics, emphasizes fundamental physical concepts and details of the mechanics of fluid flow. A good knowledge of general mechanics and mathematics as well as introductory lectures in fluid mechanics covering hydrostatics and broad definitions are assumed. Coverage in Part I is suitable up to the end of the second year (3-year degree courses) or third year (4-year degree courses) of civil and mechanical engineering undergraduate studies. Part II on Specialized Topics in Civil Engineering is meant mainly for final-year civil engineering degree students. Treatment is concentrated on discussions of the physics and concepts which have led to certain mathematical results. Equations are generally not derived but discussions centre on the merits and limitations of the equations.

The general aim of the book is to emphasize the physical concepts of fluid flow and hydraulic engineering processes with the hope of providing a foundation which is suitable for both academic and non-academic postgraduate work. Towards this end, serious efforts have been made to steer a middle course between the thorough mathematical approach and the strictly down-to-earth empirical approach.

Chapter 11 gives an introduction to the fundamental economics of water resources development which is a very important topic at postgraduate level. I feel that economics and decision theory must be given more prominence in undergraduate engineering curricula especially in countries where young graduates soon find themselves propelled to positions of responsibility and decision making.

In an attempt to make this book comprehensive and yet not too bulky and expensive, I have resorted to a literary style which uses terse but scientific words with the hope of putting the argument in a short space. I have also followed rather the classroom 'hand-out' approach than the elaborate and sometimes long-winded approach found in many books.

'The author of any textbook depends largely upon his predecessors' – Francis. Existing books and other publications from which I have benefited are listed at the end of each chapter in acknowledgement and as further references for the interested reader. The tutorial problems have been derived from my own class exercises, homework and class tests at M.I.T. and from other sources, all of which are gratefully acknowledged. In the final chapter, problems 3.18, 4.23, 4.24, 5.18, 5.19, 6.3, 6.14, 8.1, 8.2 and 8.3 have been included with the kind permission of the University of London. All statements in the text and answers to problems, however, are my responsibility.

I wish to thank Prof. J. R. D. Francis of Imperial College, London and Dr. J. O. Sonuga of Lagos University and other colleagues who read the manuscript in part or whole and made many useful suggestions. The encouragement of Prof. Francis, a former teacher with continued interest in his student and the external examiner in Fluid Mechanics and Hydraulics as well as the moderator for Civil Engineering courses at U.S.T., has been invaluable. Mr. D. W. Prah of the Department of Liberal and Social Studies, U.S.T., made some useful comments on the use of economic terms in Chapter 11. The services of the clerical staff and the draughtsmen of the Faculty of Engineering, U.S.T., especially of Messrs. S. K. Gaisie and S. F. Dadzie during the preparation of the manuscript and drawings are also gratefully acknowledged.

Finally, I wish to express my sincere gratitude to the University of Science and Technology, Kumasi, whose financial support has made the production of this book possible.

University of Science and Technology, Kumasi, Ghana Jonas M. K. Dake
1972

List of Principal Symbols

A_j	cross-section area of a jet (L^2)
A	area (L^2)
a	acceleration (L/t^2), area (L^2), wave amplitude (L)
a_b	amplitude of wave beat envelope (L)
B	top width of a channel (L)
b	bottom width of a channel (L)
C	Chezy coefficient ($L^{\frac{1}{2}}/t$); wave velocity (L/t)
C_G	group velocity (waves) (L/t)
C_p	specific heat at constant pressure (L^2/Tt^2)
C.R.F.	Capital Recovery Factor
C_v	specific heat at constant volume (L^2/Tt^2)
c	concentration of mass, surge wave speed (L/t), speed of sound (L/t)
C_D	coefficient of drag
C_d	coefficient of discharge
c_f	coefficient of drag (friction)
c.s.	control surface
c.v.	control volume
C_v	coefficient of velocity
D	molecular mass conductivity (M/Lt), pipe diameter (L), drag (ML/t^2)
D_N	sieve diameter which pass N% of soil sample (L)
D_{50}	median sand particle size (L)
d	geometric mean size (sand) (L), depth (L), drawdown (L)
d_g	geometric mean size (sand) (L)
d_n	diameter of a nozzle (L)
E	energy (ML^2/t^2), specific energy (L), Euler number, rate of evaporation (L/t); wave energy (M/t^2), modulus of elasticity (M/Lt^2)
$\dot{E}$	rate of transmission of wave energy (ML/t^3)
E_h	thermal eddy diffusivity (L^2/t)
E_m	mass eddy diffusivity (L^2/t)
E_k	kinetic energy (ML^2/t^2)
E_p	potential energy (ML^2/t^2)
e	exponential constant (=2.71828)
e	vapour pressure (mmHg), void ratio
F	force (ML/t^2), Froude number, fetch (L)
F'	densimetric Froude number
f	friction factor
f_p	infiltration capacity (L/t or L^3/t)
f_s	silt factor
g	acceleration due to gravity (L/t^2)
g_0	constant – 32.174 lbm/slug
H	enthalpy (L^2/t^2), total head (L), wave height (L)
H_m	head developed or consumed by a rotodynamic machine (L)
H_p	pump head (L)
H_r	theoretical head of a rotodynamic machine (L)

H_s	static lift (L)
H_{sv}	net positive suction head (NPSH) (L)
H_T	turbine head (L)
h	head of water above spillway crest (L), hydraulic head (L)
h_f	friction head loss (L)
I	moment of inertia (ML^2), infiltration amount (L^3), rate of interest
i	seepage (hydraulic) gradient (L/L)
i_0	rainfall intensity (L/t)
J	mechanical equivalent of heat, (ML^2/t^2)
K	thermal molecular conductivity (ML/Tt^3), coefficient of hydraulic resistances, modulus of compressibility (M/Lt^2)
K_n	nozzle (loss) coefficient
K_r	coefficient of wave refraction
k	coefficient of permeability (superficial) (L/t); wave number $(2\pi/L)$
k_s	size of roughness (L)
L	length (L), wavelength (L)
L_0	wavelength in deep water (L)
l	length (L)
lbm	pound mass (M)
lbf	pound force (ML/t^2)
M	mass (M), Mach number
MB	marginal benefit
MC	marginal cost
MP	marginal productivity
MRS	marginal rate of substitution
MRT	marginal rate of transformation
m	mass (M), mass rate of flow (M/t), hydraulic mean depth or radius (L)
N	speed of rotation (rev/min)
N_s	specific speed (turbines)
N_u	unit speed (rotodynamic machines)
n	porosity, ratio of wave group velocity to phase velocity (C_G/C)
n_s	specific speed (pumps)
O	outflow (L^3/t)
$\bar{O}$	average outflow (L^3/t)
OMR	operation, maintenance and repairs
P	force (ML/t^2), wetted perimeter (L), power (ML^2/t^3), principal investment, precipitation (rainfall)
P_u	unit power
p	pressure (M/Lt^2)
ppm	parts per million
p_{at}	atmospheric pressure (M/Lt^2)
p_v	vapour pressure (M/Lt^2)
Q	discharge rate (L^3/t), heat (ML^2/t^2)
Q_u	unit discharge
q	discharge per unit width (L^2/t)
$\mathbf{q}$	velocity vector (L/t)
q_b	rate of bed load transport per unit width (L^2/t)
q_s	rate of suspended load transport per unit width

List of Principal Symbols

R	universal gas constant ($L^2/t^2 T$), Reynolds number, rainfall amount (L^3)
R'	Reynolds number based on shear velocity ($v'd/\upsilon$)
R_i	Richardson number
°R	degrees Rankine
r	radius (L), (suffix) ratio of model quantity/prototype quantity
S	specific gravity, storage (L^3), degree of saturation, storage constant
S_f	slope of energy grade line (L/L)
SF	shape factor (sand grains)
S_F	flow net shape factor
S_0	bed slope (L/L)
S_s	specific gravity of solids
T	temperature (T), wave period (t), torque (ML^2/t^2), transmissibility (L^2/t)
t	time (t), wind duration (t)
t_c	time of concentration (t)
t_p	recurrence interval (t)
t_r	duration of rainfall (t)
U	internal energy (ML^2/t^2), wind speed (L/t)
u	specific internal energy (L^2/t^2)
V	volume (L^3)
V_v	volume of voids (L^3)
v	velocity (L/t)
$\bar{v}$	time average velocity (turbulent flow) (L/t) (sections 1.5.4, 4.1.3, 7.2.3); sectional average velocity (sections 3.1, 3.2, 3.3, 3.4.1, 3.4.2)
v_f	radial (flow) component of velocity in a rotodynamic machine (L/t)
v_n	velocity of nozzle jet (L/t)
v_s	seepage velocity (ch. 9) (L/t)
v'	shear velocity $\sqrt{(T_0/\rho)}$ (L/t)
v_w	absolute surge wave speed (L/t), rotodynamic whirl component of velocity (L/t)
W	weight (ML/t^2), Weber number, work (ML^2/t^2)
W_p	work against pressure (ML^2/t^2)
W_s	work against shear (ML^2/t^2)
W_{sh}	shaft work (ML^2/t^2)
w	settling velocity (sand particles) (L/t)
y	distance measured from wall (L)
y_c	critical depth (L)
y_n	uniform (normal) depth (L)
y_0	depth (generally) in an open channel (L)
$\bar{y}$	centroid of section measured from water surface (L)
y'	centre of pressure measured from water surface (L)
ΔZ or Z_0	height of weir (L)
Σ	summation of
$\rightarrow$	approaches (equivalent or equal to)
α	thermal molecular diffusivity (L^2/t), angle
α_m	mass molecular diffusivity (L^2/t)
β	angle, constant of proportionality in $\epsilon_s = \beta\epsilon$
γ	specific (unit) weight (M/L^2t^2)
γ_s	specific weight of solid matter (M/L^2t^2)
δ	boundary layer thickness (L)
ϵ	eddy kinematic viscosity (L^2/t)
ϵ_s	eddy diffusity for suspended load (L^2/t)
θ	angle, temperature (T)

List of Principal Symbols

η	efficiency, small amplitude wave form (L)
η_h	hydraulic efficiency
η_m	mechanical efficiency
μ	dynamic molecular viscosity (M/Lt), discharge factor
υ	kinematic molecular viscosity (L^2/t)
ρ	density (M/L^3)
ρ_s	density of solid matter (M/L^3)
σ	surface tension (M/t^2), standard deviation, wave number ($2\pi/\tau$)
σ_c	critical cavitation number
σ_g	geometric standard deviation
τ	shear stress (M/Lt^2)
τ_c	critical shear stress (M/Lt^2)
τ_0	wall shear stress (M/Lt^2)
ω or Ω	angular velocity (rad/t)
$\mathcal{C}$	Cauchy number

PART ONE:
Elementary Fluid Mechanics

1 Fundamental Concepts of Fluid Mechanics

1.1 Introduction

Matter is recognized in nature as solid, liquid or gas (or vapour). When it exists in a liquid or gaseous form, matter is known as a *fluid*. The common property of all fluids is that they must be bounded by impermeable walls in order to remain in an initial shape. If the restraining walls are removed the fluid flows (expands) until a new set of impermeable boundaries is encountered. Provided there is enough fluid or it is expandable enough to fill the volume bounded by a set of impermeable walls, it will always conform to the geometrical shape of the boundaries. In other words, a fluid by itself offers no lasting resistance to change of shape. The essential difference between a liquid and a gas is that a given mass of the former occupies a fixed volume at a given temperature and pressure whereas a fixed mass of a gas occupies any available space. A liquid offers great resistance to volumetric change (compression) and is not greatly affected by temperature changes. A gas or a vapour, on the other hand, is easily compressed and responds markedly to temperature changes.

The above definitions and observations indicate that the ultimate shape and size of a fixed mass of a fluid under a deforming force depend on the geometry of the container and on the compressibility of the fluid. The rate at which a fluid assumes the new shape is governed by the property known as viscosity. Viscosity is a molecular property of a fluid which enables it to resist rapid deformation and this is discussed more fully in Section 1.4.

1.2 The Continuum

Fluids are composed of discretely spaced molecules in constant motion and collision. An exact analysis of fluid motion would therefore require knowledge of the behaviour of each molecule or group of molecules in motion. In most engineering problems, however, average measurable indications of the general behaviour of groups of molecules are sufficient. These indications can be conveniently assumed to arise from a continuous distribution of molecules, referred to as the *continuum*, instead of from the conglomeration of discrete molecules that exists in reality.

Thus in fluid mechanics generally, the terms density, pressure, temperature, viscosity, velocity, etc. refer to the average manifestation of these quantities at a point in the fluid as opposed to the individual behaviour of individual molecules or particles. The adoption of the continuum model implies that all dimensions in the fluid space are very large compared to the molecular mean free path (the average distance traversed by the molecules between collisions). It also implies that all properties of the fluid are continuous from point to point throughout a given volume of the fluid.

1.3 Units of Measurement

Two main systems of measurement are used in engineering; the metric system and the Imperial system. The metric system mainly will be used in this book but examples based on the Imperial system are also included. The general analytical principles are the same and the student should be familiar with both systems. The SI system is basically a refined metric (mks) system.

As shown in Table 1.1 both the metric system of measurement and the imperial system of measurement have two subdivisions. The difference lies in the units used for measuring mass and sometimes length. In the metric cgs (centimetre-gramme-second) system the gramme is used as a unit of mass and the centimetre as a unit of length. In the mks (metre-kilogramme-second) system the kilogramme and the metre are used respectively. Both use the second as the basic

Table 1.1
Units of measurement

The metric system

	cgs	*SI (mks)*
Mass	gramme (g)	kilogramme (kg)
Length	centimetre (cm)	metre (m)
Time	second (s)	second (s)
Force	dyne (dyne)	newton (N)
Temperature	degree Kelvin (K)	degree Kelvin (K)

The Imperial system

	British absolute	*Engineers'*
Mass	pound mass (lbm)	slug (slug)
Length	foot (ft)	foot (ft)
Time	second (s)	second (s)
Force	poundal (pdl)	pound force (lbf)
Temperature	degree Rankine (°R)	degree Rankine (°R)

unit of time. In the Imperial system the pound mass is the unit of mass when the so-called British absolute system is adopted and the slug is the unit in the so-called engineers' system. In both cases the unit of length is the foot and the unit of time is the second.

Newton's second law of motion states that a mass moving by virtue of an applied force will accelerate so that the product of the mass and acceleration equals the component of force in the direction of acceleration. In symbols,

$$F = ma \tag{1.1}$$

where F is force, m is mass and a is acceleration.

By definition if the mass is 1 gramme and the acceleration is 1 cm/s^2, the force is 1 dyne. Similarly 1 newton of force produces an acceleration of 1 m/s^2 in 1 kg mass; 1 poundal of force accelerates a pound mass, 1 ft/s^2 and a mass of 1 slug will require 1 lbf to produce an acceleration of 1 ft/s^2.

Supposing one slug is equivalent to g_0 pound mass. From equation (1.1)

$$1 \text{ lbf} = 1 \text{ slug} \times 1 \text{ ft/s}^2$$

or

$$1 \text{ lbf} = g_0 \text{ lbm} \times 1 \text{ ft/s}^2 \tag{1.2}$$

But the pound force is defined in terms of the pull of gravity, at a specified (standard) location on the earth, on a given mass of platinum. One pound mass experiences a pull of 1 lbf due to standard gravitational acceleration of $g = 32{\cdot}174$ ft/s^2. Thus

$$1 \text{ lbf} = 1 \text{ lbm} \times 32{\cdot}174 \text{ ft/s}^2 \tag{1.3}$$

Since equations (1.2) and (1.3) define the same quantity, 1 lbf, it is obvious that $g_0 = 32{\cdot}174$ lbm/slug. This development shows that g_0 is a constant and is not necessarily equal to g which varies from location to location. This should be expected since g_0 relates two mass units and a given mass is the same anywhere in the universe. The gravitational pulling force on the same mass however varies from place to place. For instance, a 10 lbm will weigh 10 lbf under standard gravitational pull. The weight of the same 10 lbm under any other gravitational field, g, is given by

$$W(\text{lbf}) = \frac{10 \text{ (lbm)}}{g_0 \text{ (lbm/slug)}} g(\text{ft/s}^2)$$

where $g = 31{\cdot}0$ ft/s^2

$$W = \frac{10}{32{\cdot}174} \times 31{\cdot}0 = 9{\cdot}635 \text{ lbf}$$

Table 1.2 lists the SI equivalents of some of the units commonly used in fluid mechanics.

Table 1.2
SI equivalents of other Units

Physical quantity	*Unit*	*Equivalent*
Length	angstrom	10^{-10} m
	inch	0·0254 m
	foot	0·3048 m
	yard	0·9144 m
	mile	1·609 34 km
	nautical mile	1·853 18 km
Area	square inch	645·16 mm^2
	square foot	0·092 903 m^2
	square yard	0·836 127 m^2
	square mile	2·589 99 km^2
Volume	cubic inch	1·638 71 × 10^{-5} m^3
	cubic foot	0·028 316 8 m^3
	U.K. gallon (imperial)	0·004 546 092 m^3
	U.S. gallon	0·003 785 4 m^3
Mass	pound	0·453 592 37 kg
	slug	14·593 9 kg
Density	pound/cubic foot	16·0185 kg m^{-3}
Force	dyne	10^{-5} N
	poundal	0·138 255 N
	pound-force	4·448 22 N
	kilogramme-force	9·806 65 N
Pressure	atmosphere	101·325 kN m^{-2}
	torr	133·322 N m^{-2}
	pound(f)/sq. in.	6894·76 N m^{-2}
Energy	erg	10^{-7} J
	calorie (I.T.)	4·1868 J
	calorie (15°C)	4·1855 J
	calorie (thermochemical)	4·184 J
	B.t.u.	1055·06 J
	foot poundal	0·042 140 1 J
Energy	foot pound (f)	1·355 82 J
Power	horse power	745·700 W
Temperature	degree Rankine	5/9 K
	degree Fahrenheit	$t/°F = \frac{9}{5}T/°C + 32$

There are two convenient ways of converting units of measurement from one system to another. One is the method of *dimensional representation* and the other a technique of forming the ratio of a unit and the proper numerical value of another unit such that there is physical equivalence between the quantities. There are four basic dimensions and in fluid mechanics the properties of fluids can be expressed in terms of these basic dimensions of mass (M), length (L), time (t) and temperature (T). In Table 1.3 some important properties commonly met in fluid mechanics are listed together with their appropriate dimensions and the relevant engineers' units and SI units.

Table 1.3
Dimensions and units of physical quantities

Property	*Symbol*	*Dimensions*	*Imperial (engineers') units*	*SI units*
Length	Primary dimensions	L	ft	m
Mass		M	slug	kg
Time		t	s	s
Temperature		T	°F or °R	K
Velocity	v	L/t	ft/s	m/s
Acceleration	a	L/t^2	ft/s²	m/s²
Force	F	ML/t^2	lbf	N
Pressure	p	$\frac{M}{Lt^2}$	lbf/ft² or lbf/in²	N/m²
Density	ρ	$\frac{M}{L^3}$	slug/ft³	kg/m³
Specific weight	γ	$\frac{M}{L^2t^2}$	lbf/ft³	kg/(m² s²)
Viscosity (dynamic)	μ	$\frac{M}{Lt}$	slug/ft s or lbf s/ft²	N s/m²
Viscosity (kinematic)	υ	$\frac{L^2}{t}$	ft²/s	m²/s
Modulus of elasticity	E	$\frac{M}{Lt^2}$	lbf /ft² or lbf/in²	N/m²
Surface tension	σ	$\frac{M}{t^2}$	lbf/ft	N/m
Universal gas constant	R	$\frac{L^2}{t^2T}$	ft lbf/(mass °R)	J/(kg K)
Specific heat at: constant pressure, constant volume	C_p, C_v	$\frac{L^2}{t^2T}$	Btu/(mass °R)	J/(kg K)
Specific internal energy	u	$\frac{L^2}{t^2}$	Btu/mass	J/kg

As an example, the conversion factor between the engineers' units and SI units of pressure will be derived both ways. First the conversion factor is obtained by writing pressure dimensionally, substituting basic units of the Imperial system and changing these units to equivalent metric units. Using Tables 1.2 and 1.3

$$(p) = \left(\frac{M}{Lt^2}\right) = \left(\frac{\text{slug}}{\text{fts}^2}\right) \equiv \frac{32{\cdot}174\ (\text{lbm})}{2{\cdot}20\ (\text{lbm/kg})} \times \frac{1}{0{\cdot}305\ (\text{m})} \times \frac{1}{s^2}$$

Hence $$1\left(\frac{\text{lbf}}{\text{ft}^2}\right) \equiv 47{\cdot}9\ \frac{\text{kg}}{\text{ms}^2} = 47{\cdot}9\left(\frac{\text{newton}}{\text{m}^2}\right)$$

Alternatively $$(p) = \left(\frac{\text{lbf}}{\text{ft}^2}\right) \equiv \frac{(\text{lbf})(4{\cdot}45\ \text{newton/lbf})}{(\text{ft}^2)(0{\cdot}305\ \text{m/ft})^2}$$

Hence $$1\left(\frac{\text{lbf}}{\text{ft}^2}\right) \equiv 47{\cdot}9\left(\frac{\text{newton}}{\text{m}^2}\right)$$

The latter method is generally more intuitive and simpler to apply.

1.4 Some Important Fluid Properties

It has been stressed in discussing the concept of the continuum in Section 1.2 that the properties of materials normally employed in solving engineering

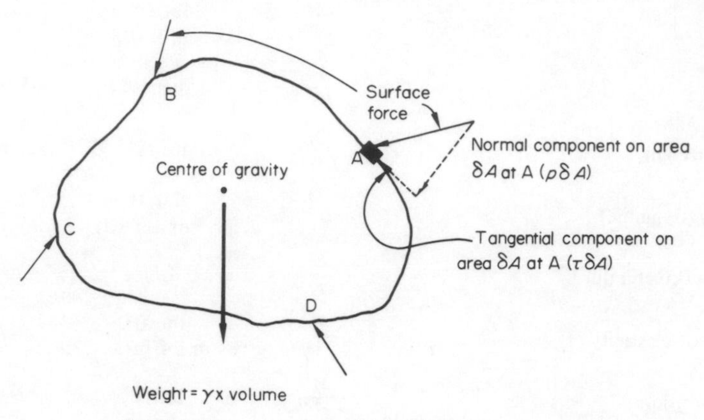

Fig. 1.1 Forces on an isolated free body of matter

problems are average manifestations of general molecular activity within the matter and these are considered to be continuous in material space. If a volume of matter is isolated as a free body (see Fig. 1.1) the force system acting on the volume is made up of a *body force* due to the gravitational pull on the mass contained in the volume and *surface forces* over every element of area bounding the volume. The gravitation pull per unit volume of the matter is known as the *specific (unit) weight* γ (force/unit volume). Thus specific weight depends on the gravitational field in contrast to *density or specific mass* ρ (mass/unit volume) which is invariant so long as the volume of the given mass is not increased or reduced.

Surface forces, in general, will have components normal and tangential to the bounding surface. The normal component per unit of area is called the normal stress. In fluids the stress is always considered compressive and is called *pressure*. Pressure is a scalar quantity but the force associated with it is a vector quantity which is always directed normal to the surface over which it acts. Pressure when measured relative to atmospheric pressure is called gauge pressure but relative to absolute zero it is called absolute pressure. The tangential component of the surface force per unit area makes up what is known as *shear stress.*

1.4.1 Viscosity and Shear Stress

Shear stress is evident in the structure of any substance whose successive layers are shifted laterally over each other. The existence of velocity changes in a particular direction, therefore, produces shear in a direction perpendicular to the direction of velocity changes. In certain fluids commonly known as *Newtonian fluids*, the shear stress on the interface tangential to the direction of flow is proportional to the gradient (rate of change with distance) of velocity in a direction normal to the interface. In mathematical terms,

$$\tau_{ns} = \mu \frac{\partial v}{\partial n} \tag{1.4}$$

where

τ_{ns} = the shear stress acting in the s direction in a plane normal to the n direction
v = velocity
μ = the coefficient of proportionality called *dynamic viscosity.*

In Fig. 1.2 the relationship expressed in equation (1.4) is illustrated. The use of the partial derivative $\partial v/\partial n$ emphasizes the point that the velocity v may be variable in all directions of space and with time but it is its gradient normal to a particular plane which produces shear on that plane. The direction of the velocity

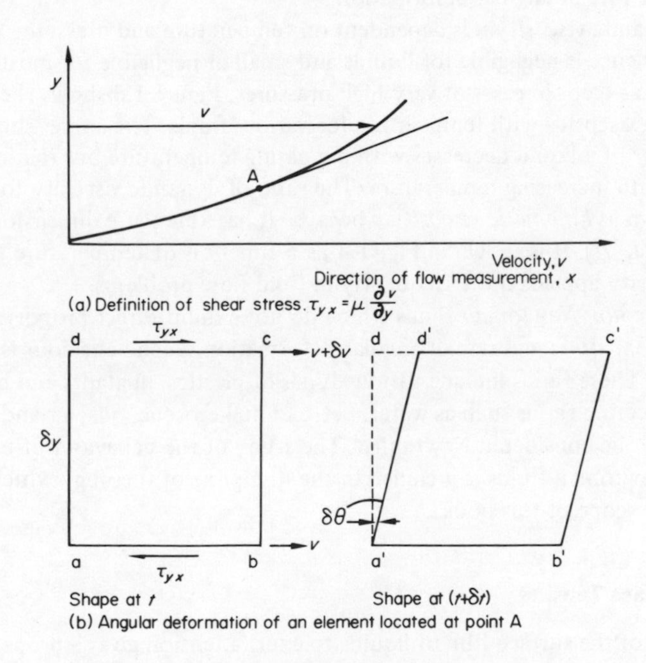

Fig. 1.2

determines the direction of the shear. Shear is, therefore, a vector quantity with magnitude and direction. Figure 1.2(a) explains the relationship expressed in equation (1.4) for a two-dimensional fluid flow in the xy plane.

Figure 1.2(b) which represents a magnification of the distortion of a fluid element located at A of Fig. 1.2(a) may be used to show that $\partial v/\partial y$ is equal to the time rate of angular deformation or displacement. A shear stress τ acts on the top and the bottom of an infinitesimally small element of fluid, abcd, in the directions shown. The relative velocity between the top cd and the bottom ab is δv. In a small interval of time δt, abcd is distorted into a'b'c'd'. The angular distortion is given by

$$\delta\theta = \frac{\mathrm{dd}'}{\delta y}$$

for a small displacement. Thus

$$\delta\theta = \frac{\delta v}{\delta y}\,\delta t$$

since

$$\mathrm{dd}' = \delta v \delta t$$

In the limit as δt and $\delta y \to 0$,

$$\frac{\partial \theta}{\partial t} = \frac{\partial v}{\partial y}$$

which is the rate of angular deformation.

The dynamic viscosity μ is dependent on temperature and pressure. The pressure dependence is negligible for liquids and small or negligible for most gases and vapours except in cases of very high pressures. Figure 1.3 shows the variation of dynamic viscosity with temperature for various fluids. The curves show that the viscosity of a liquid decreases with increasing temperature but that of a gas increases with increasing temperature. The ratio of dynamic viscosity to density μ/ρ is known as *kinematic viscosity* υ because it has kinematic dimensions and units only (L^2/t). It is shown in Fig. 1.4 as a function of temperature. Kinematic viscosity appears quite frequently in fluid flow problems.

There are *non-Newtonian* fluids which do not exhibit direct proportionality between shear stress and rate of angular deformation. Such behaviour is illustrated in Fig. 1.5. These fluids include various types of plastics, dilatants and blood. Most engineering fluids such as water, petroleum, kerosene, oils, air and steam can however be considered Newtonian. The study of the behaviour of plastics and non-Newtonian fluids is included in the discipline of rheology which is beyond the scope of this book.

1.4.2 Surface Tension

The ability of the surface film of liquids to exert a tension gives a property known as *surface tension*. It is commonly observed that small objects such as

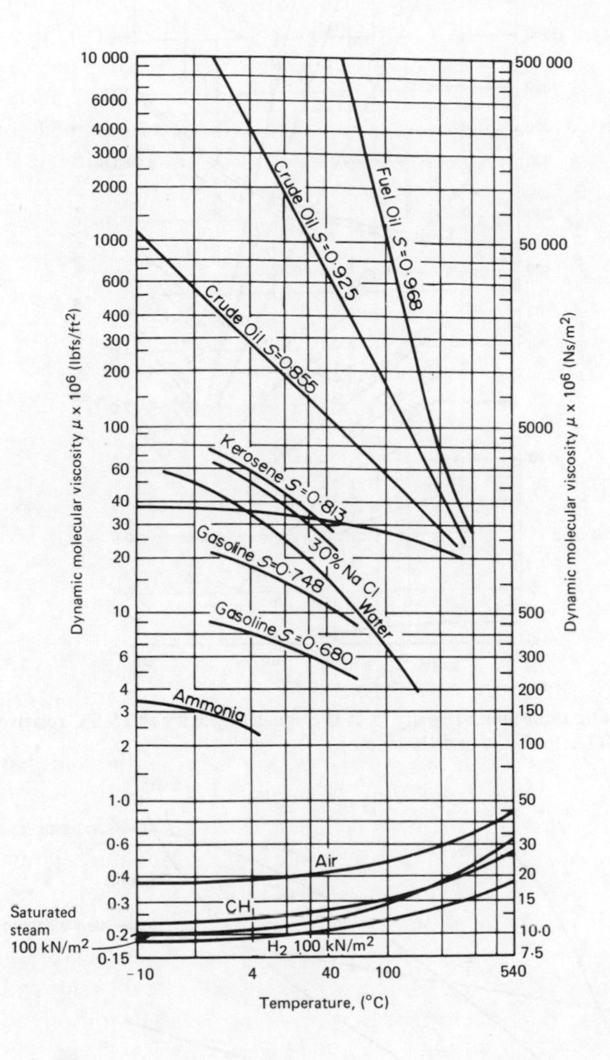

Fig. 1.3 Dynamic molecular viscosity (*s* is the specific gravity at 15.5°C relative to water at 15.5°C) (After Daily and Harleman)

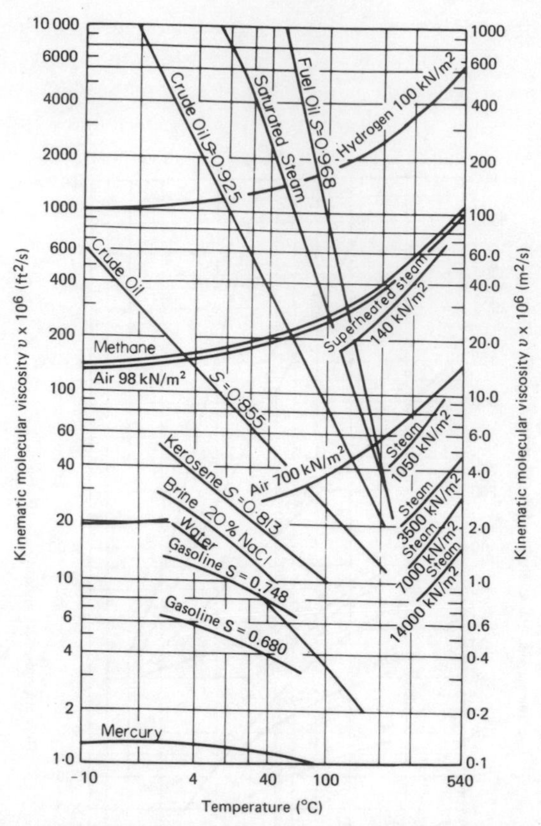

Fig. 1.4 Kinematic molecular viscosity (S is the specific gravity at 15.5°C relative to water at 15.5°C) (After Daily and Harleman)

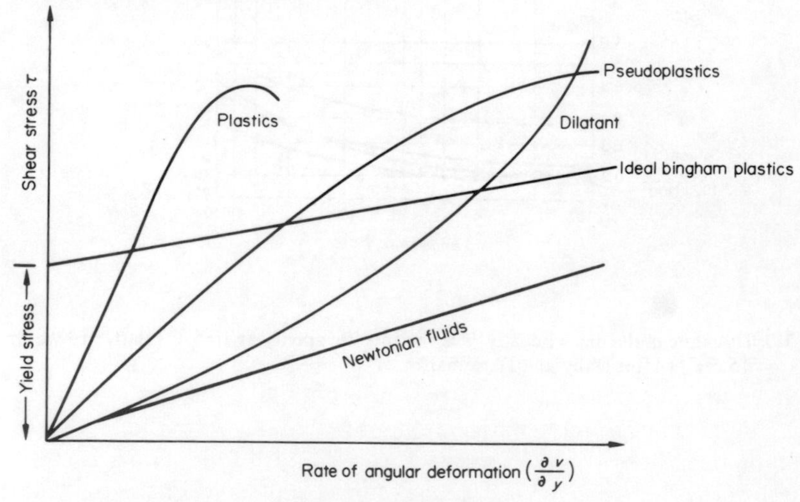

Fig. 1.5 Rheological diagram

ants can be supported on the surface of water even though the object may be much denser than water. At the interface between a liquid and a gas a film forms, apparently due to the molecular attraction of liquid molecules below the surface. Surface tension is defined as the force required to maintain a unit length of the film in equilibrium. In Table 1.4 the values of surface tension for some common liquids are listed.

Table 1.4
Surface tension of liquids

	Temperature		*Surface tension*	
	(°F)	*(K)*	*σ (lbf/ft)*	*(N/m)*
Water–air	32	273	0·00518	0·0755
	50	283	0·00509	0·0742
	59	288	0·00504	0·0735
	68	293	0·00499	0·0727
	86	303	0·00488	0·0711
	122	323	0·00464	0·0677
	150	340	0·00442	0·0645
Mercury–air	68	293	0·03192	0·466
Water–mercury	68	293	0·0257	0·375
Water–carbon tetrachloride	68	293	0·00308	0·0449
Benzene–mercury	68	293	0·0257	0·375
Water–benzene	68	293	0·0024	0·0335

1.5 Transfer Phenomena

The movement of fluids involves the transport of mass, energy and momentum. The study of the dynamic behaviour of fluids is therefore concerned with some aspect of the behaviour of the fluid to convey these materials and properties from place to place.

There are two distinct ways in which fluid properties are transported; one is the actual physical transport of fluid masses from one point to another and the other is a molecular process of diffusion. Generally the hydraulics engineer is concerned with the former and this is the main subject of this text. However, hydraulic engineers need a clear understanding of diffusional processes not only for the solution of pollution problems but also to appreciate the mechanics of turbulent processes, evaporation and related phenomena. In this section we deal with the diffusional modes of transport as transfer phenomena.

The basic premises of transfer phenomena have in recent years evolved from a rather loose collection of theories and experimental data in diverse branches of engineering. The transfer process is typified by a diffusion process which tends to establish equilibrium. For example, when a small amount of perfume is sprayed into a room, the mass transfer process causes the perfume vapour to diffuse throughout the room until its concentration is uniform and an equilibrium condition is reached. A transfer process involves the net flow of a property under the influence of a driving force. The rate of transfer of a property is the *flux* and the intensity of the driving force is the *potential gradient*. The basic transfer principle is based on the assumption that the flux in any direction is directly proportional to the potential gradient in that direction. This is assuming that the flow is well ordered (laminar) and adjacent layers of fluid move parallel to each other at all times. Thus the only mode of transfer of the property is through the interaction of the molecules of fluid under the influence of the driving force. The constant of proportionality between the flux and the potential gradient is therefore referred to as a *molecular coefficient*. Random motion of fluid particles in a turbulent flow brings about additional transport of properties. By analogy this is referred to as eddy (turbulent or convective) diffusion and will be discussed more fully in another section.

1.5.1 Momentum Transfer (Laminar)

By definition momentum is the product of mass and velocity, and according to Newton's second law of motion the rate of change of momentum of an element in any direction is equal to the sum of the external forces acting on the element in that direction. Consider a fluid confined between two wide parallel plates as shown in Fig. 1.6. If the upper plate is moved at a constant velocity v_0 relative to the lower plate, the fluid elements in contact with the upper plate will move with relative velocity v_0 while those in contact with the lower plate will remain relatively stationary because of adhesion between the viscous fluid and the plate. The latter condition is referred to in fluid flow as the 'no-slip' effect.

On the continuum model we expect that there will be a continuous variation of fluid velocity within the space, varying from v_0 at the top to zero at the bottom. Figure 1.6 shows the forms of velocity profile from the instant the top plate is set in motion until a steady-state linear velocity distribution is attained (a long time afterwards). This will occur provided the plates are sufficiently close. It is obvious that there is a velocity gradient (momentum gradient) at every point within the fluid space at all times. The momentum gradient constitutes a momentum potential which drives momentum from the upper layers of the fluid towards the lower layers. If the flow is laminar, momentum flux in a direction normal to the plates will be proportional to the (momentum) potential gradient.

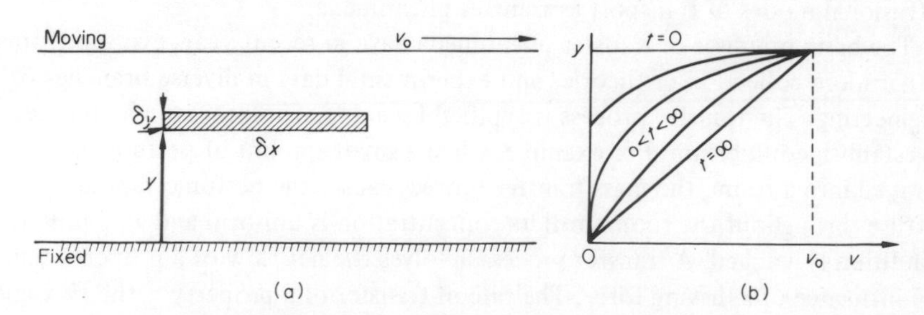

Fig. 1.6 Momentum transfer between two parallel plates

With reference to Fig. 1.6, let the momentum of an elemental strip, δy thick be increased by δM in a small time δt. This can be brought about only by an 'external' force. The only external force acting on the element is shear. For a strip of length δx and unit depth normal to the plane of the paper,

$$\tau_{yx} \cdot \delta x = \frac{\delta M}{\delta t} \tag{1.5}$$

It may therefore be inferred that the shear stress τ_{yx} is in fact a measure of the rate of momentum transfer per unit area. If the fluid velocity at y is v, the momentum per unit volume of fluid is ρv and the corresponding momentum potential (driving force) is $\partial(\rho v)/\partial y$ in the y direction. From the transfer principle, momentum flux is proportional to the momentum potential in the particular direction.

Thus

$$\tau_{yx} = -\upsilon \frac{\partial(\rho v)}{\partial y} \tag{1.6}$$

where υ is a momentum diffusion coefficient known as the *molecular kinematic viscosity*. The negative sign shows that the transfer is in the direction of decreasing momentum. By convention however the sign can be dropped without giving rise to confusion. For a homogeneous fluid medium

$$\tau_{yx} = \mu \frac{\partial v}{\partial y} \tag{1.7}$$

where $\mu = \rho \upsilon$, is the molecular dynamic viscosity. Equation (1.7) is identical to (1.4) but the development in this section stresses the point that viscosity is, in fact, a coefficient of momentum transfer. Equation (1.7) is known as Newton's equation of viscosity and is applicable to Newtonian fluids.

1.5.2 Mass Transfer (Laminar)

The fundamental law for mass transfer can be illustrated as for momentum transfer. Consider a mass of dye or salt injected continuously along the upper plate such that the concentration of the substance is maintained constant at c_0 on the upper plate and at zero on the lower plate. The distribution of the matter (salt or dye) throughout the fluid space at various times is illustrated in Fig. 1.7. If the concentration c at any point is the mass of diffusing substance per unit mass of fluid, the mass of diffusing substance per unit volume of fluid will be ρc. The flux w of the diffusing substance is given by

$$w = -\alpha_m \frac{\partial(\rho c)}{\partial y}$$

where α_m is a constant known as *molecular mass diffusivity* or coefficient of mass diffusion and has the same units as kinematic viscosity. For a homogeneous fluid medium,

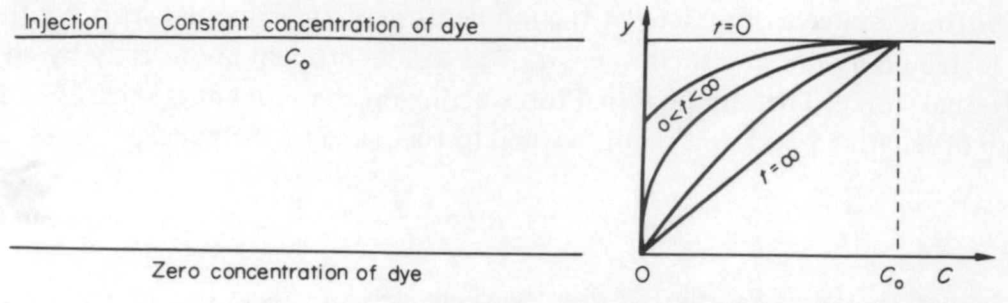

Fig. 1.7 Mass transfer between two surfaces

$$w = -D\frac{\partial c}{\partial y} \tag{1.8}$$

where $D = \rho\alpha_m$ is called the *molecular mass conductivity* and has the same units as dynamic viscosity. Equation (1.8) expresses Fick's first law of mass diffusion. The theory of mass transfer finds application in many diverse fields such as mixing and absorption processes in chemical technology, the pollution of air and contamination of ground and surface waters and intrusion of saline sea water in estuaries.

1.5.3 Heat Transfer (Laminar)

Heat transfer is like mass transfer. If the upper plate is maintained at a constant temperature T_0 and the lower plate at zero temperature, the distribution of heat (temperature) is similar to the mass distribution of Fig. 1.7. The heat quantity per unit volume of fluid at any point of temperature θ is $\rho C_p \theta$ where C_p is specific heat of the fluid at constant pressure. The heat flux q is given by

$$q = -\alpha\frac{\partial(\rho C_p \theta)}{\partial y}$$

where α is a constant called *thermal molecular diffusivity* having the same units as kinematic viscosity or molecular mass diffusivity.

For a homogeneous fluid medium

$$q = -K\frac{\partial\theta}{\partial y} \tag{1.9}$$

where $K = \rho C_p \alpha$ is the *thermal molecular conductivity*. Equation (1.9) is known as Fourier's law of heat conduction.

1.5.4 Mass, Momentum and Heat Transport in Turbulent Flow

Most natural fluid flows are not well enough ordered to be regarded as laminar. A distinctly irregular flow called *turbulent flow* occurs more frequently in nature.

In a turbulent flow, fluid particles move randomly. This is illustrated in Fig. 1.8 for flow in two dimensions. A particle only maintains a mean path and a mean

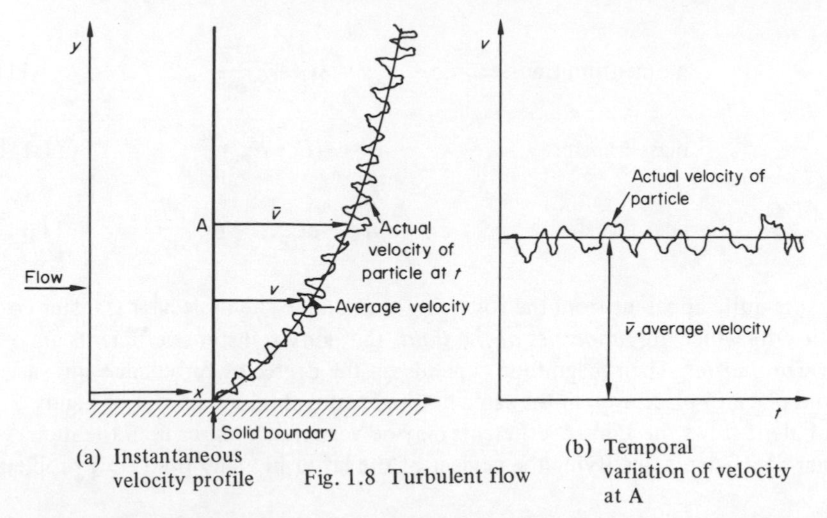

(a) Instantaneous velocity profile

Fig. 1.8 Turbulent flow

(b) Temporal variation of velocity at A

velocity over a long period of time. In addition to the mean transport of matter, momentum and energy due to molecular diffusion there is a transport effect arising from particle fluctuations. The measurable effects of mass, momentum and heat fluxes are considerably increased. For analytical convenience it is conventional to express the turbulent transport in a way analogous to molecular transfer.

$$\tau'_{yx} = \rho\epsilon \frac{\partial \bar{v}}{\partial y} \tag{1.10a}$$

$$w' = -E_{\mathrm{m}} \frac{\partial \bar{c}}{\partial y} \tag{1.10b}$$

and

$$q' = -\rho C_{\mathrm{p}} E_{\mathrm{h}} \frac{\partial \bar{\theta}}{\partial y} \tag{1.10c}$$

where the dash denotes the flux of a property due to turbulent transport, ϵ is Eddy kinematic viscosity, E_{m} is Eddy mass diffusivity and E_{h} is Eddy thermal diffusivity. The bars represent the average quantities over a large interval of time T. These are easier to measure than the corresponding parameters for individual molecules.

$$\bar{v} = \frac{1}{T}\int_{t}^{t+T} v \, \mathrm{d}t$$

$$\bar{c} = \frac{1}{T}\int_{t}^{t+T} c \, \mathrm{d}t$$

and

$$\bar{\theta} = \frac{1}{T}\int_{t}^{t+T} \theta \, \mathrm{d}t$$

The total transport in any fluid is given by the sum of the molecular transport and the turbulent transport. Thus

momentum transfer: $$\tau_{xy} = \rho(\nu + \epsilon)\frac{\partial \bar{v}}{\partial y} \tag{1.11a}$$

mass transfer: $$w = -(D + E_{\mathrm{m}})\frac{\partial \bar{c}}{\partial y} \tag{1.11b}$$

heat transfer: $$q = -\rho C_{\mathrm{p}}(\alpha + E_{\mathrm{h}})\frac{\partial \bar{\theta}}{\partial y} \tag{1.11c}$$

It is quite apparent from the foregoing that unlike the molecular transfer coefficients which are *properties of the fluid*, the Eddy transfer coefficients are *flow properties.* Their magnitude depends on the degree of turbulence and varies from place to place even in the same homogeneous fluid medium. For highly turbulent flows the Eddy coefficients may be very much bigger than the molecular coefficients, justifying the neglect of the latter in many fluid flow problems.

1.6 Types of Flow

In hydrodynamic theory fluids are classified as *ideal or real.* An ideal fluid is a hypothetical one; it has no viscosity (it is frictionless) and is incompressible. The concept of an ideal fluid has led to soluble mathematical formulations which with little or no modifications have yielded results corresponding satisfactorily to real fluid behaviour. This assumption is particularly helpful in analysing flow situations involving large expanses of fluids, as in the motion of an aeroplane or a submarine.

A real fluid flow may be *laminar* or *turbulent*. Turbulent flow situations are most prevalent in engineering practice. In turbulent flow the fluid particles move in very irregular paths, causing exchange of momentum, mass and energy from one portion of the fluid to another as discussed in the preceding section. A laminar flow is an ordered one and the fluid particles move along smooth paths in layers (laminae) with one layer gliding smoothly over an adjacent layer. Laminar flow for most common fluids is governed by Newton's law of viscosity, equation (1.4), and the effect of viscosity is to damp out turbulence. Laminar flow is observed in cases of high viscosity and low velocities. In situations combining low viscosity, high velocity and large flow passages the flow becomes unstable.

A flow may also be *steady or unsteady*. In a steady flow, conditions at any point within the regime of flow do not vary with time. The flow is unsteady when conditions at any point change with time. An unsteady flow must not be confused with turbulent flow. The latter can be steady or unsteady as demonstrated in Fig. 1.9 which depicts observation of velocity at a point. In steady flow the

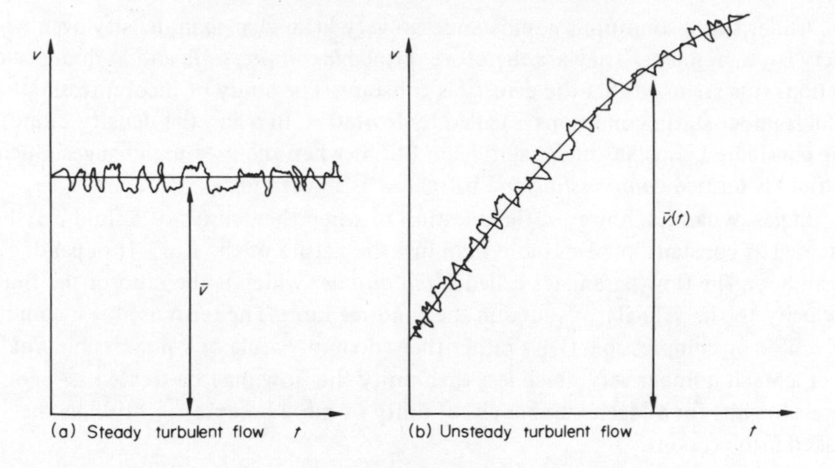

Fig. 1.9 Steady and unsteady turbulent flows

average values (at a point) of velocity ($\bar{v}$), density ($\bar{\rho}$), pressure ($\bar{p}$), or temperature ($\bar{\theta}$) do not change with time; thus

$$\frac{\partial \bar{v}}{\partial t} = 0, \quad \frac{\partial \bar{\rho}}{\partial t} = 0, \quad \frac{\partial \bar{p}}{\partial t} = 0, \quad \frac{\partial \bar{\theta}}{\partial t} = 0$$

The use of partial differentials implies that these values can be variable in space. A river discharging water at a fixed rate is an example of steady flow. The same river discharging water at an increasing rate because of a rain storm in its catchment area is an example of unsteady flow.

A steady or unsteady flow may be described as *uniform* or *non-uniform*. The flow is uniform if the average velocity vector $\bar{v}$ is identical (in magnitude and direction) at every point for any given instant. If a displacement δs is taken in any direction, a uniform flow must satisfy $\partial \bar{v}/\partial \bar{s} = 0$ at all times. In the case of a real fluid flowing in an open or closed conduit, the definition of uniform flow may usually also be applied even though the velocity vector at the boundary is always zero. When all parallel cross sections through the conduit are identical, that is when the conduit is prismatic, and if the average velocity at each cross section is the same at any given instant, the flow is said to be uniform. For a non-uniform flow $\partial \bar{v}/\partial \bar{s} \neq 0$.

A constant rate of flow of liquid through a long pipeline of a uniform cross sectional area constitutes a *steady*, *uniform* flow. The same constant rate of flow of the liquid through a pipeline of decreasing or increasing cross sectional area gives an example of *steady*, *non-uniform flow*. If the flow rate increases or decreases with time in a constant cross sectional area or a changing cross sectional area pipeline the result is an *unsteady*, *uniform* and an *unsteady*, *non-uniform* flow respectively.

Under static conditions liquids undergo very little change in density even under very large pressures. They are therefore termed *incompressible* and in doing calculations it is assumed that the density is constant. The study of incompressible fluids under static conditions is called hydrostatics. In a gas, the density cannot be considered constant under static conditions when the pressure changes. Such a fluid is termed *compressible* and is treated as aerostatics.

In gas dynamics, however, the question of when the density of a fluid may be treated as constant involves more than just the nature of the fluid. It depends mainly on the flow parameter called *Mach number* which is the ratio of the fluid velocity to the velocity of sound in the fluid medium. The term used is incompressible or compressible *flows* rather than incompressible or compressible *fluids*. For a Mach number very much less than unity the flow may be treated as incompressible but for a Mach number about unity or more compressibility must be taken into account.

1.7 Boundary Layer Concepts and Drag

A full discussion of boundary layer concepts is not given in this text for reasons of size and economy. It falls within the realm of classical hydrodynamic theory dealing with real fluids. The aim here is only to introduce the subject to enable the student to appreciate the concept of resistance or drag to fluid motion by bounding surfaces as applied in the following sections. The mathematical development which follows is quite straightforward and simple but the student whose background in mechanics is weak may be advised to read the mathematical development again after studying the next chapter.

Resistance to flow arising from objects situated in or enclosing a moving fluid or objects which are moving relative to a stationary or moving fluid has two components. The first is due to the friction between the fluid and the surface of the object and the second is due to the shape of the object and its alignment with respect to flow. The former is referred to as skin frictional drag and the latter as form drag or form resistance or pressure drag.

If an ideal fluid flows along a solid surface, no forces are exerted, for there cannot be any shear forces in an ideal fluid. In a real fluid flow, however, the fluid close to the solid surface is retarded by the shear forces due to viscosity. That region of fluid in which the velocity of flow is affected by the boundary shear is called the *boundary layer* (Fig. 1.10).

In 1904 Prandtl developed a concept of the boundary layer which provides an important link between ideal fluid flow and real fluid flow. For fluids having relatively small viscosity, the effect of internal friction in the fluid is appreciable only in a narrow region surrounding the fluid boundaries. This hypothesis allows the flow outside the narrow region near the solid boundaries to be treated as ideal fluid flow.

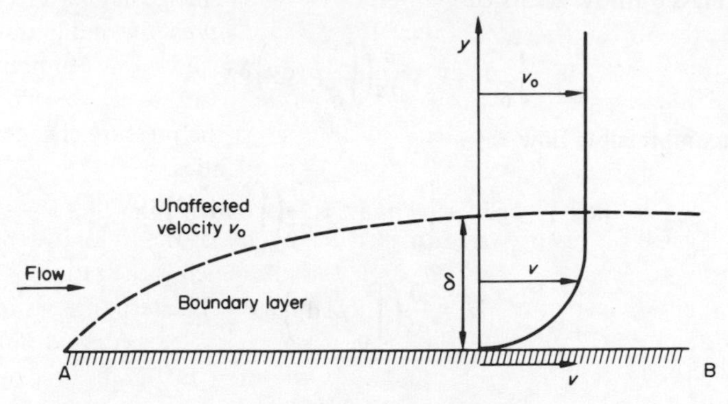

Fig. 1.10 Boundary layer

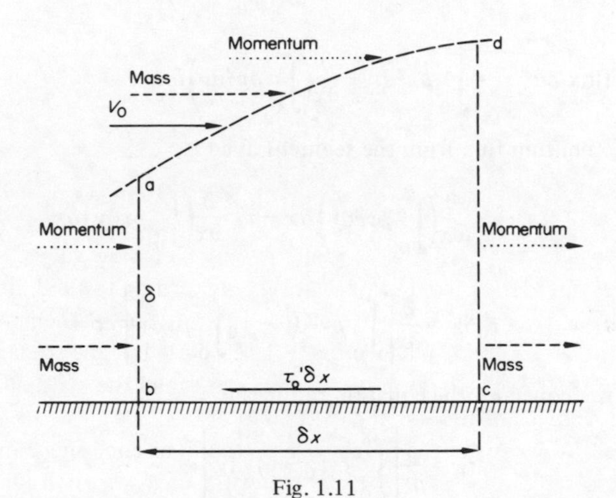

Fig. 1.11

Figure 1.10 depicts the boundary layer development on an extensive flat solid surface AB over which a fluid flows. A is the leading edge of the surface and a boundary layer has been formed which in general becomes thicker toward B. For a flat surface it can be assumed that the variation of pressure in the direction of flow is zero. The principle of momentum may be applied directly in order to determine the shear stress exerted on the fluid by the solid boundary.

Providing the flow is steady in a small segment of the boundary layer of fixed boundaries abcd (Fig. 1.11), the resultant force in the direction of motion x must balance the net momentum flux across the surfaces of the segment. The mass inflow across ad is m, v is the velocity in the boundary layer and v_0 is the undisturbed velocity. The mass inflow across ab is thus

$$\int_0^\delta \rho v \mathrm{d}y$$

And the mass outflow across cd is

$$\int_0^\delta \rho v \mathrm{d}y + \frac{\partial}{\partial x}\left(\int_0^\delta \rho v \mathrm{d}y\right)\delta x$$

For an incompressible flow,

$$m + \int_0^\delta \rho v \mathrm{d}y = \int_0^\delta \rho v \mathrm{d}y + \frac{\partial}{\partial x}\left(\int_0^\delta \rho v \mathrm{d}y\right)\delta x$$

i.e.

$$m = \frac{\partial}{\partial x}\left(\int_0^\delta \rho v \mathrm{d}y\right)\delta x$$

$$\text{Momentum flux in} = \int_0^\delta \rho v^2 \mathrm{d}y + v_0 \frac{\partial}{\partial x}\left(\int_0^\delta \rho v \mathrm{d}y\right)\delta x$$

$$\text{Momentum flux out} = \int_0^\delta \rho v^2 \mathrm{d}y + \frac{\partial}{\partial x}\left(\int_0^\delta \rho v^2 \mathrm{d}y\right)\delta x$$

Excess of momentum flux from the segment abcd is

$$\frac{\partial}{\partial x}\left(\int_0^\delta \rho v^2 \mathrm{d}y\right)\delta x - v_0 \frac{\partial}{\partial x}\left(\int_0^\delta \rho v \mathrm{d}y\right)\delta x$$

Thus

$$-\tau_0'\delta x = \frac{\partial}{\partial x}\left[\int_0^\delta \rho v^2 \mathrm{d}y - v_0 \int_0^\delta \rho v \mathrm{d}y\right]\delta x$$

Since v_0 is a constant, the equation reduces to

$$\tau_0' = \frac{\partial}{\partial x}\left[\int_0^\delta \rho(v_0 - v)v\mathrm{d}y\right]$$

or

$$\tau_0' = \rho v_0^2 \frac{\partial}{\partial x}\left[\int_0^\delta \left(1 - \frac{v}{v_0}\right)\frac{v}{v_0}\,\mathrm{d}y\right] \tag{1.12}$$

As an example for the application of equation (1.12) assume

$$v/v_0 = (y/\delta)^{1/7} \tag{1.13}$$

This is Blasius' 1/7th power law which is known to be reasonably true for turbulent flow over a smooth flat surface. Substituting in equation (1.12) and integrating

$$\tau_0' = \frac{7}{72}\rho v_0^2 \frac{\partial \delta}{\partial x} \tag{1.14}$$

The boundary layer thickness δ depends on the roughness of the solid surface, on the fluid properties and the distance x from the leading edge. In most fluid flow problems the average shear stress

$$\tau_0 = \frac{1}{l}\int_0^l \tau_0' \mathrm{d}x$$

over the solid boundary whose length is l can conveniently be used. It is conventional to express the average shear stress in the form

$$\tau_0 = \frac{1}{2} c_f \rho v_0^2$$

where c_f is an average coefficient of drag. Experiments have indicated that for large distances from the leading edge the variation in the coefficient of drag (and τ_0) becomes minor and it may be treated as a constant provided flow conditions are not severely altered.

Form resistance arises from uneven or non-symmetrical pressure distribution due to flow around an object. The shape of the object and the direction of flow are therefore very important. Take for example flow round a circular cylindrical

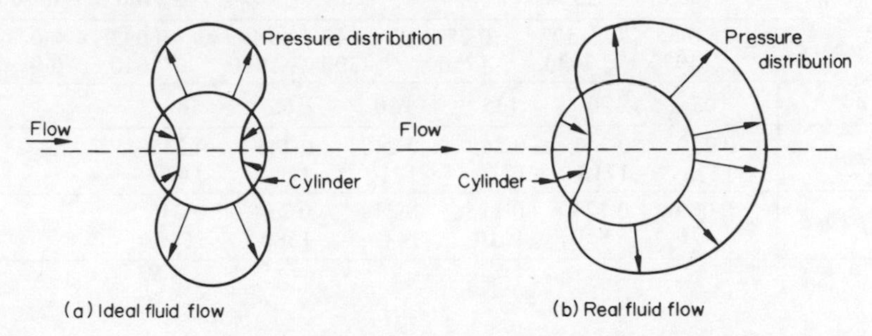

Fig. 1.12 Pressure distribution around a cylinder

object. An ideal fluid would produce a symmetrical pressure distribution about all planes containing a diameter of the circle of the cylinder (Fig. 1.12(a)). The net pressure force in the direction of motion is accordingly zero. A real fluid, however, produces a pressure distribution which is symmetrical only about the diameter parallel to the flow direction (Fig. 1.12(b)) and as a result there is a net pressure force in the direction of flow. This is form drag. Two objects, one streamlined (needle-like) and the other bluff (e.g. cylindrical), having the same surface area and the same relative roughness when placed in the same stream of a fluid, may have very nearly the same value of skin frictional drag but widely different values of form drag. Indeed the form drag of the streamlined object may be negligible compared with the frictional drag whereas the frictional drag may be negligible compared with form drag on a bluff body.

It is also important to realize at this stage that resistance problems arising from pressure changes are not confined to objects around which there is flow. They arise also in the case of conduits. For example a pipeline which suddenly changes in diameter introduces a pressure change across the junction. This therefore introduces a form resistance which results in some loss of energy. A similar situation arises when an open channel through which fluid is flowing experiences a sudden change in its bed configuration. These two cases will be discussed more fully in Chapters 3 and 4.

EXAMPLE 1.1

The following pressure measurements are for a 0.61 m long, 5.1 cm diameter laboratory cylinder standing in a wide 0.61 m deep stream of water. Estimate the drag on the cylinder assuming symmetrical pressure distribution.

Angular distance θ (degrees)	0	15	30	45	60	75	90
Pressure p (lbf/in^2)	0·652	0·521	0·293	−0·061	−0·294	−0·360	−0·260
(N/m^2)	4990	3590	2020	−420	−2020	−2480	−1790
$p \cos\theta$	0·652	0·507	0·254	−0·043	−0·147	−0·093	0·0
	4990	3490	1750	−300	−1010	−640	0·0

θ	105	120	135	150	165	180
p	−0·250	−0·248	−0·248	−0·248	−0·245	−0·245
	−1720	−1710	−1710	−1710	−1690	−1690
$p \cos\theta$	0·069	0·124	0·175	0·214	0·238	0·245
	470	850	1210	1440	1680	1690

SOLUTION

This is an example of case (b) of Fig. 1.12, see sketch in Fig. 1.13(a).

Pressure is always normal to the solid surface. If the pressure corresponding to angle θ is p, the pressure force on a small segment made by $\delta\theta$ is $(pa\delta\theta)l$ where a is the radius of the cylinder and l is the immersed length of the cylinder. The component of the pressure force in the direction of motion is $(pal\delta\theta)\cos\theta$. The total drag on the cylinder is the sum of the pressure force on all such elemental segments

$$\therefore \qquad \text{drag} = \int_0^{2\pi} pal \cos\theta \, d\theta$$

From symmetry,

$$\text{drag} = 2al \int_0^{\pi} p \cos\theta \, d\theta$$

Now $\int p \cos\theta \, d\theta$ is the area under a plot of $p \cos\theta$ against θ. Such a plot is shown in Fig. 1.13(b).

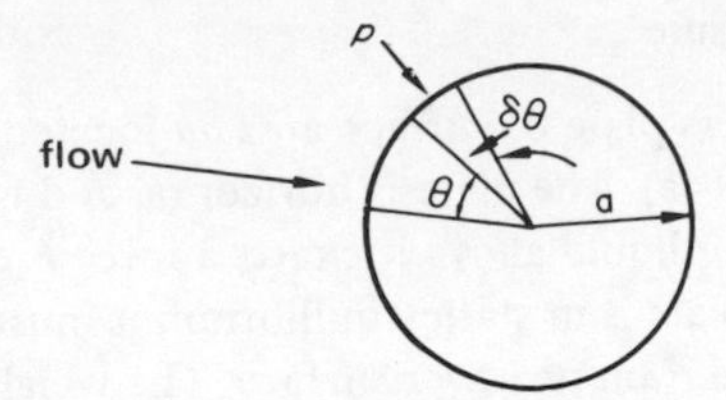

Fig. 1.13(a)

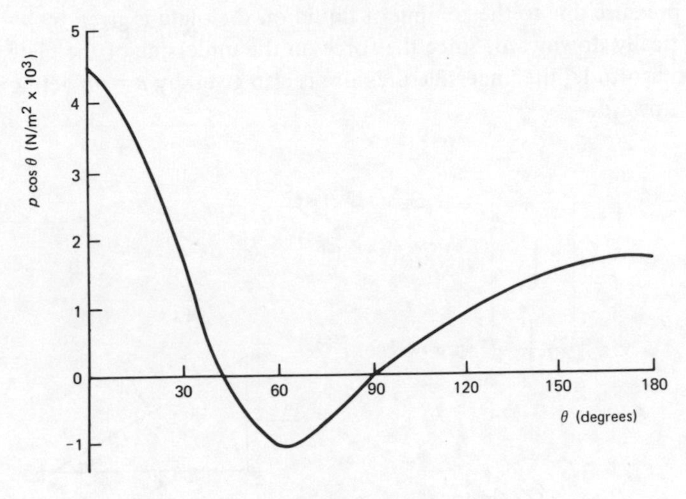

Fig. 1.13(b)

Net area under the curve = 168 000 N degrees/m^2

$$\therefore \qquad \text{drag} = \frac{2\pi}{180}(0.025)(0.61)(168000)\text{ N}$$

$$= \underline{90.6\text{N } (20.4 \text{ lbf})}$$

This is form drag. Frictional drag can be calculated only if the friction factor (c_f) is known. The former normally dominates in the case of cylinders unless the flow is well ordered (laminar).

1.8 Fluids in Static Equilibrium

The study of the behaviour of liquids under static conditions is known as hydrostatics, which is a branch of fluid mechanics. Fluids in motion is studied in hydrodynamics. In this section the general conditions of static equilibrium of liquids is reviewed. The remainder of the book covers primarily fluids in motion.

1.8.1 Hydrostatic Pressure

Consider a thin weightless plate of surface area δa located at P in a stagnant pool of a liquid (Fig. 1.14a). The plate is horizontal and is of infinitesimal thickness. The column of liquid above it exerts a force F on the upper surface of the plate. Since the plate is in static equilibrium it must be balanced by an equal and opposite force F on its lower surface. The weight of the column of liquid above the plate is given by the product of its volume and the specific weight. Thus $F = \gamma h \delta a$. By definition, pressure is a force acting on a unit area. Thus the pressure due to the column of liquid on the plate is given by $p = \gamma h$ acting vertically downward. Since the force on the underside of the plate is also F in magnitude, the underside pressure is also given by $p = \gamma h$ acting vertically upward.

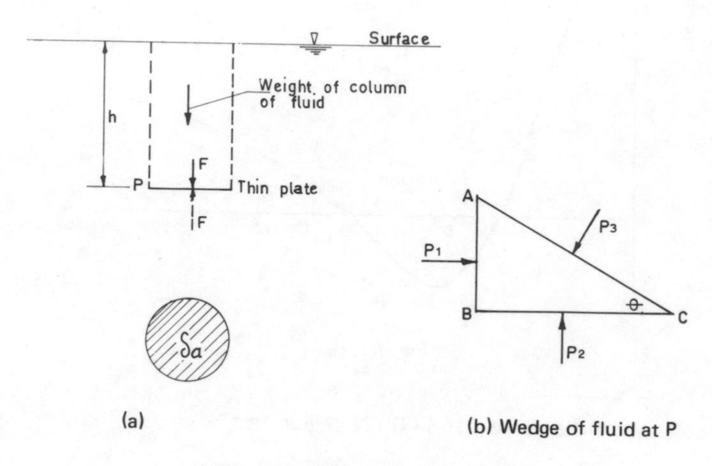

Fig. 1.14 Hydrostatic Pressure

It is shown using the prism in Fig. 1.14b, that the pressure at a point in a fluid has the same magnitude in all directions. Let the prism be of unit thickness normal to the plane of paper. The side AB is taken vertical and BC horizontal for convenience, although the desired result can be obtained using any orientation. Let the fluid pressure on AB be p_1, on BC be p_2 and on AC be p_3. Angle ACB is θ. For static equilibrium of the prism, we resolve forces horizontally and vertically.

Resolving horizontally,

$$p_1 \text{ (AB)} = p_3 \text{ (AC)} \sin \theta$$

But (AB) = (AC) sin θ, by simple trigonometry.

Thus

$$p_1 = p_3$$

Resolving vertically

$$p_2\,(BC) = p_3\,(AC)\cos\theta + \gamma/2\,(AB)\,(BC)$$

where the last term on the right-hand side represents the weight of fluid in the prism.

But again (BC) = (AC) cos θ, thus

$$p_2 = p_3 + \gamma/2\,(AB).$$

As the prism shrinks to a point AB → 0, thus

$$p_2 = p_3$$

Therefore $$p_1 = p_2 = p_3 = p \qquad (1.15)$$

The conclusion is arrived at that:

1. *the pressure at a point in a column of a fluid* (liquid) *equals the product of the depth from the liquid surface to the point and the specific weight of fluid* (liquid); and
2. *the pressure at a point in a fluid has the same magnitude in all directions. Thus pressure acts normally on any chosen plane through a point.*

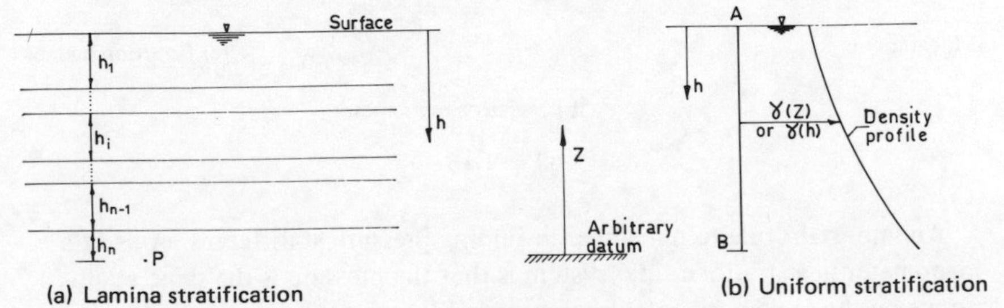

Fig. 1.15 Stratification of Liquids

Using elementary calculus, a small increment of depth δh produces a corresponding increase in pressure given by $p = \gamma\delta h$ or $p = -\gamma\delta z$ where the z-axis is taken vertically upward (Fig. 1.15). The hydrostatic pressure in a layered fluid is given by

$$p = \sum_{i=1}^{n} \gamma_i h_i \qquad (1.16)$$

starting from the exposed surface (Fig. 1.15a). γ_i is the specific weight of the ith layer of thickness h_i. In a uniformly stratified fluid system (Fig. 1.15b) it is necessary to express γ as a function of depth h or z and integrate

$$p = \int_A^B \gamma dh = -\int_B^A \gamma dz$$

to obtain the difference in pressure between two points A and B.

1.8.2 Pressure Measurement

There are three basic devices for measuring pressure in fluids; a piezometer, a manometer or a mechanical pressure gauge like the Bourdon gauge. The three are shown in Fig. 1.16. A piezometer is a simple tube connected to the fluid system. The column of fluid in the tube provides a pressure which balances the pressure at that point in the water system. A U-tube manometer is generally filled with a denser liquid and the displacement h_m of the liquid column in the two limbs of the manometer provides the necessary pressure balance.

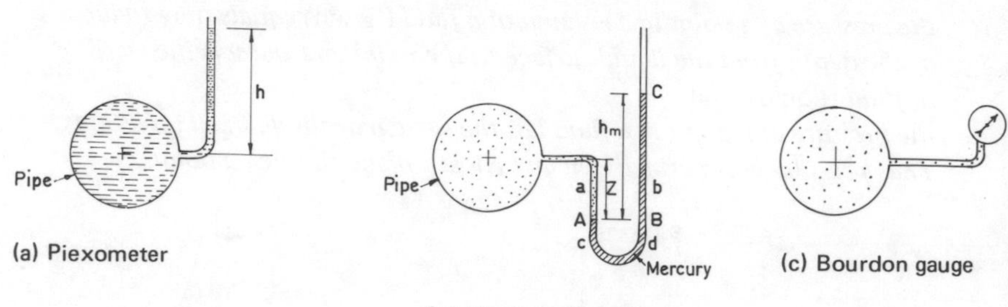

Fig. 1.16

An important rule to remember in finding pressure at different levels of a manometer in a stratified fluid system is that the pressure is the same at all points on the same horizontal level provided the points are connected by a continuous column of the same fluid. For example in Fig. 1.16b, pressure at A is equal to pressure at B and pressure at c is equal to pressure at d because each pair is connected by a continuous column of mercury. On the other hand the pressure at a is not the same as pressure at b. Although they are at the same level they are not in the same continuous fluid column.

In solving a manometric problem it is advisable to start from a point at which the pressure is known and work toward the point at which the pressure is desired. Add to this pressure in appropriate steps the changes in pressure calculated from $\pm\gamma h$, using positive for movement downward and negative for upward movement. Be consistent with units. Continue until the point of interest is reached and equate the expression thus far obtained to the pressure at that point.

For example, in Fig. 1.16(b) the pressure in the pipe is to be determined. Start at C where the pressure is known to be atmospheric. Let specific weight of mercury be γ_m and of the fluid in the pipe be γ_w.

Thus

$$p_{at} + \gamma_m h_m - \gamma_w z = p \text{ (in pipe).}$$

Note that it is not necessary in this case to consider the mercury column below plane AB since pressure at B equals pressure at A. The pressure in the pipe as given above is absolute. The corresponding gauge pressure is given by $p = \gamma_m h_m - \gamma_w z$.

The above procedure may also be applied to the case shown in Fig. 1.17 where the difference in pressures at sections 1 and 2 are required. Starting from 1

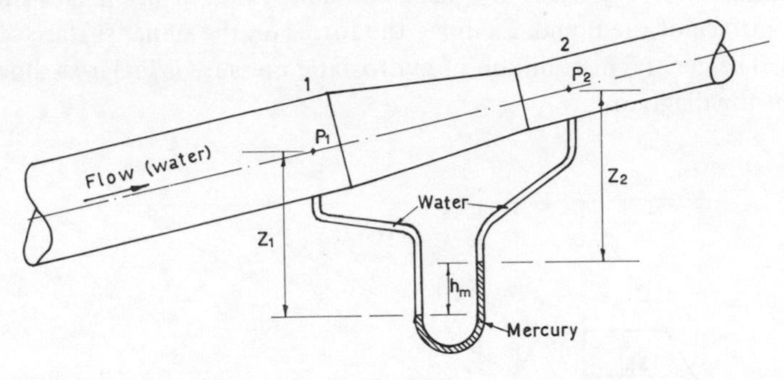

Fig. 1.17

$$p_1 + \gamma_w z_1 - \gamma_m h_m - \gamma_w z_2 = p_2$$

$$\therefore \qquad p_1 - p_2 = \gamma_m h_m - \gamma_w (z_1 - z_2)$$

For a horizontal pipe $z_1 - z_2 = h_m$

$$\therefore \qquad p_1 - p_2 = h_m (\gamma_m - \gamma_w) = \gamma_w h_m \left(\frac{\gamma_m}{\gamma_w} - 1\right) \tag{1.17}$$

This shows that the water equivalent of one unit of mercury column in a mercury–water manometer is given by 12.6 (i.e. specific gravity of mercury – 1.0).

In order to improve its sensitivity a manometer for measuring small pressures may be inclined as shown in Fig. 1.18. The scale reading is along the inclined arm such that the relevant pressure head is given by $r \sin \theta$ where θ is the angle of inclination of the arm. The reservoir R also increases the sensitivity by providing relatively large changes in r for small vertical deflections in R. The smaller θ is the better the sensitivity.

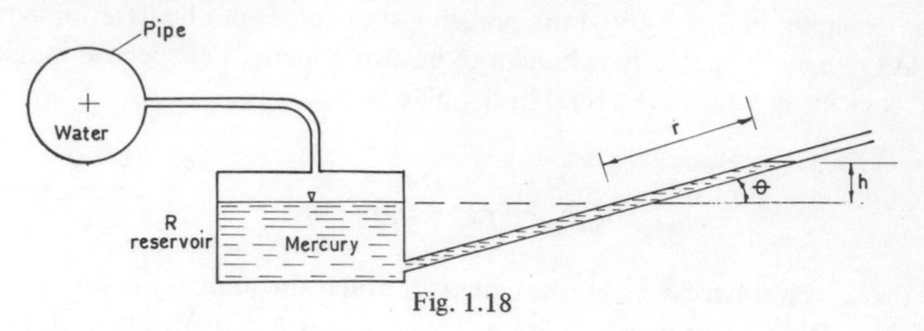

Fig. 1.18

1.8.3 Forces on Submerged Surfaces

PLANE SURFACES

Let us now examine hydrostatic forces operating on a plane surface immersed in a liquid as shown in Fig. 1.19. The plane containing the surface makes angle θ with the surface of the liquid. Examine the forces on the upper surface of the plate AB. The general distribution of hydrostatic pressure is linear as shown on the left of the diagram.

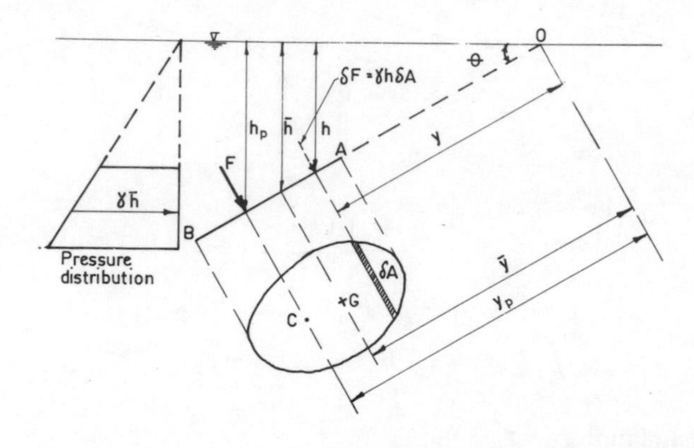

Fig. 1.19

The hydrostatic force δF on an elemental area δA running horizontally at a vertical distance h from the liquid surface is given by:

$$\delta F = \gamma h \, \delta A = \gamma y \sin \theta \, \delta A$$

where y is the distance along the plate measured from the liquid surface at O. The total force on the upper plane is given by

$$F = \int \mathrm{d}F = \int \gamma y \sin \theta \, \mathrm{d}A = \gamma \sin \theta \int y \, \mathrm{d}A.$$

But $\int y \mathrm{d}A = A\bar{y}$, the first moment of area about O. $\bar{y}$ is the distance of the centroid G from O measured along the plate and the total surface area is A

Thus
$$F = \gamma \bar{y} \sin\theta\, A = \gamma \bar{h}\, A. \tag{1.18}$$

But $\gamma \bar{h}$ is the hydrostatic pressure acting on a horizontal line through the centroid G, where $\bar{h}$ is the vertical distance of G below the fluid surface. Thus the conclusion is made that:

> *the magnitude of the hydrostatic force on a plane surface holding a column of a liquid is given by the hydrostatic pressure at the centroid of the surface multiplied by its area. This force acts at right angles to the plane surface.*

However, the hydrostatic force F acts through a point C known as the centre of pressure and not through G. Let the distance OC be y_{p}. The moment of the elemental force δF acting on the elemental area δA at the distance y from O is given by ($\gamma y^2 \sin\theta\, \delta A$). Thus the total moment about O is given by

$$F y_{\mathrm{p}} = \int \gamma y^2 \sin\theta\, \mathrm{d}A$$

and
$$y_{\mathrm{p}} = \frac{1}{F} \int \gamma y^2 \sin\theta\, \mathrm{d}A = \frac{\int y^2\, \mathrm{d}A \sin\theta}{A\bar{y}\sin\theta} \tag{1.19}$$

But $\int y^2\, \mathrm{d}A = I_{\mathrm{O}}$, the second moment of area about O.

Using the parallel theorem, I_{O} is related to the second moment I_{G} about a horizontal axis through the centroid G by

$$I_{\mathrm{O}} = I_{\mathrm{G}} + A\bar{y}^2 = A\,(k^2 + \bar{y}^2) \tag{1.20}$$

where k is the radius of gyration of the area about the horizontal axis through G. Thus the centre of pressure is located at

$$y_{\mathrm{p}} = \frac{k^2}{\bar{y}} + \bar{y} = \frac{I_{\mathrm{G}}}{A\bar{y}} + \bar{y} \tag{1.21}$$

Take for example the special case of a rectangular surface of width B standing vertically and holding back water to a depth h (Fig. 1.20a).

$$I_{\mathrm{G}} = \frac{1}{12} B h^3,\ \bar{y} = h/2$$

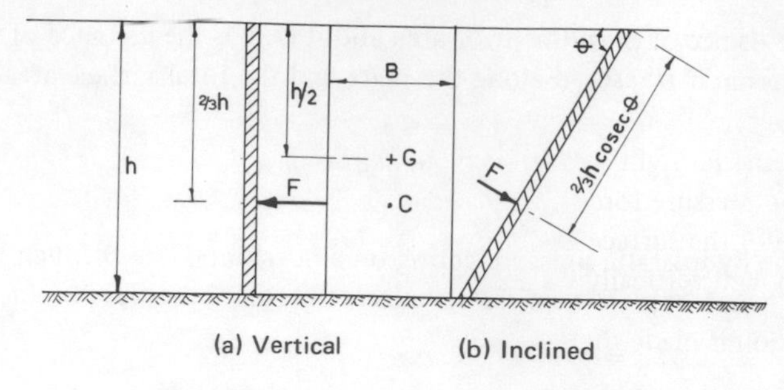

Fig. 1.20 Forces on a Rectangular Surface

The hydrostatic force

$$F = \gamma \frac{h}{2}(Bh) = \frac{1}{2}\gamma Bh^2 \tag{1.22}$$

The centre of pressure:

$$y_p = h_p = \frac{2}{3}h \tag{1.23}$$

The corresponding values of the forces and centre of pressure on an inclined rectangular plate (Fig. 1.20b) are $\frac{1}{2}\gamma Bh^2$ cosec θ and $\frac{2}{3}h$ cosec θ respectively.

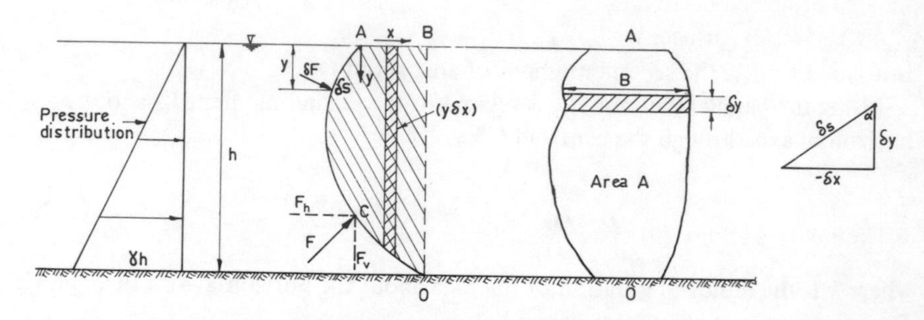

Fig. 1.21 Hydrostatic Forces on a Curved Surface

CURVED SURFACES

Many hydraulic structures holding back liquids are not plane in surface but curved. We now want to examine hydrostatic forces acting on such curved surfaces. With reference to Fig. 1.21 the curved surface ACO holds up a liquid to a depth h. The hydrostatic pressure distribution is still linear with depth as shown on the left of the diagram. The pressure at each point on the surface acts normally to the surface. Consequently the resultant force F which acts through the centre of pressure C also acts normally to the surface at C. Therefore it has horizontally (F_h) and vertical (F_v) components as shown.

Consider a small elemental length δs on the surface at depth y as shown. The diagram on the far right magnifies this length which is inclined at angle α to the vertical. The pressure force on the area $B\delta s$ is given by $\delta F = \gamma y\ B\delta s$, where B is the width of the surface at depth y. This elemental force may be resolved horizontally and vertically giving respectively,

$$\delta F_h = \gamma y B\delta s\,(\cos\alpha) \text{ and } F_v = \gamma y B\delta s\,(\sin\alpha)$$

But $\delta s\,(\cos\alpha) = \delta y$ and $\delta s\,(\sin\alpha) = -\delta x$. The signs are due to the fact that y is measured positively downward and x is measured positively to the right.

$$\therefore \quad \delta F_h = \gamma By\,\delta y \quad \text{and} \quad \delta F_v = -\gamma By\,\delta x$$

The total horizontal force,

$$F_h = \gamma \int_0^h By\mathrm{d}y = \gamma \int_0^h y\mathrm{d}A = \gamma\bar{y}A \tag{1.24}$$

since $\int_0^h yB\mathrm{d}y = \int_0^h y\mathrm{d}A = \bar{y}A$, the first moment about the liquid surface of the area projected on to a vertical plane. $\bar{y}$ is the centroid of the project area below the water surface. We conclude therefore that:

> *the horizontal component of the hydrostatic (pressure) force on a curved surface is equal to the hydrostatic (pressure) force on the horizontally projected area of the surface onto a vertical plane.*

The line of action of the horizontal component of the force still passes through the centre of pressure of the projected area.

The vertical component of the force is given from the equation

$$F_v = -\gamma\int By\mathrm{d}x.$$

But $\int By\mathrm{d}x = V$, the volume of the shaded region ACOBA. It is equal to the volume of fluid displaced by the curved surface. We therefore also conclude that:

> *the vertical component of the pressure force on a curved surface is equal to the weight of liquid (real or imaginary) vertically above the curved surface and extending up to the piezometric surface.*

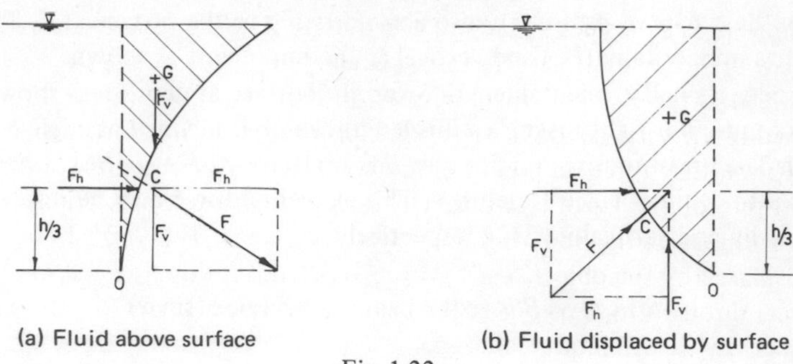

(a) Fluid above surface (b) Fluid displaced by surface

Fig. 1.22

Two cases, one in which the liquid is really held up above the surface and the other when it is imagined to be held up, are shown in Fig. 1.22. The vertical component acts through the centroid of the volume of fluid held up or displaced.

The location of the lines of action of the horizontal and vertical components of the forces enables the line of action of the resultant itself to be determined graphically or otherwise. The magnitude of the resultant force is given by $(F_h^2 + F_v^2)^{\frac{1}{2}}$.

BUOYANT FORCES

The results derived above in relation to hydrostatic pressure forces on curved surfaces can be used readily to establish Archimedes' principle which states that

> *the buoyant force on a body submerged in a fluid is equal to the weight of the volume of fluid displaced by the body.*

The buoyant force or upthrust is the vertical force exerted on a body by a static fluid in which it is submerged or floating. With reference to Fig. 1.23, the

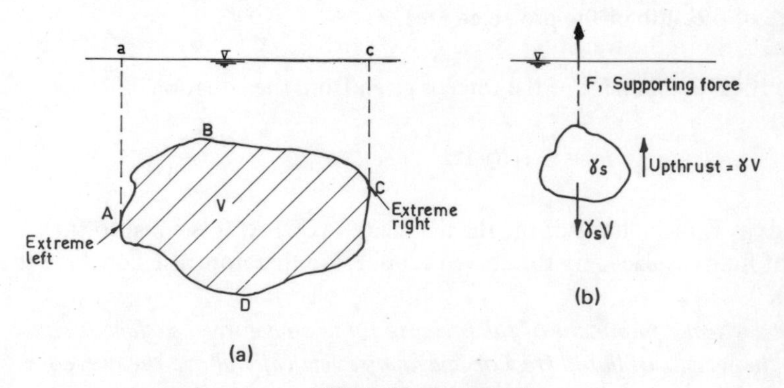

Fig. 1.23 Buoyancy

horizontal thrusts on sides of the immersed body ABCDA cancel out by virtue of the law on horizontal pressure force stated above.

The vertical force on the upper curve ABC is given by the weight of liquid contained in the space aABCca of volume V_1. This force acts vertically downward. The vertical force on the lower surface ADC is given by the weight of liquid in the volume V_2 of aADCca. This acts vertically upward. Thus the resultant upthrust on the body is $\gamma(V_2 - V_1) = \gamma V$ where V is the volume of fluid displaced by the object ABCDA. The resultant upthrust or the buoyant force acts through the centroid of the displaced volume. If the specific weight of the object is γ_s it will require a force $F = (\gamma_s - \gamma)\ V$ to maintain its static equilibrium (Fig. 1.23b).

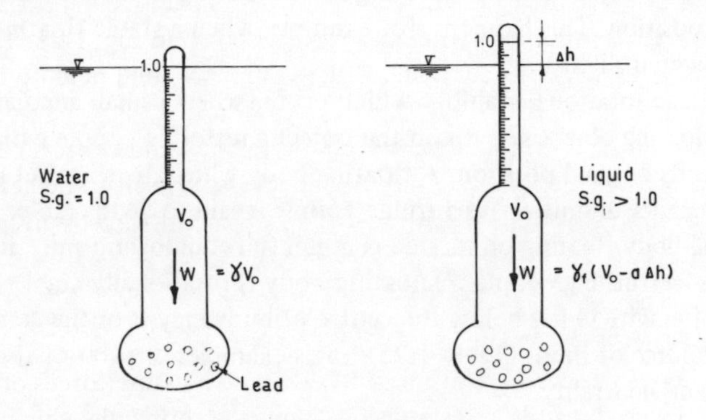

Fig. 1.24 Hydrometer

The principle of buoyancy is used to determine the specific gravity of liquids using a hydrometer (Fig. 1.24). Let the hydrometer read 1.0 when floating in water. The corresponding weight of water displaced will be $\gamma_w V_0$, where V_0 is the volume of displaced water. In a liquid of higher density (specific weight = γ_f) than water the hydrometer will pop up by an amount Δh. If the stem of the hydrometer is of cross sectional area a, the reduction in volume of fluid displaced will be $a\Delta h$. Since the weight of the hydrometer is equal to the weight of the volume of fluid displaced in each case,

$$W = \gamma V_0 = \gamma_f\ (V_0 - a\Delta h)$$

Thus

$$\Delta h = \frac{V_0}{aS_f}\ (S_f - 1)$$

where S_f is the specific gravity of the second fluid. This equation enables the stem to be calibrated in terms of specific gravities.

In a lighter liquid than water the hydrometer will sink more than in water. If it sinks by Δh, the corresponding calibration equation is

$$\Delta h = \frac{V_0}{aS_f}(1 - S_f)$$

1.8.4 Stability of Floating Bodies

1.8.4.1 *Definitions*

There are two types of stability conditions when dealing with objects which float in liquids. There is linear stability which occurs when a small linear displacement of an object sets up restoring forces tending to return the object to its original position. This happens, for example, when a stable floating body is raised or lowered slightly.

There is also rotational stability which occurs when a small angular displacement of a floating object sets up on the object a restoring couple which tends to restore it to its original position. A floating body which does not set up a restoring couple to balance an initial overturning couple is said to be unstable. A sphere or a cylindrical body floating on its side is in neutral equilibrium since it does not develop an overturning couple. A floating body is rotationally stable only when its centre of gravity G lies below the centre of buoyancy B or the centroid of the displaced volume of fluid. Figure 1.25 shows examples of rotational stability, instability and neutrality.

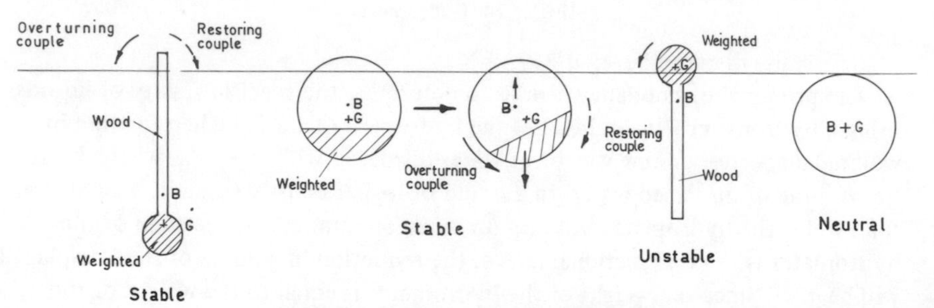

Fig. 1.25 Stability of Floating Bodies

1.8.4.2 *Floating bodies with* B *below* G

Many floating bodies, like ships loaded on the deck and pontoons, have their centre of gravity G above the centre of buoyancy B. Such bodies are, by definition, unstable. They, however, can function without capsizing provided an angular displacement from their equilibrium position remains small. One case is shown in Fig. 1.26.

The figure shows a vessel loaded such that its centre of gravity G lies above B. In static equilibrium position (Fig. 1.26(b)) the volume of water displaced gives an upthrust which balances the total weight W of the vessel and its load. In Fig.

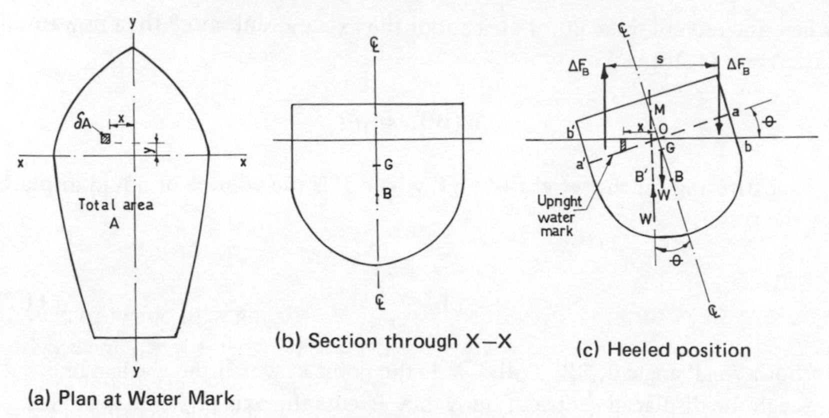

Fig. 1.26 Stability of a Vessel

1.26c the vessel is displaced by a small angle θ. The centre of gravity relative to the vessel remains in the same position but the centre of buoyancy shifts from B to B′ in relation to the structure of the vessel. The displaced volume of liquid also remains unchanged since the weight of the loaded vessel remains constant.

In relation to Fig. 1.26c the original stable water line a′Oa is tilted through θ. Liquid which occupies region aOb in the stable position now fills region a′Ob′. Thus, while the buoyancy on the right of the vessel axis is reduced by an amount $\Delta F_{\mathbf{B}}$ due to reduced volume of liquid displaced by the right side of the vessel it is increased on the left by an amount $\Delta F_{\mathbf{B}}$ due to an increase in volume of liquid displaced by the left side of the vessel. These set up a restoring moment $\Delta F_{\mathbf{B}}s$, where s is the distance between the centroids of the wedges aOb and a′Ob′. Accepting, for the sake of shortness, the proof by others that the vessel will be stable provided the restoring couple balances the imbalance $W(\mathrm{BB'})$ created by shifting the centre of buoyancy from B to B′,

$$W(\mathrm{BB'}) = \Delta F_{\mathbf{B}}s \tag{1.25}$$

where W is weight of the vessel.

The couple due to the change in geometry of the volume of fluid displaced is determined as follows: In plan (Fig. 1.26a) a small element at a distance x from the axis of symmetry yy has an area δA. In Fig. 1.26c the corresponding depth of wedge is $x\theta$. Thus volume of liquid displaced is $\delta A x\theta$ and the corresponding buoyant force is $\gamma\delta A x\theta$. Accordingly the total unbalanced moment about O of the buoyant force when the vessel is heeled is given by $\int \gamma x^2 \theta \, \mathrm{d}A$. The integration is taken over the surface of vessel corresponding to the liquid mark as in Fig. 1.26a. Thus

$$\Delta F_{\mathbf{B}}s = \gamma\theta \int x^2 \mathrm{d}A = \gamma\theta I_{\mathrm{y}} \tag{1.26}$$

where the second moment of area about the axis yy is $I_y = \int x^2 \mathrm{d}A$. From equations (1.25) and (1.26),

$$W(\mathrm{BB}') = \gamma\theta I_y$$

Substituting for the weight $W = \gamma V$ where V is the volume of liquid displaced by the vessel,

$$\mathrm{BB}' = \frac{I_y}{V}\theta \tag{1.27}$$

For a small angle θ, $\mathrm{BB}' = \overline{\mathrm{MB}}\theta$. M is the point at which the vertical line through the displaced centre of buoyancy B′ cuts the axis of symmetry. M is known as the *metacentre* and $\overline{\mathrm{MG}}$ is known as the *metacentric height*. For stability of the vessel $\overline{\mathrm{MG}}$ must be positive, that is M must lie above G, and vice-versa. The position of M depends on θ. M lies above G only when θ is small. $\overline{\mathrm{MB}}$ is given from equation (1.27) as

$$\overline{\mathrm{MB}} = \frac{I_y}{V}$$

Thus the metacentric height

$$\overline{\mathrm{MG}} = \overline{\mathrm{MB}} - \overline{\mathrm{GB}} = \frac{I_y}{V} - \overline{\mathrm{GB}} \tag{1.28}$$

when G lies above B, and

$$\overline{\mathrm{MG}} = \overline{\mathrm{MB}} + \overline{\mathrm{BG}} = \frac{I_y}{V} + \overline{\mathrm{BG}} \tag{1.29}$$

when G lies below B, in which case the vessel is permanently stable.

When the vessel oscillates about the yy axis the movement is known as rolling. Pitching occurs when the movement is about the xx axis and the corresponding second moment of area $I_x = \int y^2 \mathrm{d}A$ must be used in the above equations.

FURTHER READING

Daily, J. W. and Harleman, D. R. F., *Fluid Dynamics,* Addison-Wesley, Reading, U.S.A.

Francis, J. R. D., *A Textbook of Fluid Mechanics for Engineering Students*, Edward Arnold, London.

Rohsenow, W. M. and Choi, H. Y., *Heat, Mass and Momentum Transfer*, Prentice-Hall, New Jersey, U.S.A.
Shames, I., *Mechanics of Fluids*, McGraw-Hill, New York, U.S.A.
Streeter, V. L., *Fluid Mechanics*, McGraw-Hill, New York, U.S.A.

2 Methods of Analysis

In this chapter the general principles of fluid motion are discussed. Broadly speaking they are described as the observational principles of mass, momentum and energy. It cannot be over stressed that the concepts treated here in relation to mass, momentum and energy conservation constitute the main tool for the solution of most problems in fluid dynamics. In compressible flow, equations of state or process are also required for a complete solution of the problem.

2.1 Control Volume Concepts

There are two convenient ways of observing the behaviour of fluid particles in a regime of flow. One way is to fix one's attention on a particular mass of fluid and observe its behaviour as it moves from place to place. As it moves its volume, velocity, energy and other properties may change depending on the nature of the fluid and the surrounding conditions. This approach may be considered a material description in the sense that it describes the behaviour of a fixed quantity of matter. The other way is to fix one's attention on a particular volume within the regime of flow and observe the changes that take place as the fluid moves through it. The changes may include those of mass, temperature, velocity, pressure and others depending on the nature of fluid and the surrounding conditions. This represents a spatial description of fluid behaviour. The latter approach gives rise to control volume concepts which are discussed below. The control volume will be used quite extensively in this book. The chosen volume may be fixed in space relative to the earth, translated in space or even rotated.

The choice of a control volume depends on observational interest. In studies involving description of the entire flow field from point to point a *differential control volume* is desired. A Cartesian coordinate frame, a polar coordinate frame, a cylindrical coordinate frame or any other suitable frame depending on the physical nature of the fluid flow space may be chosen. In the Cartesian coordinate frame consideration is usually given to an elemental volume of fluid of dimensions Δx, Δy and Δz which in the limit as Δx, Δy and Δz approach zero expresses conditions at a point x, y, z. This approach gives the most general three-dimensional form of equations of the fluid motion. The equations may be solved using appropriate boundary and initial conditions to yield expressions for the various properties (velocity, pressure, temperature, density, etc.) applicable to the entire regime of flow. In polar (spherical) coordinates one chooses the radial distance r and the angular distances defined by θ and ϕ. Similarly the radial distance r, the angular distance θ and the distance z relative to a fixed plane define the cylindrical coordinate frame.

However in many fluid flow problems this approach is quite unnecessary. Indeed, the differential equations and the boundary conditions may be so complex that solutions are impossible. If the nature of the problem permits gross descriptions of the flow rather than a description of the variation from point to point a *finite control volume* is adopted. The control volume is chosen to include the entire fluid-filled space of interest. The conservation laws of mass, energy and momentum (Newton's second law) can then be applied to the mass of fluid contained in the finite volume at an instant. For example, the venturimeter is used in measuring fluid flow through a pipe. By choosing a finite control volume bounded by the solid surfaces of the meter and the planes normal to the axis of flow at its inlet and throat ends it is possible to apply the energy and mass conservation principles which yield an expression relating average velocity of flow at either end to the pressure drop across the venturimeter. The rate of discharge can thus be calculated. The finite control volume approach is particularly useful in one-dimensional flow analyses since the primary interest is generally in the variations in flow characteristics in the direction of flow rather than in the variations across a section.

2.2 The Basic Physical Laws of Mass, Energy and Momentum Transport

The transport of mass, energy and momentum associated with the motion of fluids is observed in nature to obey a basic physical law which Daily and Harleman have called an 'observational' law. Mass transport follows the principle of conservation of matter, heat energy transport obeys the conservation of energy or the first law of thermodynamics and momentum transport is according to Newton's second law of motion.

To summarize the laws of conservation, let X be any scalar quantity being transported (mass or energy). For any control volume fixed in space the law of conservation of X states that the rate of change of X within the control volume is given by the total rate of influx of X into the control volume plus the total rate of production of X inside the control volume minus the total rate of efflux of X from the control volume. It must be remembered that a rate of absorption is a negative rate of production. In symbols

$$\frac{\partial X}{\partial t} = \Sigma I_X + \Sigma P_X - \Sigma E_X \tag{2.1}$$

where the left-hand side represents the rate of increase, I_X denotes influx, P_X denotes rate of production and E_X denotes efflux, and the subscript X denotes the property being considered.

The principle of equation (2.1) may also be applied to linear momentum $\boldsymbol{M}$ (a vector quantity) associated with a control volume provided that the equation is applied in a fixed direction of motion and that proper account is taken of all external forces acting on the fluid in the control volume resolved in the direction of interest. This arises because a rate of change of momentum is invariably associated with impulsive force according to Newton's second law. In fluid mechanics the external forces are normally surface forces of pressure and shear and body forces due to gravity.

The expression for the conservation of linear momentum in a control volume fixed in space is

$$\frac{\partial \boldsymbol{M}}{\partial t} = \Sigma \boldsymbol{F} + \Sigma \boldsymbol{I}_M + \Sigma \boldsymbol{P}_M - \Sigma \boldsymbol{E}_M \tag{2.2}$$

where the first term on the right-hand side represents the summation of all external forces. The use of vector notation emphasizes that the equation is valid only if components are taken in a fixed direction.

For a condition of steady flow the left-hand side is zero. Thus

$$\Sigma \boldsymbol{F} = \Sigma \boldsymbol{E}_M - \Sigma \boldsymbol{I}_M - \Sigma \boldsymbol{P}_M \tag{2.3}$$

The principles enunciated above will be applied quite extensively in this book. The employment of a finite control volume in most cases presupposes that gross descriptions of flow conditions across a section of the control volume are permissible. This in effect implies one-dimensional consideration which stipulates that property variations are unimportant across the direction of flow. A semi-finite control volume results when the limiting case of an infinitesimally wide control volume with a finite length is adopted as for example is in Fig. 1.11 as $\partial x \rightarrow 0$.

2.3 Conservation of Mass

Consider a finite volume V fixed in a space through which a fluid flows (Fig. 2.1). The fluid enters and leaves the volume through its surfaces. The surface bounding the control volume is the *control surface*. If there is no artificial or natural production of fluid mass m inside the control volume the ΣP_m term on the right-hand side of equation (2.1) (X = mass, m) is zero. The other two terms represent the net flux of fluid mass across the control surface. This is given by integration of the mass flux over the control surface.
This may be written as

$$\int_{\substack{\text{control}\\\text{surface}}} \rho q_{\text{n}} \, \mathrm{d}A$$

where q_{n} is the component of velocity q normal to the surface of the control volume and δA is an elemental surface area. q_n is assumed positive when the flow is out of the control volume and negative when it is into it. Thus

$$\Sigma I_m - \Sigma E_m = -\int \rho q_n \, dA$$

The mass of fluid contained in an elemental volume δV within the control volume is $\rho \delta V$. Thus the total mass in the control volume is

$$\int_{\substack{\text{control}\\\text{volume}}} \rho \, dV$$

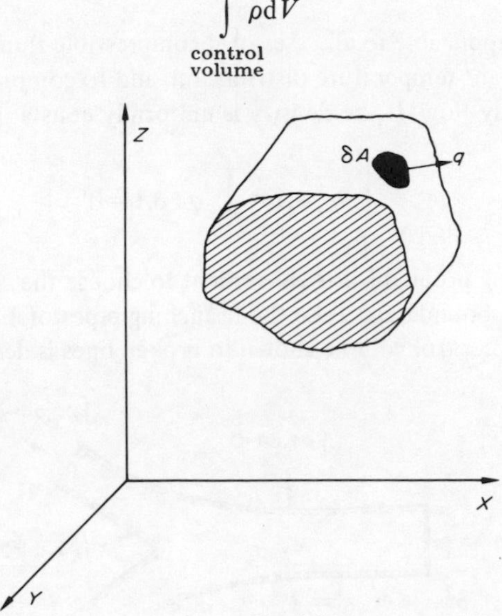

Fig. 2.1

Remembering that X is equivalent to mass in this example, equation (2.1) requires that

$$\frac{\partial}{\partial t}\int_{\substack{\text{control}\\\text{volume}}} \rho \, dV = -\int_{\substack{\text{control}\\\text{surface}}} \rho q_n \, dA \tag{2.4}$$

Since the control volume is fixed and is independent of time the differentiation can be done inside the integral sign. Thus

$$\int_{\text{c.v.}} \frac{\partial \rho}{\partial t} \, dV = -\int_{\text{c.s.}} \rho q_n \, dA \tag{2.5}$$

or in vector notation

$$\int_{\text{c.v.}} \frac{\partial \rho}{\partial t} \, dV = \int_{\text{c.s.}} \rho \boldsymbol{q} \, . \, d\boldsymbol{A} \tag{2.6}$$

where $\boldsymbol{q}$ is the absolute velocity (with a positive sign when pointing out of the control volume (c.v.)) and $d\boldsymbol{A}$ is a vector differential area pointing in the inward direction over an enclosing control surface (c.s.). Equation (2.5) or (2.6) is the general equation of continuity for a fluid of one type or of uniform mixture, that is a homogeneous fluid medium.

For the particular case of constant density and therefore of constant mass within the control volume, $\partial\rho/\partial t = 0$ and thus

$$\int_{\text{c.s.}} \rho q_n \, dA = \int_{\text{c.s.}} \rho \mathbf{q} \, . \, d\mathbf{A} = 0 \tag{2.7}$$

Equation (2.7) is applicable to all cases of incompressible fluids of one type with a reasonably constant temperature distribution, and to compressible fluids under conditions of steady flow. If the density is uniformly constant throughout the control volume

$$\int_{\text{c.s.}} q_n \, dA = \int_{\text{c.s.}} \mathbf{q} \, . \, d\mathbf{A} = 0 \tag{2.8}$$

In many engineering problems it is convenient to choose the control volume to coincide with flow boundaries. Take the branching pipes of Fig. 2.2 as an example. Here the control volume shown in broken lines is defined by the

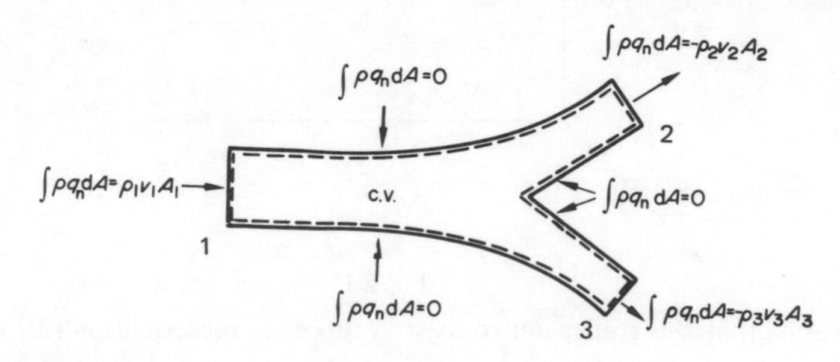

Fig. 2.2

boundaries of the pipe and the planes 1, 2 and 3. For an incompressible or a steady-state compressible flow through this volume, equation (2.7) can easily be applied. There is no flow across the wall boundaries and therefore the normal component of velocity is zero at the walls. Thus the integration gives for incompressible flow,

$$v_1 A_1 - v_2 A_2 - v_3 A_3 = 0 \tag{2.8a}$$

and for steady compressible flow,

$$\rho_1 v_1 A_1 - \rho_2 v_2 A_2 - \rho_3 v_3 A_3 = 0 \tag{2.8b}$$

where v is the average velocity across a specified section, ρ the average density and A is the cross sectional area. The same expression can be obtained directly from equation (2.1). It must be emphasized that the well-known continuity equation of vA = constant really originates from continuity of *mass* flux and reduces to *volumetric* flux only in the special case of a constant density.

In general the steady state conservation of mass m for a fixed control volume is derived from equation (2.1) as

$$\Sigma I_m + \Sigma P_m - \Sigma E_m = 0 \tag{2.9}$$

It may be observed that the derivation of equation (2.8) is general in the sense that an arbitrary volume is chosen in the fluid space. Its application to the branching pipe however assumes a finite control volume and one-dimensional conditions across the pipe sections 1, 2 and 3. The original equation (2.6) can be solved generally using Gauss' theorem

$$\int_V \nabla \cdot \phi \, \mathrm{d}V = -\int_s \boldsymbol{n} \cdot \phi \, \mathrm{d}s$$

which relates integration over a control volume to the corresponding integration over the surfaces of that control volume.* The right-hand side of equation (2.6) gives

$$\int_{\text{c.s.}} \rho \boldsymbol{q} \cdot \mathrm{d}A = \int_{\text{c.s.}} \rho \boldsymbol{q} \cdot \boldsymbol{n} \, \mathrm{d}A = \int_{\text{c.s.}} \boldsymbol{n} \cdot \rho \boldsymbol{q} \, \mathrm{d}A$$

where $\boldsymbol{n}$ is an inward unit vector at the control surface. According to Gauss' theorem,

$$\int_{\text{c.s.}} \boldsymbol{n} \cdot \rho \boldsymbol{q} \, \mathrm{d}A = -\int_{\text{c.v.}} \nabla \cdot \rho \boldsymbol{q} \, \mathrm{d}V$$

Substituting into equation (2.6)

$$\int_{\text{c.v}} \left(\frac{\partial \rho}{\partial t} + \nabla \cdot \rho \boldsymbol{q} \right) \mathrm{d}V = 0$$

Thus

$$\frac{\partial \rho}{\partial t} + \nabla \cdot \rho \boldsymbol{q} = 0$$

This is the general conservation of mass equation. In the cartesian coordinate system for homogeneous fluids this simplifies to

$$\frac{\partial \rho}{\partial t} + \rho \left(\frac{\partial u}{\partial x} + \frac{\partial v}{\partial y} + \frac{\partial w}{\partial z} \right) = 0$$

2.4 The Linear Momentum Principle (Newton's Second Law)

The statement of the time invariant linear momentum principle in equation (2.3) is that in a particular direction the sum of all surface and body forces acting on a fluid contained in a fixed control volume is equal to the sum of all components

* $\nabla = \mathrm{i}\frac{\partial}{\partial x} + \mathrm{j}\frac{\partial}{\partial y} + \mathrm{k}\frac{\partial}{\partial z}$

$\boldsymbol{q} = \mathrm{i}u + \mathrm{j}v + \mathrm{k}w$

of momentum flux across the boundaries out of the control volume less the sum of the components of momentum flux injected into the control volume.

$$\Sigma F = \Sigma E_M - \Sigma I_M - \Sigma P_M \qquad (2.10)$$

Surface and body forces	Net momentum flux out of control volume	Injected momentum flux into control volume

Equation (2.10) as it stands is applicable to compressible as well as incompressible steady flows. The following examples demonstrate the application of the linear momentum principle. The student is advised to study these examples carefully and to try as many as possible of the exercises based on chapter 2 at the end of the book.

EXAMPLE 2.1

A nozzle is a device generally employed in fluid flow to convert a relatively slow flow (high pressure) into a fast moving jet (low pressure). Figure 2.3 shows the design of a simple nozzle. If the water jet discharges into a free atmosphere, calculate the pull exerted on the bolts in the flange which couples the nozzle unit to the main pipeline for an approach velocity of 1·22 m/s. A pressure gauge indicates 103 kN/m² at the flange.

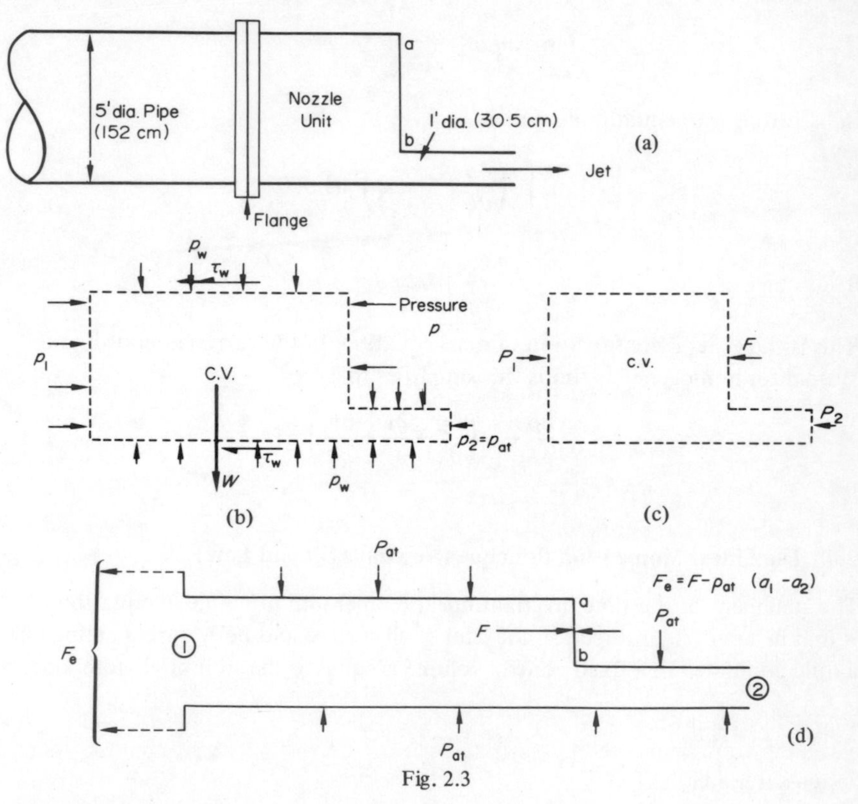

Fig. 2.3

Figure 2.3(b) shows a control volume bounded by the walls of the nozzle unit, a vertical plane through the flange and a vertical plane at the exit end of the nozzle. The forces acting on the control volume are the body force W acting vertically downward and boundary forces made up of normal (pressure) and tangential (shear) components to the various sections of the boundaries. Assuming the section of the pipe of interest is horizontal the equation of motion in the horizontal direction of flow would be sufficient to give the pull at the flange.

Pressures p_1 and p_2 are known. The unknown forces due to pressure p and shear τ_w can be combined together as a horizontal force F acting on the fluid contained in the control volume. The simplified arrangement is illustrated in Fig. 2.3(c).

Since there is no injection of momentum from external sources, applying equation (2.10) gives in the *horizontal direction*:

$$\Sigma F = \Sigma E_M - \Sigma I_M - \Sigma P_M$$

But $\quad \Sigma P_M = 0$

$$\therefore \quad \Sigma F = P_1 - P_2 - F$$

$$= \frac{\pi}{4}(1{\cdot}52)^2 \times (103 + 101) - \frac{\pi}{4}(0{\cdot}305)^2 \times 101 - F\,(\text{kN})$$

$$= 370 - 7{\cdot}4 - F = 362{\cdot}6 - F\,(\text{kN})$$

Applying the continuity equation (2.8a) the jet velocity v_2 can be obtained from

$$v_2 = \frac{v_1 A_1}{A_2} = 1{\cdot}22 \times \left(\frac{152}{30{\cdot}5}\right)^2 = 30{\cdot}30 \text{ m/s } (100 \text{ ft/s})$$

Thus

$$\Sigma E_M - \Sigma I_M = \rho Q v_2 - \rho Q v_1$$

where Q is the rate of discharge and ρ is density

$$\therefore \quad (362{\cdot}6 - F) \times 10^3 = 10^3 \times \frac{\pi}{4}(1{\cdot}52)^2 \times 1{\cdot}22 \times (30{\cdot}5 - 1{\cdot}22)$$

$$= 64{\cdot}6 \times 10^3$$

$$\therefore \quad F = 298 \text{ kN } (67600 \text{ lbf})$$

The walls of the nozzle unit exert a force of 298 kN to the left on the fluid contained in the control volume. According to Newton's third law the fluid must exert an equal amount of force to the right on the walls of the nozzle. This is made up of shear and pressure forces too complex to be defined locally. There is however a pull of 298 kN on the flange bolts.

In practice it will be necessary to design the bolts and other connections to withstand this pull. The pipe must also be designed to withstand the longitudinal and hoop stresses induced in the pipe material. The student is advised to do this as an exercise in design.

It must be pointed out that the calculations leading to the determination of F above have assumed absolute pressures. In designing the bolts, however, the equilibrium of the whole of the nozzle must be considered. Atmospheric pressure acts on all the exposed surfaces of the nozzle. The net effective hydrodynamic force which the bolts must withstand is less than obtained above by an amount due to atmospheric pressure forces. This is illustrated in Fig. 2.3(d). The relevant forces due to atmospheric pressure are those acting on the vertical right-hand solid face ab. They are given by

$$p_{at}(a_1 - a_2) = p_{at}\,a_1 - p_{at}\,a_2$$

Thus the net effective force on the nozzle, F_e, is given by

$$\begin{aligned} F_e &= F - p_{at}\,a_1 + p_{at}\,a_2 \\ &= 298000 - \frac{\pi}{4}(1{\cdot}52)^2\,(101000) + \frac{\pi}{4}(0{\cdot}305)^2\,(101000) \\ &= 123 \text{ kN } (27\,700 \text{ lbf}). \end{aligned}$$

This is the same force which would be obtained assuming atmospheric pressure as datum in the first place. For this reason the determination of reactive forces in most fluid dynamic problems assumes pressures relative to atmospheric pressure. Unless otherwise stated this practice will be followed in subsequent examples.

EXAMPLE 2.2

Figure 2.4(a) shows a tapering bend in a pipeline technically referred to as a *reducing elbow.* It is required to design a support for the elbow to take the force

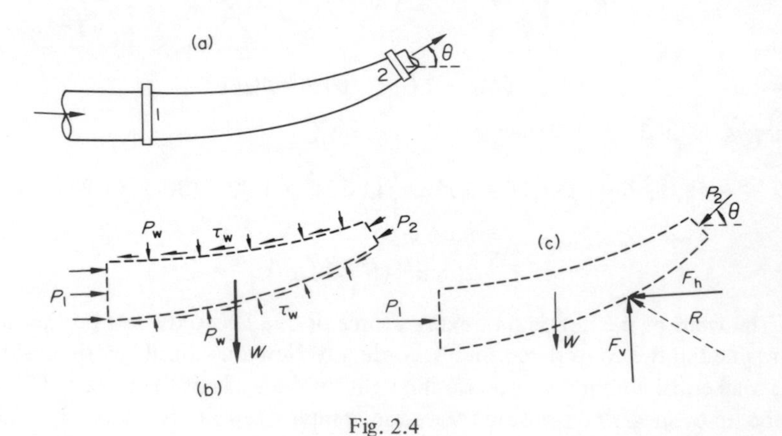

Fig. 2.4

resulting from a steady flow of fluid through the pipe. The initial direction of flow is horizontal and the bend is upward.

The control volume is isolated as shown in Fig. 2.4(*b*) and all the forces acting on the fluid are indicated. These comprise normal boundary forces (pressure), tangential boundary forces (shear) and the body force due to the weight of fluid. In the absence of a sufficiently good knowledge of the forces between the fluid and the reducer walls, the combined effects of wall pressure (p_w) and shear (τ_w) may be represented by a force R resolved horizontally and vertically into F_h and F_v (see Fig. 2.4(c)). Applying equation (2.10) without injected momentum, in the horizontal direction:

$$P_1 - P_2 \cos\theta - F_h = mv_2 \cos\theta - mv_1$$

in the vertical direction:

$$-P_2 \sin\theta - W + F_v = mv_2 \sin\theta$$

where m the constant mass flux

$$= \rho_1 A_1 v_1 = \rho_2 A_2 v_2$$

or $$F_h = \rho_1 A_1 v_1 (v_1 - v_2 \cos\theta) + p_1 A_1 - p_2 A_2 \cos\theta \quad \text{(a)}$$

and $$F_v - W = \rho_1 A_1 v_1 v_2 \sin\theta + p_2 A_2 \sin\theta \quad \text{(b)}$$

The resultant force

$$R = (F_h{}^2 + F_v{}^2)^{\frac{1}{2}}$$

inclined at

$$\tan^{-1} \frac{F_v}{F_h}$$

to the horizontal.

Changing the signs of F_h and F_v gives the force components on the elbow from the fluid. The support must be able to take the force R plus the weight of the pipe material. Knowing the magnitude and inclination of the resultant force, its point of action can be determined graphically or otherwise.

EXAMPLE 2.3

A problem quite similar conceptually to that dealt with in Example 2.1 is illustrated in Fig. 2.5. The cross sectional area of a sluice gate, a device used in controlling the flow in open channels, is shown. The sluice traps water upstream and allows flow into the downstream channel from under the gate at a high speed. The minimum depth of flow is attained at the *vena contracta* at section 2 and the problem is to determine the force exerted on the sluice gate per unit width. This force determines the structural design of the gate.

The control volume defined by the vertical planes through sections 1 and 2, the surface, the inside of the sluice gate and the channel floor is shown in Fig. 2.5(*b*). Hydrostatic pressure distribution is assumed and the pressure gauge readings are taken. The latter is justified by the fact that horizontally the atmospheric pressure forces on the left-hand side of the control volume cancel out those on the right. The same is true for the surface and floor components. A less justifiable assumption is that the shear forces on the floor are negligible. The thrust of the sluice gate on the fluid is F.

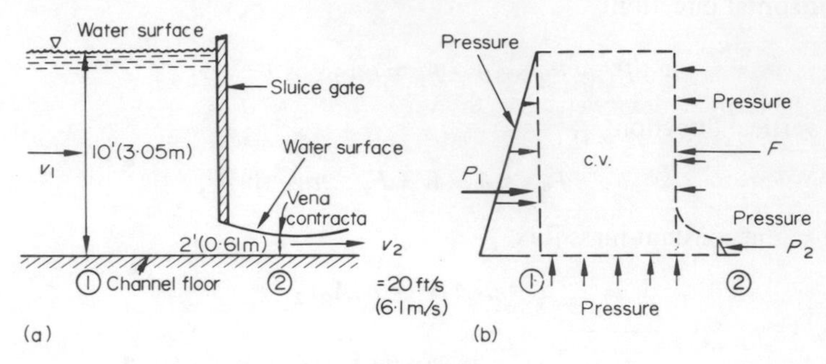

Fig. 2.5

Applying equation (2.10) horizontally,

$$P_1 - P_2 - F = E_M - I_M$$

that is
$$\frac{1}{2}\gamma\,(3{\cdot}05^2 - 0{\cdot}61^2)\,F = \rho \times 0{\cdot}61 \times 6{\cdot}1\,(6{\cdot}1 - 1{\cdot}22)$$

Since by continuity
$$3{\cdot}05 v_1 = 0{\cdot}61 v_2 = 3{\cdot}72$$

$$F = \frac{9{\cdot}81 \times 10^3}{2} \times 8{\cdot}94 - 10^3 \times 3{\cdot}72 \times 4{\cdot}88$$

$$= (43{\cdot}7 - 18{\cdot}1)\,10^3$$

$$= \underline{25{\cdot}6 \times 10^3 \text{ N/m } (1755 \text{ lbf/ft})}$$

Thus, the force on the sluice gate is 25·6 kN/m towards the right. The structural strength and rigidity of the gate must therefore be sufficient to withstand this force and the torque it produces.* However, its line of action cannot be determined precisely.

EXAMPLE 2.4

We want to analyse the performance of a liquid-liquid ejector in which the ejector liquid a enters liquid b and pumps the mixture into a delivery line. The ejector liquid a of density ρ_a and average velocity v_o and the liquid b of density ρ_b and average velocity $1/3\ v_o$ enter the cylindrical mixing chamber at section A (Fig. 2.6). The cross-sectional area of the mixing chamber is a and that of the ejector pipe is $a/3$.

* It is to be noted that if absolute pressures were used in the calculation, the value of F would be higher by p_{at} (3·05 – 0·61). This would however cancel out when the equilibrium of the gate was considered assuming that the *vena contracta* occurred under the gate.

At section B the two liquids are completely mixed; their mean velocity is v_1 and the pressure is uniform. Assuming that the flow is steady and neglecting friction on the mixing chamber walls, show that, if $\rho_b = 3\rho_a$,

$$\frac{p_a + 2p_b}{3} - p_1 = \rho_a v_0^2 \left(\frac{v_1}{v_0} - \frac{5}{9}\right)$$

where p_a and p_b are the pressures of liquids a and b at A.

In a laboratory demonstration the velocity of the mixture is 12·3 m/s, $p_a = 0{\cdot}276$ N/mm², $p_b = 0{\cdot}172$ N/mm² and $p_1 = 0{\cdot}172$ N/mm². What is the volumetric discharge ratio between the denser and lighter fluid? Take $\rho_a =$ 515 kg/m³. (Dip. IV Mech., U.S.T., 1967).

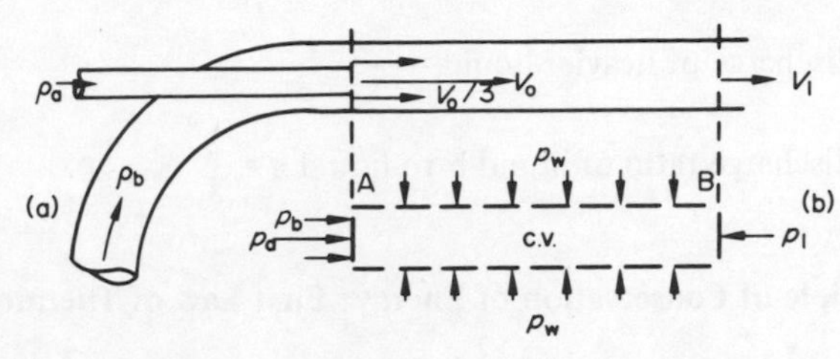

Fig. 2.6

SOLUTION

The appropriate control volume between sections A and B neglecting wall shear, is sketched in Fig. 2.6(*b*). Applying equation (2.10) in the longitudinal (horizontal) direction

F_1	+ F_2	− F_3	= E_M	− I_{M1}	− I_{M2}
liquid a at A	**liquid b at A**	**at section B**	**at B**	**liquid a at A**	**liquid b at A**

that is

$$\frac{a}{3} p_a + \frac{2}{3} a\, p_b - p_1 a = \rho_m a v_1^2 - \rho_a \frac{a}{3} v_0^2 - \rho_b \frac{2}{3} a \left(\frac{v_0}{3}\right)^2$$

i.e.

$$\frac{p_a + 2p_b}{3} - p_1 = \rho_m v_1^2 - \rho_a \frac{v_0^2}{3} - \frac{2\rho_b}{27} v_0^2 \qquad \text{(a)}$$

where ρ_m is density of mixture at B.

Applying equation (2.9) since there is no production of mass m into the control volume ($\Sigma P_m = 0$)

$$I_{m1} + I_{m2} - E_m = 0$$

$$\therefore \qquad \rho_m v_1 = \frac{\rho_a v_0}{3} + \frac{2}{3} \rho_b \frac{v_0}{3} = v_0 \rho_a \qquad \text{(b)}$$

Substitute (b) in (a)

$$\frac{p_a + 2p_b}{3} - p_1 = \rho_a v_0 v_1 - v_0^2 \rho_a (1/3 + 2/9)$$

$$\therefore \qquad \frac{p_a + 2p_b}{3} - p_1 = \rho_a v_0^2 \left(\frac{v_1}{v_0} - \frac{5}{9}\right)$$

With the given data

$$67{\cdot}6 = 12{\cdot}3v_0 - \frac{5}{9} v_0^2$$

i.e. $$v_0^2 - 22v_0 + 122 = 0$$

Thus $v_0 = 11{\cdot}0$ m/s, the other root being unrealistic. And the

volumetric discharge of lighter liquid $= \frac{a}{3} \cdot 11$

volumetric discharge of heavier liquid $= \frac{2a}{3} \cdot \frac{11}{3}$

volumetric discharge ratio of liquid b to liquid a $= \frac{2}{3}$

2.5 The Principle of Conservation of Energy: First Law of Thermodynamics

The energy of a system or a control volume may conveniently be classified into inherent or *stored energy* and *energy in transit.* The former refers to energy primarily associated with a given mass and the latter to energy moving from one system or control volume into another. In classical mechanics the inherent energy E of an element of mass is made up of:

(1) *kinetic energy*, E_k: energy acquired by virtue of the local velocity of the mass

(2) *potential energy*, E_p: energy associated with the local position of the mass and

(3) *internal energy*, U: energy associated with the molecular and atomic activity within the mass

$$E = E_k + E_p + U$$

Heat energy and *work* are the two types of transitional energy usually met in classical mechanics. Heat is the energy in transit from one mass to another as a result of a temperature difference. Work is the energy transferred to or from a system as a result of the application of an external force through a distance.

In thermodynamics the concept of work includes energy transferred to or from a system by an action of the system such that the total effect outside the system of the given action can be reduced by hypothetical frictionless mechanisms equivalent to that of raising a weight in a gravitational field. The work W done on a mass contained in a fluid system or a control volume may include:

(1) *pressure work*, W_p: due to normal stresses (pressure) acting on the boundaries of the system or control volume;

(2) *shear work*, W_s: due to tangential (shear) stresses at the boundaries of the system or control volume and

(3) *shaft work*, W_{sh}: due to a rotating element in the system or control volume and transmitted from outside into the system or control volume through a rotating shaft

$$W = W_p + W_s + W_{sh}$$

Heat energy Q and work W may be considered as energy added or generated within the system or control volume. Equation (2.1) may be applied (remembering that the rate of change of energy within the system or control volume is entirely reflected in the rate of change of its inherent or stored energy) to give

$$\frac{\partial E}{\partial t} = \Sigma I_e + \frac{\partial Q}{\partial t} + \frac{\partial W}{\partial t} - \Sigma E_e \tag{2.11}$$

where
$$\Sigma P_e = \frac{\partial Q}{\partial t} + \frac{\partial W}{\partial t}$$

Equation (2.11) is a general statement of the first law of thermodynamics which stipulates that energy must at all times be conserved. In a static system the energy flux terms do not exist and the increase in stored energy in a mass is equal to the heat transferred to the mass plus the work done on the mass

$$\Delta E = \Delta Q + \Delta W \tag{2.12}$$

As discussed above, the term E may be given as the sum of the following specific types of stored energy for a fixed mass m of fluid:

(1) *Kinetic energy*, E_k. From basic classical mechanics the kinetic energy of mass m moving with a velocity v is $m\,v^2/2$. It must be emphasized that for a fluid mass of a finite volume the velocity v is that of the centroid of the mass.

(2) *Potential energy*, E_p. If the only external force field is that of gravity, classical mechanics shows that the potential energy is related to acceleration due to gravity g and the height z of the centre of gravity of the mass above an *arbitrary* datum by mgz.

(3) *Internal energy*, U. The internal energy of a substance is the result of activity of the component molecules and of the atoms comprising the molecules as well as of forces existing between individual molecules.

The molecular activity gives rise to kinetic energy which is considered stored within the molecule, and the inter-molecular activity represents a potential energy which is also stored within the molecule.

The molecular kinetic energy is dependent primarily on temperature but the inter-molecular potential energy is determined by the change of molecular structure of the substance, A significant change of molecular structure involves a change in phase (for example, from liquid to gas) which is beyond the scope of this book. All changes of internal energy will thus be considered due entirely to molecular activity which is known to be dependent primarily on temperature changes.

The rate at which work is done by pressure forces on fluid associated with the control volume is given by the product of the pressure force and the velocity component in the direction of the force. Thus the net rate at which the surroundings of the control volume do work on the fluid through pressure is

$$\frac{\partial W_\mathrm{p}}{\partial t} = \Sigma p A v$$

where p is the average pressure at the section exposed to the surroundings, A is the cross sectional area and v is the average velocity

or

$$\frac{\partial W_\mathrm{p}}{\partial t} = \Sigma \left(p \frac{m'}{\rho} \right)$$

where the mass flux through the surface is $m' = \rho A v$.

Substituting into equation (2.11)

$$\frac{\partial E}{\partial t} = \Sigma m \left(\frac{v^2}{2} + gz + \frac{U}{m} \right)_{\text{influx}} - \Sigma m \left(\frac{v^2}{2} + gz + \frac{U}{m} \right)_{\text{efflux}} + \frac{\partial Q}{\partial t} + \frac{\partial W_\mathrm{s}}{\partial t} + \frac{\partial W_\mathrm{sh}}{\partial t} + \Sigma \left(\frac{p}{\rho} m' \right) \tag{2.13}$$

The last term of equation (2.13) can be split up according to whether the flow is into or out of the control volume. At inlets to the control volume work is done on the fluid in the control volume and fluid from the control volume does work on the surrounding at outlets.

Thus

$$\Sigma \left(\frac{p}{\rho} m' \right) = \Sigma \left(\frac{p}{\rho} m \right)_{\text{influx}} - \Sigma \left(\frac{p}{\rho} m \right)_{\text{efflux}} \tag{2.14}$$

And so, combining (2.13) and (2.14) we get

$$\frac{\partial E}{\partial t} = \Sigma mg \left(\frac{v^2}{2g} + z + \frac{p}{\rho g} + \frac{u}{g} \right)_{\text{influx}} - \Sigma mg \left(\frac{v^2}{2g} + z + \frac{p}{\rho g} + \frac{u}{g} \right)_{\text{efflux}} + \frac{\partial Q}{\partial t} + \frac{\partial W_\mathrm{s}}{\partial t} + \frac{\partial W_\mathrm{sh}}{\partial t} \tag{2.15}$$

where u is the internal energy per unit mass.

For the steady flow in which there is no increase in stored energy the general energy equation for the finite control volume is

$$-\underset{\text{heat power}}{\frac{\partial Q}{\partial t}} - \underset{\text{shear power}}{\frac{\partial W_s}{\partial t}} - \underset{\text{shaft power}}{\frac{\partial W_{sh}}{\partial t}} = \underset{\text{inlet power}}{\Sigma\, mg\left(\frac{v^2}{2g} + z + \frac{p}{\rho g} + \frac{u}{g}\right)_{\text{influx}}}$$

$$-\underset{\text{outlet power}}{\Sigma\, mg\left(\frac{v^2}{2g} + z + \frac{p}{\rho g} + \frac{u}{g}\right)_{\text{efflux}}} \qquad (2.16)$$

BERNOULLI'S EQUATION

The special case of an insulated (or negligible heat transfer) control volume in which no shaft work is done is commonly encountered in liquid flow problems. The appropriate energy equation is

$$-\frac{\partial W_s}{\partial t} = \Sigma\, mg\left(\frac{v^2}{2g} + z + \frac{p}{\rho g} + \frac{u}{g}\right)_{\text{influx}} - \Sigma\, mg\left(\frac{v^2}{2g} + z + \frac{p}{\rho g} + \frac{u}{g}\right)_{\text{efflux}}$$

For a flow through a duct

$$-\frac{1}{mg}\frac{\partial W_s}{\partial t} = \left(\frac{v^2}{2g} + z + \frac{p}{\rho g} + \frac{u}{g}\right)_{\text{influx}} - \left(\frac{v^2}{2g} + z + \frac{p}{\rho g} + \frac{u}{g}\right)_{\text{efflux}}$$

since influx mass equals efflux mass *m*.

Putting

$$h_f = -\frac{1}{mg}\frac{\partial W_s}{\partial t} = \text{work done \textit{per unit weight} of flowing fluid against wall resistance}$$

$$\left(\frac{v^2}{2g} + z + \frac{p}{\rho g} + \frac{u}{g}\right)_{\text{influx}} = \left(\frac{v^2}{2g} + z + \frac{p}{\rho g} + \frac{u}{g}\right)_{\text{efflux}} + h_f \qquad (2.17)$$

The term $(p/\rho + u)$ is called enthalpy in thermodynamics. For liquids, the variation of u with temperature is so small that $u_{\text{influx}} = u_{\text{efflux}}$ for all practical purposes. Thus for a liquid flow and denoting influx conditions by 1 and efflux conditions by 2 we get

$$\frac{v_1^2}{2g} + z_1 + \frac{p_1}{\gamma} = \frac{v_2^2}{2g} + z_2 + \frac{p_2}{\gamma} + h_{f(1\to 2)} \qquad (2.18)$$

Note that each element of equation (2.17) or (2.18) has the dimension of a length. This may be verified by the fact that each element represents energy per unit weight of the fluid, or

$$\frac{\text{Nm}}{\text{N}} \equiv \text{m} \quad \text{and} \quad \frac{\text{ft} \times \text{lbf}}{\text{lbf}} \equiv \text{ft}$$

The application of equation (2.18) to liquid flow through a pipeline is illustrated in Fig. 2.7. An arbitrary horizontal datum is chosen and the following points must be carefully noted:

(1) The total (energy) head is made up of the kinetic energy, the pressure energy and the potential energy heads plus all head losses in friction and in some cases energy additions up to the point of interest.

(2) The energy head is made up of the kinetic energy, the pressure energy and the potential energy heads. It is equal to the total energy head only when the point in question is used as the reference point as (1) is in Fig. 2.7 or in cases

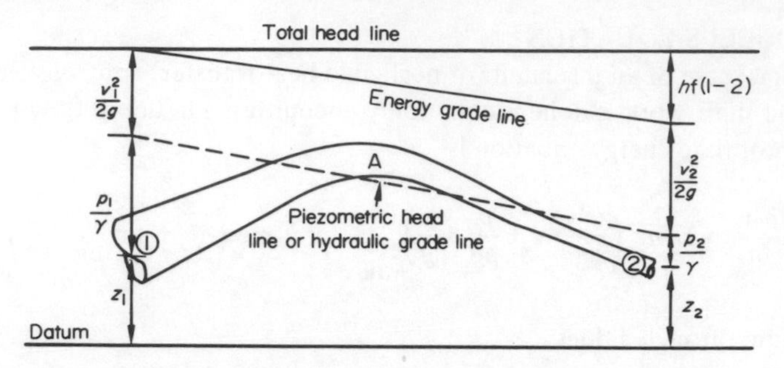

Fig. 2.7

where there is no loss of energy. The energy grade line may never be horizontal or slope upward in the direction of flow if the fluid is real and no energy is being added. Some people refer to it as the total (energy) head.

(3) The piezometric head is everywhere an amount $(p/\rho g + z)$ vertically above the datum. The line showing the variation of piezometric head is known as the hydraulic grade line. The vertical distance of the piezometric head above a point in the pipe gives the pressure head. Whenever the line falls below a point in the system as at A, the local pressure will be less than atmospheric (reference) pressure.

(4) The total energy line, the energy grade line and the hydraulic grade line are all coincident and lie horizontal (in line with the liquid level in the system) only when the liquid is at rest.

In the ideal flow case in which there are no energy losses due to shear, we get

$$\frac{v_1^2}{2g} + z_1 + \frac{p_1}{\rho g} = \frac{v_2^2}{2g} + z_2 + \frac{p_2}{\rho g} = H \text{ (a constant)} \tag{2.19}$$

This is *Bernoulli's Equation*. All the implications in quoting Bernoulli's equation must be clearly appreciated. Conditions must be steady, there must be no energy transfer from or to the outside, and the flow must be frictionless. If the flow is rotational it can be applied only along a specified streamline. The modified Bernoulli Equation (2.18) is very powerful in solving one-dimensional fluid flow problems.

EXAMPLE 2.5

Air enters a 35·6 cm (14 inch) diameter horizontal pipe at an absolute pressure of 0·102 N/mm^2 (14·8 lbf/in^2) and a temperature of 15·5°C (60°F). At the exit the pressure is 0·100 N/mm^2 (14·5 lbf/in^2) absolute. The entrance velocity is 10·7 m/s (35 ft/s) and the exit velocity is 15·3 m/s (50 ft/s). The gas constant is 292 J/kg K (54·5ft lbf/lbm °R). C_p = 10^3 J/kg K (0·24 Btu/lbm °R). Determine the quantity and direction of heat flow.

Under steady conditions the appropriate equation is derived from equation (2.16) as

$$-\frac{\partial Q}{\partial t}-\frac{\partial W_s}{\partial t}=m\left[\left(\frac{v_1^2}{2}+gz_1+\frac{p_1}{\rho_1}+u_1\right)-\left(\frac{v_2^2}{2}+gz_2+\frac{p_2}{\rho_2}+u_2\right)\right]$$

Assuming negligible shear power,

$$\frac{\partial W_s}{\partial t}=0$$

then from the perfect gas law

$$p_1=\rho_1 RT_1$$

$$102\times 10^3=\rho_1\times 292\times(273+15{\cdot}5)$$

$$\therefore \qquad \rho_1=1{\cdot}21\ \text{kg/m}^3\ (0{\cdot}00234\ \text{slug/ft}^3)$$

Continuity of mass from equation (2.8b) gives

$$\rho_2=\frac{v_1}{v_2}\rho_1=\frac{10{\cdot}7}{15{\cdot}3}\times 1{\cdot}21$$

$$=0{\cdot}85\ \text{kg/m}^3\ (0{\cdot}00164\ \text{slug/ft}^3)$$

The exit temperature

$$T_2=\frac{p_2}{\rho_2\times 292}=403\ \text{K}\ (730°\text{R})$$

The mass rate of flow

$$m=\frac{\pi}{4}(0{\cdot}356)^2\rho_1 v_1=1{\cdot}28\ \text{kg/s}\ (0{\cdot}0875\ \text{slug/s})$$

If $\qquad z_1=z_2$

$$-\frac{\partial Q}{\partial t}=m\left[\left(\frac{p_1}{\rho_1}+u_1\right)-\left(\frac{p_2}{\rho_2}+u_2\right)+\frac{v_1^2}{2}-\frac{v_2^2}{2}\right]$$

$$=m\left(h_1-h_2+\frac{v_1^2}{2}-\frac{v_2^2}{2}\right)$$

where h (enthalpy) = C_pT

$$-\frac{\partial Q}{\partial t} = m\left[C_p\,(T_1 - T_2) + \frac{v_1^2 - v_2^2}{2}\right]$$

$$= 1{\cdot}28\,[10^3\,(288{\cdot}5 - 403) - \frac{1}{2}(114{\cdot}5 - 234)]\ \text{Nm/s}$$

or $$\frac{\partial Q}{\partial t} = \underline{0{\cdot}15 \times 10^6\ \text{J/s (142 Btu/s)}}$$

Thus 0·15 MW is transferred from outside into the pipe.

EXAMPLE 2.6

A perfect gas flows steadily through the machine shown in Fig. 2.8. The gas constant for the gas is 214 J/kg K (40 ft lb/lbm °R) and 1356 J (1000 ft lbf) of heat is added per second. Calculate the shaft work of the machine.
$v_1 = v_2 = 9$ m/s (30 ft/s)

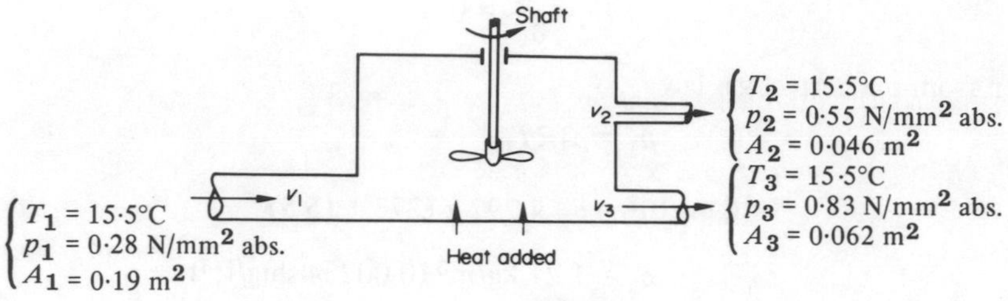

Fig. 2.8

From equation (2.16) neglecting friction

$$-\frac{\partial Q}{\partial t} - \frac{\partial W_{sh}}{\partial t} = m_1\left(\frac{v_1^2}{2} + gz_1 + \frac{p_1}{\rho_1} + u_1\right)$$

$$- m_2\left(\frac{v_2^2}{2} + gz_2 + \frac{p_2}{\rho_2} + u_2\right)$$

$$- m_3\left(\frac{v_3^2}{2} + gz_3 + \frac{p_3}{\rho_3} + u_3\right)$$

From the continuity equations

$$m_1 = m_2 + m_3$$

Also $$h_1 = h_2 = h_3$$

since $$T_1 = T_2 = T_3$$

$$z_1 = z_2 = z_3$$

Thus

$$-\frac{\partial Q}{\partial t} - \frac{\partial W_{sh}}{\partial t} = m_1\left(\frac{v_1^2}{2}\right) - m_2\left(\frac{v_2^2}{2}\right) - m_3\left(\frac{v_3^2}{2}\right)$$

$$\rho_1 = \frac{p_1}{(R)T_1} = \frac{275\,000}{214 \times 288{\cdot}5} = 4{\cdot}44 \text{ kg/m}^3$$

and $\rho_2 = 8{\cdot}88 \text{ kg/m}^3$

and $\rho_3 = 13{\cdot}32 \text{ kg/m}^3$

$$m_1 = \rho_1 A_1 v_1 = 7{\cdot}55 \text{ kg/s}$$

$$m_2 = \rho_2 A_2 v_2 = 3{\cdot}78 \text{ kg/s}$$

$$m_3 = 3{\cdot}78 \text{ kg/s}$$

and $$v_3 = \frac{378}{13{\cdot}32 \times 0{\cdot}062} = 4{\cdot}58 \text{ m/s}$$

$$-1356 - \frac{\partial W_{sh}}{\partial t} = 7{\cdot}55\left(\frac{83{\cdot}6}{2}\right) - 3{\cdot}78\left(\frac{83{\cdot}6}{2}\right) - 3{\cdot}78\left(\frac{21}{2}\right)$$

$$-1356 - \frac{\partial W_{sh}}{\partial t} = 198{\cdot}4 \text{ J/s}$$

Thus the shaft work of the machine is approximately 1·55 kW.

EXAMPLE 2.7

Calculate the rate of frictional energy dissipation for the sluice gate arrangement discussed in Example 2.3. Calculate the coefficient of discharge for an approach velocity of 1·22 m/s (4 ft/s).

Apply equation (2.18) to the control volume shown in Fig. 2.5 and note that if hydrostatic pressure is assumed:

$$(z_1 + p_1/\gamma) = 3{\cdot}05 \text{ m}$$

and $$(z_2 + p_2/\gamma) = 0{\cdot}61 \text{ m}$$

Thus $$h_{f(1\to 2)} = 0{\cdot}62 \text{ m}$$

$\therefore$ $$\frac{\partial W_s}{\partial t} = -\gamma Q h_{f(1\to 2)} = -22{\cdot}6 \text{ kW/m}$$

Thus the water expends 22·6 kW per metre in overcoming friction.

Applying Bernoulli's equation (2.19)

$$\frac{v_2^2}{2g} = \frac{v_1^2}{2g} + \left(z_1 + \frac{p_1}{\gamma}\right) - \left(z_2 + \frac{p_2}{\gamma}\right)$$

$\therefore$ $$v_2 = 7{\cdot}03 \text{ m/s}$$

The ideal discharge $= 4{\cdot}3 \text{ m}^2/\text{s}$

Thus the coefficient of discharge

$$c_d = \frac{\text{Measured discharge}}{\text{Ideal discharge}} = 0{\cdot}87$$

2.6 The Moment of Momentum Concept

The two most powerful methods employed in solving problems in classical mechanics involve the resolution of forces and taking moments of forces about a fixed axis. In the former method the sum of the components of external forces acting on a body in a particular direction is put equal to the rate of change of momentum produced. This of course is Newton's second law of motion which has been applied to cases of linear fluid motions in preceding sections. Using the other method which has been found particularly convenient for dynamic problems involving curvilinear and revolving systems, the vector sum of all external moments (torques) acting on an object is made equal to the time rate of change of angular momentum (moment of momentum) of the object. This is a direct corollary to Newton's second law since

$$\text{Torque} = \text{Force} \times r = r \times \frac{\mathrm{d}M}{\mathrm{d}t} = \frac{\mathrm{d}}{\mathrm{d}t}(r \times M)$$

where r is the perpendicular distance between the line of action of the force and the axis about which moments are taken.

In line with the linear momentum equation (2.2) the statement for a finite rotating control volume may be put as

$$\frac{\partial M_{\mathrm{a}}}{\partial t} = \Sigma T_{\mathrm{e}} + \Sigma I_{MM} - \Sigma E_{MM} \tag{2.20}$$

where M_{a} is the angular momentum, T_{e} the external torque, I_{MM} is the moment of momentum influx into the control volume and E_{MM} is the moment of momentum efflux out of the control volume.

The case of no injected momentum flux is assumed for simplicity and for steady situations

$$\Sigma T_{\mathrm{e}} = \Sigma E_{MM} - \Sigma I_{MM} \tag{2.21}$$

The external moments arise from the external forces on the boundary (shear and pressure) and from gravity (body force) or field (electric, magnetic, etc.) forces.

EXAMPLE 2.8

The arrangement shown in Fig. 2.9 illustrates the principles of a lawn sprinkler. Water is let in at the rate of q. It flows out radially in the two arms of the sprinkler and forms two symmetrical jets, The momentum efflux makes the arms rotate at a speed ω and thereby sprinkle water into the air which falls back on the lawn. Derive an expression for the speed of rotation of the rotor for a flow of q under a total head H relative to the base of the sprinkler. Neglect friction.

A convenient control volume is chosen coincident with the rotating arms. The jet velocity $v_{\mathrm{j}} = q/2A_{\mathrm{j}}$ (from continuity). Since the arms rotate in a horizontal plane, only forces and momentum fluxes whose components contribute to torques in a horizontal plane are of interest. The external torque consists of that due to pressure forces at the jet outlets and that exerted on the control volume by the walls of the rotating arm. The former is negative since the control volume does work in overcoming atmospheric pressure but the latter is positive since the rotor walls do work on the fluid in the control volume. However, taking

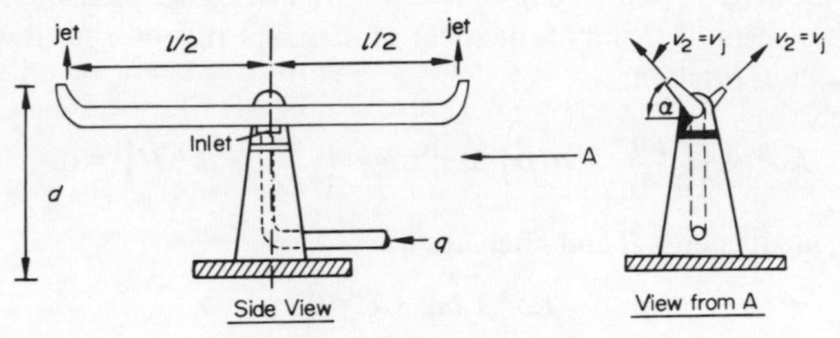

Fig. 2.9 The lawn sprinkler

atmospheric pressure as datum the torque effect due to atmospheric pressure (which will in any case cancel out when the equilibrium of the whole sprinkler is considered) is taken as zero.

Thus
$$\Sigma T_e = T$$

where T is due to reactive pressure and resistive forces only.

The inlet moment of momentum is zero since the inlet radius r is zero. Component of outlet moment of momentum flux in a horizontal plane

$$= 2\,(\rho v_j^2\, A_j \cos\alpha)\frac{l}{2}$$

$$= \rho\,\frac{q^2}{4A_j}\; l\cos\alpha$$

From equation (2.21)

$$T = \frac{\rho q^2}{4A_j}\; l\cos\alpha$$

It is significant that the body force (weight of fluid in the rotor) does not appear in the equation for torque since it acts parallel to the axis of rotation. In order to keep the rotor stationary, a torque equal and opposite to T must be applied.

The power required to produce a speed ω rad/s in the rotor is

$$T\omega = \frac{\omega\rho q^2 l\cos\alpha}{4A_j}$$

From the principle of the conservation of energy the excess of energy input over the energy output of the system equals the energy consumed in the system

plus the kinetic energy of the rotating system. The energy consumed is expended in driving the rotor.
Thus

$$\rho g q H - \rho g q \left(\frac{v_j^2}{2g} + d\right) = \frac{\omega \rho q^2}{4A_j}\, l \cos\alpha + \frac{1}{2} I\omega^2$$

where d is the height of outlet above base of sprinkler, gauge pressure is assumed in the measurement of H and I is moment of inertia of the rotor filled with water about the axis of rotation.
Alternatively

$$I\omega^2 + \left(\frac{1}{2}\frac{\rho q^2}{A_j}\, l \cos\alpha\right)\omega + 2\rho g q \left(\frac{q^2}{8gA_j^2} + d - H\right) = 0$$

For a fixed input energy H and discharge q,

$$I\omega^2 + b\omega + C = 0$$

giving the speed of rotation as

$$\omega = \frac{-b + \sqrt{(b^2 - 4IC)}}{2I}$$

where

$$b = \frac{1}{2}\rho q^2 \frac{l \cos\alpha}{A_j}$$

and

$$C = 2\rho g q \left(\frac{q^2}{8gA_j^2} + d - H\right)$$

FURTHER READING

Daily, J. W. and Harleman, D. R. F., *Fluid Dynamics*, Addison Wesley, U.S.A.
Prandtl, L., *The Essentials of Fluid Dynamics*, Blackie and Son, London.
Shames, I., *Mechanics of Fluids*, McGraw-Hill, New York.

[*Note*. See Appendix p. 401 for application of energy and continuity concepts to flow measuring devices.]

3 Steady Incompressible Flow Through Pipes

3.1 Introduction

Apart from natural circulation of air and other gases in the atmosphere and liquids of one form or another in the oceans, lakes and other reservoirs, fluid masses are transported from one locality to another through ducts. A duct may be in a closed sectional form or an open form. Flow of fluids through a closed duct under some form of pressure is referred to as a closed conduit or pressure flow. Both liquids and gases can be transported through closed conduits. Flow in an open duct exposed to an approximately constant pressure at the surface is known as a free-surface or an open-channel flow. Liquids are commonly transported through open channels and this form of transportation will be treated in Chapters 4 and 7. This chapter is devoted to the study of the steady incompressible flow of fluids through closed conduits, particularly those of circular cross-sectional area.

In the discussions which follow it is assumed that the boundary layer is fully developed, that is, the flow is fully established and, therefore, changes in the velocity distribution across two different sections along the pipeline are brought about only by changes in the size of the pipe. The establishment of flow in a pipe, due to progressive development of the boundary layer, the origin of turbulence and the associated laminar sublayer in turbulent flow is outside the scope of this book. Emphasis is nevertheless placed on the total effects of these phenomena on the capacity of conduits to transport fluids. Flows are supposed to be fully established over a range from a pipe entrance of about 130 times the pipe diameter when the flow is laminar and between 50 and 100 times the pipe diameter when the flow is turbulent.

The concepts of conservation of mass, energy and linear momentum are readily applied in most pipe-flow problems. Turbulence, which involves a varying intensity across a pipe section, introduces complications in the analysis of flows where the Reynolds number ($\bar{v}D/\nu$) (based on the average pipe velocity and diameter) is high. This makes exact solutions complicated as they require three-dimensional considerations. Fortunately, however, the use of one-dimensional concepts gives very satisfactory results for most practical applications. Discussions in the present chapter (except in Section 3.2 on flows at low Reynolds number) will therefore adopt the one-dimensional approach using the average velocity across a particular cross section. Exact solutions are possible under conditions of laminar flow since momentum transfer through shear for most fluids is Newtonian and obeys equation (1.7) to a reasonable approximation.

Two types of resistance between solids and flowing fluids are usually encountered; resistance due to friction (shear or skin friction) and resistance due to the shape of the solid object (form resistance). The concepts of skin friction and form resistance are valid whether the fluid is flowing through a duct or round an object. The concepts of resistances have already been introduced in Section 1.7. A change in the shape of a duct or an object brings about changes in pressure distribution in the duct or around the object. The net pressure force in the direction of flow gives rise to form resistance. Cases in which there is an enlargement of the flow section (decreasing velocity) result in an increased pressure (see Fig. 3.6). Increases in pressure in the downstream direction give an adverse pressure gradient in which the flow is against a higher pressure. This generally introduces a return flow which causes eddies and through which much energy may be lost. Thus, just as a loss of energy is always associated with shear, a loss of energy is always associated with form resistance especially that due to flow expansion. In closed conduits form losses are associated with changes in pipe section, flow around bends, through valves and other fittings. Some of these will be treated more fully in Subsection 3.4.2.

3.2 Enclosed Flow at a Low Reynolds Number

3.2.1 Introduction

In the 19th century Osborne Reynolds, an English scientist working at the University of Manchester, demonstrated through experiments that the flow of a liquid through a pipe would be well ordered (laminar) so long as a dimensionless parameter $R = \bar{v}D/\upsilon$ ($\bar{v}$ is the average pipe velocity, D is the pipe diameter, and υ is the molecular kinematic viscosity) was less than 2300. The dimensionless number, subsequently named after Reynolds, has since become an important parameter in the classification of fluid flows as laminar or turbulent. It is generally accepted that turbulent flow through a pipe cannot exist if $R < 2300$ and this number is known as the lower critical Reynolds number for pipe flow (the equivalent value for open channels is about 4000). Reynolds proposed that instability resulting in disorderly (turbulent) flow would occur if R were greater than 12 000. Subsequent well-controlled experiments have however proved that laminar flow can exist in a pipe at $R > 40\,000$ although the least disturbance (perturbation) will initiate fully turbulent conditions. The upper limit, that is the limit to which laminar flow can be stretched, is therefore undefined.

The practical application of laminar flow theory is limited to very viscous fluids at low velocities such as in lubrication systems and dashpots. The flow of air or water through pipes is hardly ever laminar because of the low viscosities involved. It would require extremely small velocities in small pipe diameters to make $\bar{v}D/\upsilon < 2300$. However, the fact that laminar flow is one of the few areas in fluid mechanics which yield to exact mathematical solution is sufficient to keep academic interest in the theory alive even in civil engineering programmes.

3.2.2 Laminar Flow Through Closed Conduits

Consider the dynamics of laminar fluid flow through an elemental control volume located in a circular pipe discharging a viscous fluid, as shown in Fig. 3.1.

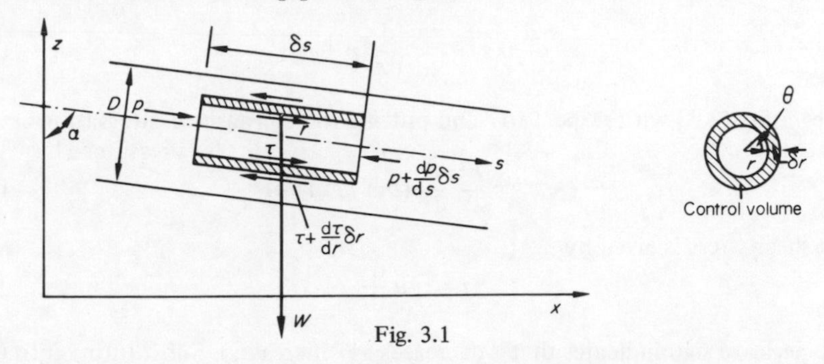

Fig. 3.1

The control volume is a differential one in a cylindrical coordinate system. It is defined by an elemental length δs in the direction of flow and radii r and $(r + \delta r)$. It is assumed that conditions are the same in all lengthwise planes passing through the centre of the pipe and therefore independent of angle θ. It is also assumed that pressure is uniform on any plane normal to the pipe axis.

The direction of shear on the boundaries of the control volume must be carefully noted. It is assumed that the flow on the inside boundary of the cylindrical control volume tends to drag the fluid element in the control volume along while the flow on the outside tends to slow it down. This implies that the fluid velocity increases inwards. An increasing pressure p in the s-direction is assumed and for a constant pipe diameter the velocity is independent of s. Continuous functions are assumed for shear stress τ, velocity v and pressure p. The above assumptions may be summed up in a mathematical language as

$$\tau = \tau(r) \text{ only}; \; v = v(r) \text{ only and } p = p(s) \text{ only},$$

where the brackets mean 'function of'.

The net momentum flux out of the control volume in the s-direction is zero since v is assumed to be independent of s. Thus from equation (2.10) (remember $\Sigma P_m = 0$),

$$\Sigma F = p(2\pi r\delta r) - \left(p + \frac{dp}{ds}\delta s\right)(2\pi r\delta r) + \tau(2\pi r\delta s)$$

$$-\left(\tau + \frac{d\tau}{dr}\delta r\right)[2\pi(r + \delta r)\delta s] + W\cos\alpha = 0 \quad (3.1)$$

where W is the weight of fluid in the control volume and α is the angle of inclination of the pipe to the vertical. Equation (3.1) can be simplified to equation (3.2) below by putting

$$W\cos a = (2\pi r\delta r\delta s)\rho g\left(-\frac{\delta z}{\delta s}\right) \quad \text{as } \delta r, \delta s \to 0$$

and neglecting terms containing incremental powers higher than two,

$$r\frac{dp}{ds} + \tau + r\frac{d\tau}{dr} + \rho g r\frac{dz}{ds} = 0 \quad (3.2)$$

or

$$\frac{d}{dr}(r\tau) = -r\frac{d}{ds}(p + \rho g z) \quad (3.3)$$

Integrating (3.3) with respect to r and putting the constant of integration as A

$$r\tau = -\frac{r^2}{2}\frac{d}{ds}(p + \rho g z) + A \quad (3.4)$$

But shear stress is given by

$$\tau = -\mu\frac{dv}{dr}$$

(the negative sign indicates that v decreases as r increases). Substituting into (3.4) and integrating once more,

$$v = \frac{r^2}{4\mu}\frac{d}{ds}(p + \rho g z) - A\log_e r + B \quad (3.5)$$

where B is another constant of integration. Two boundary conditions are required for a complete solution of the problem.

For a full pipe, the velocity must be finite on the centre line $r = 0$ and zero on the pipe wall $r = r_0$. Thus

$$A = 0 \text{ and } B = -\frac{r_0^2}{4\mu}\frac{d}{ds}(p + \rho g z)$$

$$\text{and } v = -\frac{\rho g}{4\mu}\frac{d}{ds}\left(\frac{p}{\rho g} + z\right)[r_0^2 - r^2] \quad (3.6)$$

The laminar velocity distribution in a circular pipe is paraboloidal according to equation (3.6) and it is determined by the *hydraulic* (*piezometric*) *gradient* and not the pressure gradient alone. For a horizontal pipe however $dz/ds = 0$. Discharge

$$Q = \int_0^{r_0} 2\pi r v\,dr = -\frac{\rho g}{\mu}\frac{d}{ds}\left(\frac{p}{\rho g} + z\right)\pi\frac{r_0^4}{8} \quad (3.7a)$$

The average velocity $\bar{v} = \dfrac{Q}{\pi r_0^2}$ is

$$\bar{v} = -\frac{\rho g}{8\mu}\frac{\mathrm{d}}{\mathrm{d}s}\left(\frac{p}{\rho g} + z\right) r_0^2 \qquad (3.7\text{b})$$

Thus the average velocity is half the velocity on the centre line (the maximum velocity) where $r = 0$. The point of average velocity may be obtained by substituting equation (3.7b) into (3.6), as $r = r_0/\sqrt{2}$. Writing the piezometric gradient in a finite difference form as $-\Delta h/\Delta L$ (head drop over length) and with $2r_0 = D$, equation (3.7a) becomes

$$Q = \frac{\rho g \pi}{128\mu}\frac{\Delta h}{\Delta L} D^4 \qquad (3.8)$$

and this is known as the Hagen–Poiseuille equation. This equation is used quite extensively in the determination of the viscosity of liquids. The head drop and discharge are measured and used in (3.8) to calculate $\upsilon = \mu/\rho$.

For annular sections with outer radius $r_0 = a$ and inner radius $r_i = b$, the constants of integration can be determined from equation (3.5) and the velocity distribution established as

$$v = -\frac{\rho g}{4\mu}\frac{\mathrm{d}}{\mathrm{d}s}\left(\frac{p}{\rho g} + z\right)\left[a^2 - r^2 + \frac{a^2 - b^2}{\log_e b/a}\log_e\frac{a}{r}\right] \qquad (3.9)$$

Another class of laminar flow problems deals with flow through parallel plates. Imagine flow through a small space between two infinitely wide and long parallel

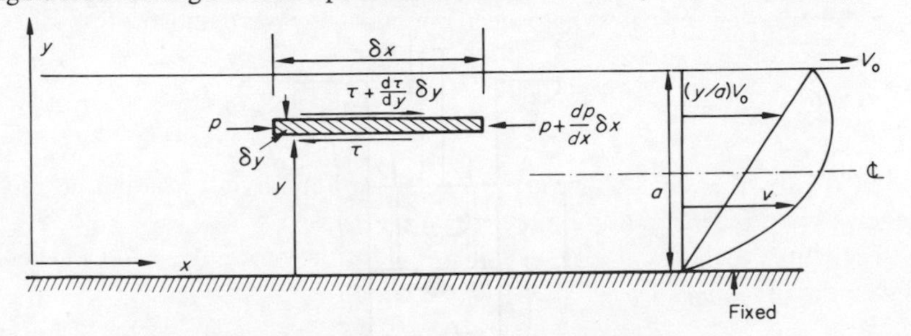

Fig. 3.2 Laminar flow between parallel plates

plates. If the lower plate is imagined stationary and the upper plate moved parallel to it at a constant velocity V_0 the no-slip boundary conditions with reference to Fig. 3.2 are

$$v = 0 \text{ at } y = 0 \text{ and } v = V_0 \text{ at } y = a.$$

Taking an elemental control volume at a distance y from the inner plate the dynamic equation (following steps similar to those above) can be shown to be

$$-\frac{\mathrm{d}\tau}{\mathrm{d}y} = -\frac{\mathrm{d}p}{\mathrm{d}x} \qquad (3.10\text{a})$$

or

$$\mu \frac{d^2 v}{dy^2} = \frac{dp}{dx} \tag{3.10b}$$

since

$$\tau = \mu \frac{dv}{dy}$$

Integration with respect to y gives

$$v = \frac{y^2}{2\mu} \frac{dp}{dx} + Ay + B$$

where A and B are constants of integration which can be eliminated using the boundary conditions stated above

$$B = 0 \text{ and } A = \frac{V_0}{a} - \frac{a}{2\mu} \frac{dp}{dx}$$

Thus

$$v = \frac{y}{a} V_0 - \frac{a^2}{2\mu} \frac{dp}{dx} \frac{y}{a} \left(1 - \frac{y}{a}\right) \tag{3.11}$$

The particular case of $dp/dx = 0$ is known as Couette flow and shows a linear distribution of velocity arising out of shear action alone introduced by the moving upper plate. Equation (3.11) may be considered as a description of the general Couette flow. Couette flow is the steady-state case of Fig. 1.6.

EXAMPLE 3.1

The sketch below shows the arrangement for a shock absorber (dashpot). There are four holes of diameter $D = 0{\cdot}08$ cm and the piston is 2·54 cm long and

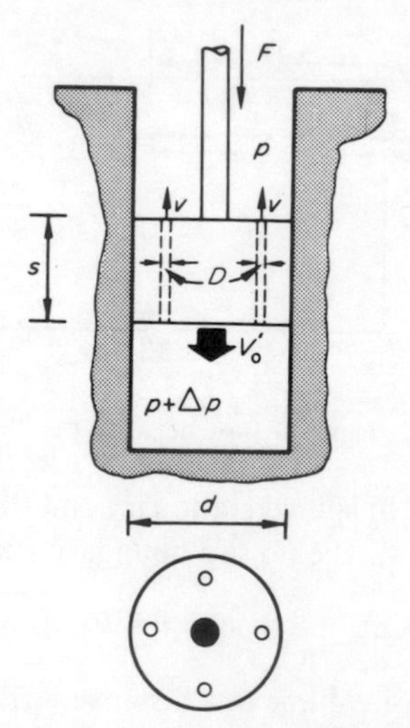

2.54 cm in diameter. Calculate the maximum shock the arrangement can sustain if the piston should not move faster than 0·305 m/s. The viscosity of the oil is 48×10^{-3} Ns/m^2.

SOLUTION

For the piston to move downward at velocity V_0 a force F must be applied to it. The force creates a pressure difference Δp on both sides of the piston which causes oil to flow through the holes at velocity v thus permitting the piston to move down. Neglect flow through the annular space between the piston and the cylinder wall. For n holes continuity requires that

$$d^2 V_0 = nD^2 v \tag{a}$$

From equation (3.8) the head drop Δh is given by

$$\Delta h = \frac{32\mu}{\rho g D^2} \Delta L \frac{4Q}{\pi D^2} = \frac{32\mu s v}{\rho g D^2}$$

Assuming $\quad s \ll \Delta p/\rho g, \; \rho g \Delta h \to \Delta p$

Substituting for v in (a) and putting $F = \pi \dfrac{d^2}{4} \Delta p$

$$F = \frac{8\pi}{n} \left(\frac{d}{D}\right)^4 \mu s V_0$$

Substituting the given data

$$F = \underline{2{\cdot}44 \text{ kN}} \text{ (550 lbf)}$$

Check the Reynolds number

From (a) $\quad v = \dfrac{1}{n}\left(\dfrac{d}{D}\right)^2 V_0 = 7{\cdot}75 \text{ m/s}$

Assuming the specific gravity of oil is 0·9

$$R = \frac{vD}{\upsilon} = \frac{7{\cdot}75 \times 0{\cdot}0008}{48 \times 10^{-3}/900}$$

$$= \underline{116}$$

And thus the flow is laminar.

EXAMPLE 3.2

The power dissipation due to friction in a cylindrical bearing of a pump motor rotating at 1500 rev/min (157 rad/s) has been estimated as 6 watts. If the bearing is 7·61 cm long and 3·81 cm in diameter calculate the clearance between the shaft and its journal. The oil viscosity is 48×10^{-3} Ns/m^2.

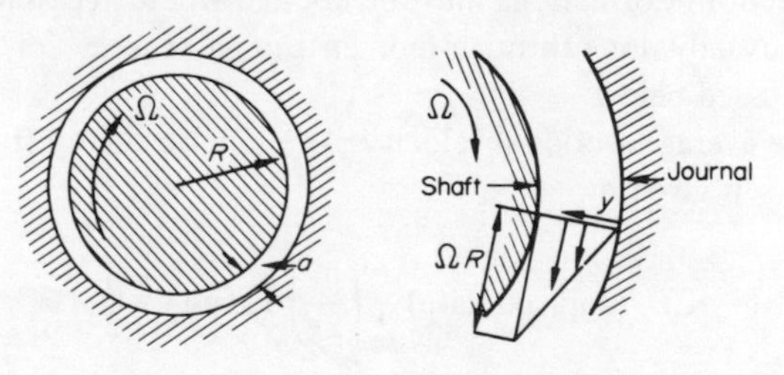

SOLUTION

Let Ω be the angular speed of the bearing shaft. If the clearance a is very small compared with the diameter $2R$ of the shaft, the flow is very similar to viscous flow through two parallel plates. Also for a uniform pressure in the clearance, $dp/dx = 0$ in equation (3.11) and the velocity distribution is linear. The shear stress on the shaft

$$\tau_0 = \mu \frac{dv}{dy}\bigg|_{y=a} = \mu \frac{R}{a} \Omega$$

Total shear force on shaft of length L

$$= (2\pi R) L \mu \frac{R}{a} \Omega$$

Torque applied to shaft to resist shear force

$$= \frac{2\pi R^2 L \Omega R \ \mu}{a}$$

Rate of energy dissipation against shear

$$= \text{Torque} \times \Omega$$

$$= \frac{2\pi R^3 L \mu \Omega^2}{a}$$

$\therefore$ The clearance $\quad a = \dfrac{2\pi\,(1{\cdot}905)^3 \times 7{\cdot}61 \times 48 \times 10^{-3} \times (157)^2}{10^8 \times 6}$

$$= \underline{0{\cdot}065 \text{ cm}}$$

3.3 Momentum and Energy Correction Factors

In adopting a one-dimensional approach to flow situations which are not strictly one-dimensional (as will be done in the following sections on turbulent flow through pipes) one must be fully aware of the errors so introduced. One must, for example, evaluate the errors introduced by basing momentum and energy calculations on average flow velocities instead of on integrations of the actual velocities across a section. Corrections must be applied where necessary to remove these errors. This is usually done through momentum and energy correction factors which are discussed below.

Let $\bar{v}$ be the average sectional velocity. The kinetic energy flux based on this average velocity is given by

$$\text{K.E. (approximate)} = \left(\frac{\bar{v}^2}{2g}\right)(\rho g \bar{v} A) = \frac{\rho}{2} A \bar{v}^3$$

where A is the cross-sectional area. The correct energy flux is however given by the summation of the energy fluxes through elemental areas δA. Thus

$$\text{K.E. (actual)} = \Sigma\left(\frac{v^2}{2g}\right)(\rho g v \mathrm{d}A) = \rho\int_0^A \frac{v^3}{2}\,\mathrm{d}A$$

We may define a ratio

$$\alpha = \frac{\text{actual energy flux}}{\text{approximate energy flux}} = \frac{\int_0^A v^3\,\mathrm{d}A}{A\bar{v}^3} \tag{3.12}$$

α is known as the *energy correction factor* and it represents the factor by which the kinetic energy flux based on average sectional velocity must be multiplied to make it equal to the correct energy flux.

Similarly the *momentum correction factor* β is given by

$$\beta = \frac{\text{actual momentum flux}}{\text{approximate momentum flux}}$$

$$= \frac{\int_0^A (\rho v \mathrm{d}A)v}{\rho(\bar{v}A)\bar{v}} = \frac{\int_0^A v^2 \mathrm{d}A}{A\bar{v}^2} \tag{3.13}$$

Unlike the parabolic form of the velocity profile in the case of laminar flow through closed conduits the velocity profile in the case of turbulent flow is like that sketched in Fig. 3.3(b). In turbulent flow the average velocity is not very different from the maximum velocity and both α and β have been calculated to be

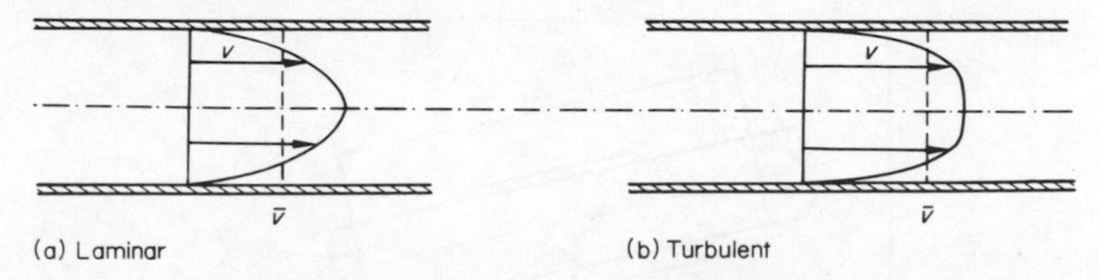

Fig. 3.3 Velocity profiles in pipes

very nearly unity. In laminar flow, however, α and β are about 2 and must therefore be considered when using average velocities. However, with laminar flow exact mathematical solutions can in many cases be found and approximations are unnecessary. In the discussions of turbulent flow which follow α and β are assumed to be unity unless otherwise stated.

3.4 Pipe Flow at a High Reynolds Number

3.4.1 Formulation of the Problem

In cases where the Reynolds number is high the flow is not well ordered and therefore in view of the existence of momentum transfer through turbulence,

Newton's law of viscosity does not hold. The appropriate energy equation for a suitably chosen finite control volume is given by equation (2.16) as

$$-\frac{1}{mg}\left[\underset{\text{heat power}}{\frac{\mathrm{d}Q}{\mathrm{d}t}}+\underset{\text{shear power}}{\frac{\mathrm{d}W_\mathrm{s}}{\mathrm{d}t}}+\underset{\text{shaft power}}{\frac{\mathrm{d}W_\mathrm{sh}}{\mathrm{d}t}}\right]=\Sigma\left(\frac{v^2}{2g}+z+\frac{p}{\rho g}+\frac{u}{g}\right)_\mathrm{influx}-\Sigma\left(\frac{v^2}{2g}+z+\frac{p}{\rho g}+\frac{u}{g}\right)_\mathrm{efflux} \quad (3.14)$$

In this section we will apply this equation to various conditions of pipe flow. The heat power exchange is assumed negligible and for liquids the internal energy terms (u/g) will always cancel out. Thus the conditions to be examined in this section involve the equation

$$-\frac{1}{mg}\left(\frac{\mathrm{d}W_\mathrm{s}}{\mathrm{d}t}+\frac{\mathrm{d}W_\mathrm{sh}}{\mathrm{d}t}\right)=\Sigma\left(\frac{v^2}{2g}+z+\frac{p}{\rho g}\right)_\mathrm{influx}-\Sigma\left(\frac{v^2}{2g}+z+\frac{p}{\rho g}\right)_\mathrm{efflux} \quad (3.15)$$

where m = mass flow per unit time.

3.4.2 Losses Against Resistances

Following the idea expressed in Section 3.1, losses arising from the two types of resistance to flow through pipes will be estimated. The concept had already been

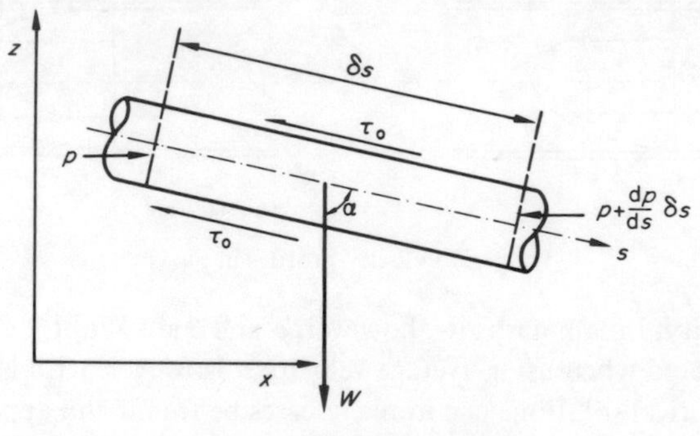

Fig. 3.4

introduced in deriving (2.17). The work done per unit weight of the flowing fluid against resistance is given by the expression

$$h_\mathrm{f}=-\frac{1}{mg}\frac{\mathrm{d}W_\mathrm{s}}{\mathrm{d}t}$$

In the discussions here however it is necessary to distinguish between pipe wall resistance and form resistance. Losses due to the former are referred to as pipe losses. Both losses were tacitly put together in Fig. 2.7.

(1) PIPE LOSSES–The Darcy–Wiesbach formula
Isolate an element δs long of a uniform pipe of cross-sectional area A (Fig. 3.4). Assume an average wall shear stress τ_0. The equilibrium equation for steady fluid flow through the control volume is given by

$$-\frac{dp}{ds} A + \rho g A \cos\alpha - \pi D \tau_0 = 0$$

or

$$-\rho g A \frac{d}{ds}\left(\frac{p}{\rho g} + z\right) - \pi D \tau_0 = 0$$

since

$$\cos\alpha = -\frac{dz}{ds} \tag{3.16}$$

Thus

$$\frac{dh}{ds} = -\frac{\pi D \tau_0}{\rho g A} \tag{3.17}$$

Equation (3.17) can also be derived using energy concepts. If the head loss in overcoming shear resistance per unit weight of flowing fluid is δh, the rate of doing work against shear

$$\frac{dW_s}{dt} = -mg\delta h = -\rho g A \bar{v} \delta h$$

But

$$\frac{dW_s}{dt} = (\pi D \tau_0 \delta s)\bar{v}$$

∴

$$-\rho g A \bar{v} \delta h = \pi D \tau_0 \delta s \bar{v}$$

In the limit as $\delta s \to 0$

$$\frac{dh}{ds} = -\frac{\pi D \tau_0}{\rho g A}$$

Making use of the convention (see Section 1.7) which expresses shear stress in terms of velocity and a friction (drag) factor f (f is conventionally used for pipes in place of c_f) as $\tau_0 = \frac{1}{2}\rho f \bar{v}^2$, equation (3.17) gives the head loss per unit length in overcoming wall resistance as

$$\frac{dh}{ds} = -\frac{1}{2}\frac{\rho f \bar{v}^2}{\rho g m} = -\frac{f\bar{v}^2}{2mg} \tag{3.18}$$

where $m = (A/\pi D) = (D/4)$ is known as the hydraulic mean radius. The slope $-(dh/ds) = S_f$ is the hydraulic gradient. Integrating equation (3.18) over a length L of the pipe gives

$$h_f = h_0 - h_L = \frac{fL\bar{v}^2}{2gm} \tag{3.19}$$

Equation (3.19) is generally referred to as the Darcy–Weisbach formula.

It must be emphasized that the friction factor f is empirical and can only be determined through experiments. Nikuradse performed a series of such experiments using uniform artificial roughness in pipes. His original results were similar to those of the modern form shown in Fig. 3.5 which show that f is dependent on Reynolds number ($\bar{v}D/v$) and the relative roughness (k_s/D) of the pipe. k_s is related to the size of the roughening element. In the laminar range, however, f is independent of pipe roughness as can be shown from equations (3.8) and (3.19). From (3.8)

$$h_f = \int_0^L dh = \frac{32\mu L}{\rho g D^2}\left(\frac{4Q}{\pi D^2}\right)$$

$$= \frac{32 v L \bar{v}}{g D^2}$$

Substituting into (3.19) and simplifying gives

$$f = 16/\left(\frac{\bar{v}D}{v}\right) = 16/R \tag{3.20}$$

Thus the friction factor is analytically defined for laminar flows. The general pipe friction diagram (frequently referred to as the Moody diagram in American literature) is however the result of semi-analytical derivations. f tends to be independent of Reynolds number when the flow is fully turbulent. There is a zone of transition in which f is not well defined with respect to R and k_s/D. Note, by the way, that the value of f from charts in most American books is four times the value as defined above. This is because equation (3.19) in terms of the pipe diameter is $h_f = (4fL\bar{v}^2)/(2gD)$ while the Americans prefer to incorporate the factor 4 into f.

The variation of friction factor f with time must also be emphasized. A pipe deteriorates with age due to the growth of roughness elements k_s and correspondingly f increases and the capacity of the pipe reduces. The rate of deterioration of a pipe does not depend only on the pipe material but also on the quality of the fluid it conveys and on maintenance. The roughness sizes given in Fig. 3.5 are therefore appropriate for young pipes only. The growth rate for roughness elements is sometimes assumed to follow a law such as

$$k_s = k_0 + \alpha t$$

where k_0 is the initial roughness, t is time and the factor α depends on the degree of corrosive attack. It ranges from 0·025 mm/yr for slight attack to 0·75 mm/yr for severe attack, by domestic water supply for example.

(2) LOSSES DUE TO FORM RESISTANCE

Consider a fluid-carrying pipe which terminates in another, larger, pipe as shown in Fig. 3.6. As the flow expands the pressure tends to increase causing separation of the flow as shown in the sketch. As a result energy is lost. Assuming

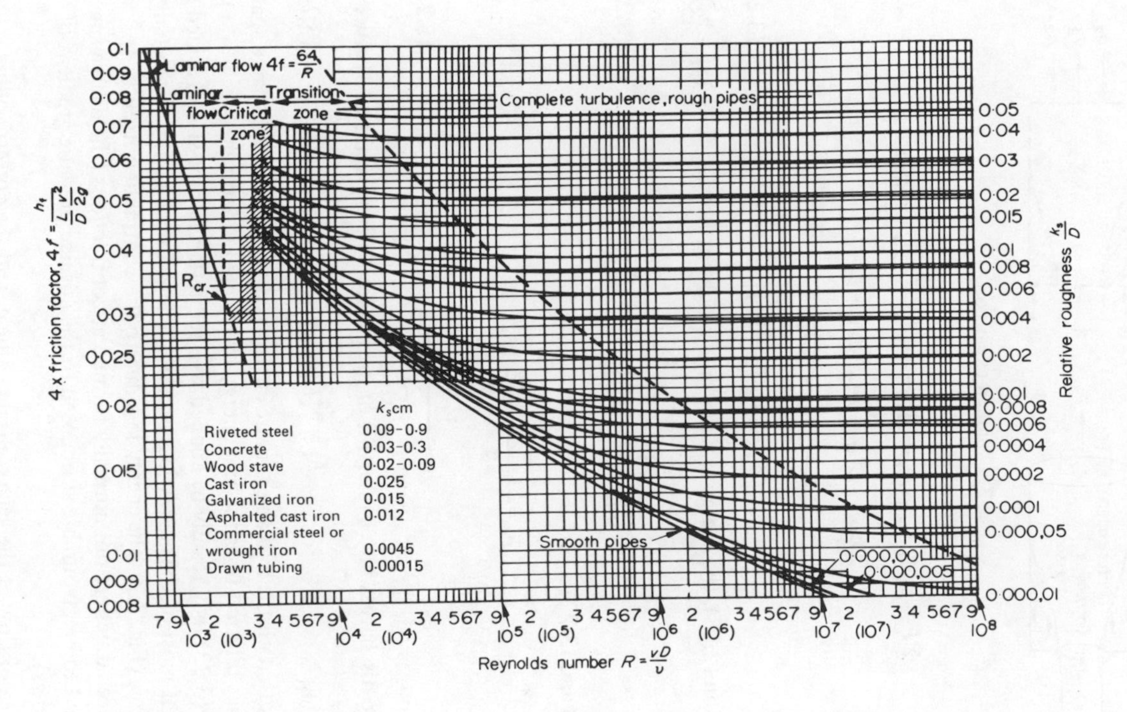

Fig. 3.5 Pipe friction diagram

that pressure is constant across all sections of the pipes, the energy loss in resisting the pressure forces slowing down the flow can be estimated. Considering the control volume as shown,

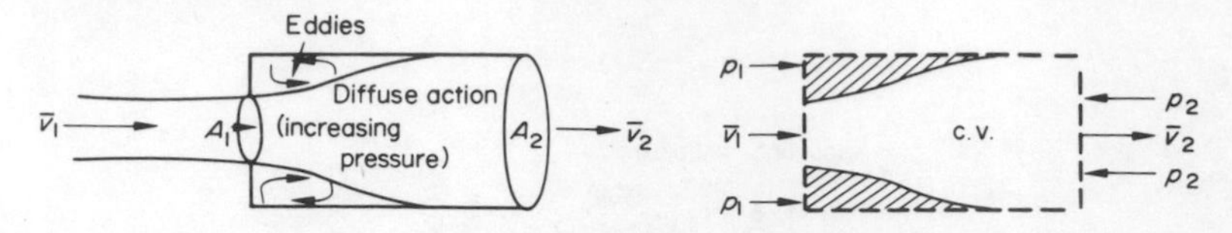

Fig. 3.6 Sudden expansion

$$\Sigma F = p_1 A_1 + p_1(A_2 - A_1) - p_2 A_2 = A_2(p_1 - p_2)$$

$$\Sigma E_M - \Sigma I_M = \rho A_2 \bar{v}_2(\bar{v}_2 - \bar{v}_1)$$

Since

$$\Sigma F = \Sigma E_M - \Sigma I_M$$

$$p_1 - p_2 = \rho \bar{v}_2(\bar{v}_2 - \bar{v}_1) \tag{3.21}$$

From equation (2.18), the energy loss per unit weight of flowing fluid ($z_1 = z_2$) is given by

$$h_f = -\frac{1}{mg}\frac{dW_s}{dt} = \frac{\bar{v}_1^2 - \bar{v}_2^2}{2g} + \frac{p_1 - p_2}{\rho g} \tag{3.22}$$

From (3.21) and (3.22)

$$h_f = \frac{\bar{v}_1^2 - \bar{v}_2^2}{2g} + \frac{\bar{v}_2}{g}(\bar{v}_2 - \bar{v}_1)$$

or

$$h_f = \frac{(\bar{v}_1 - \bar{v}_2)^2}{2g} \tag{3.23}$$

If the pipe discharges into a reservoir or into space, $\bar{v}_1 \gg \bar{v}_2$ and (3.23) gives

$$h_f = \bar{v}_1^2/2g \tag{3.24}$$

which means that all the available kinetic energy is dissipated into the large space. It is sometimes referred to as the exit energy or rejected kinetic energy.

Equation (3.23) may readily be applied to a reduction in pipe size provided it is realized that most of the energy loss occurs at the re-expansion of the initially contracted flow (Fig. 3.7). The energy loss due to the contraction itself is negligible since eddies are not normally formed at contractions (the decreasing pressure is said to be favourable as opposed to the adverse pressure gradient of expanding flows).

Let the area of flow at the vena contracta a be a_c and the corresponding velocity be $\bar{v}_c$. The head loss equation (3.23) becomes

$$h_f = \frac{(\bar{v}_c - \bar{v}_2)^2}{2g}$$

But
$$\bar{v}_c = \frac{A_2\bar{v}_2}{a_c} = \frac{\bar{v}_2}{C_c}$$

where
$$C_c = a_c/A_2$$

$$\therefore \qquad h_f = K\frac{\bar{v}_2^2}{2g} \qquad (3.25)$$

where the coefficient of hydraulic resistance $K = ([1/C_c] - 1)^2$. The value $K = 0{\cdot}5$ (corresponding to $C_c = 0{\cdot}585$) is usually adopted for sudden contractions when A_1/A_2 is large. Three typical cases of entry into a pipe from a large reservoir are illustrated in Fig. 3.8 together with their approximate values for K.

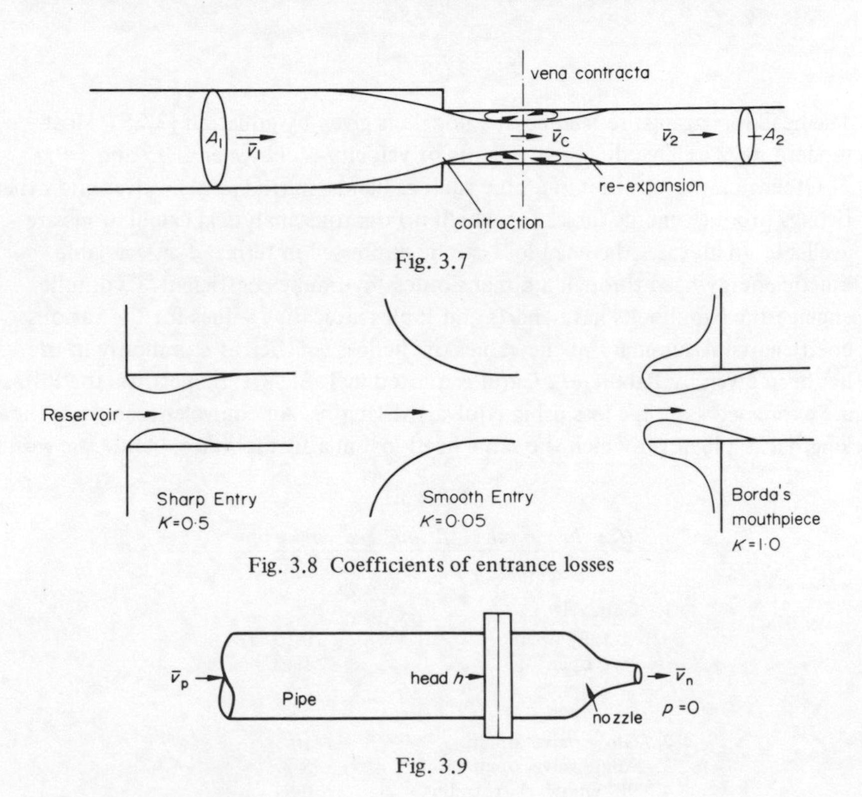

Fig. 3.7

Fig. 3.8 Coefficients of entrance losses

Fig. 3.9

(3) LOSSES IN A NOZZLE

A nozzle is essentially a constriction at the end of a pipe which converts pressure energy into kinetic energy. It is used to provide high-velocity jets, for example, for impulse turbines and from fire water hoses. With reference to Fig. 3.9, let the pressure head just before the nozzle be h. The energy equation gives

$$h + \frac{\bar{v}_p^2}{2g} = \frac{\bar{v}_n^2}{2g} + h_f \qquad (3.26)$$

where $\bar{v}_p$ and $\bar{v}_n$ are the pipe and jet velocities respectively and h_f is the head loss in the nozzle. The orifice equation which expresses the jet velocity in terms of the total head upstream of the orifice (nozzle) and a coefficient of velocity C_v may also be applied to give

$$\bar{v}_n = C_v \sqrt{[2g(h + \bar{v}_p^2/2g)]} \tag{3.27}$$

Eliminating $\bar{v}_p/2g$ from (3.26) using (3.27)

$$h_f = \frac{\bar{v}_n^2}{2g}\left(\frac{1}{C_v^2} - 1\right)$$

or

$$h_f = K_n \frac{\bar{v}_n^2}{2g} \tag{3.28}$$

where

$$K_n = \frac{1}{C_v^2} - 1$$

The head loss against resistance in a nozzle is given by equation (3.28). Most modern nozzles have their coefficients of velocity C_v between 0·95 and 1·0.

Other non-uniformities in a pipe such as bends, mitred joints, valves and other fittings produce energy losses for which no rigorous analytical calculations are available. In all cases the head loss can be expressed in terms of an available kinetic energy head through a dimensionless hydraulic coefficient. Hydraulic engineering handbooks have charts and tables indicating values for the various coefficients. A summary of the values of the loss coefficient commonly in use has been given by Babbit, *et al.* and is quoted in Table 3.1. Sometimes the losses are expressed as a pipe loss using equivalent lengths. An equivalent length is the length of a pipe over which the same head lost in a fitting, valve, bend, etc. would

Table 3.1

Head loss in valves, fittings and service pipes

	Values of K in $h = K\frac{v^2}{2g}$
1. Gate valve:	
fully open	0·19
$\frac{3}{4}$ open	1·15
$\frac{1}{2}$ open	5·6
$\frac{1}{4}$ open	24
2. Globe valve, open	10
3. Angle valve, open	5
4. 90° elbow, short radius	0·9
medium radius	0·75
long radius	0·6
5. Return bend (180°)	2·2
6. 45° elbow	0·42
7. $22\frac{1}{2}$° elbow (45 cm)	0·13
8. Tee joint	1·25
9. Reducer (valve at small end)	0·25
10. Increaser	0·25 $(v_1^2 - v_2^2)/2g$
11. Bellmouth reducer	0·10
12. Open orifice	1·80

be produced. For convenience the bars from the average velocities are dropped for the remainder of the chapter. Since only average velocities are used there is no danger of confusion.

3.4.3 Applications

(1) PIPES IN SERIES

Figure 3.10 shows two different sizes of pipes in series delivering water from reservoir A to reservoir B. Applying the energy equation (2.18) or (3.15) between the liquid levels in reservoirs A and B (there is no shaft work),

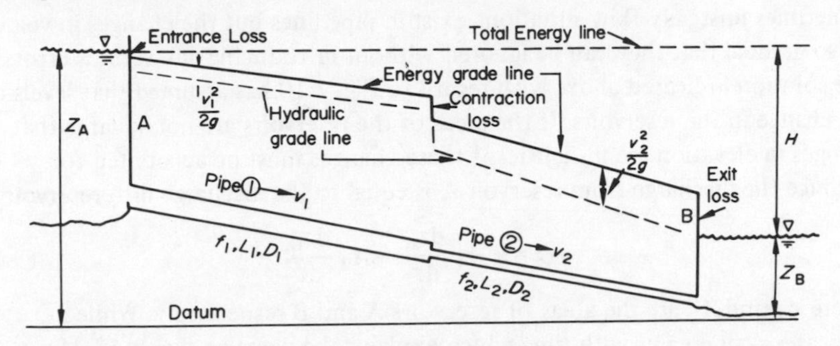

Fig. 3.10

$$\frac{v_A^2}{2g} + z_A + \frac{p_A}{\rho g} = \frac{v_B^2}{2g} + z_B + \frac{p_B}{\rho g} + \text{losses (A to B)}.$$

If both reservoirs are large $v_A^2/2g$ and $v_B^2/2g$ are negligible compared with the other terms. Also $p_A = p_B$ since both are atmospheric pressures. Thus

$$z_A - z_B = H = \text{losses}$$

i.e.

$$H = 0{\cdot}5\,\frac{v_1^2}{2g} + \frac{4f_1L_1v_1^2}{2gD_1} + K\frac{v_2^2}{2g} + \frac{4f_2L_2v_2^2}{2gD_2} + \frac{v_2^2}{2g} \tag{3.29}$$

entrance, friction pipe (1), contraction, friction pipe (2), exit

From continuity

$$v_2 = \left(\frac{D_1}{D_2}\right)^2 v_1 \tag{3.30}$$

and equation (3.30) can be used to eliminate v_2 from (3.29). After solving for v_1, the discharge

$$Q = \pi\frac{D_1^2}{4}v_1$$

This approach can be extended to any number of pipes in series and to bend, valve and other losses. For long pipes the pipe friction losses are so large compared with the others which are called minor losses that the latter can be neglected. Except in very accurate work the pipe friction term

$$h_f = \frac{fLQ^2}{3D^5} \text{ approx (SI units)} = \frac{fLQ^2}{10D^5} \text{ approx (Imperial units)}$$

$$h_f = \frac{64fLQ^2}{2g\pi^2 D^5} = \frac{fLQ^2}{10D^5} \text{ approx (Imperial units)}$$

(2) SLOW UNSTEADY FLOW THROUGH PIPES

Sometimes unsteady flow situations exist in pipe lines but the changes in velocity are so gradual that they can be ignored without introducing any serious errors. The solution indicated above with regard to Fig. 3.10 has assumed that levels do not change in the reservoirs. If the areas of the reservoirs are not so large that changes in elevation are insignificant these changes must be accounted for.

Since the discharge from reservoir A is equal to the discharge into reservoir B,

$$Q = -A_1 \frac{dz_A}{dt} = A_2 \frac{dz_B}{dt} \tag{3.31}$$

where A_1 and A_2 are the areas of reservoirs A and B respectively. While z_A decreases z_B increases with time which explains the negative sign in (3.31)

$$z_A - z_B = H$$

$$\therefore \quad \frac{dz_A}{dt} - \frac{dz_B}{dt} = \frac{dH}{dt} \tag{3.32}$$

From (3.31) and (3.32)

$$\left(1 + \frac{A_1}{A_2}\right) \frac{dz_A}{dt} = \frac{dH}{dt}$$

or

$$\left(1 + \frac{A_1}{A_2}\right)\left(-\frac{Q}{A_1}\right) = \frac{dH}{dt} \tag{3.33}$$

Neglecting minor losses,

$$H = \frac{f_1 L_1 Q^2}{3D_1^5} + \frac{f_2 L_2 Q^2}{3D_2^5} \text{ very nearly}$$

$$\therefore \quad Q = H^{\frac{1}{2}} \left(\frac{f_1 L_1}{3D_1^5} + \frac{f_2 L_2}{3D_2^5}\right)^{-\frac{1}{2}} \tag{3.34}$$

Substituting (3.34) into (3.33) gives

$$\frac{dH}{dt} = -\frac{A_1 + A_2}{A_1 A_2} \mathscr{K} H^{\frac{1}{2}} \tag{3.35}$$

where

$$\mathscr{K} = \left(\frac{f_1 L_1}{3D_1^5} + \frac{f_2 L_2}{3D_2^5}\right)^{-\frac{1}{2}}$$

Integration of equation (3.35) gives the time interval for the available head to change from H_1 to H_2 as

$$\Delta t = 2\frac{A_1A_2}{A_1 + A_2}\left(\frac{f_1L_1}{3D_1^5} + \frac{f_2L_2}{3D_2^5}\right)^{\frac{1}{2}}(H_1^{\frac{1}{2}} - H_2^{\frac{1}{2}}) \qquad (3.36)$$

In the integration of equation (3.35) we assumed that the friction factor is constant. This assumption is valid as long as the flow is fully turbulent (see Fig. 3.5). This requires that H must be quite large. Equation (3.36) however ceases to be accurate if H becomes small since f will then depend on the flow velocity (Reynolds number). Equation (3.35) should then be solved some other way (such as iterative methods).

(3) POWER TRANSMISSION THROUGH A NOZZLE

Let reservoir B in Fig. 3.10 be replaced by a nozzle of diameter d_n at an elevation H below the liquid surface level in A as sketched in Fig. 3.11.

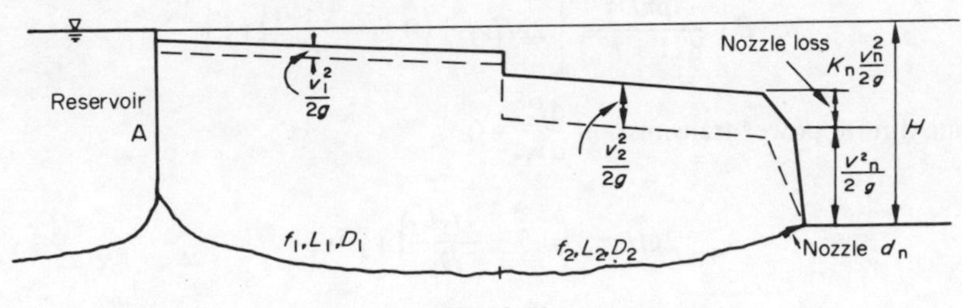

Fig. 3.11

There are three unknowns v_1, v_2 and v_n requiring three equations for solution. The appropriate equations are as follows.

From energy considerations:

$$H = 0{\cdot}5\frac{v_1^2}{2g} + \frac{4f_1L_1v_1^2}{2gD_1} + K\frac{v_2^2}{2g} + \frac{4f_2L_2v_2^2}{2gD_2} + \frac{K_nv_n^2}{2g} + \frac{v_n^2}{2g} \qquad (3.37)$$

From continuity:

$$v_1D_1^2 = v_2D_2^2 \qquad (3.38)$$

and

$$v_2D_2^2 = v_nd_n^2 \qquad (3.39)$$

Substituting (3.38) and (3.39) in (3.37)

$$H = \frac{v_n^2}{2g}\left[0{\cdot}5\left(\frac{d_n}{D_1}\right)^4 + \frac{4f_1L_1}{D_1}\left(\frac{d_n}{D_1}\right)^4 + K\left(\frac{d_n}{D_2}\right)^4 + \frac{4f_2L_2}{D_2}\left(\frac{d_n}{D_2}\right)^4 + (1 + K_n)\right]$$

or

$$v_n = \sqrt{(2gH/\mathscr{K})} \qquad (3.40)$$

where $\mathscr{K}$ = terms in the square brackets.

The discharge

$$Q = \pi \frac{d_n^2}{4} v_n$$

and the power transmitted is given by

$$P = \frac{\rho g Q v_n^2}{2g} = \rho g \frac{\pi d_n^2}{4 \times 2g} v_n^3 \tag{3.41}$$

or

$$P = \frac{\rho \pi d_n^2}{8} \left(\frac{2gH}{\mathscr{K}}\right)^{\frac{3}{2}} \tag{3.42}$$

If $L_2 = 0$ equation (3.37) gives

$$v_n^2 = \frac{2gH - \left(0{\cdot}5 + \frac{4f_1L_1}{D_1}\right) v_1^2}{1 + K_n}$$

Substituting into (3.41) and using continuity ($v_n d_n^2 = v_1 D_1^2$),

$$P = \frac{\rho \pi D_1^2}{8(1 + K_n)} \left[2gHv_1 - \left(0{\cdot}5 + \frac{4f_1L_1}{D_1}\right) v_1^3\right]$$

For maximum power transmission $\frac{dP}{dv_1} = 0$

∴

$$2gH - 3\left(0{\cdot}5 + \frac{4f_1L_1}{D_1}\right) v_1^2 = 0$$

i.e.

$$\left(0{\cdot}5 + \frac{4f_1L_1}{D_1}\right) \frac{v_1^2}{2g} = H/3 \tag{3.43}$$

The left-hand side is the total energy loss in the pipe before the nozzle. This must be equal to a third of the available head for maximum energy to be transmitted by the system. It is necessary to combine the size and roughness of the pipe (over a fixed distance) to achieve as near as possible this requirement for the optimum use of the available head. The corresponding nozzle diameter for maximum power transmission is given as follows

$$\left(0{\cdot}5 + \frac{4f_1L_1}{D_1}\right) \frac{v_1^2}{2g} = \frac{H}{3} = \frac{1}{3}\left[\left(0{\cdot}5 + \frac{4f_1L_1}{D_1}\right) \frac{v_1^2}{2g} + \left(1 + K_n\right) \frac{v_n^2}{2g}\right]$$

∴

$$\frac{1 + K_n}{3} \frac{v_n^2}{2g} = \frac{2}{3}\left(0{\cdot}5 + \frac{4f_1L_1}{D_1}\right) \frac{v_1^2}{2g}$$

or

$$\left(\frac{v_n}{v_1}\right)^2 = \left(\frac{D_1}{d_n}\right)^4 = \frac{2}{1 + K_n}\left(0{\cdot}5 + \frac{4f_1L_1}{D_1}\right)$$

∴

$$d_n = \left(\frac{1 + K_n}{D_1 + 8f_1L_1}\right)^{\frac{1}{4}} D_1^{\frac{5}{4}} \tag{3.44}$$

(4) PIPES IN PARALLEL

Pipes which carry fluid from and/or to the same points but without the same particles of fluid passing through the pipes in succession are said to be in parallel. Two examples of parallel pipes are given in Fig. 3.12. The common points from

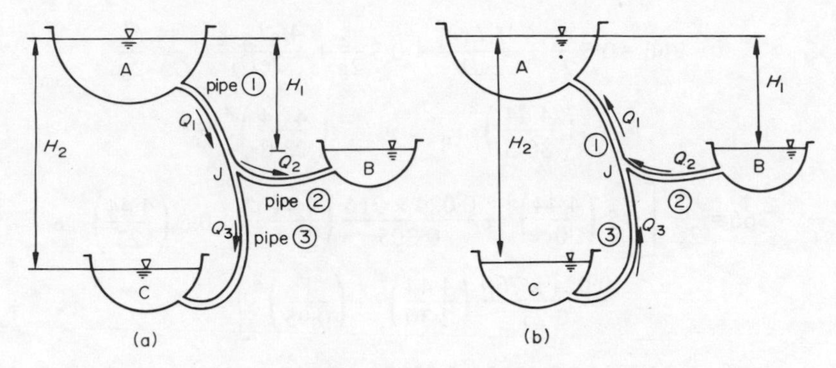

Fig. 3.12 Parallel pipes

or to which fluid is being carried are reservoir A and the junction J. Reservoir A supplies water to B and C in Fig. 3.12(a) but A takes water from B and C in Fig. 3.12(b). JB and JC are in parallel in each case; the directions of flow are however reversed. In case (a) the surface elevation in A is higher than those of B and C, but in case (b), the reverse is true. Applying the energy equation for AJB and AJC for case (a) and for BJA and CJA for case (b) gives identical equations:

$$H_1 = \frac{f_1 L_1 Q_1^2}{3D_1^5} + \frac{f_2 L_2 Q_2^2}{3D_2^5} \tag{3.45}$$

$$H_2 = \frac{f_1 L_1 Q_1^2}{3D_1^5} + \frac{f_3 L_3 Q_3^2}{3D_3^5} \tag{3.46}$$

Considerations of continuity gives the third equation (3.47) from which the three unknowns Q_1, Q_2, Q_3 can be determined

$$Q_1 = Q_2 + Q_3 \tag{3.47}$$

The important hydraulic difference between pipes in series and pipes in parallel is that in the first case the discharge is the same for the pipes at the junction whereas in the second case the discharge is different but the piezometric head is assumed common at the junction.

EXAMPLE 3.3

A pipeline to an impulse turbine consists of a 0·305 m diameter pipe 914 m long followed by a 0·229 m diameter pipe 760 m long. At the end of the pipe is a nozzle 44·4 mm in diameter. If the nozzle is 366 m below reservoir level calculate the power contained in the jet. What is the system efficiency? Take $f = 0{\cdot}006$ and $C_v = 0{\cdot}95$.

Assume the inlet to the 0·305 m pipe to be sharp and the contraction in pipe sections to be sudden. Using equation (3.37) and assuming K for the pipe contraction to be 0·5,

$$366 = 0{\cdot}5\,\frac{v_1^2}{2g} + \frac{4f_1L_1v_1^2}{2gD_1} + 0{\cdot}5\,\frac{v_2^2}{2g} + \frac{4f_2L_2v_2^2}{2gD_2} + \frac{1}{C_v^2}\,\frac{v_n^2}{2g}$$

$$v_1 = \left(\frac{4{\cdot}44}{30{\cdot}5}\right)^2 v_n; \quad v_2 = \left(\frac{4{\cdot}44}{23{\cdot}0}\right)^2 v_n$$

$$\therefore \quad 366 = \frac{v_n^2}{2g}\left[0{\cdot}5\left(\frac{4{\cdot}44}{30{\cdot}5}\right)^4 + \frac{0{\cdot}024 \times 914}{0{\cdot}305}\left(\frac{4{\cdot}44}{30{\cdot}5}\right)^4 + 0{\cdot}5\left(\frac{4{\cdot}44}{23}\right)^4 + \frac{0{\cdot}024 \times 760}{0{\cdot}23}\left(\frac{4{\cdot}44}{2{\cdot}30}\right)^4 + \left(\frac{1}{0{\cdot}95}\right)^2\right]$$

i.e. $\mathscr{K} = (0{\cdot}00023 + 0{\cdot}0325 + 0{\cdot}00072 + 0{\cdot}117 + 1{\cdot}10) = 1{\cdot}251$

$$\therefore \quad \frac{v_n^2}{2g} = 366/\mathscr{K} = 292 \text{ m}$$

Power of jet using equation (3.42)

$$= 10^3\left(\frac{\pi}{8}\right)(0{\cdot}0444)^2\left(19{\cdot}62 \times \frac{366}{1{\cdot}251}\right)^{\frac{3}{2}}$$

$$= \underline{330 \text{ kW}}$$

Efficiency $= \dfrac{v_n^2/2g}{H} = \dfrac{292}{366} = \underline{80\%}$

Note: The calculations for $\mathscr{K}$ show how negligible minor losses sometimes are compared with pipe friction losses.

EXAMPLE 3.4

In the pipe system shown in the figure, $L_1 = 914$ m, $D_1 = 0{\cdot}305$ m $f_1 = 0{\cdot}005$; $L_2 = 608$ m, $D_2 = 0{\cdot}204$ m, $f_2 = 0{\cdot}0045$; $L_3 = 1216$ m, $D_3 = 0{\cdot}408$ m and $f_3 = 0{\cdot}0043$. The gauge pressure at A is 0·550 N/mm², its elevation is 30·5 m and that of B is 24·3 m. For a total flow of 0·34 m³/s, determine the flow through each pipe and the pressure at B.

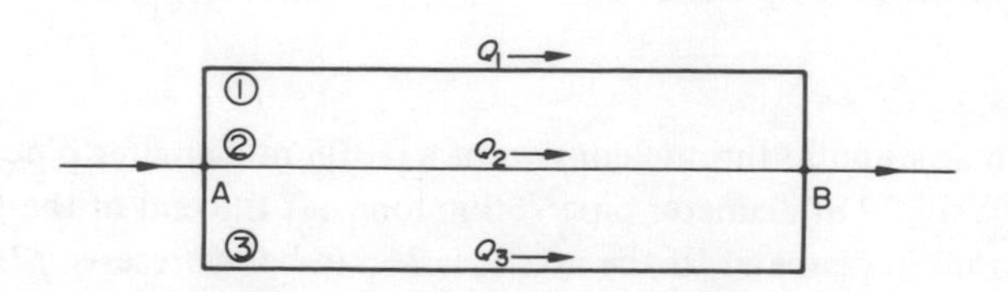

SOLUTION

The head drop H from A to B is the same no matter which pipe forms the path. Neglecting minor losses,

$$H = \frac{f_1 L_1 Q_1^2}{3D_1^5} = \frac{f_2 L_2 Q_2^2}{3D_2^5} = \frac{f_3 L_3 Q_3^2}{3D_3^5}$$

$$\therefore \qquad \frac{Q_1}{Q_2} = \sqrt{\left(\frac{f_2}{f_1}\,\frac{L_2}{L_1}\,\frac{D_1^5}{D_2^5}\right)} = 2{\cdot}14 \qquad \text{(a)}$$

$$\frac{Q_1}{Q_3} = \sqrt{\left(\frac{f_3}{f_1}\,\frac{L_3}{L_1}\,\frac{D_1^5}{D_3^5}\right)} = 0{\cdot}526 \qquad \text{(b)}$$

From continuity: $$Q = 0{\cdot}34 = Q_1 + Q_2 + Q_3 \qquad \text{(c)}$$

From (a), (b) and (c), $0{\cdot}34 = Q_1\,[1 + 0{\cdot}467 + 1{\cdot}90] = 3{\cdot}37Q_1$

$$\therefore \qquad \underline{Q_1 = 0{\cdot}101 \text{ m}^3/\text{s}}$$

$$Q_2 = 0{\cdot}467Q_1 = \underline{0{\cdot}047 \text{ m}^3/\text{s}}$$

$$Q_3 = 1{\cdot}90Q_1 = \underline{0{\cdot}192 \text{ m}^3/\text{s}}$$

$$\frac{p_A}{\rho g} + \frac{v_A^2}{2g} + z_A = \frac{p_B}{\rho g} + \frac{v_B^2}{2g} + z_B + \text{losses along any path}$$

$$p_B = p_A + \rho g(z_A - z_B) - \frac{\rho g f_1 L_1 Q_1^2}{3D_1^5}$$

since $v_A = v_B$

$$\therefore \qquad \underline{p_B = 0{\cdot}555 \text{ N/mm}^2 = 555 \text{ kN/m}^2}$$

3.4.4 Pump and Turbine Works

The last term on the left-hand side of equation (3.15) dealing with shaft work will now be discussed. The shaft work is considered to be transmitted from outside into the system or control volume through a rotating shaft (see Section 2.5). The rate of shaft work per unit weight of the flowing fluid is

$$\left(\frac{1}{mg}\,\frac{\partial W_{sh}}{\partial t}\right)$$

Negative work is involved if the fluid does work on the shaft. The usual ways of transferring mechanical energy into a fluid system are through pumps, compressors, fans or paddles. A turbine is the reverse of a pump and therefore extracts energy from the fluid system for conversion into the mechanical energy of a rotating shaft. The inclusion of a pump in a pipe line means a rise in the energy content through a rise in pressure across the pump. The pressure drops across a turbine.

If the pressure head rise across a pump is H_p, the fluid power P_w transferred to a discharge Q is given by

$$P_w = \rho g Q H_p \tag{3.48}$$

If the pump works at an efficiency η, the total power required from the pump motor is

$$P = \rho g Q H_p/\eta \tag{3.49}$$

Similarly if a turbine in a pipe consumes a head H_T across itself, its power extraction from the fluid system is

$$P_w = \rho g Q H_T \tag{3.50}$$

The effective output power is however given by

$$P = \eta \rho g Q H_T \tag{3.51}$$

where η is the turbine efficiency.

The one-dimensional energy equation derived from (3.15) incorporating a hydraulic machine (pump or turbine) head H_m and energy losses is

$$\frac{v_1^2}{2g} + z_1 + \frac{p_1}{\rho g} \pm H_m = \frac{v_2^2}{2g} + z_2 + \frac{p_2}{\rho g} + \text{losses} \tag{3.52}$$

The positive sign applies to a pump and the negative sign to a turbine. The flow is from point (1) toward point (2) with the machine located between the two points.

EXAMPLE 3.5

0·708 m³/s of water is transported from a reservoir on one side of a 1220 m high mountain range to a town on the other side through a 1·53 m diameter pipe. The gauge pressure just before a booster pump P is 689 kN/m² and the pipe pressure

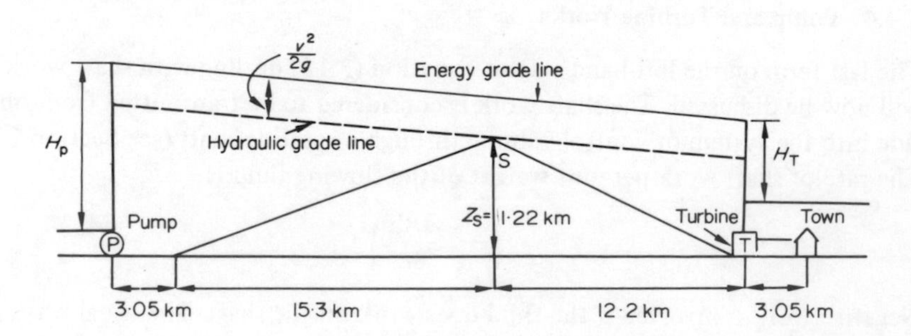

at the summit of the mountain is atmospheric. Determine the power of the booster pump motor for 80% efficiency. Neglect minor losses and take f = 0·006. If the delivery pressure at the town should be 0·689 N/mm² how much power is developed by a turbine T at 80% efficiency?

SOLUTION

Since the pressure at the summit S is atmospheric, the hydraulic grade line must pass through the centre of the pipe at the summit. The general form of the energy and hydraulic grade lines are indicated in the sketch. Apply equation (3.52) between the point just before the pump and the summit

$$\frac{p_1}{\rho g} + H_p = \frac{p_s}{\rho g} + z_s + \frac{fL_1Q^2}{3D^5}$$

since $v_1 = v_s$

$$\therefore \quad H_p = \frac{(p_s - p_1)}{\rho g} + 1220 + \frac{0{\cdot}006 \times (3050 + 15\,300)}{3 \times (1{\cdot}53)^5}\,(0{\cdot}708)^2$$

$$= -\frac{689}{9{\cdot}81} + 1220 + 2{\cdot}3 = 1152\text{ m} = 1{\cdot}15\text{ km}$$

$$\text{Pump motor h.p.} = \frac{10^3 \times 9{\cdot}81\,(0{\cdot}708) \times 1152}{0{\cdot}80} = \underline{10\text{MW}}$$

Between the summit and the town,

$$\frac{p_s}{\rho g}\!\!\!\!\!\!/\ 0 + 1220 - H_T = \frac{689}{9{\cdot}81} + \frac{fL_2Q^2}{3D^5} = 70 + 2{\cdot}5 \times \frac{1{\cdot}53}{1{\cdot}835}$$

$$H_T = 1148\text{ m}$$

$$\text{Power of turbine} = 0{\cdot}8 \times 9{\cdot}81\,(0{\cdot}708) \times 1148 \times 10^3$$

$$= \underline{7{\cdot}9\text{ MW}}$$

3.5 Analysis of Pipe Systems

Pipe systems with which the practising engineer has to deal may not be as simple as illustrated above. A city supply network is often complex and to design an

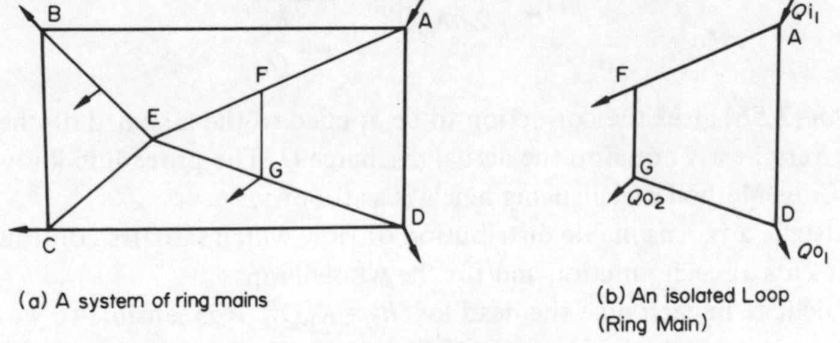

Fig. 3.13

effective water distribution system throughout the city it is necessary to calculate the pressure and discharge at all points in the network. Another example is a system of three or more reservoirs (or other pressure points) with a common pipe junction for which it is desired to calculate the flow in each pipe. A simplified isolated portion of a city network is shown in Fig. 3.13(a) and an example of a number of reservoirs whose pipe connections have a common joint J is shown in Fig. 3.14 (p. 91).

3.5.1 Ring Mains

A city network can be broken up into a convenient number of loops or 'rings'. The two theoretical requirements are that the net head drop around a loop must be zero and that the net flow toward a junction must be zero.

Let the head loss against friction and others in each pipe be expressed in the form

$$h_f = K_p Q^n \tag{3.53}$$

where K_p and the index n are assumed constant and Q is the discharge through the pipe. Let

$$Q = Q_0 + \Delta Q \tag{3.54}$$

where Q_0 is an assumed discharge (satisfying continuity conditions) which underestimates the actual discharge by a small value ΔQ.

Substituting (3.54) into (3.53) and expanding by the binomial theorem (neglecting terms containing $(\Delta Q)^2$ and higher powers)

$$h_f = K_p (Q_0^n + nQ_0^{n-1} \Delta Q) \text{ approximately}$$

In moving around a loop, $\Sigma h_f = 0$, thus

$$\Sigma n K_p Q_0^{n-1} \Delta Q = -\Sigma K_p Q_0^n \tag{3.55}$$

In order to satisfy continuity requirements at each junction (for fixed inflows and outflows into the particular loop), the value of ΔQ must be the same in each pipe. It can therefore be removed from the summation sign. Thus equation (3.55) gives

$$\Delta Q = \frac{-\Sigma K_p Q_0^n}{\Sigma n K_p Q_0^{n-1}} = \frac{-\Sigma h_f}{\Sigma n \dfrac{h_f}{Q_0}} \tag{3.56}$$

Equation (3.56) gives the correction to be applied to the assumed discharge Q_0 to make it very nearly equal to the actual discharge Q. The procedure known as the Hardy Cross Method of balancing heads is as follows.

1. Assume any reasonable distribution of flow which satisfies continuity requirements at each junction and for the whole loop.

2. Calculate in each pipe the head loss $h_f = K_p Q_0^n$. It is sensible to work in a fixed direction (clockwise or counter-clockwise). The head loss is positive if the flow is in the fixed direction and is negative (i.e. head rise) if the flow is against this direction. Summing up the head losses algebraically,

$$\Sigma h_f = \Sigma K_p Q_0^n$$

3. Calculate $\Sigma nh_f/Q_0 = \Sigma nK_pQ_0^{n-1}$.

The value of h_f/Q_0 is always positive since head loss ($+h_f$) is associated with flow in the positive sense of progress ($+Q_0$) and vice versa.

4. Set up in each loop a counter-balancing flow according to equations (3.56) and (3.54) to obtain a new and better approximate flow distribution in the ring main. Use the revised flows to repeat 1, 2, and 3, until the desired precision is attained with Σh_f approximately equal to zero.

EXAMPLE 3.6 (Imperial units)

Analyse the flow through the various pipes of the following network. If the pressure head at A is 200 ft what is the pressure at C, 15 ft below the elevation of A? Use the Hazen–William formula for head loss

$$h_f = \frac{15LQ^{1.85}}{C^{1.85}D^{4.87}}$$

where Q is the discharge (in gall/min), L is the length of the pipe (in ft) and D is the pipe diameter (in inches). $C = 100$ for all pipes.

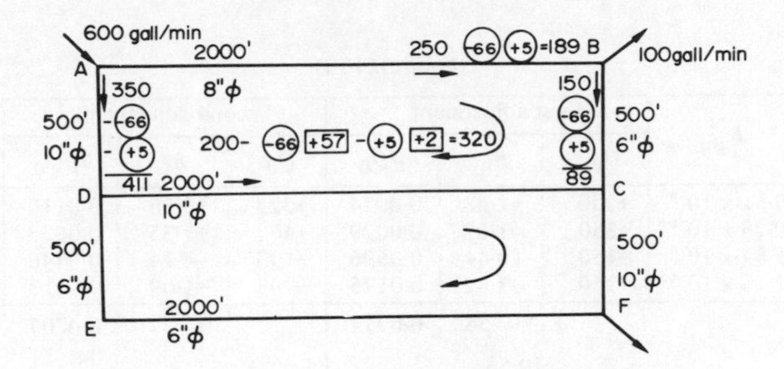

SOLUTION

Since $n = 1.85$

$$h_f = K_pQ^{1.85}$$

where

$$K_p = \frac{15L}{(100)^{1.85}D^{4.87}}$$

The outgoing discharge at F is given from continuity, as (600 − 100) = 500 gall/min for the whole loop. Moving in the clockwise direction, the first estimates of discharge through the various pipes are given in column 3 of the table below in gall/min. Check that continuity is satisfied at all junctions. While adjustment of flow for all other pipes is relatively straightforward, the exercise for pipe CD which is common to both loops needs special attention. The corrections arising from both loops must be appropriately applied. Examples of application of the corrections are indicated on the figure. The residual head loss after the second adjustment is negligible for each loop.

To find ΔQ we use

$$\Delta Q = -\Sigma h_f/(n\Sigma h_f/Q_0)$$

Loop ABCDA

Pipe	K_p	First adjustment			Second adjustment			Q
		Q_0	h_f	h_f/Q_0	Q_0	h_f	h_f/Q_0	
AB	2.38×10^{-4}	+250	+6.47	0.0260	+184	+3.66	0.0199	+189
BC	2.46×10^{-4}	+150	+2.62	0.0175	+ 84	+0.89	0.0106	+ 89
CD	0.81×10^{-4}	−200	−1.48	0.0074	−323	−3.56	0.0110	−320
DA	0.20×10^{-4}	−350	−1.02	0.0029	−416	−1.41	0.0034	−411
		Σ	+6.59	0.0538		−0.42	0.0449	

$$\Delta Q = \frac{-6.59}{1.85 \times 0.0538} = -66 \qquad \Delta Q = \frac{0.42}{1.85 \times 0.0449} = +5$$

(indicated by O on the diagram)

Loop DCFED

Pipe	K_p	First adjustment			Second adjustment			Q
		Q_0	h_f	h_f/Q_0	Q_0	h_f	h_f/Q_0	
DC	0.81×10^{-4}	+200	+1.48	0.0074	+323	+3.56	0.0110	+320
CF	0.20×10^{-4}	+350	+1.02	0.0029	+407	+1.35	0.0033	+409
FE	9.84×10^{-4}	−150	−10.44	0.0696	− 93	−4.14	0.0446	− 91
ED	2.46×10^{-4}	−150	−2.62	0.0175	− 93	−1.09	0.0118	− 91
		Σ	−10.58	0.0974		−0.32	0.0707	

$$\Delta Q = \frac{10.58}{1.85 \times 0.0974} = +57 \qquad \Delta Q = +2$$

(indicated by □ on the diagram)

The flows indicated in the last column are considered to be satisfactorily correct. Neglecting the difference in velocity heads at A and C, the pressure at C is given by

$$p_C/\rho g = p_A/\rho g + z_A - z_C - \text{losses (A} \rightarrow \text{C)}$$

$$= 200 + 15 - \left[\frac{(189)^{1.85}}{29 \times 145} + \frac{(89)^{1.85}}{29 \times 140}\right]$$

$$= 210.3 \text{ ft of water}$$

$$\therefore \qquad p_C = 210.3 \times \frac{62.4}{144} = \underline{91 \text{ lbf/in}^2}$$

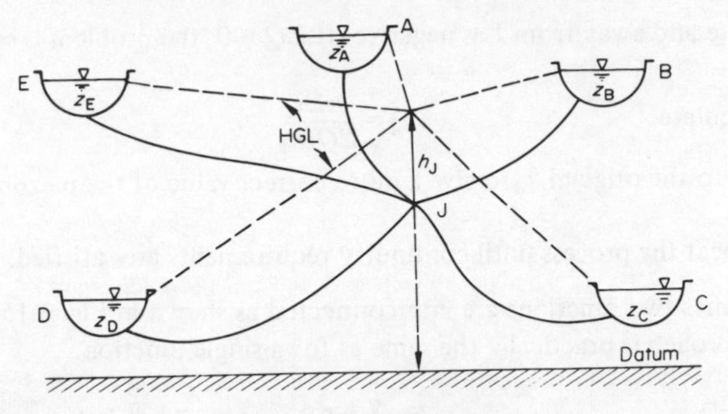

Fig. 3.14 Branching pipes

3.5.2 Branching Pipes

The theory behind the analysis of branching pipes hinges on the principle that there should be no accumulation of liquid at a junction. Assuming that the hydraulic grade line (HGL) coincides with the water surface in each reservoir (i.e. the velocity head in the reservoir is negligible) and that there is a definite single value for the hydraulic head at a junction, the head loss between the junction and a reservoir whose surface elevation is z_R above a common datum is

$$h_J - z_R = h_f = K_p Q^n \tag{3.57}$$

where h_J is the piezometric head at the junction above datum. A small error δh_J in the piezometric head at J introduces a corresponding error in the discharge through each pipe given by

$$\delta h_f = \delta h_J = n K_p Q^{n-1} \delta Q \tag{3.58}$$

From (3.57) and (3.58),

$$\delta h_f = \delta h_J = n h_f \delta Q / Q \tag{3.59}$$

The small error in the assumed position of the hydraulic grade line at J produces an error in the discharge through each pipe which therefore means that the algebraic sum of the flows toward the junction will not be zero. The residual ΔQ will be equal to the algebraic summation of δQ for all the pipes

i.e. $$\Delta Q = \Sigma \delta Q = \delta h_f / n \; \Sigma Q / h_f$$

or $$\delta h_f = \delta h_J = \frac{n \Delta Q}{\Sigma Q / h_f} \tag{3.60}$$

Equation (3.60) thus gives an estimate by which the value of the piezometric head must be *increased* to satisfy $\Delta Q = \Sigma \delta Q = 0$ considering flows toward the junction as positive. The procedure for estimating the discharge through branching pipes is as follows.

(1) Assume an elevation h_J of the hydraulic grade at the junction.

(2) Calculate the discharges between reservoir and junction taking flow toward

J as positive and away from J as negative. If $\Sigma Q = 0$, the problem is solved, otherwise

(3) Calculate
$$\delta h_J = \frac{n\Sigma Q}{\Sigma Q/h_f}$$
and add it to the original h_J to give a more correct value of the piezometric head at J.

(4) Repeat the process until continuity requirements are satisfied.

Sometimes two junctions are interconnected as shown in Fig. 3.15(a). The approach is practically the same as for a single junction.

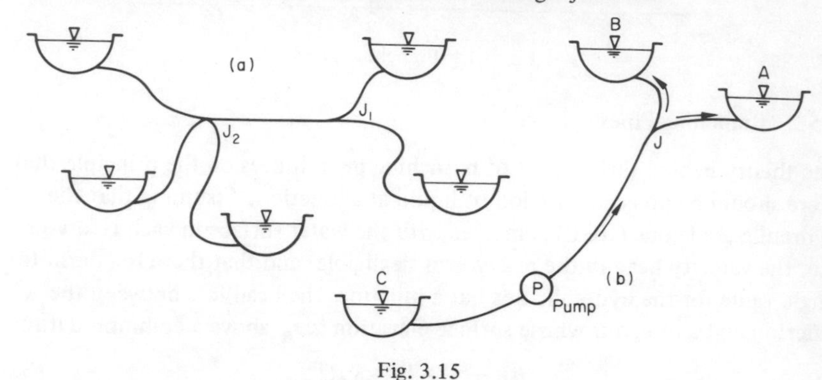

Fig. 3.15

(1) Assume the hydraulic grade at one junction J_1.
(2) Calculate the flow in the pipes directly connecting J_1 to reservoirs.
(3) Use continuity, $\Sigma Q = 0$, to determine the flow in the common pipe J_1J_2 and determine the hydraulic grade at J_2.
(4) Calculate flows in the other pipes branching at J_2. If $\Sigma Q = 0$ at J_2, the problem is solved, otherwise
(5) Calculate δh_{J2} using equation (3.60).
(6) With the new value $h_{J2} = h_{J2} + \delta h_{J2}$, repeat 2 to 4, working toward J_1.
(7) Repeat the process until satisfactory precision is attained.

Another variation of the branching pipe system is illustrated in Fig. 3.15(b). Pump P whose characteristics are assumed known delivers water toward junction J. The pump may be assumed to run at a constant speed. The procedure is as follows:

(1) Assume a discharge through the pump. From the pump characteristics read the pump head H_P.
(2) Calculate the hydraulic head at the pump and add to it H_P.
(3) Calculate the hydraulic head at J.
(4) Calculate flows into or from the adjoining reservoirs. If $\Sigma Q = 0$, the problem is solved, otherwise
(5) Assume a new discharge through the pump and repeat the process until satisfactory precision is attained.

EXAMPLE 3.7

The pump P in Fig. 3.15(b) delivers water from reservoir C. Its characteristics are

H (m)	21	18	16·5	15	13·5	12	10·5	9
Q (m^3/s)	0	0·057	0·072	0·086	0·098	0·108	0·116	0·124

Find the discharge into or from reservoirs A and B. Use the British form of the Darcy–Weisbach equation.

It is necessary that the withdrawal of water from C be limited to 0·043 m^3/s in the dry season. In which of the pipes leading to A and B should a throttle valve be fitted? What is the rate of energy dissipation in kW across the valve in the dry season?

Pipe Properties:

	L (m)	D (cm)	f
AJ	305	20	0·008
BJ	1015	20	0·008
CJ	610	30	0·015

Elevation of reservoirs are A (24 m), B (30 m) and C (15 m). J is at 0 m.

(U.S.T. Part III, 1967)

SOLUTION

$$h_f = \frac{4fLv^2}{2gD} = K_p Q^2 \text{ and } Q = \sqrt{(h_f/K_p)}$$

where

$$K_p = \frac{64fL}{2g\pi^2 D^5}$$

$$K_p\,(AJ) = 2{\cdot}56 \times 10^3;\; K_p\,(BJ) = 7{\cdot}68 \times 10^3;\; K_p\,(CJ) = 1{\cdot}25 \times 10^3$$

Assume flow Q_1 from C, Q_2 into A and Q_3 into B. From continuity:

$$Q_1 = Q_2 + Q_3$$

Q_1 (m^3/s)	H_p (m)	h_f(C→J) (m)	h_J (m)	h_f(J→A) (m)	h_f(J→B) (m)	Q_2 (m^3/s)	Q_3 (m^3/s)	$Q_2 + Q_3$ (m^3/s)
0·057	18	4·06	28·84	4·84	−1·16	0·044	−0·012	0·032
0·042	19	2·22	31·78	7·78	1·78	0·055	0·015	0·070
0·051	18·6	3·25	30·85	6·35	0·35	0·050	0·007	0·057
0·052	18·5	3·38	30·12	6·12	0·12	0·049	0·004	0·053

The flow rate into reservoir A is 0·049 m^3/s and into B, 0·004 m^3/s. With Q_1 = 0·042 m^3/s, head at J = 31·78 m; Q_2 = 0·055 m^3/s; Q_3 = 0·015 m^3/s. The throttle valve should cut down Q_2 such that $Q_2 + Q_3$ = 0·042 m^3/s.

∴ Flow required into A = 0·042 − 0·015 = 0·027 m^3/s.

Let head loss across valve be h_v.

$$h_J = h_A + h_f (J \rightarrow A) + h_v$$

i.e. $$31{\cdot}78 - 24 = 2{\cdot}56 \times 10^3 \, (0{\cdot}027)^2 + h_v$$

$\therefore$ $$h_v = 5{\cdot}9 \text{ m}$$

Power lost $= \rho g h_v Q_2 = \underline{1{\cdot}56 \text{ kW}}$

4 Flow in Non-Erodible Open Channels

4.1 Introduction

4.1.1 Definitions

Open channel flow or, more appropriately, *free surface flow* involves flow through a conduit with the liquid surface exposed to a constant pressure (atmosphere). Thus there can be open channel flow through a closed conduit (a

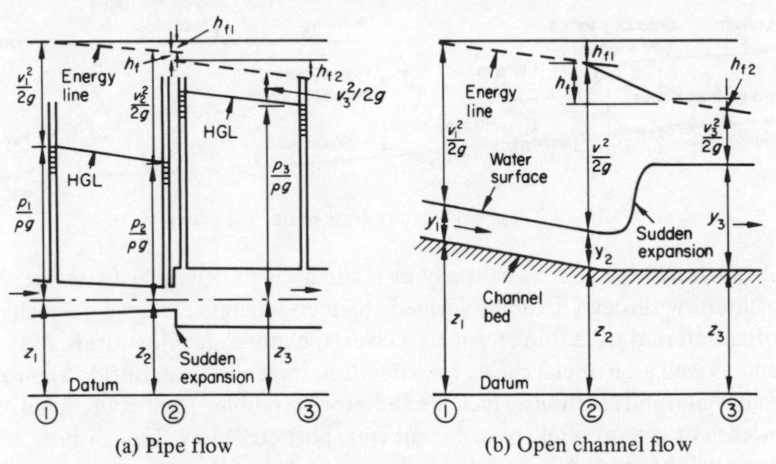

(a) Pipe flow (b) Open channel flow

Fig. 4.1 Comparison between pipe flow and open channel flow

pipe) provided the flow is not full bore. Full-bore closed conduit flow is referred to as pressure flow since pressure is the major controlling factor. In contrast, gravity and therefore conduit slope and elevation changes are important factors determining the nature of flow in an open channel. Nevertheless, all the fundamental principles of conservation of momentum, energy and mass (discussed in Chapter 2) are applicable to flow through open channels as they are to closed conduit pressure flows. Figure 4.1 shows the similarity of the energy concepts in both cases. It may be observed from the figure that depth (y) changes in an open channel follow the pattern of pressure (p) changes in a pipe.

All definitions and descriptions in Section 1.6 may be applied to open channel flows. However, open channel flows invariably involve liquids which are normally treated as incompressible. The flow may be laminar or turbulent. Laminar open channel flows are however rare in practice. Examples of unsteady open channel flows are the propagation of flood waves or the attenuation of discharge due to the termination of the supply. Rapid changes in flow will be treated in Section 4.5 under the heading of open channel surges. A steady open channel flow may either be uniform or non-uniform. By definition a uniform open channel flow is one in which there are no changes in depth (and average velocity) along the conduit. This is normally true in prismatic (constant geometry) channels. The liquid surface is always parallel to the bed for a uniform free surface flow. Changes in the geometry of a channel give rise to non-uniform flow. Changes in the depth and other characteristics in such a case may be abrupt or gradual (smooth). The latter produces draw-down and backwater effects. A rapid change in a steady open channel flow produces what is known as a *hydraulic jump*. The unsteady equivalent of the hydraulic jump is the *hydraulic surge or bore.* Waves are unsteady conditions which may be produced in an open channel due to a disturbance. Various steady-state flow conditions are illustrated in Fig. 4.2.

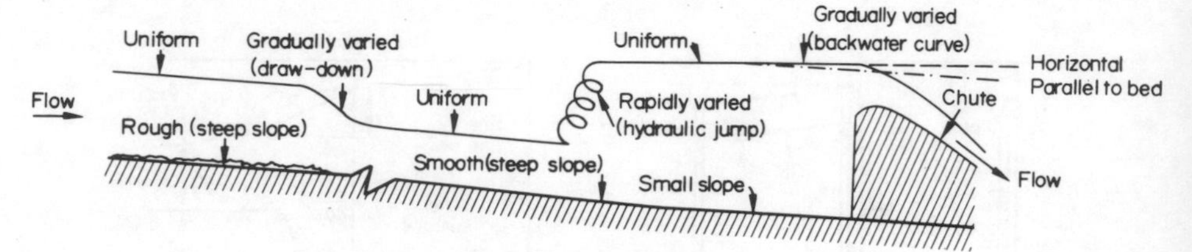

Fig. 4.2 Types of steady flow in open channels

Channels may be formed naturally or made by man. Rivers and streams normally flow through naturally formed channels whereas artificial open channels are often referred to as flumes, canals, culverts, chutes, etc. Most rivers and streams as well as artificial canals for irrigation, transportation or flood control have alluvial (sand) sections which are therefore erodible. Their study involves a knowledge of geomorphology, sediment transport etc. Flow through such channels will be treated separately in Chapter 7. This chapter is devoted to flow through channels whose geometry is stable and whose beds are non-erodible.

4.1.2 Influence of Channel Characteristics on Flow

The geometry, roughness and permeability of a channel affect the flow through it. In what follows we will assume the channel wall to be impermeable. The channel geometry includes its shape, slope and alignment. A natural channel usually has an irregular shape and a longitudinal profile. In the interests of analysis it is convenient to smooth out the irregularities and reduce the section to a smooth geometry. The longitudinal slope of a channel is taken as the average slope, and the bends are replaced by their equivalent lengths. The slope S_0 is defined as the net drop z_0 in elevation divided by the longitudinal distance L (Fig. 4.3a). Irregular cross sections may be smoothed into convenient shapes.

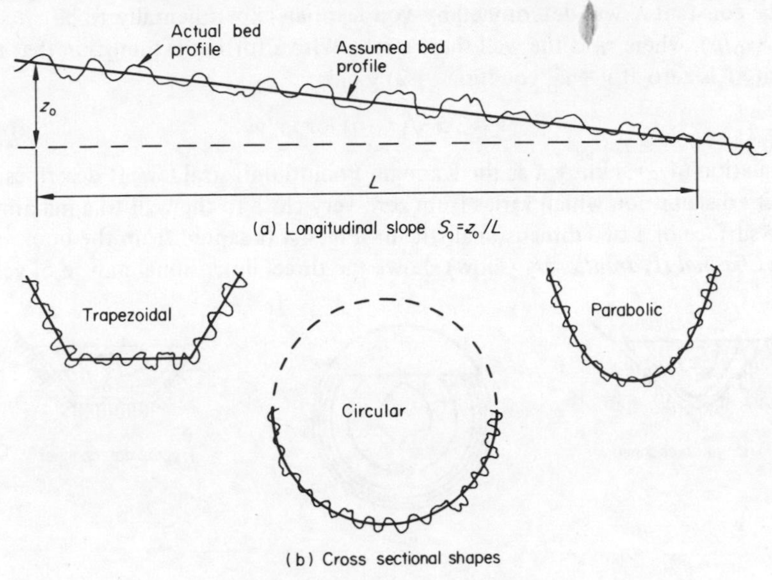

Fig. 4.3 Channel geometry

The geometrical shapes usually adopted for design and computational purposes are the trapezium (including the rectangle and triangle), the parabola and the circle (or part of it). These are illustrated in Fig. 4.3(b). Irregularities and other minor obstacles to flow (stones, vegetation etc) are represented by a friction factor for each particular reach of channel. These approximations are valid only when the irregularities are very small compared with a principal channel dimension, usually the depth of flow.

4.1.3 Channel Roughness (Velocity Distribution)

Roughness in a channel resists flow. Since most open channel flows are turbulent, equation (1.11a) which relates shear stress to the time average velocity $\bar{v}$ at a distance y from the wall through the apparent coefficient of viscosity $(\upsilon + \epsilon)$ may be applied. Except very close to the wall $(y < y')$ in the region known as the laminar sublayer where turbulence is insignificant, eddy viscosity ϵ dominates molecular viscosity υ. Thus (assuming a steady state) shear stress is given by

$$\tau_{yx} = \rho\epsilon \frac{d\bar{v}}{dy} \quad \text{for } y \geqslant y' \tag{4.1}$$

Based on Prandtl's mixing length concepts and experiments, von Karman has hypothesized that the ratio of shear stress to dynamic eddy viscosity is inversely proportional to the distance y from the surface. Thus

$$\frac{\tau_{yx}}{\rho \epsilon} = \frac{d\bar{v}}{dy} = \frac{K}{y} \text{ for } y \geqslant y' \tag{4.2}$$

The constant K was determined by von Karman experimentally to be $2{\cdot}5\sqrt{(\tau_0/\rho)}$, where τ_0 is the wall shear stress. With a further assumption that the velocity $\bar{v}$ is zero at $y = y'$, equation (4.2) yields:

$$\bar{v} = 2{\cdot}5\sqrt{(\tau_0/\rho)} \log_e y/y' \tag{4.3}$$

Equation (4.3) is known as the Karman–Prandtl universal law. It describes a velocity distribution which varies from zero very close to the wall to a maximum at the surface of a two-dimensional stream. Fig. 4.4 (adapted from the book *Open Channel Hydraulics* by Chow) shows the three-dimensional nature of velocity

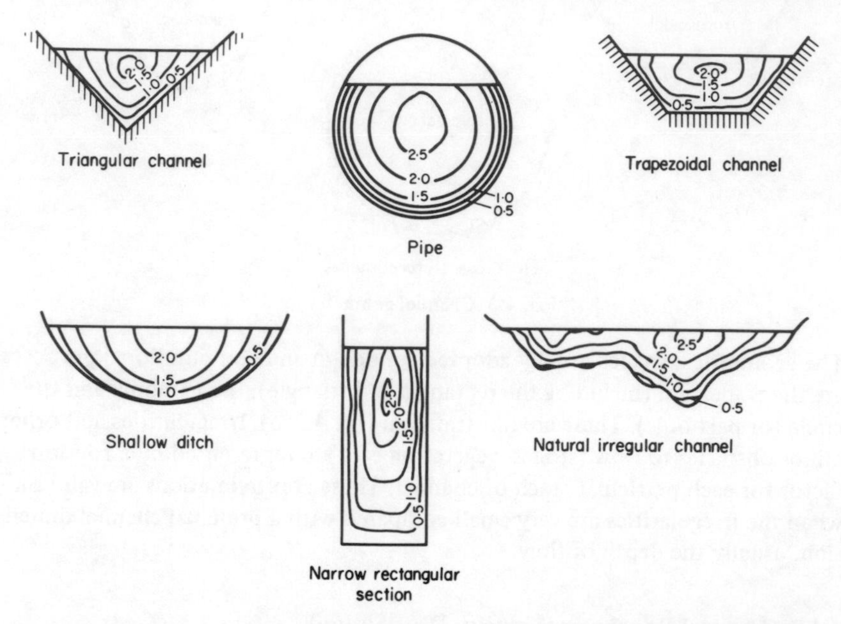

Fig. 4.4 Typical curves of equal velocity in various channel sections (After V.T. Chow)

profiles in open channels. The need for an approximation to two-dimensional conditions in wide open channels is well illustrated. The reduction in velocity near the water surface in most cases is due to wind resistance which the above theory ignores.

Assuming the two-dimensional case (a wide open channel) of depth y_0, the maximum velocity is given by

$$\bar{v}_{max} = 2{\cdot}5\sqrt{(\tau_0/\rho)} \log_e y_0/y' \tag{4.4}$$

From (4.3) and (4.4)

$$\frac{\bar{v}_{max} - \bar{v}}{v'} = 2{\cdot}5 \log_e y_0/y \tag{4.5}$$

where the *shear velocity*

$$v' = \sqrt{(\tau_0/\rho)}$$

Equation (4.5) which gives the difference between the velocity at any depth relative to the maximum velocity is commonly referred to as the *velocity defect law.* The average velocity of flow is given by

$$\bar{v}_{av} = \frac{1}{y_0} \int_0^{y_0} \bar{v} \, dy \tag{4.6}$$

Substituting (4.5) into (4.6) and integrating by parts gives

$$\frac{\bar{v}_{av}}{\bar{v}_{max}} = 1 - 2{\cdot}5 \frac{v'}{\bar{v}_{max}} \tag{4.7}$$

From (4.5) and (4.7)

$$\bar{v}_{av} + 2{\cdot}5\, v' = \bar{v} + 2{\cdot}5\, v' \log_e y_0/y$$

$$\therefore \quad \bar{v} = \bar{v}_{av} + 2{\cdot}5\, v'(1 - \log_e y_0/y)$$

or

$$\frac{\bar{v}}{\bar{v}_{av}} = 1 + \frac{2{\cdot}5\, v'}{\bar{v}_{av}}(1 - \log_e y_0/y) \tag{4.8}$$

If $\bar{v}/\bar{v}_{av} = 1$ in equation (4.8), $\log_e y_0/y = 1$

or

$$y/y_0 = \frac{1}{e} = 0{\cdot}368 \tag{4.9}$$

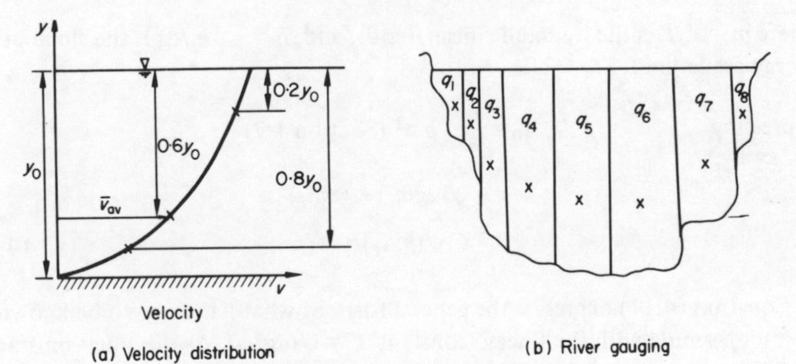

Fig. 4.5 Flow measurement in channels

Equation (4.9) which has been verified in practice shows that the average velocity of a stream occurs at approximately 0·6 y_0 from the surface. This relationship is used in the measurement of discharge through rivers. The river is divided into a convenient number of segments as illustrated in Fig. 4.5(b). The discharge q through each segment is given by the measured velocity at 0·6 times the depth multiplied by the area of the segment. The total discharge is then given by the sum of the q's. To minimize errors, velocities at 0·2 and 0·8 times the depth for each segment may be measured. The average of the two measurements is known to be very close to the velocity at 0·6 of the depth. (See Section A 1.2).

ROUGHNESS (FRICTION) FACTORS

It has been stated or implied in Section 1.7 and elsewhere that it is more convenient in practice to express shear stress in terms of a measurable factor c_f or f. Other empirical factors exist for open channel flows and these will be introduced here. For a small length δs along a channel, there is a small head loss δh_f due to friction on the bed and sides of the channel. This head loss represents the rate of doing work against shear forces by a unit weight of the flowing fluid. For an average wetted perimeter for the elemental length δs of P, the shear force is $\tau_0 \delta s P$. Using average velocities which are for convenience simply designated as v,

rate of doing work against shear force $= \tau_0 P \delta s v$

rate of shear work per unit weight of flowing fluid $= \tau_0 P \delta s v / \rho g A v$

where A is the cross sectional area of flow.

$$\therefore \qquad -\delta h_f = \frac{\tau_0 \delta s}{\rho g} \frac{P}{A}$$

and

$$\tau_0 = \rho g m S_f \tag{4.10}$$

where $m = A/P$ is the hydraulic mean depth, and $S_f = -\,\mathrm{d}h_f/\mathrm{d}s$ is the slope of the energy grade line.

Expressing

$$\tau_0 = \tfrac{1}{2} c_f \rho v^2 \text{ (Section 1.7)}$$

$$v = \sqrt{(2g/c_f)}\sqrt{(mS_f)}$$

or

$$v = C\sqrt{(mS_f)} \tag{4.11}$$

Equation (4.11) expresses the general form of what has come to be known as *Chezy's formula* with the Chezy 'constant' $C = \sqrt{(2g/c_f)}$. As the equation stands, C and S_f may be variable but for fully established flows the friction factor c_f and therefore C are fairly constant. For the special case of a uniform open channel flow, the bed, the water surface and the energy grade line have an equal slope, $S_f = S_0$. Thus equation (4.11) for a *uniform flow* becomes,

$$v = C\sqrt{(mS_0)} \tag{4.12}$$

With reference to equation (4.3) the average velocity occurs at $y = 0{\cdot}4\,y_0$, and experiments have established that $y' = k_s/33$ for a hydraulically rough bed, where k_s is the average size of the roughness protrusions.

$$\therefore \quad v = 2{\cdot}5\sqrt{(\tau_0/\rho)}\,\log_e \frac{13{\cdot}2}{k_s}\,y_0 \tag{4.13}$$

Equating (4.12) and (4.13), changing the base of the logarithm and remembering that $\tau_0 = \rho g m S_0$,

$$C = 5{\cdot}75\sqrt{g}\,\log_{10}\frac{13{\cdot}2\,y_0}{k_s} \tag{4.14}$$

Equation (4.14) shows that the Chezy factor is not really constant for a particular roughness of a channel but depends on the depth of flow. C increases with decreasing c_f and k_s and is therefore a measure of smoothness as opposed to roughness.

Another common empirical formula used in calculating the discharge through open channels is the *Manning formula*. In its modern form (see Chow, 1959 edition, p. 98) this is given as

$$v = \frac{1}{n}\,m^{\frac{2}{3}}\,S_0^{\frac{1}{2}} \tag{4.15}^*$$

where n is Manning's friction factor which is not dimensionless as it stands. The form given in (4.15) is different from Manning's original form given in 1889. The value 1·486 arising from the conversion from the metric system ($m^{\frac{1}{3}}$) to Imperial units ($ft^{\frac{1}{3}}$) is believed to have a dimension which satisfies dimensional homogeneity of the equation. Despite or because of the many conversions over the years the numerical value of n is widely taken to be the same in both systems. Manning's n is related to Chezy's C as

$$C = \frac{1}{n}\,m^{\frac{1}{6}} \tag{4.16}$$

and to channel roughness size k_s as

$$n = m^{\frac{1}{6}}/(5{\cdot}75\sqrt{g}\,\log_{10} 13{\cdot}2y_0/k_s) \tag{4.17}$$

A table of ranges of n for natural channels and excavations with or without lining is given in Tables 5.5 and 5.6 on pages 109–113 of *Open Channel Hydraulics* by Chow. Some typical ranges are quoted here in Table 4.1 for easy reference. Many other empirical formulae due for instance to Bazin (1897), Ganguillet and Kutter (G–K formula) (1870) and Powell (1950) seek to relate flow conditions and nature of the channel surface to the Chezy C but none has so far been found to be universally valid or superior to Manning's formula which is extensively used.

* The Manning equation in Imperial units is $v = \dfrac{1{\cdot}486}{n}\,m^{\frac{2}{3}}\,S_0^{\frac{1}{2}}$

Table 4.1
Typical values for Manning n

Channel material	*Range of n*
Metals	0·010 – 0·024
Lucit, glass, cement	0·009 – 0·013
Concrete	0·011 – 0·017
Wood	0·012 – 0·017
Clay	0·013 – 0·016
Gravel bottom canal	0·020 – 0·033
Masonry lining	0·025 – 0·032
Asphalt	0·013 – 0·016
Vegetal lining	0·030 – 0·500

SUMMARY
The following points must be carefully noted.

(1) Both Chezy's C and Manning's n depend on the depth of flow and also on Reynolds number although for most practical problems, the assumption of appropriate constant values does not introduce any serious error.

(2) The local values of C and n (friction factors) may be different from the average values.

(3) The Chezy and Manning formulae are generally applied to uniform flow computations. They are however valid for non-uniform flows provided local values of C or n and energy slope S_f are used.

4.1.4 Best Hydraulic and Economic Sections

From Manning's formula (the same results can be deduced from Chezy's formula), the discharge Q through an open channel of cross-sectional area A is

$$Q = \frac{1}{n} A\, m^{\frac{2}{3}} S_0^{\frac{1}{2}} \tag{4.18}$$

The discharge increases with increasing hydraulic mean depth for a fixed cross sectional area (assuming constant n). The maximum discharge occurs when the wetted perimeter P is a minimum. In other words, the minimum possible wetted perimeter gives the best hydraulic performance if all other factors remain constant. Also for a particular discharge, slope and roughness,

$$A = c\, m^{-\frac{2}{3}}$$

where c is a constant

$$\therefore \quad A = c\, P^{\frac{2}{5}}$$

Thus A is a minimum if P is a minimum, that is when the section is hydraulically most efficient. This means that the best hydraulic geometry is also the most economical since it involves the least amount of excavations and lining.

(1) TRAPEZOIDAL SECTIONS

Consider a trapezoidal section with sides sloping at $1/k$ as shown in Fig. 4.6. The depth of flow is y_0, the width at the bottom is b and the surface width is B.

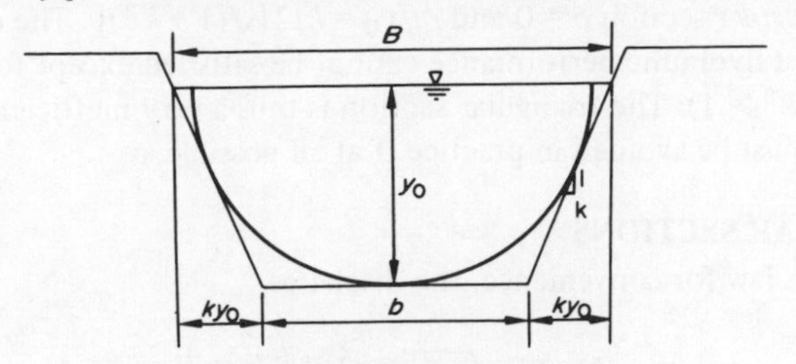

Fig. 4.6 Trapezoidal section

$$m = \frac{A}{P} = \frac{y_0(b + ky_0)}{b + 2y_0(1 + k^2)^{\frac{1}{2}}} \tag{4.19}$$

$$b = \frac{A}{y_0} - ky_0 \tag{4.20}$$

Substituting (4.20) into (4.19)

$$m = \frac{Ay_0}{A + C_1 y_0^2} \tag{4.21}$$

where $$C_1 = 2\sqrt{(1 + k^2)} - k$$

For the best hydraulic performance, m must be maximum

i.e. $$\frac{dm}{dy_0} = \frac{A(A + C_1 y_0^2) - 2C_1 y_0(Ay_0)}{(A + C_1 y_0^2)^2} = 0$$

or $$A + C_1 y_0^2 - 2C_1 y_0^2 = 0$$

$\therefore$ $$C_1 y_0^2 = A \tag{4.22}$$

Substituting (4.22) into (4.21) for conditions of most efficient performance

$$\frac{m_e}{y_0} = \frac{1}{2} \tag{4.23}$$

and from (4.20)

$$b/y_0 = 2\sqrt{(1 + k^2)} - 2k \tag{4.24}$$

The condition obtained in equation (4.23), namely that the hydraulic mean depth should be half the depth of flow, can also be derived from elementary geometry. The three sides of the trapezoidal section are considered tangential to a semi-circle described on the water surface. Equation (4.24) determines the relationship of bottom width to depth of flow for fixed side slopes which are generally determined by stability conditions.

For the special case of a *rectangular* section, $k = 0$ and $b/y_0 = 2$. Thus for best hydraulic performance the depth is half the width of the channel and the hydraulic mean depth a quarter.

For a *triangular* section $b = 0$ and $m/y_0 = k/2\,[\surd(1 + k^2)]$. The condition (4.23) for best hydraulic performance cannot be satisfied except for very large values of k, ($k^2 \gg 1$). The triangular section is thus a very inefficient hydraulic section and must be avoided in practice if at all possible.

(2) CIRCULAR SECTIONS

Using Chezy's law for convenience, the discharge

$$Q = AC\sqrt{mS_0} = C\sqrt{\left(\frac{A^3}{P}S_0\right)}$$

For maximum discharge, A^3/P must be maximum.
This requires that at best performance,

$$\frac{\mathrm{d}(A^3/P)}{\mathrm{d}y_0} \quad \text{or} \quad \frac{\mathrm{d}(A^3/P)}{\mathrm{d}\theta} = 0$$

where 2θ is the angle subtended at the centre by the water surface (Fig. 4.7(a))

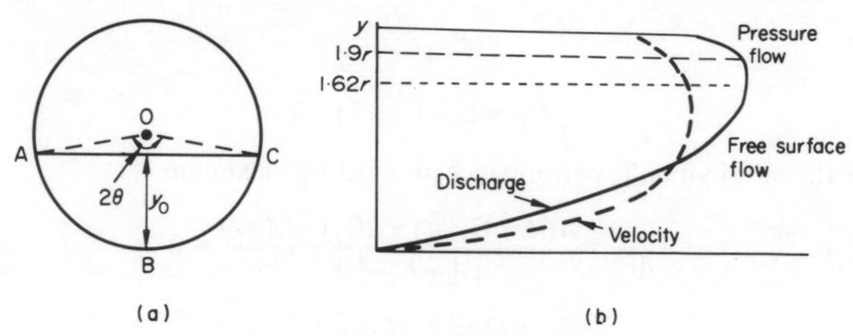

Fig. 4.7 Flow in circular sections

i.e.
$$P(3A^2)\frac{\mathrm{d}A}{\mathrm{d}\theta} - A^3\frac{\mathrm{d}P}{\mathrm{d}\theta} = 0$$

∴
$$3P\frac{\mathrm{d}A}{\mathrm{d}\theta} - A\frac{\mathrm{d}P}{\mathrm{d}\theta} = 0 \tag{4.25}$$

A = area of sector $ABCOA$ – area of triangle AOC.

$$= r^2\theta - \frac{r^2 \sin 2\theta}{2}$$

$$= r^2\left(\theta - \frac{\sin 2\theta}{2}\right)$$

and $P = 2r\theta$

where r is the radius of the circular section.

Thus from equation (4.25)

$$6r^3\theta\,(1-\cos 2\theta) - 2r^3\left(\theta - \frac{\sin 2\theta}{2}\right) = 0$$

or

$$4\theta - 6\theta\cos 2\theta + \sin 2\theta = 0$$

The solution of this equation gives $\theta = 154°$ approximately. The depth corresponding to the best performance is point $y_0 = r - r\cos\theta = 1{\cdot}9r$ approximately. Thus a pipe discharges most efficiently when it is flowing 95% full bore. A sketch of the variation of discharge with depth is given in Fig. 4.7(b). This fact is very important for the design of culverts. Flood waters, which tend to produce flow at a depth higher than the optimum depth, immediately introduce pressure flow. The culvert capacity falls and the accumulated water builds up to provide the head necessary to maintain a pressure flow. An approach similar to that above can be used to show that the maximum average velocity occurs when the pipe is flowing at 81% of its diameter.

Table 4.2

Section	Area A	Wetted perimeter P	Hydraulic mean depth $m=A/P$	Top width B	Average depth A/B
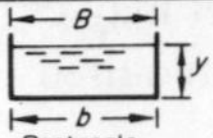 Rectangle	by	$b+2y$	$\frac{by}{b+2y}$	b	y
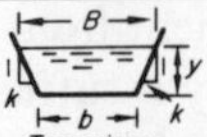 Trapezium	$(b+ky)y$	$b+2y\sqrt{(1+k^2)}$	$\frac{(b+ky)y}{b+2y\sqrt{(1+k^2)}}$	$b+2ky$	$\frac{(b+ky)y}{b+2ky}$
 Triangle	ky^2	$2y\sqrt{(1+k^2)}$	$\frac{ky}{2\sqrt{(1+k^2)}}$	$2ky$	$(1/2)y$
 Circle	$1/8\,(2\theta-\sin 2\theta)d_0^2$	θd_0	$1/4\left(1-\frac{\sin 2\theta}{2\theta}\right)d_0$	$(\sin\theta)d_0$ or $2\sqrt{(y(d_0-y))}$	$1/8\left(\frac{2\theta-\sin 2\theta}{\sin\theta}\right)d_0$
 Parabola	$2/3\,By$	$B+\frac{8}{3}\frac{y^2}{B}$*	$\frac{2B^2y}{3B^2+8y^2}$*	$\frac{3}{2}\frac{A}{y}$	$(2/3)y$
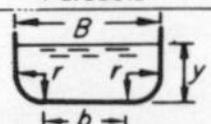 Round-cornered rectangle ($y>r$)	$(\frac{\pi}{2}-2)r^2+(b+2r)y$	$(\pi-2)r+b+2y$	$\frac{(\pi/2-2)r^2+(b+2r)y}{(\pi-2)r+b+2y}$	$b+2r$	$\frac{(\pi/2-2)r^2}{b+2r}+y$
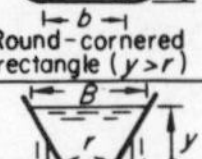 Round-bottomed triangle	$\frac{B^2}{4k}-\frac{r^2}{k}(1-k\cot^{-1}k)$	$\frac{B}{k}\sqrt{(1+k^2)}-\frac{2r}{k}(1-\cot^{-1}k)$	$\frac{A}{P}$	$2[k(y-r)+r\sqrt{(1+k^2)}]$	$\frac{A}{B}$

*Approximation for the interval $0 < 4y/B \leqslant 1$.

In Table 4.2 the important geometric properties of the more commonly employed sections are summarized. (In the table and elsewhere in this chapter, B is the top width, b is the bottom width, y or y_0 is the depth of flow, $\bar{y}$ is the distance of the centroid and y' the distance of the centre of pressure from the top and the other terms are defined or illustrated in column 1.)

4.1.5 Factors Determining the Choice of Channel Geometry

Having discussed the hydraulic influence of various geometrical properties of channels it is now opportune to try to summarize the principal considerations which influence the choice of shape and slope of channels.

(1) Hydraulic effectiveness. Hydraulically the most effective section is one which has the minimum perimeter for a particular cross-sectional area. The circular section is thus the most efficient. Excavation and lining costs decrease as the section approaches the hydraulically most efficient shape.

(2) Stability of side slopes. This is generally determined by the material through which the channel runs (soil mechanics considerations).

(3) Evaporation. In hot climates evaporation may be a major consideration. The larger the surface width B the more the evaporation losses, all other things being equal.

(4) Contours of the land. Possible gradients between terminal points are determined by the land mark and economy. The gradient determines the velocity (Chezy or Manning formula). If the gradient is too high bed erosion may take place in moveable beds. It may be necessary to go in steps with drop structures between terminal points to avoid erosion. For very small slopes it may be desirable to make m large and the friction factor small to pass a required discharge.

(5) Sediments. If it is not possible to prevent suspended sediment from entering the channel, it is necessary to make sure that the flow velocity is sufficient to carry it away otherwise it will settle and impair the efficiency of the channel.

(6) Permeability problems. The channel material may be porous enough to allow seepage of flowing water. If loss of water is undesirable, sealing may be required. This is sometimes achieved by passing through a large amount of clay containing materials at a very slow speed to clog the soil pores.

(7) Uncontrollable natural factors. Water plants (e.g. on Lake Kariba in Zimbabwe), water buffalos and other burrowing animals and human beings.

(8) Accessibility for maintenance.

EXAMPLE 4.1

A rectangular open channel 5·5 m wide and 1·22 m deep has a slope of 1 in 1000 and is lined with rubble masonry (Manning's $n = 0{\cdot}017$). We wish to increase the amount of water discharged as much as possible without changing the channel slope or the rectangular form of the section. The dimensions of the section may be changed but the amount of excavation must not be changed.

Determine: (1) the discharge of the original channel; (2) new dimensions of a channel to give the maximum discharge; (3) the ratio of the new discharge to the original discharge. What is the new discharge? (U.S.T., Dip. IV Mech., 1967.)

SOLUTION

(1) Original cross-sectional area of flow $A = 6{\cdot}71\ m^2$. Original hydraulic mean depth $m = 6{\cdot}71/(5{\cdot}5 + 2 \times 1{\cdot}22) = 0{\cdot}84$ m. From equation (4.18)

$$Q = \frac{1}{0{\cdot}017} \times 6{\cdot}71 \times (0{\cdot}85)^{\frac{2}{3}} \left(\frac{1}{1000}\right)^{\frac{1}{2}}$$

$$= \underline{11{\cdot}1\ m^3/s}$$

(2) The best hydraulic rectangular section has a width/depth ratio of 2

i.e.
$$y_0 = \frac{1}{2} b$$

Since the cross sectional area must be unchanged,

$$b y_0 = \text{constant} = 6{\cdot}71$$

$$\therefore \quad b^2 = 13{\cdot}42$$

New width $\quad b = \underline{3{\cdot}66\ m}$

New depth $\quad y_0 = \underline{1{\cdot}83\ m}$

(3) The new
$$m = \frac{6{\cdot}71}{3{\cdot}66 + 2 \times 1{\cdot}83} = 0{\cdot}91\ m$$

$$\frac{\text{New discharge}}{\text{Original discharge}} = \left(\frac{0{\cdot}91}{0{\cdot}84}\right)^{\frac{2}{3}} = \underline{1{\cdot}055}$$

New discharge $= \underline{11{\cdot}7\ m^3/s}$

4.2 Momentum Concepts

It is apparent from the discussions on velocity distribution in an open channel (Section 4.1.3) that precise analysis would involve three-dimensional considerations. Fortunately however this has been found to be unnecessary for most problems encountered in engineering practice. Gross descriptions of the flow using a finite or semi-finite control volume have yielded satisfactory results. The use of an average flow velocity across the cross section of the channel introduces an error in the evaluation of momentum and energy fluxes which may be taken into account by means of momentum and energy correction factors (see Section 3.3). As for flow through pipes the factors are very nearly unity if the channel flow is turbulent, and this is usually the case in practice. In the following discussions therefore fully turbulent flow will be assumed and the correction factors for energy (α) and momentum (β) will be taken as unity. Again unless otherwise stated v is used for average velocity at a section throughout the rest of the chapter.

4.2.1 Force of a Stream

In the one-dimensional analysis of flow in open channels we use a control volume bounded by the walls of the channel, the free liquid surface and two cross sections upstream and downstream of the flow regime of interest. This is illustrated for the force exerted by a stream on an obstruction in Fig. 4.8. The obstruction

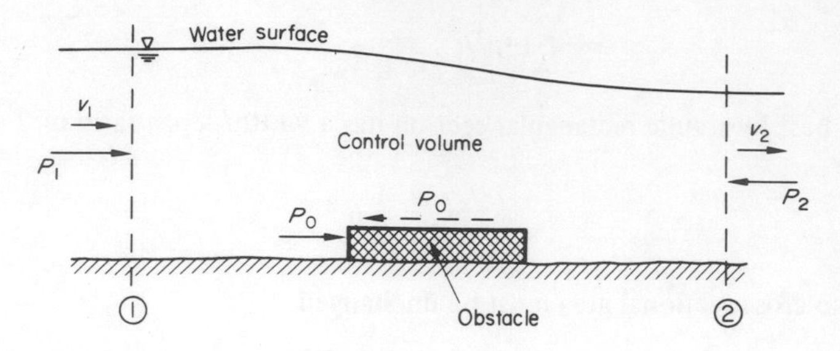

Fig. 4.8

is represented by a block in the figure and may arise from many sources including normal channel bed friction, stones, bridge piers or side walls of channel constrictions. For simplicity a horizontally aligned channel is assumed at this stage.

If the stream exerts a force P_0 to the right on the obstacle the stream in turn will experience a force P_0 to the left from the obstacle. Equation (2.10) may be applied to the control volume with average pressures p_1 and p_2 at sections 1 and 2.

$$\Sigma F = \Sigma E_M - \Sigma I_M$$

The external forces are given by

$$\Sigma F = A_1 p_1 - A_2 p_2 - P_0 = \rho g(A_1 \bar{y}_1 - A_2 \bar{y}_2) - P_0$$

and the net momentum flux by

$$\Sigma E_M - \Sigma I_M = \rho Q(v_2 - v_1)$$

where A is the cross sectional area, $\bar{y}$ is the centroid of the cross sectional area from the liquid surface, Q is the discharge and v is the velocity. The pressure terms imply an assumption of hydrostatic pressure conditions.
Thus

$$\rho g(A_1 \bar{y}_1 - A_2 \bar{y}_2) - P_0 = \rho Q(v_2 - v_1) \qquad (4.26)$$

or

$$(\rho g A_1 \bar{y}_1 + \rho Q v_1) - (\rho g A_2 \bar{y}_2 + \rho Q v_2) - P_0 = 0 \qquad (4.27)$$

Equation (4.27) may be written simply as

$$P_1 - P_2 - P_0 = 0 \qquad (4.28)$$

where the *force function* **P** is given by the pressure force and the momentum flux relevant to a section. For a frictionless, unobstructed flow through a prismatic channel, $P_0 = 0$ and $P_1 = P_2 = P$, a constant. The force function per unit width for a rectangular channel $\left(\bar{y} = \frac{1}{2} y_0\right)$ is given by

$$P = \frac{1}{2} \rho g y_0^2 + \rho q v \tag{4.29}$$

where q is the discharge per unit width, Q/b

or
$$P = \frac{1}{2} \rho g y_0^2 + \rho q^2/y_0 \tag{4.30}$$

Equation (4.30) is cubic in y_0 for a fixed value of q. The variation of the force function P with depth (for a fixed discharge through a prismatic rectangular channel) is illustrated in Fig. 4.9.

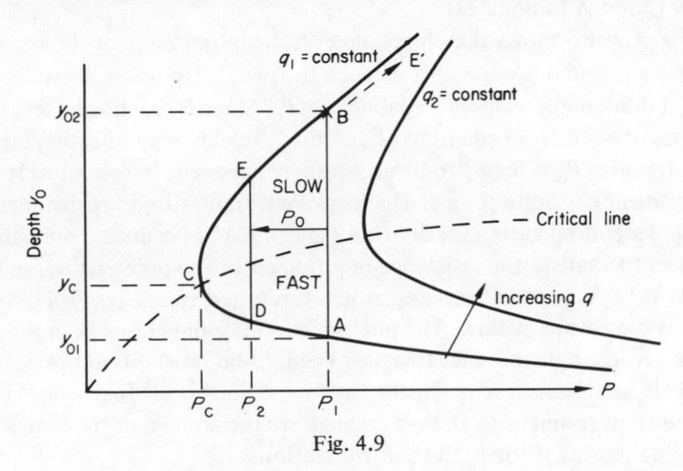

Fig. 4.9

The following points should be noted about the curves in Fig. 4.9.

(1) For a particular P a fixed flow can pass through the channel at two possible depths. At the smaller depth (corresponding to A), the flow will be fast and at the bigger depth (corresponding to B) the flow will be slow. Under normal conditions of flow the steepness and roughness of the channel bed determine whether the flow should be fast or slow as may be inferred from Chezy's equation (4.12).

(2) There is a minimum value of P at which a particular discharge can pass through the channel. The depth corresponding to the point of minimum value C is known as the critical depth, y_c. At the minimum

$$\frac{dP}{dy_0} = \rho g y_0 - \rho q^2/y_0^2 = 0$$

$$\therefore \quad y_0^3 = q^2/g = y_c^3$$

or
$$y_c = (q^2/g)^{\frac{1}{3}} \tag{4.31}$$

A flow with a depth greater than the critical depth is said to be *slow or subcritical* and one with a depth smaller than the critical depth is *fast or supercritical.* At the point of critical flow, the *Froude number F* is given by

$$F = \frac{v_c^2}{gy_c} = \frac{q^2}{gy_c^3} = 1$$

The Froude number can thus be conveniently used to define the nature of a flow in an open channel. The flow is slow if F is less than unity and is fast if F is greater than unity. F is generally defined by q^2/gy_0^3.

(3) A flow which is initially slow (corresponding to point B) will show a drop in surface when it flows past an obstacle (to point E). The reverse is true for a fast flow (point A to point D).

(4) Fig. 4.9 also shows that depending on the initial value of the force function P_1 there is a maximum value of resistance to flow P_0 (for each discharge) which will give a minimum downstream value $P_2 = P_c$. Since P_2 cannot be less than P_c it becomes obvious from equation (4.28) that the only way of satisfying the demand for large P_0 is for P_1 to be accordingly adjusted. In case of a slow flow, the phenomenon is quite simple. The message is transmitted upstream and the upstream water level automatically rises (say to E′) to produce a pressure build-up sufficient to satisfy the resistance requirements. The point corresponding to section (2) at which conditions are critical is referred to as a *control point* since it controls the nature of flow. The phenomenon is more complex in the case of a fast flow because a disturbance (message) cannot be transmitted upstream in fast flow. It may be necessary for the flow to convert itself first from fast to slow conditions (point A to B, say) through what is known as the hydraulic jump before passing through the control section.

4.2.2 The Hydraulic Jump

The hydraulic jump (Fig. 4.10) is a sudden and turbulent passage of a flowing liquid from a low stage below critical depth to a high stage above critical depth during which the velocity changes from super-critical to subcritical. Like all expanding flows which produce an increased pressure downstream, the hydraulic jump is accompanied by large and violent eddies and therefore invariably involves a loss of kinetic energy. The hydraulic jump is, therefore, a very useful instrument of energy dissipation. Equation (4.27) is applicable generally to a hydraulic jump which forms in a channel of any section provided the slope is sufficiently gentle to justify the neglect of the component of the weight of fluid in the control volume acting in the direction of flow. The subscript (1) refers to conditions just before the jump and (2) to conditions after the jump. The resistance to flow P_0 due to bed friction or blocks is generally neglected since the transition is usually

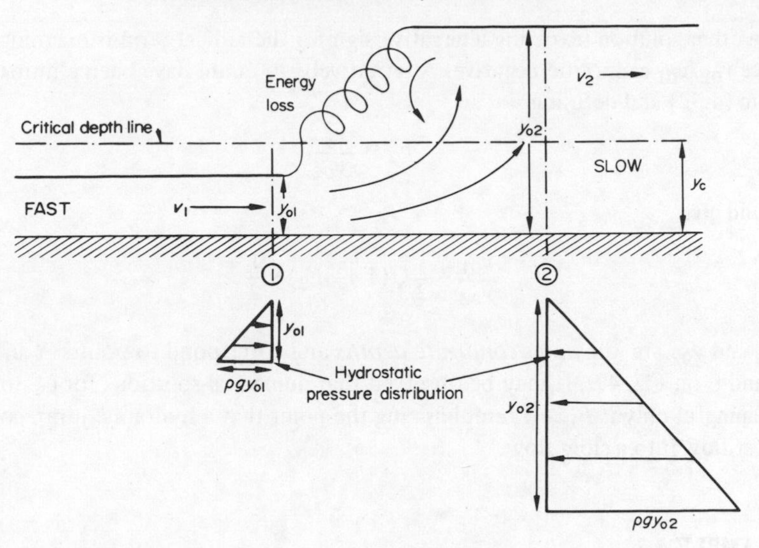

Fig. 4.10 The hydraulic jump

relatively short. For a wide channel reasonably rectangular in shape, the condition $P_1 = P_2$ gives

$$\frac{\rho}{2} g y_{01}^2 + \frac{\rho q^2}{y_{01}} = \frac{\rho}{2} g y_{02}^2 + \frac{\rho q^2}{y_{02}}$$

or

$$\frac{1}{2} g y_{01}^2 + y_{01} v_1^2 = \frac{1}{2} g y_{02}^2 + y_{02} v_2^2 \tag{4.32}$$

From continuity

$$v_2 = \frac{y_{01}}{y_{02}} v_1 \tag{4.33}$$

Substitution of (4.33) into (4.32) yields after simplification

$$\frac{1}{2} g (y_{01} + y_{02}) - \frac{y_{01}}{y_{02}} v_1^2 = 0 \tag{4.34}$$

Dividing (4.34) through by $gy_{01}/2$ and defining a Froude number $F_1 = v_1^2/gy_{01}$, we get after simplification

$$\left(\frac{y_{02}}{y_{01}}\right)^2 + \left(\frac{y_{02}}{y_{01}}\right) - 2F_1 = 0 \tag{4.35}$$

This quadratic equation (4.35) may be solved to give

$$\frac{y_{02}}{y_{01}} = \frac{1}{2}\sqrt{(1 + 8F_1)} - \frac{1}{2} \tag{4.36}$$

The other solution involving a negative sign for the radical term is meaningless since y_{02}/y_{01} cannot be negative. Alternatively, v_1 could have been eliminated from (4.32) and defining

$$F_2 = \frac{v_2^2}{g y_{02}}$$

would give

$$\frac{y_{01}}{y_{02}} = \frac{1}{2}\sqrt{(1 + 8F_2)} - \frac{1}{2} \tag{4.37}$$

y_{01} and y_{02} are known as *conjugate depths* and correspond to points A and B or D and E on Fig. 4.9. It may be observed that numerical solutions for (4.36) are meaningful only if $F_1 > 1$, emphasizing the point that a hydraulic jump converts a fast flow into a slow flow.

EXAMPLE 4.2

A 6·1 m wide horizontal stream flows over a short stretch of stones. The flow upstream of the rough stretch is 0·91 m deep and has a speed of 1·52 m/s. If the stones are estimated to experience a drag of 4448 N, determine the depth and speed of the stream downstream of the rough stretch. Sketch the water surface profile and indicate the critical depth line. Is the flow fast or slow? (U.S.T., Part I, 1968).

SOLUTION

Discharge per unit width

$$q = 1{\cdot}38 \text{ m}^2/\text{s}$$

∴ Corresponding critical depth

$$y_c = (q^2/g)^{\frac{1}{3}} = 0{\cdot}58 \text{ m}$$

Since the initial depth of flow is higher than the critical depth, the flow is initially slow. On flowing over the stretch of stones the depth will decrease (see Fig. 4.9).

Applying the momentum equation (4.26)

$$6{\cdot}1 \times 9{\cdot}81 \times 10^3 \,(y_{01}^2/2 - y_{02}^2/2) - 4{\cdot}448 \times 10^3 = 10^3 \times 6{\cdot}1 \times 1{\cdot}38\,(v_2 - v_1)$$

From continuity

$$v_2 = v_1 y_{01}/y_{02}$$

$$\therefore \quad 3{\cdot}05 \times 10^3 \times 9{\cdot}81\,(0{\cdot}83 - y_{02}^2) - 4{\cdot}448 \times 10^3 = 8{\cdot}42 \times 10^3 v_1\,(0{\cdot}91/y_{02} - 1)$$

$\therefore \qquad y_{02}^3 - 1{\cdot}1\, y_{02} + 0{\cdot}39 = 0$

Thus $\qquad y_{02} = 0{\cdot}77$ m

$$v_2 = 1{\cdot}38/0{\cdot}77 = \underline{1{\cdot}79 \text{ m/s}}$$

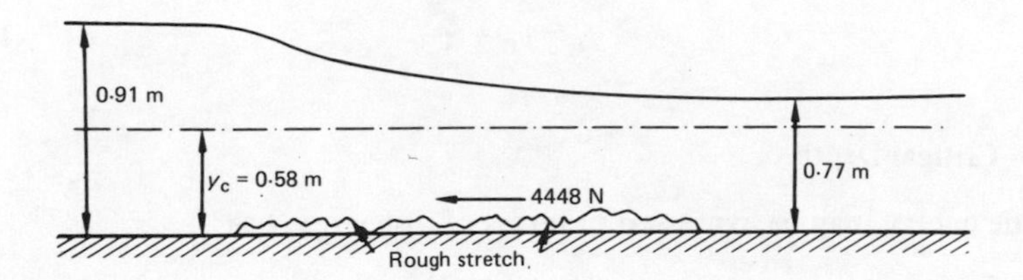

4.3 Energy Concepts

4.3.1 Specific Energy

The total energy (H) per unit weight of flowing fluid is given by the sum of the

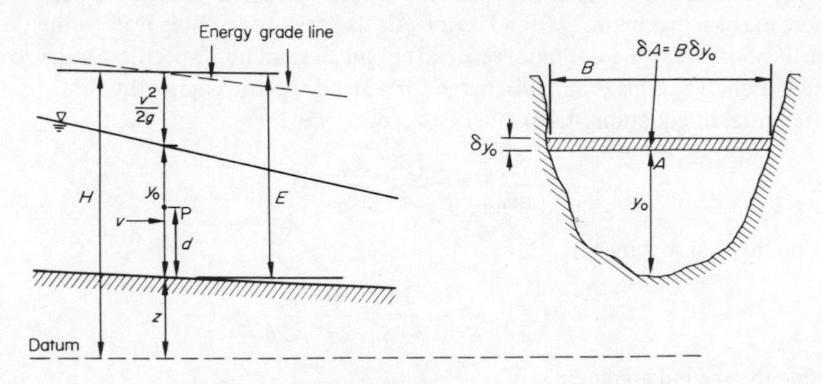

Fig. 4.11

potential head above an arbitrary datum, the pressure head and the kinetic head. With reference to Fig. 4.11 this is given for a particle located at P by

$$H = (z + d) + (y_0 - d) + \frac{v^2}{2g}$$

or

$$H = z + y_0 + \frac{v^2}{2g} \tag{4.38}$$

(assuming a hydrostatic pressure distribution). The bed is also assumed to have a gentle or no slope. z is the bed level above the chosen datum, y_0 is the depth of flow and the energy correction factor is assumed to be unity. The total head is the same for all particles in a particular sectional plane since the expression in (4.38) is independent of the location d. By definition, *specific energy E* is the total energy when the bed level is chosen as datum. Thus

$$H = z + E$$

where

$$E = y_0 + \frac{v^2}{2g} \tag{4.39}$$

4.3.2 Critical Depth

Specific energy may be expressed in terms of discharge Q as

$$E = y_0 + \frac{Q^2}{2gA^2} \tag{4.40}$$

It is apparent from equation (4.40) that a fixed discharge Q may be passed through a channel at various depths corresponding to different specific energies. There exists a minimum value of E for which the particular discharge can pass. At that point $\mathrm{d}E/\mathrm{d}y_0 = 0$. Similarly it can be argued that for a particular specific energy various discharges at different depths can be passed through the channel. There exists a maximum Q for a fixed E. Both lines of reasoning lead to the same conclusion that at the optimum (critical) point of minimum specific energy for a fixed discharge or maximum discharge for a fixed specific energy the first differential of equation (4.40) must be zero.

i.e.

$$\frac{\mathrm{d}E}{\mathrm{d}y_0} = 1 - \frac{Q^2}{gA^3}\frac{\mathrm{d}A}{\mathrm{d}y_0} = 0$$

(using the first argument)

or

$$0 = 1 - \frac{Q^2}{gA^3}\frac{\mathrm{d}A}{\mathrm{d}y_0} + \frac{Q}{gA^2}\frac{\mathrm{d}Q}{\mathrm{d}y_0}$$

(using the second argument)

From both equations, remembering that $\mathrm{d}Q/\mathrm{d}y_0 = 0$ for a maximum value of Q,

$$1 - \frac{Q^2}{gA^3}B = 0 \text{ since } \frac{\mathrm{d}A}{\mathrm{d}y_0} = B$$

(see Fig. 4.11)

or

$$1 - \frac{Q^2}{gA^2}\frac{B}{A} = 0$$

Thus the critical 'average' depth

$$y_c = \frac{A}{B} = \frac{Q^2}{gA^2} \tag{4.41}$$

Equation (4.41) gives the general expression for the critical depth in any open channel. For the special case of a *rectangular channel*

$$y_c = \frac{Q^2}{B^2 y_c^2}\frac{1}{g} = \frac{q^2}{g y_c^2}$$

$$y_c = (q^2/g)^{\frac{1}{3}} \tag{4.42}$$

which is the same as equation (4.31). Critical conditions in open-channel flow therefore refer to conditions at which the channel passes a particular discharge at the minimum energy and with a minimum 'force' or passes the maximum possible discharge with a particular energy and 'force'.

The specific energy for a rectangular section is

$$E = y_0 + \frac{v^2}{2g} = y_0 + \frac{q^2}{2g y_0^2} \tag{4.43}$$

At critical conditions (substituting from (4.42))

$$E = y_c + \frac{q^2}{2g y_c^2} = y_c + \frac{y_c^3}{2 y_c^2} = \frac{3}{2} y_c$$

Thus the critical depth is two-thirds the specific energy. Plots of the cubic equation (4.43) for fixed values of q are illustrated in Fig. 4.12. The curves are asymptotic to the E-axis and the line $y_0 = E$. There are two possible depths

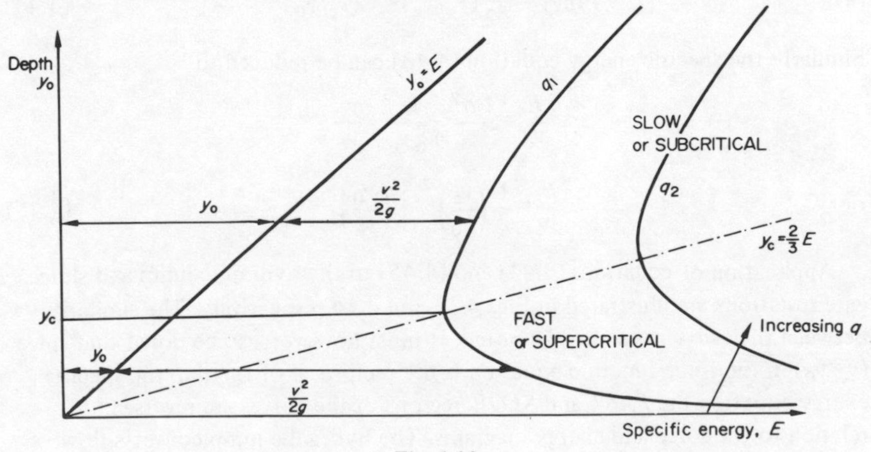

Fig. 4.12

known as *alternate depths* for any E. The particular depth of flow is determined by the slope and roughness of the channel (see Manning or Chezy equation).

The points of minimum specific energy (critical points) fall on the $y_0 = \frac{2}{3}E$ line. Flow conditions with depths greater than $\frac{2}{3}E (= y_c)$ are slow or subcritical and with depths less than $\frac{2}{3}E$ they are fast or supercritical.

4.3.3 Use of the Generalized 'Force' and Energy Equations

A thorough understanding of the curves in Figs. 4.9 and 4.12 simplifies considerably the application of momentum and energy concepts to open channel flows. Before applying the concepts to certain types of channel transitions it is advisable for the student to study carefully the similarities and differences of both sets of curves. It is appropriate to point out that although both concepts are valid for any type of flow conditions the choice of which to employ is dependent on known factors and on which assumptions can be made. The case of the hydraulic jump in sub-section 4.2.2 is a good example. Since the energy loss in a jump is an unknown factor which cannot be ignored, we use momentum concepts in the analysis since the assumptions (negligence of friction) necessary in this case can more easily be justified. Having determined the downstream depth after the jump, the energy loss can be calculated or read from curves similar to Fig. 4.12. Conversely in flow over a weir or through constrictions, the forces arising from the obstruction are unknown but quite significant. Energy concepts may therefore be used to estimate flow conditions from which the resistance to flow can then be calculated.

It is possible to reduce the curves of Figs 4.9 and 4.12 to one general curve in each case by plotting them on dimensionless axes. Substituting y_c^3 for q^2/g in equation (4.30) gives

$$P/\rho g = \frac{1}{2}y_0^2 + y_c^3/y_0$$

or

$$P/\rho g y_c^2 = \frac{1}{2}(y_0/y_c)^2 + (y_c/y_0) \tag{4.44}$$

Similarly the specific energy equation (4.43) can be reduced to

$$\frac{E}{y_c} = \frac{1}{2}\frac{q^2}{g}\frac{1}{y_0^2 y_c} + \frac{y_0}{y_c}$$

or

$$\frac{E}{y_c} = \frac{1}{2}\left(\frac{y_c}{y_0}\right)^2 + \left(\frac{y_0}{y_c}\right) \tag{4.45}$$

Application of equations (4.44) and (4.45) to the hydraulic jump and sluice gate situations are illustrated in Figs 4.13 and 4.14 respectively. The similarity between the two curves must be noted. It must however also be noted that since (y_c/y_0) in the force function equation is the reciprocal of y_0/y_c in the specific energy equation the *FAST* and *SLOW* regions of the curves are reversed in relation to the force and energy diagrams. The hydraulic jump converts flow from point A to point B and since the external forces due to friction and other obstacles are ignored AB is parallel to the y_c/y_0 axis. The corresponding energy loss ΔE can be read from the energy diagram. A sluice gate exerts an unknown force P_0 on the flow in converting it from slow (A′) to fast (B′) conditions: It is however reasonable to assume that the energy loss in flowing under the sluice is negligible. Thus the line A′B′ on the energy diagram is parallel to the y_0/y_c axis. Determination of point B′ makes the determination of the force possible from the force diagram.

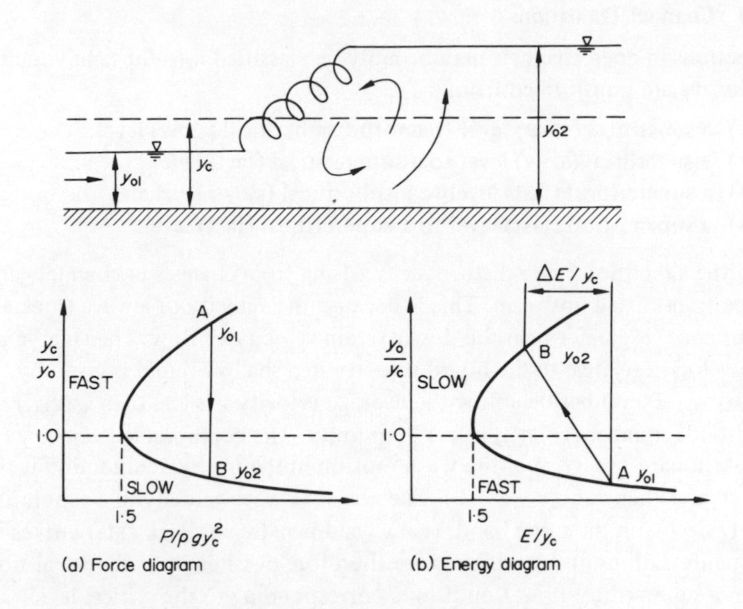

Fig. 4.13 The hydraulic jump on the generalized force and energy diagrams

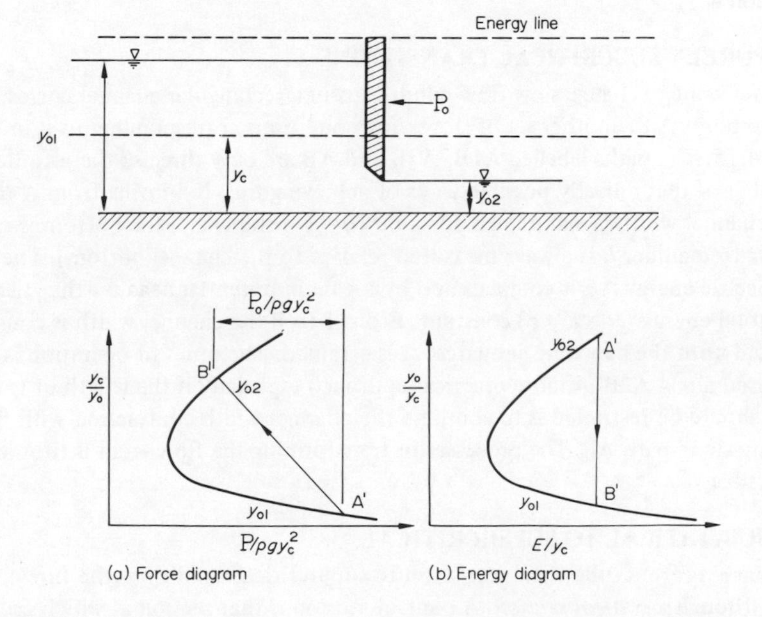

Fig. 4.14 Flow under a sluice gate on the generalized force and energy diagrams

4.3.4 Channel Transitions

Transitions in open channels may broadly be classified into those in which flow conditions are transformed from

(1) a subcritical (slow) level to another subcritical (slow) level
(2) a subcritical (slow) level to a supercritical (fast) level
(3) a supercritical (fast) level to a subcritical (slow) level and
(4) a supercritical (fast) level to a supercritical (fast) level.

In the subcritical range disturbances arising from changes in channel geometry can be transmitted upstream. This is because the velocity of a wave (message of disturbance) is greater than the downstream velocity of flow. The small-amplitude wave velocity relative to the liquid velocity in a shallow liquid is given by $c = \sqrt{(gy_0)}$. In a subcritical flow the average velocity v is less than $\sqrt{(gy_0)}$ since the Froude number (v^2/gy_0) is less than unity. The compounded velocity (relative to a stationary observer) of the wave motion in the upstream direction is therefore greater than zero $(c - v > 0)$. The absolute wave velocity in a supercritical flow $(F > 1)$, on the other hand, is always downstream. Backwater curves which are manifestations of disturbances are therefore possible in a subcritical flow but not in a supercritical flow. Conditions corresponding to the critical level produce standing waves since the absolute velocity is zero. Large disturbances (large amplitude waves) have wave velocities higher than $\sqrt{(gy_0)}$ and can therefore be transmitted upstream for some flows in the supercritical range. This gives rise to the so-called hydraulic bore which is a moving hydraulic jump (see Section 4.5).

(1) PURELY SUBCRITICAL TRANSITIONS

Say we want to change slow flow conditions in a rectangular channel corresponding to point A to another set of slow flow conditions corresponding to B in Fig. 4.15. The paths labelled A1B, A2B and AB are only three of the infinite number of theoretically possible ways of achieving this. In moving from A to 1 the channel width is maintained constant (q = constant) and the bottom is raised by Δz (remember E is always measured relative to the channel bottom). The loss in specific energy ΔE is compensated by a gain in potential head Δz thus keeping the total energy $(H = E + z)$ constant. From 1 to B the channel width is contracted until the new unit discharge q_2 is attained. The order of operation is reversed along A2B. A more practical approach especially if the length of transition should be restricted is to combine the channel width contraction with floor raising along path AB. The processes in transforming the flow from B to A are reversible.

(2) SUBCRITICAL TO SUPERCRITICAL

In changing from subcritical condition to supercritical conditions the flow must pass through a *control section*. A control section is that section at which critical flow conditions exist. Two of the possible ways by which flow conditions corresponding to A (Fig. 4.15) can be transformed to critical conditions at C or C′ are to raise the channel floor by an amount Δz_0 or contract the channel width. The former method gives rise to a weir and the latter to a venturi flume. The initiation and maintenance of supercritical flow downstream of the control section depends on whether conditions there are favourable. If the downstream channel slope is sufficiently large fast flow will persist, otherwise the flow will change back to slow flow through a hydraulic jump. As discussed at the end of subsection 4.2.1, if the rise in bottom Δz_0 or the width contraction is too large the message will be transmitted upstream and the liquid level will adjust itself until critical conditions are produced at the control section. Too small a value of Δz_0 or contraction will only produce effects discussed in (1) above.

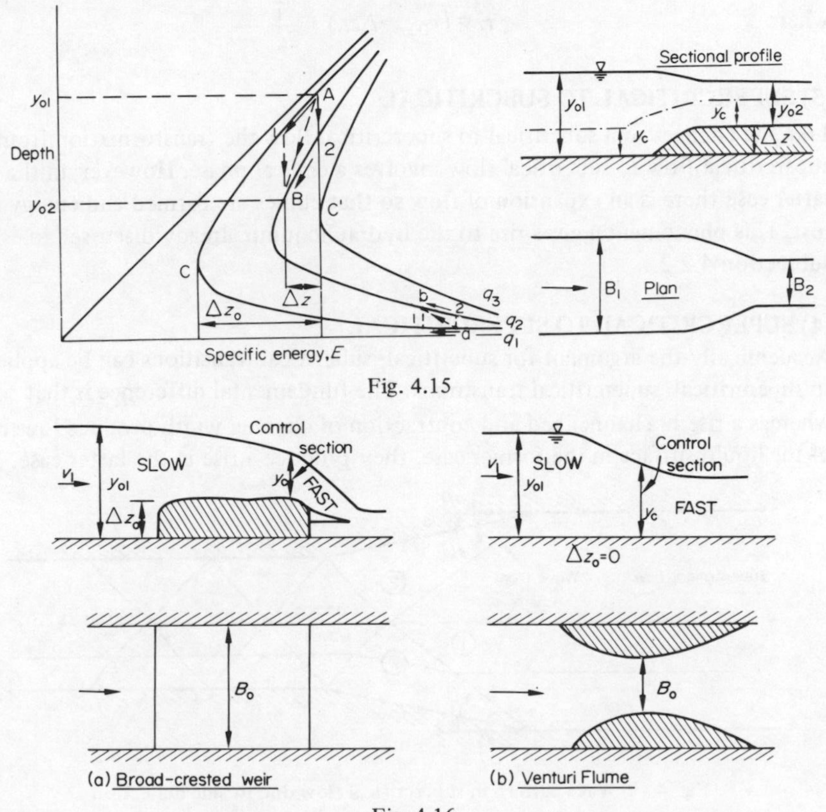

Fig. 4.15

Fig. 4.16

Broad-crested weirs (Fig. 4.16(a)) and venturi flumes (Fig. 4.16(b)) or their combinations are used extensively in measuring discharge through small rivers and canals. The same calibration equation is applicable in either case. Let the width at the control section be B_0. The discharge per unit width referred to the critical depth is

$$q^2 = gy_c^3$$

$\therefore$ Discharge $$Q = Bq = B_0\sqrt{g}(y_c)^{\frac{3}{2}}$$

But the critical depth $y_c = \frac{2}{3}E$ where E is the approach energy measured relative to the crest of the weir (bed level).

$$\therefore \quad Q = B_0\left(\frac{2}{3}\right)^{\frac{3}{2}}\sqrt{g}\,(E)^{\frac{3}{2}} = 1{\cdot}70\,B_0\,(E)^{\frac{3}{2}} \qquad (4.46a)$$

In order to account for minor losses of energy due to friction a factor C_d less than unity known as the coefficient of discharge is applied, yielding

$$Q = 1{\cdot}70\,C_d\,B_0E^{\frac{3}{2}} \qquad (4.46b)^*$$

where $$E = (y_{01} - \Delta z_0) + \frac{v_1^2}{2g}$$

(3) SUPERCRITICAL TO SUBCRITICAL

Like the change from subcritical to supercritical flow the transformation from supercritical flow to subcritical flow involves a critical phase. However, in the latter case there is an expansion of flow so that eddies are formed and energy is lost. This phenomenon gives rise to the hydraulic jump already discussed in subsection 4.2.2.

(4) SUPERCRITICAL TO SUPERCRITICAL

Academically the argument for subcritical-subcritical transitions can be applied to supercritical-supercritical transitions. The fundamental difference is that whereas a rise in channel bed and contraction of channel width produce lowering of the liquid surface in the former case, they produce a rise in the latter case. This

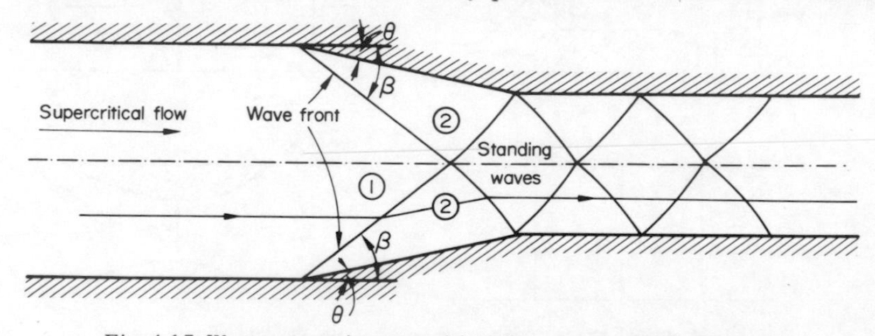

Fig. 4.17 Wave pattern in supercritical flow due to side deflection

may be inferred from Fig. 4.15 following the paths a1b and a2b. The phenomenon is however complicated by the fact that disturbances in supercritical flow produce shock waves whose continuous reflection from the side walls of the channel produces an irregular liquid surface of standing waves downstream of the source of disturbance. The situation for an inward channel deflection is illustrated in

* The Imperial Units equivalent is $Q = 3{\cdot}09\,C_dBE^{\frac{3}{2}}$.

Fig. 4.17. Open channel shocks are analogous to shock waves produced in supersonic gas flow. If the shock is strong enough subcritical conditions are produced downstream of the shock front giving rise to an oblique hydraulic jump. The wave angle β is related to the angle of deflection θ and the conjugate depth ratio y_2/y_1 by

$$y_2/y_1 = \frac{\tan \beta}{\tan (\beta - \theta)} \tag{4.47}$$

and

$$y_2/y_1 = \frac{1}{2} [\sqrt{(1 + 8F_1 \sin^2 \beta)} - 1] \tag{4.48}$$

where F_1, the approach Froude number $= v_1^2/gy_1$. Equations (4.47) and (4.48) may be solved by trial and error for a known θ. Ippen has produced charts to facilitate computations. The interested reader is referred to his chapter 'Channel Transitions and Controls' in *Engineering Hydraulics* (edited by Rouse) or to Chow's *Open Channel Hydraulics* for a detailed analysis.

EXAMPLE 4.3

An open channel of constant width has its floor raised 4·57 cm at a given section. If the depth of the approaching flow is 45·7 cm calculate the rate of flow indicated by: a 7·63 cm drop in the surface elevation over the raised bottom; a 7·63 cm increase in surface elevation over the raised bottom.

Neglect losses at the sudden change of section.

Calculate the critical depth in each case and plot it relative to the water surface profile.

SOLUTION

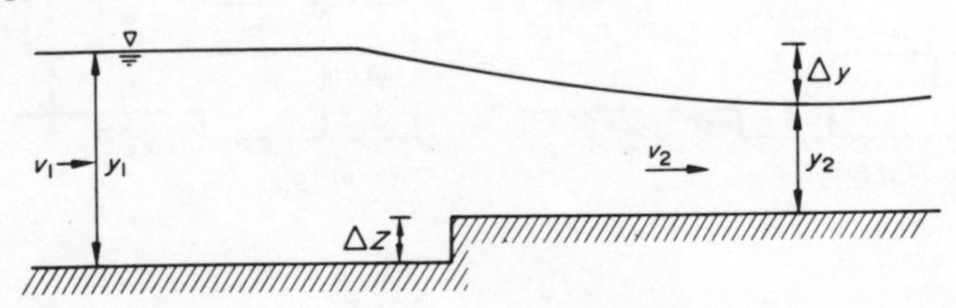

$$y_1 = y_2 + \Delta y + \Delta z$$

Neglecting losses,

$$E_1 = E_2 + \Delta z$$

$$\frac{v_1^2}{2g} + y_1 = \frac{v_2^2}{2g} + y_2 + \Delta z$$

i.e.

$$\frac{v_1^2}{2g} + y_1 = \frac{v_2^2}{2g} + y_1 - \Delta y$$

From continuity, $v_1 y_1 = v_2 y_2$

Thus $$\frac{v_1^2}{2g}\left[1 - \left(\frac{y_1}{y_2}\right)^2\right] = -\Delta y$$

In Case (a) $(\Delta y = 7{\cdot}63 \text{ cm})$

$$\frac{y_1}{y_2} = \frac{0{\cdot}457}{(0{\cdot}457 - 0{\cdot}046 - 0{\cdot}076)} = 1{\cdot}36$$

$\therefore$ $$v_1^2 = \frac{-2g \times 0{\cdot}076}{1 - (1{\cdot}36)^2} = 1{\cdot}75$$

Discharge $q = 0{\cdot}046\, v_1 = \underline{0{\cdot}60 \text{ m}^2/\text{s}}$

$$y_c = (0{\cdot}61^2/g)^{\frac{1}{3}} = \underline{33{\cdot}3 \text{ cm}}$$

In Case (b) $(\Delta y = -7{\cdot}63 \text{ cm})$

$$\frac{y_1}{y_2} = \frac{0{\cdot}457}{0{\cdot}487} = 0{\cdot}94$$

$$v_1^2 = 2g \times 0{\cdot}076/(1 - 0{\cdot}88) = 12{\cdot}4$$

$$q = 0{\cdot}046 v_1 = \underline{1{\cdot}62 \text{ m}^2/\text{s}}$$

$$y_c = (1{\cdot}62^2/g)^{\frac{1}{3}} = \underline{64{\cdot}5 \text{ cm}}$$

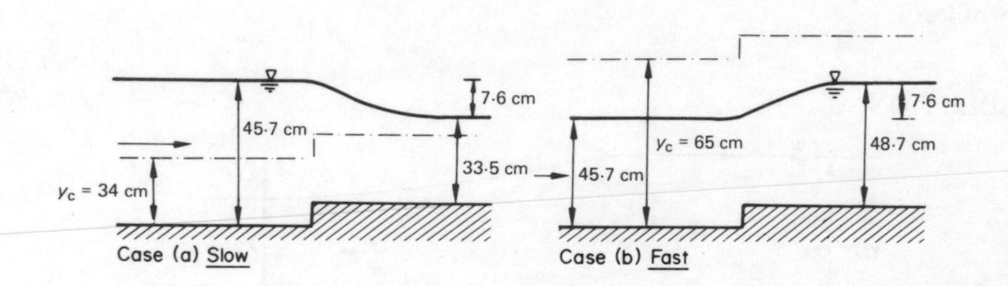

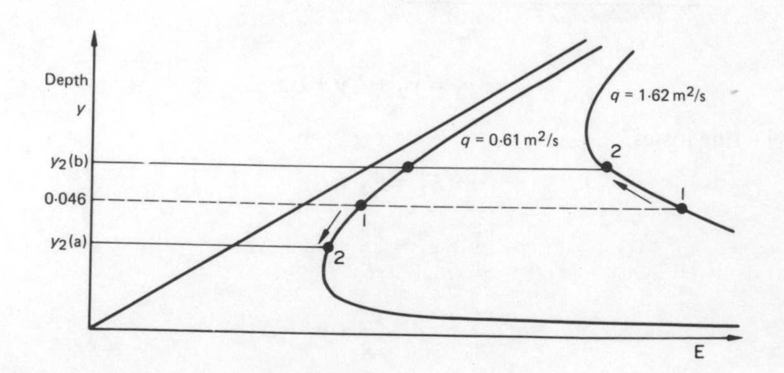

4.4 Gradually Varied Flow

4.4.1 Basic Equations

In this section we will consider the mathematical description of liquid surface transitional changes under steady flow conditions. Rapid, unsteady changes will be discussed in the next section. In the cases discussed in this section the changes are presumed so gradual and smooth that no energy is lost except by friction.

The total energy head at any section with depth y_0 is, according to equation (4.38) (assuming the energy correction factor $\alpha = 1$)

$$H = z + y_0 + \frac{v^2}{2g} = z + E$$

Differentiating with respect to x (a horizontal distance in the direction of flow),

$$\frac{\mathrm{d}H}{\mathrm{d}x} = \frac{\mathrm{d}z}{\mathrm{d}x} + \frac{\mathrm{d}E}{\mathrm{d}x} \tag{4.49}$$

$$-\frac{\mathrm{d}H}{\mathrm{d}x} = S_f, \text{ the slope of the energy grade line}$$

$$-\frac{\mathrm{d}z}{\mathrm{d}x} = S_0, \text{ the slope of the channel bottom.}$$

$$\frac{\mathrm{d}E}{\mathrm{d}x} = \frac{\mathrm{d}E}{\mathrm{d}y_0}\frac{\mathrm{d}y_0}{\mathrm{d}x} = \frac{\mathrm{d}}{\mathrm{d}y_0}\left(y_0 + \frac{Q^2}{2gA^2}\right)\frac{\mathrm{d}y_0}{\mathrm{d}x}$$

Thus for a constant discharge Q,

$$\frac{\mathrm{d}E}{\mathrm{d}x} = \left[1 - \frac{Q^2}{gA^3}\frac{\mathrm{d}A}{\mathrm{d}y_0}\right]\frac{\mathrm{d}y_0}{\mathrm{d}x} = \left[1 - \frac{Q^2}{A_c^2 g}\frac{A_c^2}{A^3}\frac{\mathrm{d}A}{\mathrm{d}y_0}\right]\frac{\mathrm{d}y_0}{\mathrm{d}x} \tag{4.50}$$

where A_c is the wetted cross-sectional area corresponding to critical flow conditions. Denoting the average depth by $\bar{y}_0$,

$$\frac{1}{A}\frac{\mathrm{d}A}{\mathrm{d}y_0} = \frac{B}{A} = 1/\bar{y}_0$$

and

$$\frac{Q^2}{gA_c^2} = \frac{v_c^2}{g} = \bar{y}_c \qquad \text{(see equation (4.41))}$$

Substituting into equation (4.50)

$$\frac{\mathrm{d}E}{\mathrm{d}x} = \left[1 - \frac{\bar{y}_c}{\bar{y}_0}\left(\frac{A_c}{A}\right)^2\right]\frac{\mathrm{d}y_0}{\mathrm{d}x} = \left[1 - \left(\frac{\bar{y}_c}{\bar{y}_0}\right)^3\left(\frac{B_c}{B}\right)^2\right]\frac{\mathrm{d}y_0}{\mathrm{d}x} \tag{4.51}$$

From (4.49) and (4.51)

$$S_0 - S_f = \left[1 - \left(\frac{\bar{y}_c}{\bar{y}_0}\right)^3\left(\frac{B_c}{B}\right)^2\right]\frac{\mathrm{d}y_0}{\mathrm{d}x} \tag{4.52}$$

Equation (4.52) describes in differential form the variation of depth y_0 with horizontal distance x for any shape of channel section. For a wide rectangular channel $y_0 = \bar{y}_0$ and $B = B_c$ and

$$S_0 - S_f = \left[1 - \left(\frac{y_c}{y_0}\right)^3\right] \frac{dy_0}{dx} \tag{4.53}$$

From Chezy's formula (wide rectangular channel)

$$S_f = q^2/(C^2 y_0^3)$$

The uniform (normal) depth y_n for the same discharge would be given by

$$S_0 = \frac{q^2}{C^2 y_n^3}$$

Assuming C to be constant,

$$\frac{S_f}{S_0} = \left(\frac{y_n}{y_0}\right)^3 \tag{4.54}$$

Substituting (4.54) into (4.53) and simplifying gives

$$\frac{dy_0}{dx} = S_0 \left[\frac{1 - (y_n/y_0)^3}{1 - (y_c/y_0)^3}\right] \tag{4.55}$$

The corresponding equation using Manning's formula is

$$\frac{dy_0}{dx} = S_0 \left[\frac{1 - (y_n/y_0)^{\frac{10}{3}}}{1 - (y_c/y_0)^3}\right] \tag{4.56}$$

The engineer's main interest lies in the determination of the depth of flow along the channel. This requires the integration of equation (4.55) or (4.56) which is possible only under some special circumstances. Bresse integrated a more general form of (4.55) for uniform channels yielding what are called backwater functions but many engineers prefer to solve the equations by iterative methods. Different numerical methods, all of which seek to find a finite length Δx over which a finite change in depth Δy_0 takes place, are available. Summation of the length segments ($\Sigma\Delta x$) for the total change in depth gives the length of the backwater curve of interest. It is quite obvious that accuracy of the estimate depends very much on the value of incremental depths chosen; the smaller the Δy_0's the better the results. A simplified step method which is known to be adequate for estimating backwater curves in uniform channels is illustrated in Example 4.4 below.

EXAMPLE 4.4

A weir is installed in a 30·5 m wide rectangular channel delivering 71 m^3/s. The bed slopes at 1 : 1000 and the depth of water just upstream of the weir is 2·0 m. If Chezy $C = 55$ may be assumed constant, estimate the length of the backwater curve.

SOLUTION

Equation (4.55) will be used.

Critical depth $$y_c = (q^2/g)^{\frac{1}{3}} = \left(\frac{5{\cdot}42}{9{\cdot}81}\right)^{\frac{1}{3}} = 0{\cdot}82 \text{ m}$$

Normal depth $$y_n = \left(\frac{q^2}{C^2 S_0}\right)^{\frac{1}{3}} = 1{\cdot}2 \text{ m}$$

The length of the backwater curve is the channel length over which the depth changes from 1·2 m to 2·0 m.

Divide the change in depth into four increments of 0·2 m.

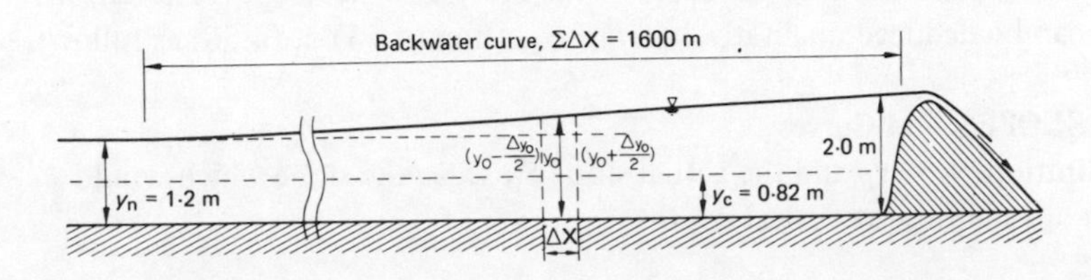

Write equation (4.55) in finite difference form.

$$\frac{\Delta y_0}{\Delta x} = \frac{S_0\,[1-(y_n/y_0)^3]}{1-(y_c/y_0)^3} = \frac{0{\cdot}001\,[1-1{\cdot}73/y_0^3]}{1-0{\cdot}55/y_0^3}$$

$$\therefore \qquad \Delta x = \left(\frac{1-0{\cdot}55/y_0^3}{0{\cdot}001-0{\cdot}00173/y_0^3}\right)\Delta y_0$$

y_0 = average depth over the segmental length Δx

= depth at beginning of segment $y_{01} + \frac{1}{2}\Delta y_0$

$y_{02} = y_{01} + \Delta y_{01}$ = depth at end of segment.

The table below shows the details of the working procedure

y_{01} (m)	Δy_0 (m)	y_0 (m)	y_0^3 (m^3)	$1-0{\cdot}55/y_0^3$	$0{\cdot}001-0{\cdot}00173/y_0^3$	Δx (m)	y_{02} (m)
1·2	0·2	1·3	2·20	0·786	0·00021	744	1·4
1·4	0·2	1·5	3·37	0·832	0·00049	340	1·6
1·6	0·2	1·7	4·93	0·889	0·00065	273	1·8
1·8	0·2	1·9	6·85	0·920	0·00075	245	2·0
					$\Sigma\Delta x$	1602	

The backwater curve is about 1602 m long.

4.4.2 Classification of Curves

A good qualitative knowledge of the transition curves facilitates computations and the student is well advised to familiarize himself with the following code of description and the principal details of all the curves. The code letters refer to the slopes: M for mild or small slopes, S for steep slopes, H for horizontal slopes and A for adverse slopes. The numbers 1, 2 and 3 refer to the location of the curve; 1 refers to depths greater than both normal and critical depths, 2 to depths lying between normal and critical depths and 3 to depths smaller than normal and critical depths. Normal depths are given by the Chezy or Manning formula for mild and steep slopes. Its value is infinite for a horizontal bed and is imaginary for an adverse slope. The code is applied to adverse slopes for convenience and because of the similarity of the curves with those of other slopes. The various curves can be deduced qualitatively from equation (4.55) or (4.56) as follows.

MILD SLOPES – M Curves
By definition, $y_c < y_n$ and the actual slope $S_0 < S_c$, the slope which would sustain normal flow at critical depth.

M1:
$$y_0 > y_n \text{ and } y_c$$

According to equation (4.55), when $y_0 \gg y_n$, dy_0/dx approaches S_0 which means that changes in depth of flow are entirely due to the sloping nature of the channel bed i.e. the liquid surface must be horizontal. As $y_0 \to y_n$ however, $dy_0/dx \to 0$, which means that the surface slope approaches that of normal flow. The M1 curve, as illustrated in Example 4.4, is asymptotic to the normal depth line and becomes horizontal as the depth becomes large relative to the normal depth.

M2:
$$y_c < y_0 < y_n$$

Again as

$$y_0 \to y_n, \frac{dy_0}{dx} \to 0,$$

i.e. the surface curve is asymptotic to the normal depth line.

But as
$$y_0 \to y_c, \frac{dy_0}{dx} \to \infty,$$

i.e. there is a sharp drawdown.

An M2 curve is produced when a mildly sloping channel discharges onto a steeply sloping channel (Fig. 4.18(a)) or into a drop pool (Fig. 4.18(b)).

M3:
$$y_0 < y_c \text{ and } y_n$$

As
$$y_0 \to 0, \frac{dy_0}{dx} \to S_0 \left(\frac{y_n}{y_c}\right)^3$$

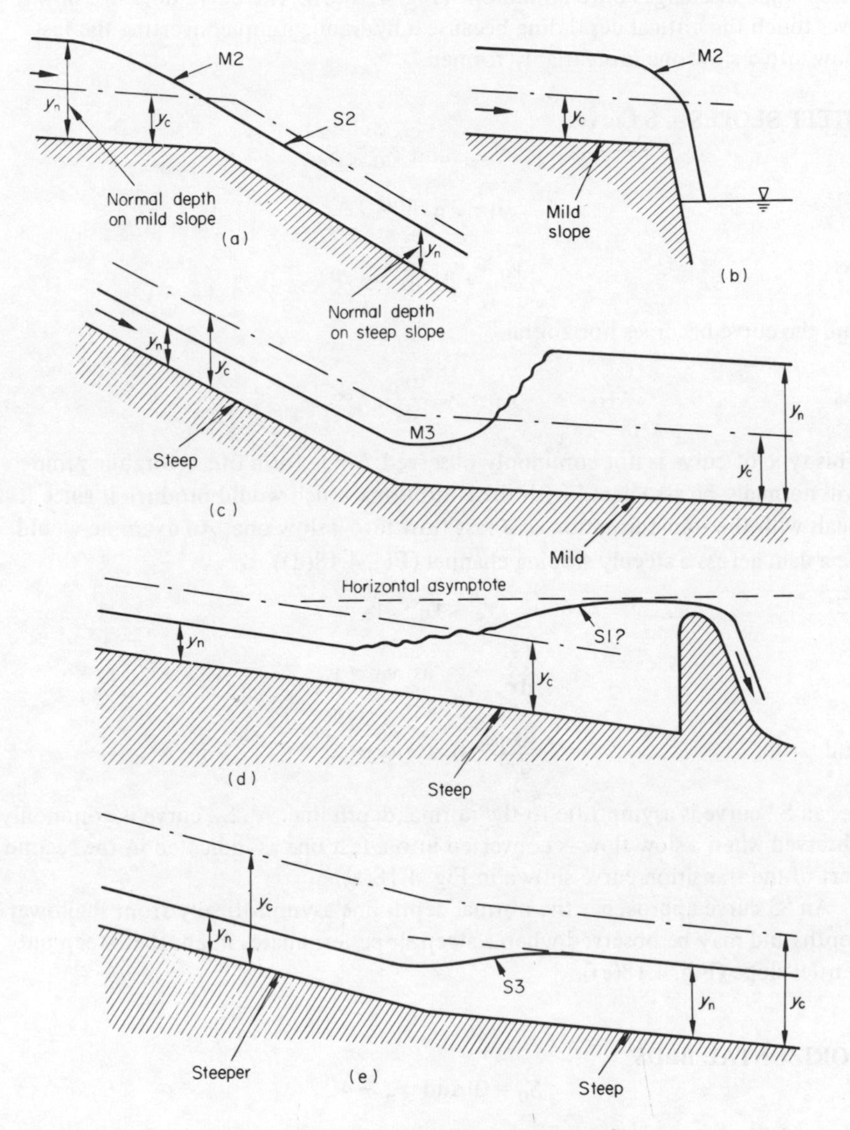

Fig. 4.18 Examples of common surface profiles

i.e. for low values of y_0 the M3 curve is asymptotic to an imaginary straight line.

As $$y_0 \to y_c, \frac{dy_0}{dx} \to \infty$$

i.e. the curve approaches the critical depth sharply. An M3 curve occurs when a steep slope discharges onto a mild one (Fig. 4.18(c)). The curve does not however touch the critical depth line because a hydraulic jump converting the fast flow into a slow one is invariably formed.

STEEP SLOPES – S Curves

$$y_c > y_n \text{ and } S_0 > S_c$$

S1: $$y_0 > y_n \text{ and } y_c$$

As $$y_0 \gg y_n, \frac{dy_0}{dx} \to S_0$$

and the curve becomes horizontal.

As $$y_0 \to y_c, \frac{dy_0}{dx} \to \infty$$

This type of curve is not commonly observed. Some form of a hydraulic jump will normally be associated with the conditions which would produce it since it deals with the transformation of a fast flow into a slow one. An example would be a dam across a steeply sloping channel (Fig. 4.18(d)).

S2: $$y_c < y_0 < y_n$$

$$\frac{dy_0}{dx} \to \infty \text{ as } y_0 \to y_c$$

and $$\frac{dy_0}{dx} \to 0 \text{ as } y_0 \to y_n$$

i.e. an S2 curve is asymptotic to the normal depth line. An S2 curve is commonly observed when a slow flow is converted into a fast one as indicated in the second part of the transition curve shown in Fig. 4.18(a).

An S3 curve approaches the normal depth line asymptotically from shallower depths and may be observed when a steep slope terminates in another steep but gentler slope (Fig. 4.18(e)).

HORIZONTAL BEDS

$$S_0 = 0 \text{ and } y_n \to \infty$$

From (4.53)

$$-S_f = [1 - (y_c/y_0)^3] \frac{dy_0}{dx}$$

Putting $$S_f = q^2/(C^2 y_0^3) = \frac{g}{C^2}\left(\frac{y_c}{y_0}\right)^3$$

$$\left[\left(\frac{y_0}{y_c}\right)^3 - 1\right]\frac{dy_0}{dx} = g/C^2$$

This may be integrated with $x = 0$ at $y_0 = y_c$ to give

$$x = \frac{C^2}{g} y_0 \left[1 - \frac{3}{4} y_c/y_0 - \frac{1}{4}(y_0/y_c)^3\right] \tag{4.57}$$

H2 and H3 curves which according to this equation are very similar to the M2 and M3 curves are shown in Fig. 4.19(c). They arise under conditions similar to those of M curves.

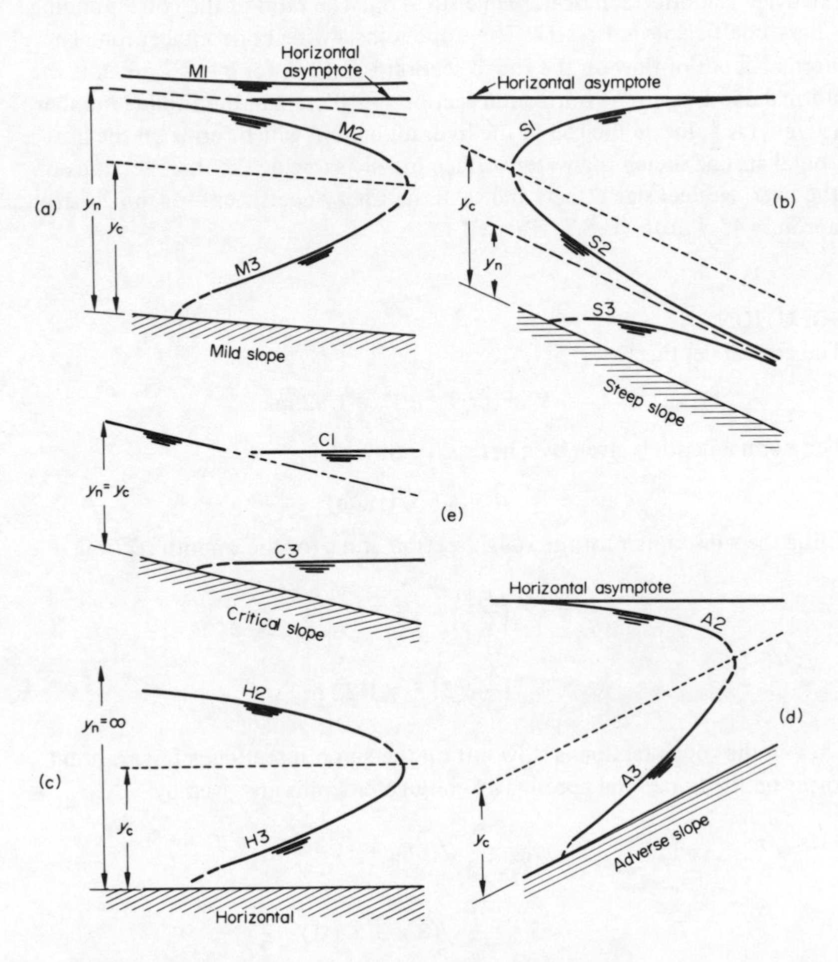

Fig. 4.19 Surface profiles (backwater curves) for gradually varied flow. Flow is from left to right; vertical scales are greatly enlarged

ADVERSE SLOPE (rising in an upstream direction)
Curves A2 and A3 are similar in appearance to H2 and H3 curves. They are also brought about by a drawdown from a flooded region covering the adverse slope (A2) or when a fast flow is forced up the slope (A3). The various curves are sketched in Fig. 4.19 and the student is advised to learn to apply them to different physical situations of flow. A C-curve occurs rarely since very few natural channels can ever be critical. Critical slopes are very sensitive and can only be defined for specific discharges.

EXAMPLE 4.5
A longitudinal section of a wide river shows a rough length followed by a long relatively smooth reach of the same slope S_0. The ratio of the corresponding Chezy coefficients is 1 : $2\sqrt{2}$. The slope joins a wide horizontal apron. The normal depth of flow on the rough section is 2·43 m for q = 7·7 m²/s. If the normal depth y_n in the horizontal section is 3·05 m and the Froude number (v^2/gy_n) is $\frac{3}{8}$, locate the toe of the hydraulic jump which forms on the horizontal apron. Sketch the water surface profile showing the three sections of the river. Neglect side effects and take the Chezy coefficient for the horizontal apron as 45·7 m$^{\frac{1}{2}}$/s, (U.S.T., Part III.)

SOLUTION
The critical depth,

$$y_c = ((7{\cdot}7)^2/g)^{\frac{1}{3}} = 1{\cdot}82 \text{ m}$$

The normal depth is given by Chezy's formula,

$$q = y_n C \sqrt{(y_n S_0)}$$

Using the subscripts r for the rough section and s for the smooth section

$$\frac{q_r}{q_s} = 1 = \left(\frac{y_{nr}}{y_{ns}}\right)^{\frac{3}{2}} \frac{C_r}{C_s} = \left(\frac{y_{nr}}{y_{ns}}\right)^{\frac{3}{2}} \frac{1}{2\sqrt{2}}$$

$$y_{ns} = y_{nr}\left(\frac{1}{4}\sqrt{2}\right)^{\frac{2}{3}} = \underline{1{\cdot}22 \text{ m}}$$

Flow in the rough section is slow but on the smooth section is fast. A jump forms on the horizontal apron. The conjugate depths are given by

$$y_{01}/y_{02} = \frac{1}{2}\sqrt{(8F_2 + 1)} - \frac{1}{2}$$

$$= \frac{1}{2}\sqrt{(8 \times 3/8 + 1)} - \frac{1}{2} = \frac{1}{2}$$

$$y_{01} = \underline{1{\cdot}52 \text{ m}}$$

The depth on the horizontal apron will rise from 4 ft to 5 ft and a jump will form (see sketch below).

$$S_0 = 0, S_f = q^2/(y_0^3C^2)$$

$$\therefore \qquad -\frac{q^2}{y_0^3C^2}\,dx = [1 - (y_c/y_0)^3]\,dy_0$$

Integrating $\qquad x_2 - x_1 = \frac{C^2}{q^2}\,[y_c^3y_0 - y_0^4/4]_{y_0 = 1{\cdot}22\text{ m}}^{y_0 = 1{\cdot}52\text{ m}}$

$$= \underline{37{\cdot}2\text{ m}}$$

The hydraulic jump will form at 37·2 m from the end of the slope.

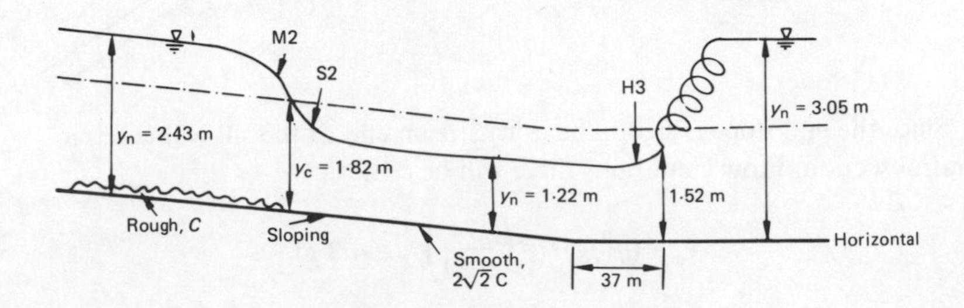

EXAMPLE 4.6

The spillway of a dam discharges 27·8 m²/s onto a horizontal stilling apron. The depth of high velocity flow at the entry point to the apron is 0·61 m. Both the apron and sill are sufficiently long to permit establishment of normal conditions. The sill is followed by a 1 to 10 sloping wide channel. (1) Calculate the height of the hydraulic jump and the height z_0 of sill necessary for its stabilization at the location shown. Neglect energy dissipation in flow over the sill. (2) Calculate the normal depth of flow in the downstream channel. (Manning's $n = 0{\cdot}015$). (3) Sketch and label the surface profile at the various sections. (4) What practical difficulties would you expect in the operation of the structure?

SOLUTION

(1) Incoming Froude no.,

$$F_1 = \frac{q^2}{gy_1^3} = \frac{(27{\cdot}8)^2}{2{\cdot}25} = 350$$

The conjugate depth ratio

$$y_2/y_1 = \frac{1}{2}(\sqrt{(1 + 8F_1)} - 1) = 26$$

$$\therefore \qquad y_2 = 26y_1 = \underline{15{\cdot}8\text{ m}}$$

The specific energy after the jump $= \frac{1}{2}\frac{q^2}{gy_2^2} + y_2$

$$= \underline{16{\cdot}0 \text{ m}}$$

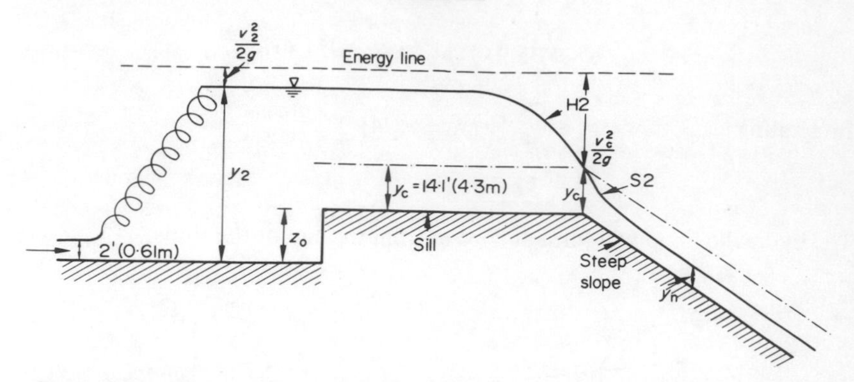

Since the end slope is steep, the downstream end of the sill will act as a control section. Flow conditions there will be critical.

$$y_c = (q^2/g)^{\frac{1}{3}} = \left(\frac{774}{9{\cdot}81}\right)^{\frac{1}{3}} = \underline{4{\cdot}3 \text{ m}}$$

Neglecting energy lost in flowing over the sill, the specific energy before the sill is equal to the specific energy on the sill plus z_0

i.e.
$$16 = \frac{3}{2}y_c + z_0 = \frac{3}{2} \times 4{\cdot}3 + z_0$$

$\therefore$
$$z_0 = \underline{9{\cdot}55 \text{ m}}$$

(2) From Manning's formula:

$$v = q/y_n = \frac{1}{n}y_n^{\frac{2}{3}}S_0^{\frac{1}{2}}$$

$\therefore$
$$y_n^{\frac{5}{3}} = \frac{nq}{S_0^{\frac{1}{2}}} = 1{\cdot}23 \text{ m}^{\frac{5}{3}}$$

$\therefore$
$$y_n = \underline{1{\cdot}14 \text{ m}}$$

(3) See sketch above.

(4) *Possible practical difficulties*

(i) The hydraulic jump may be submerged especially for relatively low flows thus reducing its energy dissipation. In such a case the fast incoming flow shoots under water in the stilling basin before dispersing (jet dispersion).

(ii) Because of high velocities, cavitation problems may arise (see Section 6.5).

(iii) The hydraulic jump for high discharges may be pushed downstream and may in fact jump the sill unless the stilling basin is sufficiently long.

4.5 Open Channel Surges

A surge is a discontinuity in the liquid (depth of flow) which is propagated along the channel. Like all other types of waves a surge is produced by a disturbance at some point along the channel. Examples of such disturbances are the partial closing or opening of a gate, the failure of a downstream hydro-plant, a landslide blocking fully or partially the channel and many others. In a precise analysis of a surge the roughness, slope, and cross-sectional area of the channel and other pertinent variables must be taken into account. The absolute velocity v_w (relative to a stationary observer) at which a surge is propagated may be evaluated using the elementary principles of momentum and continuity. The surge velocity relative to the flow velocity v_1 corresponding to the shallower depth is designated c. Thus $v_w = c \pm v_1$ (positive sign for downstream moving surges and negative for upstream moving surges). Chow has classified open channel surges into types A, B, C and D (see Fig. 4.20). A surge front which moves in the direction of smaller depth is described as *positive* and one which moves in the direction of greater depth is *negative*.

Fig. 4.20(b) shows the various types of surges as they would appear to an observer moving in the direction of the wave with the speed of the wave. It is tantamount to choosing a finite control volume which includes the wave front at all times. The control volume translates at a velocity v_w and the situation is steady relative to the control volume. The momentum equation (4.26) is thus applicable to the control volume with P_0 and the channel slope assumed to be negligible and a hydrostatic pressure distribution assumed. Using a type B surge for example,

$$g(A_1\bar{y}_1 - A_2\bar{y}_2) = Q\ [(v_w + v_2) - (v_w + v_1)] = Q(v_2 - v_1) \tag{4.58}$$

$$Q = A_1(v_w + v_1) = A_2(v_w + v_2) \tag{4.59}$$

Eliminating Q and v_2 from equation (4.58) using (4.59)

$$g(A_1\bar{y}_1 - A_2\bar{y}_2) = A_1(v_w + v_1)\left[\frac{A_1}{A_2}(v_w + v_1) - v_w - v_1\right] = A_1(v_w + v_1)^2\left(\frac{A_1}{A_2} - 1\right)$$

$$\therefore \qquad v_w + v_1 = c = \sqrt{\left[\frac{g(A_2\bar{y}_2 - A_1\bar{y}_1)}{A_1(1 - A_1/A_2)}\right]} \tag{4.60}$$

It can similarly be shown that the surge velocity c relative to the velocity in the shallow depth is given by the expression on the right-hand side of (4.60) for all the other cases. The absolute velocities are given as indicated in Fig. 4.20.

For a rectangular channel, $\bar{y}_1 = y_{01}/2, \bar{y}_2 = y_{02}/2, A_1 = By_{01}$ and $A_2 = By_{02}$. Thus equation (4.60) reduces to

$$c = \sqrt{\left[\frac{gy_{02}}{2}\left(\frac{y_{02}}{y_{01}} + 1\right)\right]} \tag{4.61}$$

or

$$c = \sqrt{(gy_{01})}\left[\tfrac{1}{2}y_{02}/y_{01}\ (y_{02}/y_{01} + 1)\right]^{\frac{1}{2}} \tag{4.62}$$

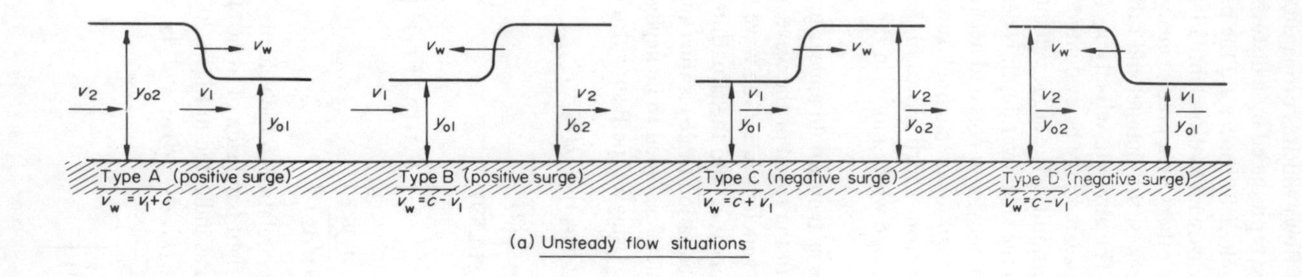

(a) Unsteady flow situations

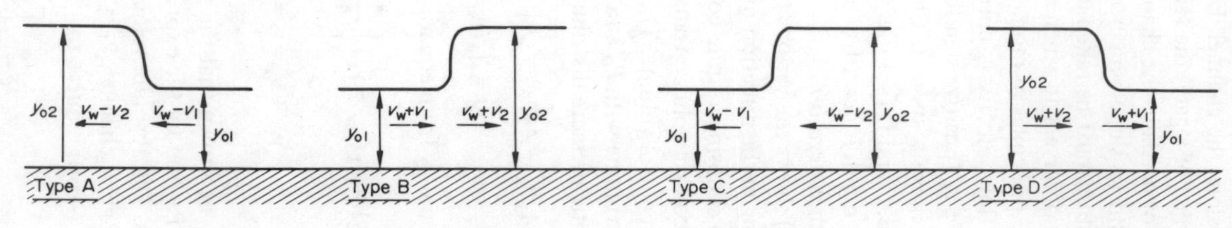

(b) Flows relative to an observer moving with the wave front

Fig. 4.20

For a small wave, $y_{02} \to y_{01}$ and $c = \sqrt{(gy_{01})}$. It is also clear from equation (4·62) that steep waves have a much higher speed of propagation than is predicted for small amplitude waves in shallow liquid. For this reason waves which propagate into deep water (Type C and D) have a tendency to spread out since particles at higher elevations tend to move faster. Tumbling over (over-running) occurs when the wave propagates into shallow depth as is manifested in hydraulic jumps. Fig. 4.21 illustrates the generation of the four types of surges through sudden-partial closing or opening of a sluice gate.

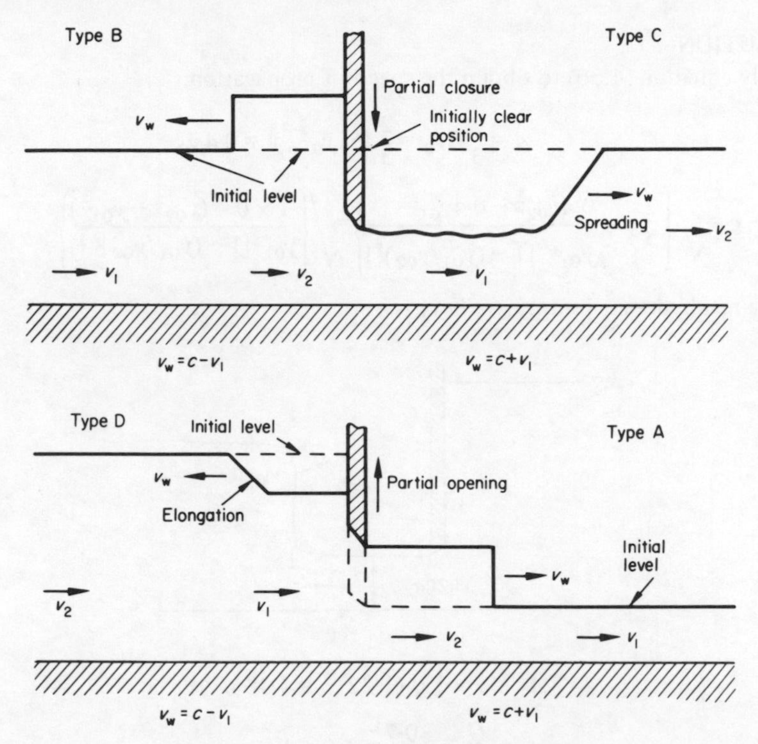

Fig. 4.21 Surges due to sudden gate operation

EXAMPLE 4.7

Establish that the speed of a hydraulic bore relative to the upstream water surface in a gently sloping parabolic open channel whose surface width B is proportional to the square root of the depth of flow y_0 is

$$c = \sqrt{\left\{0{\cdot}4g \frac{(y_{02}^{\frac{5}{2}} - y_{01}^{\frac{5}{2}})}{y_{01}^{\frac{3}{2}}\left[1 - \left(\frac{y_{01}}{y_{02}}\right)^{\frac{3}{2}}\right]}\right\}}$$

The subscripts 1 and 2 refer to the shallower and deeper depths respectively. (The centroid of a parabolic lamina is 0·4 times the height measured from the base.) Assume that the water surface is horizontal at both sections.

Water is released from under a sluice gate into a parabolic channel at the rate of 30·7 m^3/s (1085 ft^3/s) and with a normal depth of 1·22 m (4 ft). The channel is described by

$$B = 5y^{\frac{1}{2}}$$

Find the minimum right of way to avoid overtopping from the formation of a hydraulic jump. What is your criticism of the method of calculation? (U.S.T., Part III, 1967.)

SOLUTION

Apply equation (4.60) to obtain the speed of propagation.

$$A = \frac{2}{3}By_0 = \frac{2}{3}(5)y_0^{\frac{3}{2}};\ \bar{y} = 0{\cdot}4y_0$$

$$c = \sqrt{\left\{\frac{2}{3}g\,\frac{0{\cdot}4y_{02}^{\frac{5}{2}} - 0{\cdot}4y_{01}^{\frac{5}{2}}}{\frac{2}{3}y_{01}^{\frac{3}{2}}\,[1-(y_{01}/y_{02})^{\frac{3}{2}}]}\right\}} = \sqrt{\left\{\frac{g \times 0{\cdot}4\,(y_{02}^{\frac{5}{2}} - y_{01}^{\frac{5}{2}})}{y_{01}^{\frac{3}{2}}[1-(y_{01}/y_{02})^{\frac{3}{2}}]}\right\}}$$

For a real hydraulic jump, $v_w = 0$, $v_1 = c$

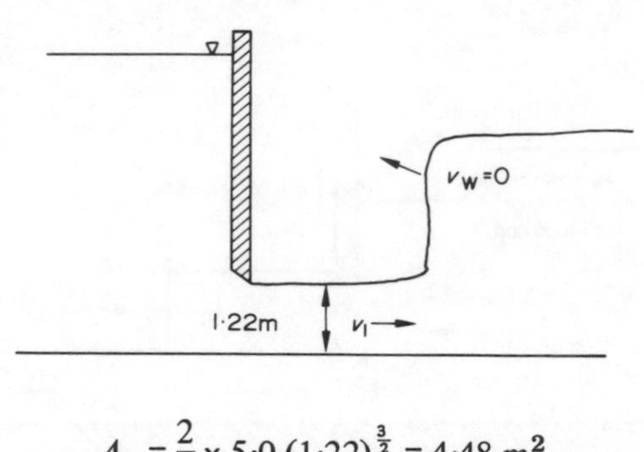

$$A_1 = \frac{2}{3} \times 5{\cdot}0\,(1{\cdot}22)^{\frac{3}{2}} = 4{\cdot}48\ \text{m}^2$$

$$v_1 = \frac{Q}{A_1} = \frac{30{\cdot}7}{4{\cdot}48} = 6{\cdot}9\ \text{m/s}$$

$$\frac{v_1^2}{gy_{01}} = 0{\cdot}4\,[(y_{02}/y_{01})^{\frac{5}{2}} - 1]/[1-(y_{01}/y_{02})^{\frac{3}{2}}]$$

$$\therefore \quad \frac{(6{\cdot}9)^2}{9{\cdot}81 \times 1{\cdot}22} = 0{\cdot}4\,[(y_{02}/y_{01})^{\frac{5}{2}} - 1]/[1-(y_{01}/y_{02})^{\frac{3}{2}}]$$

or

$$(y_{02}/y_{01})^{\frac{5}{2}} + 10\,(y_{01}/y_{02})^{\frac{3}{2}} = 11$$

Solving by trial and error:

$$y_{02}/y_{01} = 2{\cdot}3$$

$$y_{02} = 2{\cdot}8\ \text{m}$$

$$B = 5 \times \sqrt{2{\cdot}8} = 8{\cdot}34 \text{ m}$$

∴ Minimum right of way is (8·34 m)

By assuming that the water surface at each section is straight and horizontal, we have simplified the real situation considerably. The widening of the cross section involves transverse pressure gradients and velocities which thereby involve transverse wave action. The downstream water surfaces at a number of sections will be irregular, looking something like any of the sketches below. This

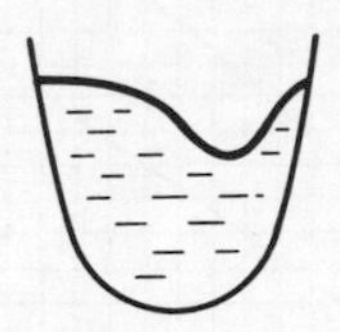

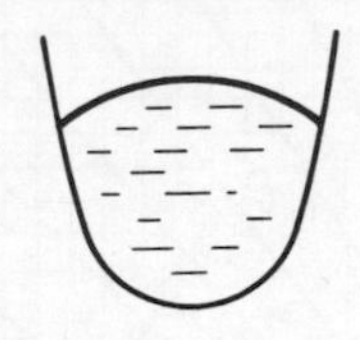

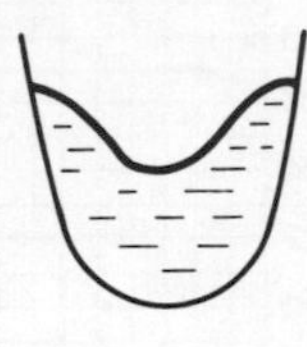

will affect the height and width of the stream. Average velocities have also been used and bed and side resistances ignored. The values obtained above can however be used as a first approximation in design; a simple model will help to improve the final design.

4.6 Miscellaneous Information

4.6.1 The Hydraulic Jump in a Steeply Sloping Channel

In considering a hydraulic jump in a channel whose slope is significant it is necessary to include the component of the weight of the fluid contained in the control volume in the dynamic equation. The major problem is connected with the determination of the volume of fluid in the jump. Straight line approximation of the surface profile of the jump in a smooth rectangular channel has led to an equation similar to equation (4.36) as

$$y_{02}/y_{01} = \frac{1}{2}\,[\sqrt{(1 + 8G)} - 1] \tag{4.63}$$

where G is a function of F_1 and the slope of the channel. Fig. 4.22 shows experimental relations between F_1 and y_{02}/y_{01} as a function of slope, S_0.

4.6.2 Flow Spreading in a Stilling Basin

A problem which frequently occurs in hydraulic engineering practice is the need to spread and dissipate the energy of a concentrated, fast flowing stream of water, so that it does not cause erosion of the valley into which it flows. Besides encouraging more energy loss through friction (the thinner the flow is the greater is the proportion of its depth subjected to boundary shear) adequate spreading and thus shallowing of a fast flow encourages the formation of a hydraulic jump which dissipates still more energy. In general the depth of flow in the channel downstream of a basin is fixed and adequate spreading of a fast flow enables the basin to be laid shallower or made shorter. This is because in many cases, the depth of flow in the downstream channel of a spillway outlet is smaller than that required to form a hydraulic jump, so the basin is laid deeper to provide the

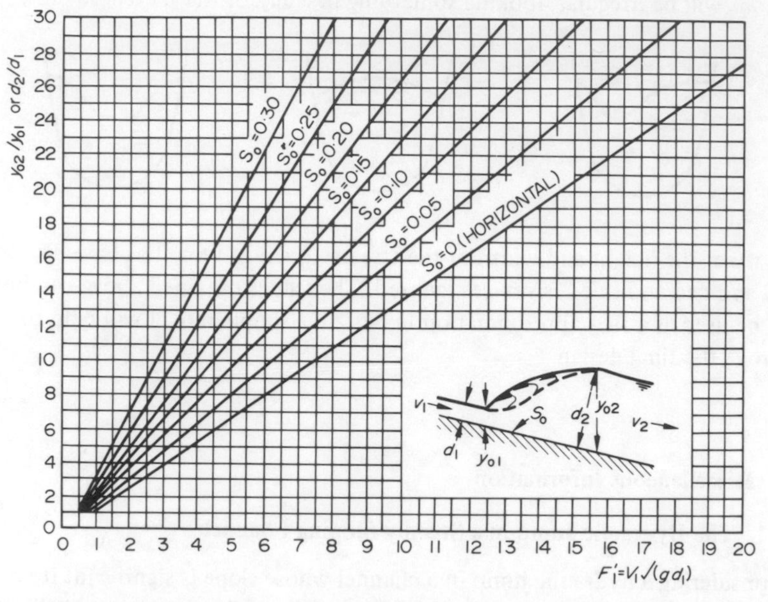

Fig. 4.22 Experimental relations between F_1 and y_{02}/y_{01} or d_2/d_1 for jumps in sloping channels. (After V. T. Chow)

necessary depth. A shallower incoming stream would provide a higher conjugate depth ratio of the jump and thereby reduce the necessity for a deep basin. Alternatively the shallow upstream depth for a given downstream depth produces a hydraulic jump in a position further upstream and, therefore, makes it possible for a shorter basin to be used.

One method of encouraging a jet to spread is to impose upon it a pressure gradient transverse to the initial direction of motion. In an open-channel flow this is tantamount to sloping the bed across the flow, and to maintain symmetry this means that a cross section of the bed has a triangular shape with a crest at the centreline. On an initially flat surface, such a device demands that the bed shall rise temporarily to afford the elevation from which the pressure gradient develops. The rise is not carried on indefinitely, so that the result is a pyramidal hump, first rising, then falling in the direction of motion with a triangular cross section across the flow (see Fig. 4.23). Such a hump, which has been experimentally shown to be effective, was successfully used at the outlet of the spillway

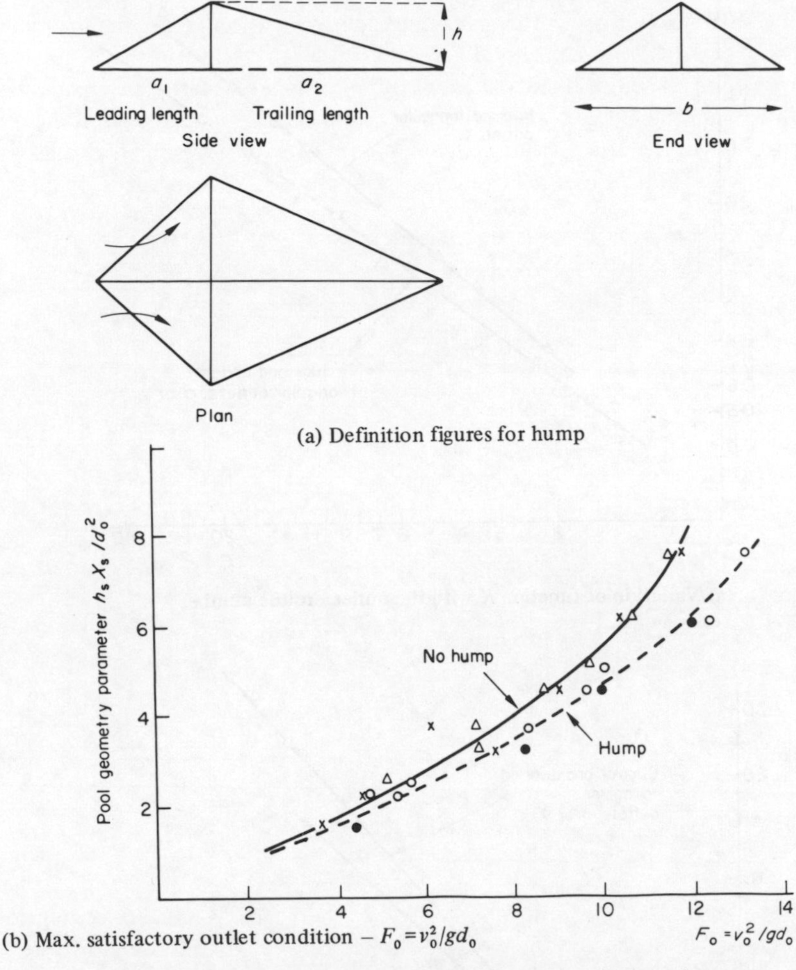

(a) Definition figures for hump

(b) Max. satisfactory outlet condition – $F_0 = v_0^2/gd_0$

Fig. 4.23 Characteristics of tunnel outflow control using a hump

tunnel at the Balderhead dam of the Tees Valley Water Board (England). Experimental results with a typical hump with 7·5 cm leading length, 15 cm trailing length, 1·5 cm width and 5 cm height are shown in Fig. 4.23. h_s is the height of end sill, X_s the position of the end sill from the outlet of a tunnel of diameter d_0 and v_0 is the tunnel outlet velocity. The ordinate represents a measure of excavations required. Table 4.3 gives the optimum position X of the hump for various ranges of Froude number.

Free expansion of flow from a tunnel on a horizontal bed may be approximated by the empirical equation

$$y = \frac{A_1 d_0^2 K}{x} \exp(-\alpha z^2/x^2) \tag{4.64}$$

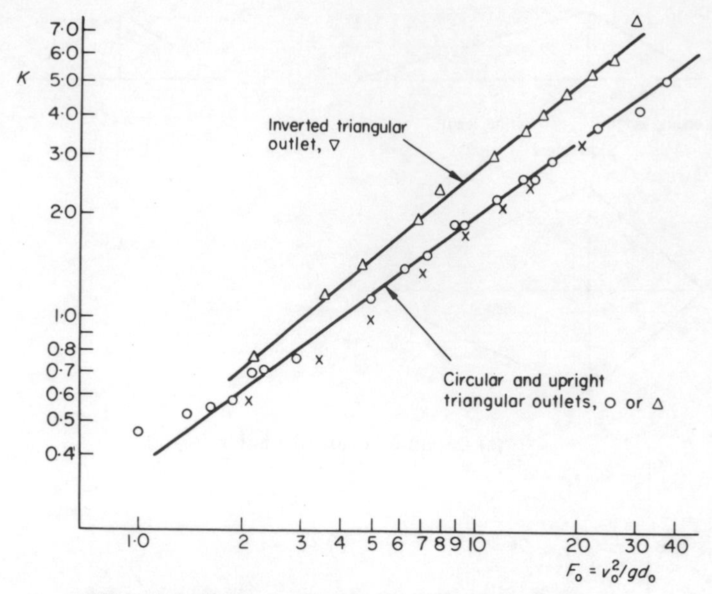

(a) Variation of function K with the outlet Froude number

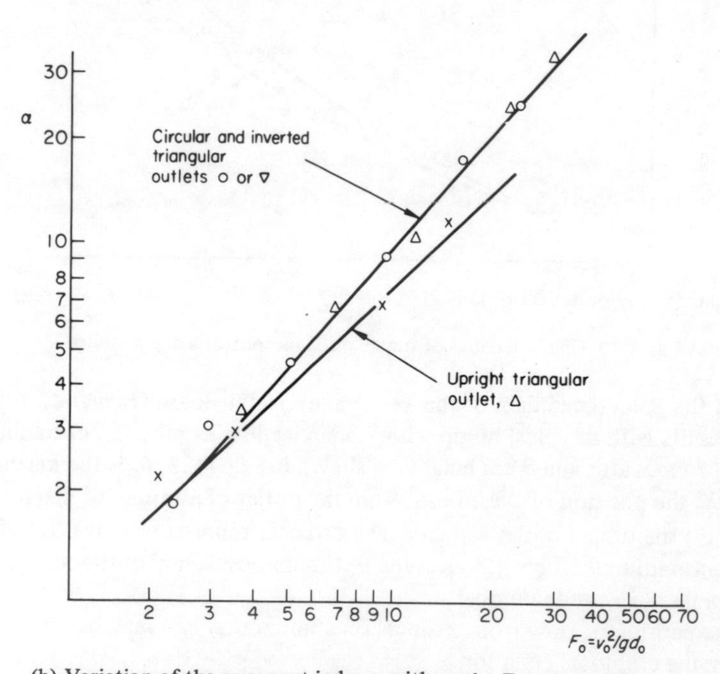

(b) Variation of the exponent index a with outlet Froude number

Fig. 4.24 Free spreading characteristics of tunnel outflow

where y is the depth at a point (x, z) from the centre of the outlet. K and α (as determined from experiments) are plotted against outlet Froude number in Fig. 4.24. A_1 depends on the shape of the outlet (0·85 for a circular outlet) and d_0 also represents the equivalent diameter for non-circular outlets.

Table 4.3

Optimum positions for a hump

[Hump geometry: $a_1/d_0 = 0{\cdot}75$; $a_2/d_0 = 1{\cdot}5$; $b/d_0 = 1{\cdot}5$; $h/d_0 = 0{\cdot}5$]

Hump position X/d_0	Approximate range of outflow $F_0 = \frac{v_0^2}{gd_0}$
1·0	up to 7
1·25	7 to 10
1·5	9 to 15
1·75	12 to 19
2·0	19 to 25
2·25	25 to 30
2·5	30 to 40
3·0	above 40

FURTHER READING

Bakhmeteff, B. A., *Hydraulics of Open Channels,* McGraw-Hill, New York.

Chow, V. T., *Open Channel Hydraulics,* McGraw-Hill, New York.

Dake, J. M. K. and Francis, J. R. D., *Outlet Works of Spillway Tunnels,* La Houille Blanche No. 6, 1965.

Elevatorski, E., *Hydraulic Energy Dissipators,* McGraw-Hill, New York.

Francis, J. R. D., *A Text Book of Fluid Mechanics for Engineering Students,* Edward Arnold, London.

Lewitt, E. H., *Hydraulics and Fluid Mechanics,* Pitmans, London.

Rouse, H. (Ed.), *Engineering Hydraulics,* John Wiley and Sons, New York, London.

5 Experimental Fluid Mechanics

5.1 Introduction

Many fluid flow problems cannot be solved mathematically because of the complex nature of the equations involved and of the boundaries of the flow regime. Many others can be solved only after certain simplifying assumptions have been made and it is then necessary to verify the solution through experiments. In the first group of problems it goes without saying that the exact behaviour of fluid in the system can be ascertained only after observations are made in the system when in operation. Such problems may be encountered in the design, construction and operation of dams, harbours, aircraft, ships, to mention only a few. The stability of these structures and their performance depend on the nature of flow in and around them. Huge sums of money, many years of work, prestige, and human lives are involved in such projects and their stability, suitability and performance cannot be left to chance. In the absence of a rigorous and reliable mathematical analysis the designer has found the technique of model tests a most valuable tool. A conveniently sized model of the structure is built and tested under controllable operating conditions and the results are supposed to predict what will happen in the real structure.

The application of model studies to prototype (or field) conditions presupposes that once a particular property (velocity, discharge, force, pressure, time, temperature, etc.) is known for the model, multiplication by a suitable factor then gives the equivalent property in the prototype. In other words, a perfect matching of model and prototype properties (qualitatively and quantitatively) is achieved by applying one or more constant factors to one of the systems. The factors are obtained from specific *scaling laws.* The theory of models seeks to specify these scaling laws.

In this chapter we will examine in detail how some of these scaling laws arise and their relative importance in some special cases of fluid flow. Emphasis is placed on correlation of experimental data for universal application to geometrically similar systems. There is a section on dimensional analysis whose principal aim is to stress the merits and limitations of the approach.

5.2 Dynamic Similarity

The basic partial differential equations governing the flow of incompressible fluids with reasonably constant viscosity are the Navier-Stokes equations. Their derivation is beyond the scope of this book but a thorough discussion of the structure of the equations is necessary for a good appreciation of the theory underlying model consideration. In the cartesian coordinate system, Newton's second law of motion applied to a unit mass of a fluid element whose components of velocity are u, v and w in the x, y and z directions respectively gives Navier–Stokes equations in the three principal directions. In the x direction

$$\underbrace{\frac{\partial u}{\partial t}}_{\text{local acceleration}} + \underbrace{u\frac{\partial u}{\partial x} + v\frac{\partial u}{\partial y} + w\frac{\partial u}{\partial z}}_{\text{convectional acceleration}} = \underbrace{-g\frac{\partial h}{\partial x}}_{\text{gravity (body) force}} \underbrace{- \frac{1}{\rho}\frac{\partial p}{\partial x}}_{\text{pressure force}} + \underbrace{\frac{\mu}{\rho}\left(\frac{\partial^2 u}{\partial x^2} + \frac{\partial^2 u}{\partial y^2} + \frac{\partial^2 u}{\partial z^2}\right)}_{\text{viscous force}} \tag{5.1}$$

Similarly in the y and z directions

$$\frac{\partial v}{\partial t} + u\frac{\partial v}{\partial x} + v\frac{\partial v}{\partial y} + w\frac{\partial v}{\partial z} = -g\frac{\partial h}{\partial y} - \frac{1}{\rho}\frac{\partial p}{\partial y} + \frac{\mu}{\rho}\left(\frac{\partial^2 v}{\partial x^2} + \frac{\partial^2 v}{\partial y^2} + \frac{\partial^2 v}{\partial z^2}\right) \tag{5.2a}$$

$$\frac{\partial w}{\partial t} + u\frac{\partial w}{\partial x} + v\frac{\partial w}{\partial y} + w\frac{\partial w}{\partial z} = -g\frac{\partial h}{\partial z} - \frac{1}{\rho}\frac{\partial p}{\partial z} + \frac{\mu}{\rho}\left(\frac{\partial^2 w}{\partial x^2} + \frac{\partial^2 w}{\partial y^2} + \frac{\partial^2 w}{\partial z^2}\right) \tag{5.2b}$$

The terms on the left-hand side of these equations are the acceleration or inertia terms which must balance the sum of all external forces operating on the fluid element as given on the right-hand side. The acceleration of a fluid particle arises from two sources; local changes of velocity because of the unsteady nature of the flow and the fact that the velocity of the fluid particle changes when it moves from one locality into another of a different velocity. In mathematical terms, the total derivative of u since it is a function of both time and space is

$$\frac{du}{dt} = \frac{\partial u}{\partial t} + \frac{\partial u}{\partial x}\frac{\delta x}{\delta t} + \frac{\partial u}{\partial y}\frac{\delta y}{\delta t} + \frac{\partial u}{\partial z}\frac{\delta z}{\delta t}$$

or

$$\underbrace{\frac{du}{dt}}_{\text{total acceleration}} = \underbrace{\frac{\partial u}{\partial t}}_{\text{local acceleration}} + \underbrace{u\frac{\partial u}{\partial x} + v\frac{\partial u}{\partial y} + w\frac{\partial u}{\partial z}}_{\text{convectional acceleration}}$$

The forces acting on the particle of fluid whose components in the various directions are shown on the right-hand side of equations (5.1) and (5.2) are gravity by virtue of the position of the fluid particle a vertical distance h above an arbitrary datum, pressure and shear or viscous forces. Gravity provides the body force while pressure and shear give normal and tangential surface forces respectively. These forces are illustrated for a particle located in a liquid flowing over a steep slope in an open channel (Fig. 5.1). The polygon of forces shows that the resultant force F must be balanced by inertia force F_i for equilibrium of the particle. This, in fact, is the diagrammatic form of the Navier–Stokes equations.

The basic concept of hydrodynamic similarity is the requirement that two systems with geometrically similar boundaries have similar flow patterns (kinematic similarity) at corresponding instants of time. This implies that the polygon

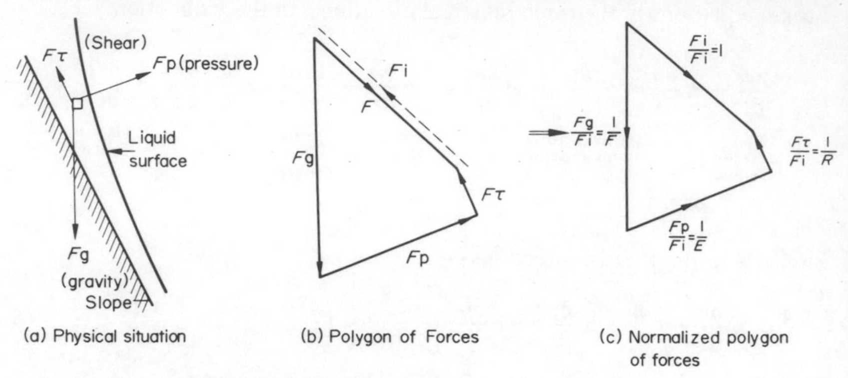

Fig. 5.1 Dynamic equilibrium of a fluid particle

of forces must become identical when a constant factor is appropriately applied to one of them. This is *dynamic similarity*. It is not possible to achieve kinematic similarity in two geometrically similar systems unless the operating forces are similarly arranged. This condition may also be deduced from the fact that the Navier–Stokes equations are applicable in both systems in which the fluid is incompressible, has a constant viscosity and surface tension is unimportant. Thus providing the boundary conditions are similar, the solution of the equations, if effected, will be qualitatively similar. Quantitative differences arise only through differences in quantitative controls of physical elements involved. The exercise is to find the combination of the physical quantities which will operate on the various elements of the external forces of the partial differential equations (5.1) and (5.2) so as to make the inertial forces identical and thus find a unique solution for all similarly controlled systems of similar geometry.

Choosing a constant geometric quantity L (length) a constant kinematic quantity V_0 (velocity) and a constant dynamic quantity ρ_0 (density), it is possible to define the following dimensionless parameters.

$$
\begin{aligned}
&x' = x/L, \quad y' = y/L, \quad z' = z/L, \quad h' = h/L \\
&u' = u/V_0, \quad v' = v/V_0, \quad w' = w/V_0, \\
&p' = p/\rho_0 V_0^2, \quad t' = t/\left(\frac{L}{V_0}\right) \text{ and } \rho' = \rho/\rho_0
\end{aligned}
\tag{5.3}
$$

Substituting (5.3) into (5.1)

$$\frac{\partial (V_0 u')}{\partial \left(\frac{L}{V_0} t'\right)} + V_0 \left[u' \frac{\partial (V_0 u')}{\partial (x'L)} + v' \frac{\partial (V_0 u')}{\partial (y'L)} + w' \frac{\partial (V_0 u')}{\partial (z'L)} \right]$$

$$= -g \frac{\partial (h'L)}{\partial (x'L)} - \frac{1}{\rho_0 \rho'} \frac{\partial (p' \rho_0 V_0^2)}{\partial (x'L)}$$

$$+ \frac{\mu}{\rho_0} \frac{1}{\rho'} \left[\frac{\partial^2 (u'V_0)}{\partial (x'^2 L^2)} + \frac{\partial^2 (u'V_0)}{\partial (y'^2 L^2)} + \frac{\partial^2 (u'V_0)}{\partial (z'^2 L^2)} \right] \tag{5.4}$$

By removing the constant terms from the differentiation brackets of equation (5.4) and by dividing through by V_0^2/L

$$\frac{\partial u'}{\partial t'} + u' \frac{\partial u'}{\partial x'} + v' \frac{\partial u'}{\partial y'} + w' \frac{\partial u'}{\partial z'}$$

$$= -\frac{gL}{V_0^2} \frac{\partial h'}{\partial x'} - \frac{1}{\rho'} \frac{\partial p'}{\partial x'} + \frac{\mu}{\rho_0 V_0 L} \frac{1}{\rho'} \left(\frac{\partial^2 u'}{\partial x'^2} + \frac{\partial^2 u'}{\partial y'^2} + \frac{\partial^2 u'}{\partial z'^2} \right) \tag{5.5}$$

The equation of motion has now been reduced to a dimensionless form. A similar structure can be obtained from (5.2a) and (5.2b) and the inferences made below are the same. The corresponding normalized polygon of forces is shown in Fig. 5.1(c). From equation (5.5), it is required for dynamic similarity between a prototype (suffix p) and a model (suffix m) that particles of fluid in both systems have the same relative acceleration given by the left-hand side of the equation. This requires that the relative forces of the right-hand side of the equation must be identically equal. That is, the two fluid motions under consideration can become similar only if the solutions, expressed in terms of the respective dimensionless variables, are identical. This requires that for both

$$\left[-\frac{gL}{V_0^2} \frac{\partial h'}{\partial x'} - \frac{1}{\rho'} \frac{\partial p'}{\partial x'} + \frac{\mu}{\rho_0 V_0 L} \frac{1}{\rho'} \left(\frac{\partial^2 u'}{\partial x'^2} + \frac{\partial^2 u'}{\partial y'^2} + \frac{\partial^2 u'}{\partial z'^2} \right) \right]_p$$

$$\equiv \left[-\frac{gL}{V_0^2} \frac{\partial h'}{\partial x'} - \frac{1}{\rho'} \frac{\partial p'}{\partial x'} + \frac{\mu}{\rho_0 V_0 L} \frac{1}{\rho'} \left(\frac{\partial^2 u'}{\partial x'^2} + \frac{\partial^2 u'}{\partial y'^2} + \frac{\partial^2 u'}{\partial z'^2} \right) \right]_m$$

Normally temperature variations of the fluid do not bring about significant density changes in incompressible fluids and $\rho_0 = \rho$ or $\rho' = 1$. To satisfy the identity

$$\left(\frac{gL}{V_0^2} \right)_p = \left(\frac{gL}{V_0^2} \right)_m \quad \text{and} \quad \left(\frac{\mu}{\rho_0 V_0 L} \right)_p = \left(\frac{\mu}{\rho_0 V_0 L} \right)_m \tag{5.6}$$

From the development above, the term V_0^2/gL, known as the Froude number F, indicates the significance of gravity forces relative to inertia forces. The Reynolds number, $R = \rho_0 V_0 L/\mu$ indicates the significance of viscous (friction) forces relative to inertia forces.

Thus provided the Froude number and the Reynolds number are equal in any geometrically similar systems of an incompressible flow (barring external forces other than gravity, pressure and shear) and provided the boundary conditions are similar, the flow patterns will be identical. The solution of equations (5.5) and (5.2) similarly normalized (plus equations of continuity and of state) will be universal for all such systems and it requires only constant factors defined in equations (5.3) to obtain quantitative values for different sizes and operating conditions for all members of the family. Dimensionless presentation of experimental data is therefore a powerful instrument in the correlation and application of experimental data in hydraulic and other branches of engineering.

5.3 Physical Significance of Modelling Laws

The above development has clearly shown that in order to achieve dynamic similarity in two geometrically similar fluid systems of constant density and viscosity, the Froude and Reynolds numbers must have the same values in both systems if gravity and friction are the only external forces operating. Denoting the ratio (model to prototype) of any quantity of interest by the suffix r, dynamic similarity requires that

$$F_r = \frac{V_r^2}{g_r L_r} = 1 \tag{5.7a}$$

$$R_r = \frac{V_r L_r}{\upsilon_r} = 1 \tag{5.7b}$$

simultaneously, where $\upsilon = \mu/\rho$.

From the Froude law, $V_r = L_r^{\frac{1}{2}}$ for $g_r = 1$

and from the Reynolds law, $V_r = \upsilon_r / L_r$

Thus for both criteria, $\upsilon_r = L_r^{\frac{3}{2}}$

Take, for an example, a model of a stretch of a river (or of a dam) which is to be built to a scale of 1:25. In order to satisfy ideal dynamic similarity, Froude and Reynolds requirements demand the ratio of the kinematic viscosity of the model fluid to that of the prototype fluid to be 1:125. Table 5.1 lists some of the natural low-viscosity fluids. No known fluid satisfies the condition. It becomes necessary therefore to seek achievement of dynamic similarity through other means. The most logical approach is to seek similarity according to the more dominant force (friction or gravity) and to correct for the effect of the other through some other means. The theoretical considerations and hypotheses guiding this approach will be discussed in the following subsections; other forces will also be considered.

5.3.1 Reynolds Models

The pressure on an object located in a flowing fluid may be considered to have two components. One part, the hydrostatic pressure p_s, is that pressure which the object would experience if the fluid was not flowing and the other, the

Table 5.1

Viscosity of some fluids (21° C)

Fluid	$\upsilon(m^2/s)$	υ_r (relative to water prototype)
Water	$0{\cdot}93 \times 10^{-6}$	1·00
Benzene	$0{\cdot}74 \times 10^{-6}$	0·80
Carbon tetrachloride	$0{\cdot}60 \times 10^{-6}$	0·65
Gasoline (petrol)	$0{\cdot}40 \times 10^{-6}$	0·47
Mercury	$0{\cdot}12 \times 10^{-6}$	0·13

dynamic pressure p_d is the increase in pressure brought about as a result of the disturbance of the flow. Thus

$$p = p_s + p_d$$

Imagine a particle P in the flow field of a fluid of an infinite expanse. If the particle is at a vertical distance h above an arbitrary datum (Fig. 5.2) the hydrostatic pressure at P is $\rho g(H - h)$, where H is the level of the fluid above the same

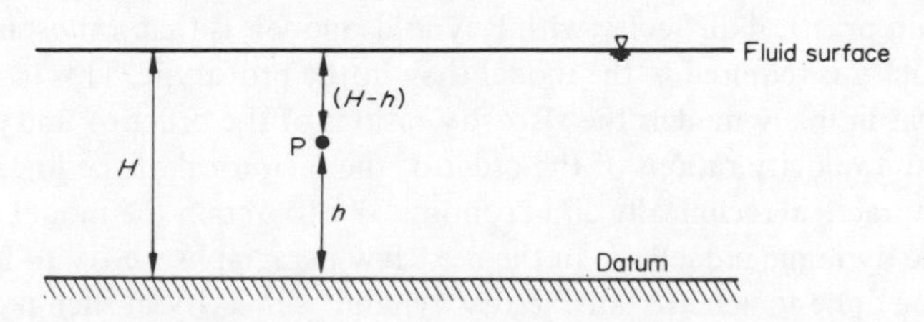

Fig. 5.2

datum. For flow conditions in which *variation of fluid surface does not exist or is unimportant* H will be constant. The total pressure may therefore be represented by

$$p = \text{constant} - \rho g h + p_d \tag{5.8}$$

Substituting (5.8) into the equations of motion (5.1), for example, gives

$$\frac{\partial u}{\partial t} + u\frac{\partial u}{\partial x} + v\frac{\partial u}{\partial y} + w\frac{\partial u}{\partial z} = -\frac{1}{\rho}\frac{\partial p_d}{\partial x} + \frac{\mu}{\rho}\left(\frac{\partial^2 u}{\partial x^2} + \frac{\partial^2 u}{\partial y^2} + \frac{\partial^2 u}{\partial z^2}\right) \tag{5.9}$$

Equation (5.9) is completely independent of the gravity term. Its normalized equation will therefore not contain a Froude number. Thus dynamic similarity for objects submerged in incompressible flows in which fluid surface variations (e.g. gravity waves) are not produced is attained by modelling according to the Reynolds criterion only. The same is true for incompressible flows in which the fluid is contained entirely within fixed boundaries and there are no free (exposed to atmosphere) surfaces. Examples of flow studies in which the Reynolds law is important are skin friction (pipe) losses, viscous drag on submerged objects, resistance to objects moving through large bodies of liquid (e.g. submarines), lubrication, flow through nozzles and orifices, flow in porous media, etc. In these examples viscous forces dominate gravity forces. Some of these may require the consideration of other dynamic conditions such as surface tension and compressibility as well as friction, as will be made clearer in subsection 5.3.3.

For $$R_r = \frac{v_r L_r}{\upsilon_r} = 1$$

velocity ratio, $$v_r = \upsilon_r / L_r$$

time ratio, $$t_r = L_r / v_r = L_r^2 / \upsilon_r$$

discharge ratio, $$Q_r = A_r v_r = L_r^2 \frac{\upsilon_r}{L_r} = \upsilon_r L_r$$

force ratio, $$P_r = \rho_r L_r^3 \frac{v_r}{t_r} = \rho_r \upsilon_r^2$$

pressure ratio, $$p_r = \frac{P_r}{A_r} = \rho_r \upsilon_r^2 / L_r^2$$

The main practical difficulty with Reynolds models is that a substantially higher velocity is required in the model than in the prototype. This is seen from the fact that in many models the viscosity ratio is of the order of unity and therefore the velocity ratio is of the order of the reciprocal of the linear ratio. It is often impractical technically and economically to obtain the model velocity to give true dynamic modelling. In the next few paragraphs we try to indicate what can be done to achieve satisfactory dynamic similarity in such models without necessarily attaining the required theoretical model velocity.

The development above leading to the establishment of the Reynolds criterion has assumed Newtonian fluid conditions. It may be argued, therefore, that Reynolds' law cannot strictly be applied to turbulent flow conditions in which the momentum transfer has become less of a fluid property, but rather more of a flow property. It is nevertheless important to appreciate that the hydrodynamic characteristics associated with shear forces (resistance to flow) are of interest. Since this resistance originates from the wall boundaries the significance of geometric similarity with regard to wall roughness cannot be over-emphasized. A very convenient way of expressing the hydrodynamic characteristics is the well-known Darcy–Weisbach formula:

$$S_f = \frac{f}{m} \frac{v^2}{2g} \tag{5.10}$$

where S_f is the slope of the energy grade line
f is an empirical *friction factor*

m is the hydraulic mean depth = $\dfrac{\text{wetted area}}{\text{wetted perimeter}}$

v is the velocity
and g is the gravitational acceleration.

In laminar flow the friction factor f is determined by purely analytical means (see Section 3.4) to be

$$f = 16/R \tag{5.11}$$

meaning that Reynolds' law automatically and completely accounts for f and wall friction. In fully turbulent flows, however, semi-analytical derivation due to Prandtl and von Karman and Nikuradse's grain experiments in pipes give

$$\frac{1}{\sqrt{f}} = 4 \log_{10} R\sqrt{f} - 1{\cdot}60 \tag{5.12}$$

for a hydraulically smooth surface and

$$\frac{1}{\sqrt{f}} = 4 \log_{10} \frac{4m}{k_s} + 2{\cdot}28 \tag{5.13}$$

for a hydraulically rough surface where k_s is the effective surface roughness. In the region of transition between laminar and turbulent flows, Colebrook and White have established that

$$\frac{1}{\sqrt{f}} = -4{\cdot}0 \log_{10}\left[\frac{1}{3{\cdot}7}\left(\frac{k_s}{4m}\right) + \frac{2{\cdot}51}{2R}\frac{1}{\sqrt{f}}\right] \tag{5.14}$$

A graphical representation of the variation of the friction factor with Reynolds number and pipe roughness is shown on the pipe friction diagram in Fig. 5.3 (which is the same as Fig. 3.5). The pipe friction diagram is used quite extensively in open channel problems as well by substituting the hydraulic mean depth for $D/4$. The following points are worth noting with special regard to hydrodynamic similitude, remembering that the R model implies the f-model and vice versa.

1. Laminar zone. For low Reynolds numbers (large viscosity and low velocities; $R < 2\,300$) frictional hydrodynamic similitude is governed entirely by the Reynolds law and departure from the stipulated prototype Reynolds number is undesirable.

2. Fully turbulent zone. For large Reynolds numbers (low viscosity and large velocities; $R > 10^5$) hydrodynamic resistance is governed by turbulent momentum transfer and only to a minor extent by molecular viscosity. The latter influence is felt only in the laminar sublayer near the walls of the surface. Since there is less tendency for separation and eddy formation near the walls of a smooth surface than there is near a rough surface with large protuberances we expect intuitively

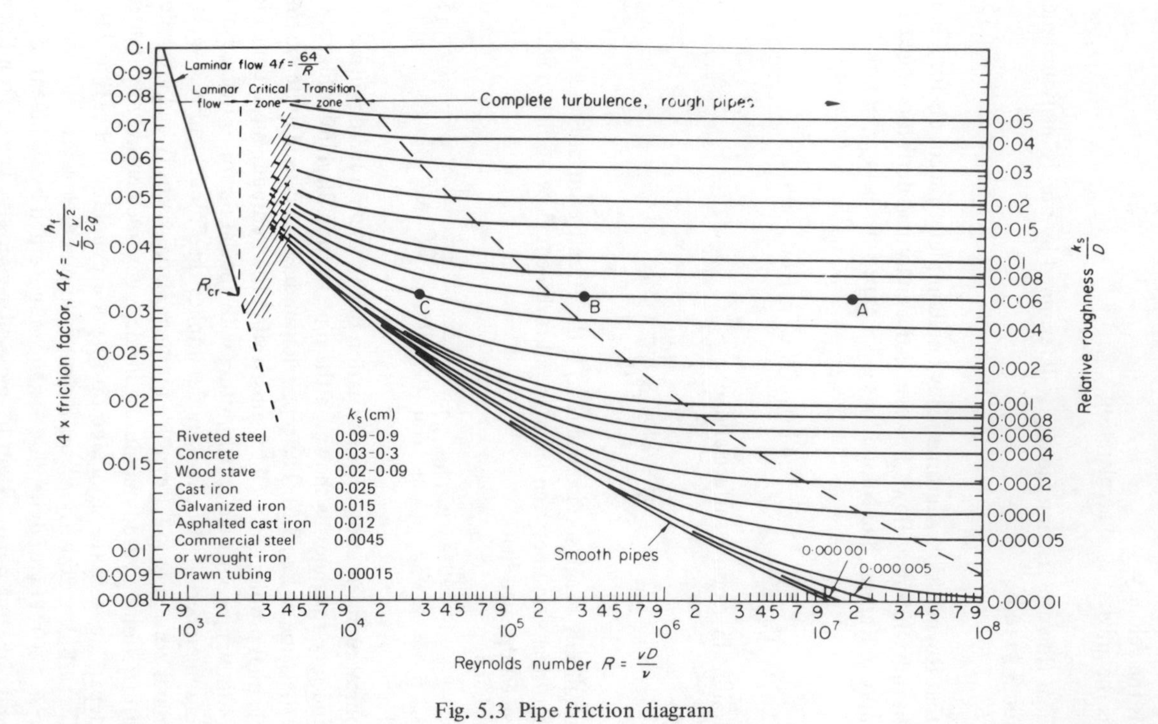

Fig. 5.3 Pipe friction diagram

that the effect of viscosity will be more pronounced on smooth surfaces that it is on relatively rough ones. This is clearly demonstrated on the pipe friction diagram. There are thus two distinct cases which must be considered in modelling a fully turbulent flow.

For rough surfaces the friction factor is independent of Reynolds number. Attaining hydrodynamical resistance similitude thus becomes a geometrical consideration of roughness $k_{sr}/L_r) = 1$. Supposing prototype conditions correspond to a point A (Fig. 5.3). Provided roughness requirements are satisfied, a model operated at point B will adequately simulate resistance conditions of the prototype. The necessary requirement in addition to the geometric conditions is that the model flow is sufficiently turbulent ($R > 10^5$). If practical considerations demand, it is even possible to operate the model at a point in the transition region such as point C which lies on a different relative roughness line. It is however clear that in such a case faulty operating conditions will produce resistance conditions in the model different from those it seeks to simulate because of the sensitivity of f to the Reynolds number. Model considerations involving relatively smooth surfaces require satisfaction of both viscous (Reynolds) and roughness effects. The model must operate at the exact point of design.

3. Transition zone. Models in this zone must satisfy both the Reynolds and relative roughness conditions. Because of the high sensitivity of the friction factor in this zone to the Reynolds number and relative roughness extreme caution is required in planning and operating the model.

EXAMPLE 5.1

The head loss across a gate valve is commonly expressed in terms of a loss coefficient C_L defined as head loss $h_L = C_L\ (v^2/2g)$ where v is the average velocity in the approach pipe. Columns 2 and 3 of the table below give the experimental results of a series of tests on a 20 cm gate valve in a 20 cm pipeline using water at 24°C. The valve disc was 7 cm off the seat in all tests.

Determine the head loss to be expected when the flow rate is 28·3 l/s in the same pipeline with the same valve and opening when water of 94°C flows through.

Test no.	Q l/s	h_L (cm)	v (m/s)	C_L	R	*Remarks*
1	6·2	0·6	0·19	3·21	$1{\cdot}48 \times 10^4$	υ of water at 24°C
2	12·7	2·4	0·38	3·20	$2{\cdot}96 \times 10^4$	$= 0{\cdot}9 \times 10^{-6}$
3	25·2	7·6	0·78	2·45	$6{\cdot}05 \times 10^4$	$R = 0{\cdot}078 \times 10^6\ V$
4	32·0	14·9	0·99	3·0	$7{\cdot}35 \times 10^4$	
5	85·0	105	2·62	3·0	$20{\cdot}4 \times 10^4$	

SOLUTION

Columns 5 and 6 show the variation of the head loss coefficient with the Reynolds number based on approach pipe velocity and size of valve opening. At a high Reynolds number C_L is constant and is independent of the Reynolds number. The variation is similar to that of f against R for pipes. In the fully turbulent region eddies are fully developed and momentum transfer is no longer governed by viscosity.

When flow is 28·3 l/s (24°C, $\upsilon = 0{\cdot}32 \times 10^{-6}$ m²/s)

$$R = 19{\cdot}2 \times 10^4, \text{ thus } C_L = 3{\cdot}0$$

$$\text{The expected head loss} = (3{\cdot}0)\frac{v^2}{2g} = 3{\cdot}0 \times \left(\frac{Q}{\pi(0{\cdot}1)^2}\right)^2 \times \frac{1}{2g}$$

$$= \underline{12 \text{ cm of water}}$$

5.3.2. Froude (Gravity) Models

In many fluid flow problems encountered in nature the fluid has a free surface, that is, a surface exposed to atmospheric conditions. The flows may also exhibit marked changes in surface elevation in response to geometric and other physical changes. These include waves in liquids, flow in rivers and over weirs, objects floating on the surface of water and multilayered (stratified) flow systems. Since the free surface level is not the same everywhere the gravity term cannot be eliminated from the equations of motion as demonstrated in subsection 5.3.1. It is necessary to consider both gravity and friction in such models. This has already been shown to be difficult. The practical approach involves construction and operation of the model according to which of the two forces is more significant and then appropriate correction for the other.

The case of fully turbulent free-surface flow is of particular interest to hydraulic engineers since most natural problems connected with rivers, spillways, energy dissipators, sailing ships, etc. fall into this category. Since the Reynolds number is usually very high in such situations, the last term of equation (5.5) is of a much smaller order than the first term on the right-hand side. In any case the discussions in subsection 5.3.1 have shown that the frictional effect is controlled essentially by the relative roughness of the surface and is insensitive to Reynolds number changes for fully turbulent flows over rough surfaces. The model conditions are therefore derived from Froude's criterion. Resistance conditions are normally satisfied by making the relative roughness (turbulence generating effects) equal in model and prototype and by choosing the model scale to ensure full turbulence. Usually, satisfaction of the Froude law will make the model operate at a much smaller Reynolds number than exists in the prototype. Practical considerations may in fact make the attainment of fully turbulent conditions in the model impossible. The problem is identical to that discussed in the preceding section. With reference to Fig. 5.3, if the model should be operated at C (thus satisfying Froude's law) instead of at A (which would satisfy Reynolds' law), the appropriate model roughness should be selected and operating conditions must not depart seriously from C.

The appropriate transfer parameters according to Froude's law are:

$$F_r = \frac{v_r^2}{g_r L_r} = 1$$

velocity ratio, $v_r = g_r^{\frac{1}{2}} L_r^{\frac{1}{2}}$

time ratio, $t_r = L_r^{\frac{1}{2}} / g_r^{\frac{1}{2}}$

discharge ratio, $Q_r = g_r^{\frac{1}{2}} L_r^{\frac{5}{2}}$

force ratio, $P_r = \rho_r g_r L_r^3$

pressure ratio, $p_r = \rho_r g_r L_r$

EXAMPLE 5.2

A 30·5 m high spillway of a dam is designed to pass 9·3 m³/s of water per metre. A model 1/25th the size of the prototype is to be tested in the laboratory. Calculate the supply of water necessary for the model. What is the minimum height of the model required for a reliable operation?

SOLUTION

The dominating force is that of gravity; thus the Froude law is the determining criterion.

$$v_r = L_r^{\frac{1}{2}} = 1/5; \; g_r = 1$$

Discharge per unit width,

$$q_r = \frac{Q_r}{L_r} = L_r^{\frac{3}{2}}$$

$$\therefore \quad q_m = q_p \left(\frac{1}{25}\right)^{\frac{3}{2}}$$

$$= \underline{0{\cdot}074 \text{ m}^2/\text{s}}$$

Check the Reynolds number.

$$R_p = \frac{v_p y_{0p}}{\upsilon_p} = \frac{q_p}{\upsilon_p} = 10^7$$

$\therefore$ the prototype flow is fully turbulent.

$$R_m = \frac{q_m}{\upsilon_m} = 0{\cdot}8 \times 10^5$$

Thus the model will just operate in the fully turbulent range

$$R_m \approx 10^5$$

For a fully turbulent flow model $R > 10^5$. That is

$$R_m = R_r R_p > 10^5$$

$$\frac{v_r L_r}{\upsilon_r} > \frac{10^5}{R_p} = 10^{-2}$$

$$L_r^{\frac{3}{2}} > 10^{-2}, \text{ since } \upsilon_r = 1$$

$$\therefore \qquad L_r > \left(\frac{1}{10}\right)^{\frac{4}{3}} = \frac{1}{21{\cdot}5} = 0{\cdot}0465$$

and minimum satisfactory height of model = 1·42 m.

5.3.3 Other Dimensionless Numbers

In discussing the Navier–Stokes equations in Section 5.2, the point was made that the only external forces supposed to be acting on the fluid element were due to gravity, pressure and sheer. The balance between these forces and inertia force is illustrated in the polygon of forces in Fig. 5.1. Although the situation depicted in the figure or described by the Navier–Stokes equation covers most flows of interest to the hydraulic engineer it leaves out external forces such as surface tension, magnetic or electrostatic force fields and others which could be quite important under special circumstances. Inclusion of these forces will increase the number of sides of the polygon and introduce other terms into equation (5.1.). The development and discussions above have also assumed that compressibility effects on the fluid are unimportant and that the fluid is homogenous. In this subsection we introduce some of the other hydrodynamic similitude laws which are commonly met in practice and discuss their significance. The Froude and Reynolds numbers are derived again by an intuitive and less sophisticated method in order to emphasize the physical significance of these and the other numbers.

From Newton's second law of motion an inertia force F_i is given by the product of mass and acceleration,

i.e. $$F_i = m \times a$$

Choosing a suitable dimension L the volume of an object is proportional to L^3. Choosing the constant of proportionality as unity, the inertia force is represented by

$$F_i = \rho L^3 \frac{v}{t} = \rho L^3 \frac{v^2}{L} = \rho L^2 v^2$$

Similarly, the gravity force = $\rho L^3 g$.

By definition, the Froude number is the ratio of the inertia force to the gravity force.

Thus $$F = \frac{\rho L^2 v^2}{\rho L^3 g} = \frac{v^2}{gL}$$

Now shear force = shear stress × area

$$\alpha\left(\frac{v}{L}\right)\mu\, L^2 = \mu v L$$

By definition,

$$\text{Reynolds number,} \qquad R = \frac{\text{inertia force}}{\text{shear force}}$$

$$= \frac{\rho L^2 v^2}{\mu v L} = \frac{\rho L v}{\mu}$$

COMPRESSIBILITY EFFECTS

Pressure force $\propto pL^2$, where p is pressure

$$\therefore \qquad \frac{\text{inertia force}}{\text{pressure force}} = \frac{\rho L^2 v^2}{pL^2} = \frac{\rho v^2}{p}$$

The dimensionless number $E = (\rho v^2/2p)$ is known as the *Euler number* or *pressure coefficient*. It is employed in high pressure flows and cavitation studies in liquids. It is generally specified together with the Reynolds number or Froude number.

It can be established from gas dynamics that the local speed of a sound wave is given by $c = \sqrt{(kp/\rho)}$ where k is the ratio C_p/C_v of the specific heats of the gas through which the wave travels. Thus for a gas the ratio of inertia force to pressure force is given by

$$M^2 = \frac{kv^2}{c^2}$$

The form $M = v/c$, known as *Mach number*, is employed in model studies for convenience since the constant k has approximately the same value (1·4 to 1·6) for all common gases.

The Mach number is the equivalent of the Euler number, the former being used for model studies of gas flows at high speed when the velocity approaches that of a sound wave through the gas. A form of Mach number suitable for liquid flows involving high pressures (e.g. the water hammer) is the so called *Cauchy number*.

$$\mathscr{C} = v/\sqrt{(K/\rho)}$$

which arises from thermodynamic considerations relating velocity of sound waves through a medium of modulus of compressibility K as $c = \sqrt{(K/\rho)}$.

The Euler and Cauchy numbers (for liquids) and the Mach number (for gases) are usually specified together with the Reynolds number for the study of drag on immersed bodies such as aircraft, torpedoes and other submarines. Compressibility effects are usually not significant unless $M \gg 0{\cdot}5$.

For cavitation studies the Froude number is specified with a modified form of Euler number first introduced by Thoma and now called the *cavitation number* (see Section 6.5).

$$\frac{1}{\sigma_c} = \frac{\rho v^2}{2(p - p_v)}$$

where p_v is the vapour pressure of the liquid of interest.

WEBER MODELS

Surface tension is the energy stored per unit area as a result of molecular activity at the interface between two fluids which do not mix. It has the units of force per unit length. The ratio of inertia force to surface tension force gives rise to a dimensionless combination known as the *Weber number*

$$W = \frac{\rho L^2 v^2}{\sigma L} = \frac{\rho v^2 L}{\sigma}$$

Since surface tension arises at the interface of two immiscible fluids it follows that its effects are noticeable only under stratified conditions, including free surface (liquid–air) situations. It is important only when the relative motion is very slow and its effect may be observed as very small amplitude waves known as

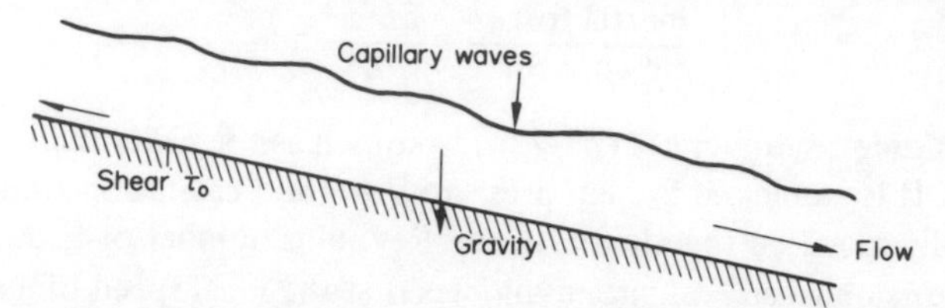

Fig. 5.4 Illustrating surface tensional effect

capillary waves. In general the Weber number is modelled with the Froude number in dealing with such problems. In some problems such as that illustrated in Fig. 5.4, which depicts a thin liquid flow over a solid surface, the Reynolds number must also be considered. The general approach, as before, is to model for W and F and correct for R. Surface tension problems normally involve laminar flow and

$$f = 16/R.$$

With $W_r = 1$ and $F_r = 1$

$$v_r^2 = \frac{\sigma_r}{\rho_r L_r} = g_r L_r$$

and

$$\frac{\sigma_r}{\rho_r} = L_r^2$$

As in the case of the Reynolds and Froude models the practical difficulty is connected with the availability of a suitable fluid. It is fortunate that very few engineering problems involve consideration of surface tension. Surface tension is insignificant if $W > 100$. It is important to avoid surface tension in a model whose prototype does not show its effect. The effect may be removed from models with detergents (e.g. soap).

STRATIFIED FLOW MODELS

The hydraulic engineer is sometimes confronted with flow problems dealing with two or more distinct layers (of different density) of fluid. One such problem which deals with the so-called arrested saline wedge in an estuary is illustrated in

Fig. 5.5. The denser sea water in the wedge displaces the fresh river water upwards. The position of the upstream end of the wedge depends on many variable parameters such as tides and the river flow conditions. Its determination is however important for fresh water supply, waste disposal and pollution studies along the river.

Another important example is the need to withdraw cold water from a thermally stratified reservoir. Deep down in the reservoir the water may be cool. Due to solar radiation absorption and discharge of cooling water from a plant

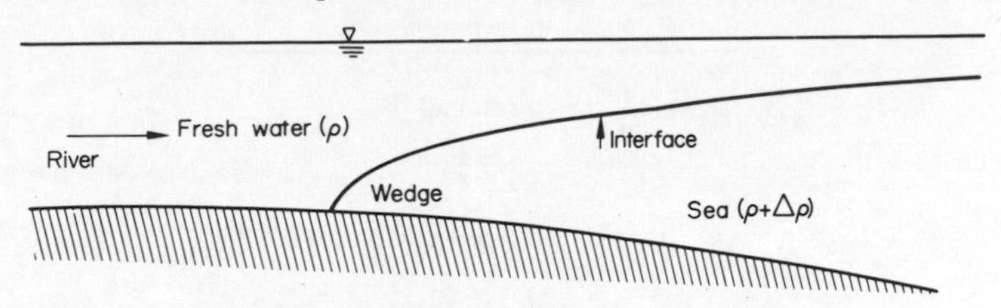

Fig. 5.5 Arrested saline wedge in an estuary

there normally exists an isothermal surface layer of lighter water. The surface mixing is aided by wind agitation and convective processes brought about by evaporative cooling. For a more detailed discussion of such processes the interested reader is referred to Dake and Harleman's report in the reference list. One consideration in deciding where to locate intakes in such stratified reservoirs is the critical condition at which warm water begins to be taken in.

A full discussion of stratified flow theory is beyond the scope of this text and the interested reader should refer to the literature, especially Harleman's chapter in the *Handbook of Fluid Dynamics* edited by V. Streeter. Unfortunately, no reliable theory yet exists for the solution of stratified flow problems. The designer has to resort to model studies. Dynamic similarity for the internal motion of stratified fluids is based on a modified form of Froude number commonly referred to as *Richardson number.* This is given by

$$R_{\mathrm{i}} = \frac{\rho g}{g\Delta\rho}\,\frac{v^2}{gL} = \frac{\gamma}{\Delta\gamma}\,\frac{v^2}{gL}$$

where $\Delta\rho$ is the difference in densities of the two layers.

The square root of the Richardson number $F' = v/\sqrt{(gL\Delta\rho/\rho)}$ is referred to, especially in American literature, as the *densimetric Froude number*. It is worth noting that for the study of a uniformly stratified fluid a more convenient Richardson number is expressed in differential form as

$$R_{\mathrm{i}} = \frac{g}{\rho}\,\frac{\partial\rho}{\partial x}\bigg/\left(\frac{\partial v}{\partial x}\right)^2$$

EXAMPLE 5.3

We want to study cavitation of a bridge pier in a pressurized water tunnel. If the model ratio is 1:25, determine the required pressure in the water tunnel for dynamic similarity to be achieved. Take p_{v} = 0·13 m of water.

SOLUTION

In the prototype the surrounding pressure is atmospheric = 10·3 m of water. Let the tunnel pressure be h_m of water.

This is a free surface flow and cavitation is involved. Thus F and the cavitation number σ_c must be modelled.

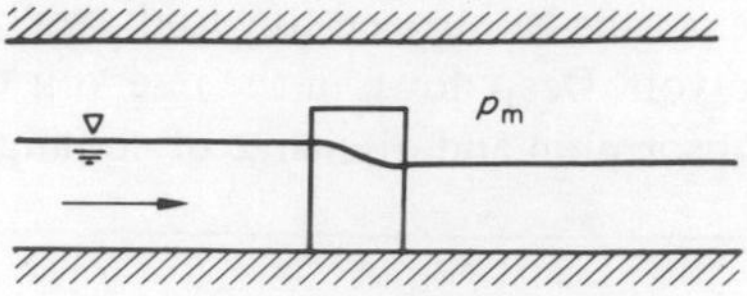

Water tunnel

$$F_r = 1, \; v_r = L_r^{\frac{1}{2}}$$

$$\sigma_{cr} = 1, \; v_r^2/(p - p_v)_r = 1$$

Thus

$$\frac{L_r}{(\rho g)_r(h - h_v)_r} = 1$$

$$\therefore \quad (h - h_v)_m = L_r(h - h_v)_p = (10{\cdot}3 - 0{\cdot}13)/25$$

$$\therefore \quad h_m = \frac{10{\cdot}17}{25} + 0{\cdot}13 \text{ m of water}$$

$$= 0{\cdot}437 \text{ m of water (absolute)}$$

Required model pressure = 0·437 m of water (absolute)
= − 9·86 m of water (gauge)

EXAMPLE 5.4

An aircraft travels at 644 km/h through air at − 18°C and a pressure of 68·9 kN/m² absolute. A model 1/20th the size of the prototype is to be tested in a pressure wind tunnel. If the tunnel operates at 21°C and 345 kN/m² absolute, what will be the necessary tunnel wind velocity? Assume an adiabatic condition with $k = 1{\cdot}40$.

This is a problem of drag. The drag is brought about through skin friction and pressure distribution (form drag) around the object. Ideally the model should simulate the two conditions through Reynolds and Mach numbers. Thus

$$R_r = 1, \qquad \therefore \frac{v_r L_r}{\upsilon_r} = 1, \qquad \therefore v_r = \frac{\upsilon_r}{L_r}$$

$$M_r = 1, \qquad \therefore v_r/c_r = 1, \qquad \therefore v_r = c_r$$

Mach conditions:

$$c = \sqrt{(kp/\rho)} = \sqrt{(kRT)}$$

$$\therefore \quad v_r = c_r = \sqrt{T_r}$$

$$T_m = 21 + 273 = 294 \text{ K}; T_p = 255 \text{ K}; \therefore v_r = 1{\cdot}08$$

Reynolds conditions: $\nu_r = v_r/L_r = \mu_r/L_r\rho_r$

From $p = \rho RT$; $\rho_r = p_r/T_r = 4{\cdot}36$

$$\mu_r = 4/3{\cdot}5 = 1{\cdot}14 \text{ (from Tables or Fig. 1.3)}$$

$$\therefore \qquad \nu_r = \frac{1{\cdot}14 \times 20}{4{\cdot}36} = 5{\cdot}24$$

To satisfy both conditions would mean a different scale ratio determined from $\nu_r = c_r = v_r/L_r$. Practical considerations obviously make this impossible. The problem now is to decide which of the two laws to use. In the prototype

$$c = \sqrt{(1{\cdot}4 \times 285 \times 255)}$$

$$= 320 \text{ m/s} = 1150 \text{ km/h}$$

$$M_p = (v/c)_p = 644/1150 = \underline{0{\cdot}56}$$

Since $M_p \ll 1{\cdot}0$ compressibility effects in the prototype are insignificant. Prototype drag is thus due mainly to skin friction. Reynolds' criterion may be employed in the model. The Mach number will however be higher than unity in the model and this represents a defect which must be considered when evaluating the results.

EXAMPLE 5.5

Drag tests on an ellipsoid with a minor diameter of 3 cm in a high speed wind tunnel have shown the following results.

Coeff. of drag, C_D	0·50	0·50	0·18	0·20	0·80	1·00
Reynolds number, R	6×10^4	2×10^5	3×10^5	5×10^5	$6{\cdot}6 \times 10^5$	1×10^6

What will be the drag on an ellipsoidal object of 15 cm in minor diameter travelling through standard air at 46 m/s? Also estimate the drag on a 30 cm diameter ellipsoidal object travelling in water at 24°C with a velocity of 3 m/s, assuming a large depth of submergence.

SOLUTION

Drag D is given by

$$D = \frac{C_D}{2} A\rho v^2$$

where C_D = coefficient of drag which is dependent on Reynolds number
and A = projected area of the object normal to the direction of notion.

Standard air has density $\rho = 1.22$ kg/m³ and
kinematic viscosity $\upsilon = 14{\cdot}6 \times 10^{-6}$ m²/s

For the ellipsoid of d = 15 cm, travelling through standard air

$$R = \frac{vd}{\upsilon} = \frac{46 \times 0{\cdot}15}{14{\cdot}6 \times 10^{-6}} = 4{\cdot}72 \times 10^5$$

From the table C_D corresponding to R = 4·7 x 10⁵ is approximately 0·20

$$\therefore \quad \text{Drag on ellipsoid} = \frac{0{\cdot}20}{2} \times \left(\frac{\pi}{4}\right) (0{\cdot}15)^2 \, (1{\cdot}22)(46)^2$$

$$= \underline{4{\cdot}6 \text{ N}}$$

For the ellipsoidal object travelling through water

$$R = 3 \times 0{\cdot}30/0{\cdot}93 \times 10^{-6} = 10^6$$

A plot of C_D against R as determined by the wind tunnel experiment is shown in the sketch below. At $R > 5{\cdot}0 \times 10^5$, C_D increases rapidly. This is the result of compressibility effects in the wind tunnel. Checking the Mach number:

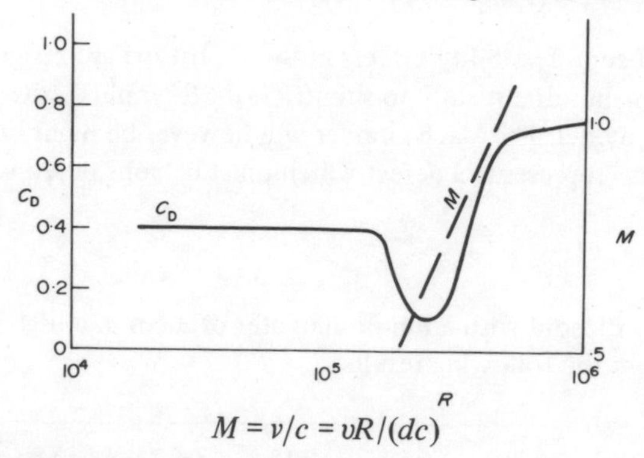

$$M = v/c = \upsilon R/(dc)$$

At standard air conditions, c = 342 m/s

$$\therefore \quad M = 1{\cdot}5 \times 10^{-6} R$$

For $R = 0{\cdot}5 \times 10^6$, M = 0·75 showing that the drag on the object in the wind tunnel has a compressibility influence (pressure drag). This effect will be negligible in water since the Cauchy number will still be small. Thus for $R \geqslant 0{\cdot}5 \times 10^6$, it will assume a constant value C_D = 0·20 for the water experiment. Then the drag to be expected

$$= \frac{0{\cdot}20}{2} \times \frac{\pi}{4} \times (0{\cdot}3)^2 \times 10^3 \times 9$$

$$= \underline{64{\cdot}7 \text{ N}}$$

5.4 Models of Rivers and Channels

Having discussed in the preceding sections of this chapter the theory behind model considerations generally and various dimensionless parameters which represent the common forces in fluid mechanics, attention is now turned to one major class of hydraulic engineering problems, namely rivers and channels. Models in this class may be considered in four main groups: (1) river systems with specific problems of flood routeing (discharge measurement), effects of river modifications and flooding and its prevention (where will overflow occur and how much of it?); (2) hydraulic effects on structures, e.g. dams, spillways and jetties; (3) calibration of measuring devices, e.g. the Venturi or the Parshall flume and (4) sediment or bed movement problems.

Three cases often arise in connection with open channel (river) models; the case in which gravity and pressure forces dominate, e.g. short lengths of river reaches at which changes are rapid; the case in which friction is important alongside gravity and pressure; and the case where friction is dominant, e.g. long straight uniform reaches of a channel, and sediment problems.

(i) *Gravity and pressure forces are dominant.* Frictional forces are small compared with gravity forces in fast flow regimes of rapid surface changes. This is particularly true for short reaches at the confluence of channels, flow over spillways or past contractions or enlargements. The Froude law is the governing criterion and scaling is as already discussed in subsection 5.3.2.

(ii) *Gravity, pressure and friction are important.* Because of practical difficulties of finding a model fluid to satisfy both the Froude and Reynolds conditions the model is designed according to the Froude criterion and suitable adjustments made for the friction factor.

These problems arise in connection with long reaches of rivers with small changes in water surface level. Friction factor correction is effected through one of the known empirical formulae, namely the Manning, Darcy–Weisbach or Chezy formula. It is hardly ever possible to have a uniform distribution of roughness in a river. In practice, therefore, it becomes necessary to resort to trial and error adjustment of the boundary roughness of the model until the water levels at corresponding locations in the model compare well with already known levels in the prototype river. This is known as model verification. The approach presupposes that if a model can be adjusted to reproduce events that have occurred in the prototype, it should also indicate what will occur under changed flow conditions. The method of trial and error is not reliable and the need for two models of different scales for a check arises in expensive and important projects.

DISTORTED MODELS

In practice models using different vertical and horizontal scales have to be used quite frequently. Models of this type are said to be distorted. The vertical scale is usually exaggerated relative to the horizontal scale. The main reasons for constructing a distorted model are connected with space availability, operating convenience and satisfactory model performance.

Space limitations normally impose a small horizontal scale for a long reach of a river. In order to be able to measure depth changes in the model a relatively large vertical scale is necessary. It may be possible to maintain fully turbulent flow in the model in accordance with conditions in the prototype river only if the vertical scale is exaggerated. This also helps to keep the relative thickness of the boundary layer nearer to that existing in the field. Relatively large vertical gradients also give freedom in the choice of friction generating technique in the model.

From Froude's law

$$v_r = g_r^{\frac{1}{2}} L_r^{\frac{1}{2}}$$

Let the model be distorted such that $L_r = X_r$ horizontally and $L_r = Y_r$ vertically. Since Froude's law involves consideration of gravity force and since gravity acts in a vertical direction it is obvious that Froude's number should be defined in terms of a vertical dimension. (Examination of Bernoulli's equation indicates that velocity is governed by changes in vertical dimensions). Thus for a distorted model

$$v_r = Y_r^{\frac{1}{2}} \text{ for } g_r = 1$$

discharge ratio, $$Q_r = A_r v_r = (X_r Y_r) Y_r^{\frac{1}{2}} = X_r Y_r^{\frac{3}{2}}$$

time ratio, $$t_r = \frac{L_r}{v_r} = \frac{X_r}{Y_r^{\frac{1}{2}}}$$

force ratio, $$P_r = (\rho_r X_r Y_r v_r) v_r = (\text{mass flux}) \times (\text{velocity}) = \rho_r X_r Y_r^2$$

pressure ratio, $$p_r = \frac{P_r}{X_r Y_r} = \rho_r Y_r$$

Manning's formula or Darcy-Weisbach's formula is used to correct for friction effects. The longitudinal slope of a river bed S_0 = vertical dimension/horizontal dimension. Thus the ratio of slopes $S_{0r} = Y_r/X_r$

Manning's formula:

$$v = \frac{1}{n} m^{\frac{2}{3}} S_0^{\frac{1}{2}}$$

where n is the friction factor, m the hydraulic mean depth and S_0 the longitudinal slope. Thus

$$v_r = \frac{v_m}{v_p} = \frac{m_r^{\frac{2}{3}} S_{0r}^{\frac{1}{2}}}{n_r}$$

substituting for v_r

$$Y_r^{\frac{1}{2}} = \frac{m_r^{\frac{2}{3}}}{n_r}\left(\frac{Y_r}{X_r}\right)^{\frac{1}{2}}$$

or $$n_r = m_r^{\frac{2}{3}} / X_r^{\frac{1}{2}}$$

For a rectangular channel,

$$m_r = \frac{y_{0m} b_m}{2y_{0m} + b_m} \bigg/ \left(\frac{y_{0p} b_p}{2y_{0p} + b_p}\right) = Y_r X_r \; \frac{2y_{0p} + b_p}{2y_{0m} + b_m}$$

where y_0 refers to depth and b is the width of channel. If the channel is very wide $b \gg 2y_0$,

$$m_r = Y_r$$

Thus

$$n_r = Y_r^{\frac{2}{3}} / X^{\frac{1}{2}}$$

If
$$Y_r^{\frac{4}{3}} / X_r > 1, \quad n_m > n_p$$

This shows the advantage of exaggerated vertical scale since it is much easier to provide roughening in models than it is to provide smoothening. Similarly, using the Darcy–Weisbach formula,

$$v_r = \sqrt{(S_{0r} m_r / f_r)}$$

and $f_r = m_r / X_r = Y_r / X_r$ for a wide rectangular channel

$$f_m > f_p \text{ if } Y_r > X_r$$

EXAMPLE 5.6

The average width of a certain river is 915 m and the average depth is 9·1 m. Calculate the discharge ratio and Manning's n ratio between model and prototype. If the model is observed to discharge a flood in 1 h, how long will an equivalent flood be observed in the prototype?

$$X_r = \frac{1}{2000} \text{ and } Y_r = \frac{1}{100}$$

SOLUTION

Distortion ratio;
$$Y_r / X_r = 20$$

Required model criteria: Froude and friction

$$Q_r = X_r Y_r^{\frac{3}{2}} = \frac{1}{2\,000} \times \left(\frac{1}{100}\right)^{\frac{3}{2}} = \frac{1}{2 \times 10^6}$$

$$m_p = \frac{9{\cdot}1 \times 915}{915 + 18{\cdot}2}; \; m_m = \frac{0{\cdot}09 \times 0{\cdot}46}{0{\cdot}46 + 0{\cdot}18}$$

$$= 8{\cdot}98 \text{ m} \qquad = 0{\cdot}065 \text{ m}$$

$$m_r = 0{\cdot}0072$$

Manning's n,
$$n_r = m_r^{\frac{2}{3}} / X_r^{\frac{1}{2}}$$
$$= (0{\cdot}0072)^{\frac{2}{3}} / (1/2000)^{\frac{1}{2}}$$

$$= \underline{1{\cdot}67}$$

$$t_r = \frac{X_r}{Y_r^{\frac{1}{2}}} = \frac{1}{200}$$

$$t_p = 200\ t_m = 200\ \text{h}$$

Prototype flood will pass in 200 h.

(iii) *Friction is dominant.* Resistance dominates in long, straight, uniform reaches of a channel. For fully turbulent flows in such channels the friction factor is virtually constant and is independent of Reynolds number. Similitude of the flow characteristics can therefore be obtained from Manning's or Darcy–Weisbach's formula. Such a model does not give a complete dynamic similarity since inertia force does not come in and it must not be used unless the solid surface is reasonably rough and the flow is fully turbulent. The main advantage of using a friction formula (apparent from the velocity ratio equation) is that one degree of freedom is gained. For a fixed laboratory flow rate different combinations of the linear scale and friction factor are possible. The transfer equations for velocity, discharge and time for a distorted model are given below.

	Manning	*Darcy–Weisbach*
v_r	$\frac{1}{n_r} m_r^{\frac{2}{3}} \frac{Y_r^{\frac{1}{2}}}{X_r^{\frac{1}{2}}}$	$\left(\frac{1}{f_r}\right)^{\frac{1}{2}} m_r^{\frac{1}{2}} \frac{Y_r^{\frac{1}{2}}}{X_r^{\frac{1}{2}}}$
Q_r	$\frac{1}{n_r} m_r^{\frac{2}{3}} Y_r^{\frac{3}{2}} X_r^{\frac{1}{2}}$	$\left(\frac{1}{f_r}\right)^{\frac{1}{2}} m_r^{\frac{1}{2}} Y_r^{\frac{3}{2}} X_r^{\frac{1}{2}}$
t_r	$\frac{n_r X_r^{\frac{3}{2}}}{m_r^{\frac{2}{3}} Y_r^{\frac{1}{2}}}$	$\frac{f_r^{\frac{1}{2}} X_r^{\frac{3}{2}}}{m_r^{\frac{1}{2}} Y_r^{\frac{1}{2}}}$

SEDIMENT TRANSPORT

Another class of problems in which friction plays a leading role is fluid flow over moveable beds. This is one of the most complex and difficult problems in hydraulic engineering. The shear of a river flowing over a sandy bed may be sufficient to cause the sand particles to move and travel downstream. As the velocity increases during flood, more sand is moved and vice versa. The different rates of movement along the stretch of the river may result in blocking of the channel at a point causing overflowing of banks and accidents and destruction of property and lives. In the absence of a sound theory and reliable empirical relationship (see Chapter 7), the hydraulic engineer's best hope of averting such catastrophes is through model studies.

Notwithstanding the encouraging efforts that have been made in the past two decades or so to promote a better understanding of the mechanics of sediment transport by streams, knowledge is still very inadequate in this field. (See the divergence of different theories as illustrated in Fig. 7.7). With present knowledge, it is possible only to make qualitative prediction of rate of sediment movement. General flow pattern and regions of scour can, however, be skilfully predicted. A fuller discussion of moveable bed models is given in Section 7.4.

5.5 Dimensional Approach to Experimental Analysis

Every physical quantity representing a characteristic property of a body (weight, mass, length, velocity, temperature, etc.) has both magnitude (quantity) and character (quality). The magnitude depends on the system of measurement used whereas the character is uniquely determined by the basic dimensions of length (L), mass (M), time (t) and temperature (T). For example, the distance between two points while having a unique characteristic of length may be 3·0 ft using an Imperial system of measurement, 91·44 cm in the cgs system or 0·9144 m in the mks system. A mathematical equation which relates physical quantities must therefore be correct numerically as well as in character. This demands that the units of all the terms of any physically correct equation must all be the same.

The basic dimensions of a physical quantity may be measured in a variety of units of measurement. Since however, these units differ from one another only by a constant factor, a dimensionally homogeneous equation can be reduced to the same quantitative value for all units of measurement by dividing through by an appropriate combination of the basic dimensions, thus making the equation dimensionless. This concept is based on the fact that if a quantity is a ratio, (such as the slope of a hill or a hydraulic gradient), and if all the basic dimensions disappear, it will be numerically the same irrespective of what (consistent) system of units is used. Take for example the familiar parabolic equation,

$$y = a + bx + cx^2 \tag{5.15}$$

The equation relates a dependent variable (physical quantity) y to the independent variable (physical quantity) x, through the constants a, b and c. Dimensional homogeneity demands that the three elements on the right-hand side of equation (5.15) must have the same dimensions as y on the left-hand side.

If say y has the units of length, the equation as it is will give different quantitative values depending on which system of measurement is employed but will preserve its basic character (dimensions). By dividing through by a physical quantity having the basic dimension of length the equation will become dimensionless (a ratio) and, therefore, will be numerically the same for all consistent units. As a further explanation; y may be considered as representing force in fluid flow and the elements on the right-hand side of equation (5.15) are made up of the physical quantities of the fluid and its kinematics. The equation may be made dimensionless by dividing through by a combination of geometric, kinematic and dynamic properties of the flow having dimensions of force. A commonly used combination of density ρ (dynamic), velocity v (kinematic) and a geometric length, l is $\rho v^2 l^2$.

In very many fluid flow problems the relationship between the various physical quantities is not explicitly known and cannot be determined by mathematical analysis. While dimensional analysis can be used to show the various combinations of the physical quantities which control the flow characteristics, it does not give an explicit combination or mathematical relationship. Experimental data have to be used to find a form of an equation. This constitutes a limitation on the use of dimensional analysis in engineering fluid mechanics.

Suppose a series of data had been collected on the motion of a ship on the surface of water and the investigator wished to derive a dimensionally correct formula to represent them. The force P on the ship may be assumed to be determined by gravity and viscous forces, speed of ship, the roughness of the body of the ship and the size of the ship. The general functional relationship may thus be written as

$$P = f(v, l, g, \rho, \mu, k_s)$$

The function f may be a linear function, a parabolic function of equation (5.15) type, a hyperbolic or any other type of function. Whatever type the function is it can be expanded in a series, and dimensional homogeneity demands that each element (made up from suitable combinations of the physical quantities listed) must have the dimensions of force. That is

$$P = A_1 (v^{a_1} l^{b_1} g^{c_1} \rho^{d_1} \mu^{e_1} k_s^{h_1}) + A_2 (v^{a_2} l^{b_2} g^{c_2} \rho^{d_2} \mu^{e_2} k_s^{h_2}) + \ldots.$$

where $A_1, A_2, \ldots.$ are pure numbers and $a_1, b_1, c_1, d_1, e_1, h_1$, etc. are indices such that the combinations in each set of brackets have the dimension of force (ML/t^2).

Thus

$$\left[\frac{ML}{t^2}\right] \equiv \left[\left(\frac{L}{t}\right)^{a_1} \left(L\right)^{b_1} \left(\frac{L}{t^2}\right)^{c_1} \left(\frac{M}{L^3}\right)^{d_1} \left(\frac{M}{Lt}\right)^{e_1} L^{h_1}\right]$$

where the square brackets indicate identity of dimensions.

M-terms: $\quad 1 = d_1 + e_1, \quad \therefore\ d_1 = 1 - e_1$

t-terms: $\quad -2 = -a_1 - 2c_1 - e_1; \quad \therefore\ a_1 = 2 - 2c_1 - e_1$

L-terms: $\quad 1 = a_1 + b_1 + c_1 - 3d_1 - e_1 + h_1$

$$1 = (2 - 2c_1 - e_1) + b_1 + c_1 - (3 - 3e_1) - e_1 + h_1$$

$$\therefore\ b_1 = 2 + c_1 - e_1 - h_1$$

Similar results are obtained for d_2, a_2 and b_2, etc.

$$\therefore \qquad P = \Sigma A v^a l^b g^c \mu^e \rho^d k_s^h$$

$$= \Sigma A(v^2 l^2 \rho)(v^{-2c-e} l^{c-e-h} g^c \rho^{-e} \mu^e k_s^h)$$

or
$$\frac{P}{\rho v^2 l^2} = \Sigma A\left(\frac{lg}{v^2}\right)^c \left(\frac{\mu}{\rho v l}\right)^e \left(\frac{k_s}{l}\right)^h = \Sigma A\left(\frac{1}{F}\right)^c \left(\frac{1}{R}\right)^e \left(\frac{k_s}{l}\right)^h$$

where Reynolds number
$$R = \frac{\rho v l}{\mu}$$

and Froudes number
$$F = \frac{v^2}{lg}$$

Thus the force on the ship

$$P = \rho v^2 l^2 \phi(R, F, k_s/l) \tag{5.16}$$

where ϕ is some function.

The actual function has to be determined from experimental results. For surface ships the contribution from Froude number (gravity effect) can be quite significant. For deeply submerged bodies, however, the Reynolds number term (viscous effect) dominates. Equation (5.16) is generally written in terms of a coefficient of drag $C_D = 2\phi(R, F, k_s/l)$ as

$$\text{Drag, } P = \frac{1}{2} C_D \rho v^2 l^2 \tag{5.17}$$

FLOW OVER NOTCHES

As a further example, the discharge over a V-notch will be established as a function of the fluid properties and geometrical characteristics of the notch.

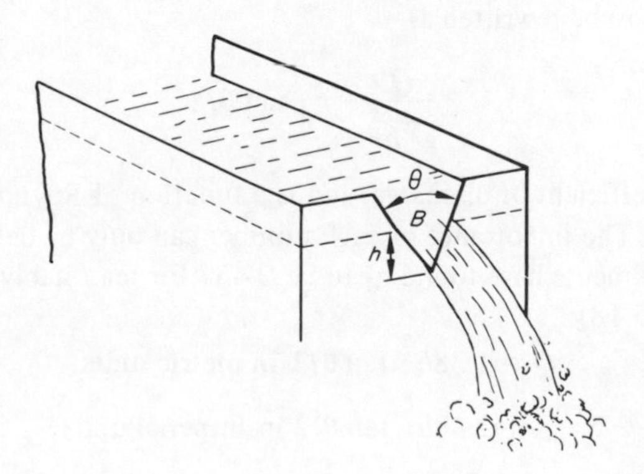

Fig. 5.6 Flow over a V-notch

With reference to Fig. 5.6, the factors which will most likely influence the discharge Q over a V-notch may be listed as: h, head of fluid above the vertex of notch; ρ, density of the fluid; μ, viscosity of the fluid; σ, surface tension of the fluid; g, gravitational acceleration; and B, width of water surface over the notch.

Thus

$$Q = f(h, B, \rho, \mu, \sigma, g)$$

or

$$Q = \Sigma A(h^a\, B^b\, \rho^c\, \mu^d\, \sigma^e\, g^f)$$

where A is a pure number and the indices $a, b, c \ldots f$, are such that the terms on the right-hand side of the equation have dimensions of discharge.

$$\therefore \qquad \left[\frac{L^3}{t}\right] = \left[(L)^a (L)^b \left(\frac{M}{L^3}\right)^c \left(\frac{M}{Lt}\right)^d \left(\frac{M}{t^2}\right)^e \left(\frac{L}{t^2}\right)^f\right]$$

M-terms: $\quad 0 = c + d + e, \quad \therefore\ c = -d - e$

t-terms: $\quad -1 = -d - 2e - 2f; \quad \therefore\ f = \frac{1}{2} - d/2 - e$

L-terms: $\quad 3 = a + b - 3c - d + f; \quad \therefore\ a = 5/2 - b - 3d/2 - 2e$

Thus

$$Q = \Sigma A(h^{\frac{5}{2}} g^{\frac{1}{2}}) \left(\frac{B}{h}\right)^b \left(\frac{\mu}{\rho h^{\frac{3}{2}} g^{\frac{1}{2}}}\right)^d \left(\frac{\sigma}{\rho g h^2}\right)^e$$

$$\therefore \qquad \frac{Q}{g^{\frac{1}{2}} h^{\frac{5}{2}}} = \phi\left(\frac{B}{h},\ \frac{\rho (gh)^{\frac{1}{2}} h}{\mu},\ \frac{\rho (gh) h}{\sigma}\right)$$

The first term of the function ϕ is a statement of geometric similarity ($B/h = 2 \tan \theta/2$); the second term is a form of Reynolds number and the last term is the Weber number since in both cases $(gh)^{\frac{1}{2}}$ is proportional to velocity. The equation may be rewritten as

$$\frac{Q}{g^{\frac{1}{2}} h^{\frac{5}{2}}} = c_{\mathrm{d}} \tan \theta/2 \tag{5.18}$$

where c_{d} is a coefficient of discharge, and is a function of Reynolds and Weber numbers. The importance of each number can only be determined experimentally. Experiments have found c_{d} to be 0·437 for reasonably large values of h, giving from (5.18)

$$Q = 1{\cdot}38 h^{\frac{5}{2}} \tan \theta/2 \text{ in metric units} \tag{5.19}$$

$$= 2{\cdot}48 h^{\frac{5}{2}} \tan \theta/2 \text{ in Imperial units.}$$

If $\theta = 90°, Q = 1{\cdot}38 h^{\frac{5}{2}}$ or $2{\cdot}48 h^{\frac{5}{2}}$ in Imperial units

and if $\theta = 60°, Q = 0{\cdot}79 h^{\frac{5}{2}}$ or $1{\cdot}43 h^{\frac{5}{2}}$ in Imperial units

The method for establishing dimensional homogeneity by the process of indices as outlined above is often referred to as the Rayleigh method. A much quicker and therefore more convenient method of establishing dimensionless quantities is attributed to Buckingham and is illustrated below.

THE BUCKINGHAM π-THEOREM

The π-theorem attributed to E. Buckingham (1915) states that in a physical problem involving n quantities in which there are k fundamental dimensions, the n quantities may be arranged into $(n - k)$ independent dimensionless parameters. Let $q_1, q_2 \ldots q_n$ be n quantities involved in a physical problem and let them be connected by the function $f(q_1, q_2, \ldots q_n) = 0$. The π-theorem states that if the q's contain k fundamental dimensions, then the function may be replaced by a new function $\phi(\pi_1, \pi_2 \ldots \pi_{n-k})$, where the π's are independent dimensionless combinations of the q's.

The procedure for determining the π's is as follows.

(1) List the various variable quantities q and their dimensions. Do not list any quantities that are functions of the other quantities, e.g. list length l with either velocity v or discharge Q, but not with both.

(2) Note the number of fundamental dimensions k. For mechanical problems these will be L, M, t or L, P (force), t and if heat flow is involved L, M, t and T (temperature).

(3) Select k of the quantities q such that the k fundamental dimensions are appropriately involved.

(4) Form the $(n - k)$ π's by combining the remaining q's with the quantities selected.

(5) If convenient the π's can be rearranged by multiplying or dividing, e.g. we can replace π_3 by $\pi_3 \,.\, \pi_6/\pi_2$.

The procedure is best illustrated by examples.

EXAMPLE 5.7

Establish the equation for a discharge over a V-notch by the Buckingham π-theorem.

As listed above the discharge Q may be a function of h, B, ρ, μ, σ, and g. Thus

$$f(Q, h, B, \rho, \mu, \sigma, g) = 0$$

There are 7 quantities (q's) and 3 fundamental dimensions (L, M, t) and therefore 4 π's. Thus the function may be replaced by

$$\phi(\pi_1, \pi_2, \pi_3, \pi_4) = 0$$

Selecting h, ρ and Q (implying L, M and t) the π's are given as

$$\begin{aligned} \pi_1 &= h^a \rho^b Q^c B \\ \pi_2 &= h^a \rho^b Q^c \mu \\ \pi_3 &= h^a \rho^b Q^c \sigma \\ \pi_4 &= h^a \rho^b Q^c g \end{aligned}$$

in which the indices a, b and c are such that the π's are dimensionless.

$$\left[\pi_1\right] = \left[L^a \left(\frac{M}{L^3}\right)^b \left(\frac{L^3}{t}\right)^c L\right] \equiv 0$$

$\therefore$ $b = 0$, $c = 0$ since there are no M's and t's, and $a + 1 = 0$, $\therefore$ $a = -1$

$$\therefore \qquad \pi_1 = B/h$$

$$[\pi_2] = \left[L^a \frac{M^b}{L^{3b}} \frac{L^{3c}}{t^c} \frac{M}{Lt}\right] \equiv 0$$

M-terms: $\quad b + 1 = 0; \quad \therefore b = -1$

t-terms: $\quad c + 1 = 0; \quad \therefore c = -1$

L-terms: $\quad a - 3b + 3c - 1 = 0, \quad \therefore a = 1$

$$\therefore \qquad \pi_2 = \frac{h\mu}{\rho Q} = \frac{\mu}{h\rho(Q/h^2)}$$

Similarly,

$$\pi_3 = \frac{h^3\sigma}{\rho Q^2} = \frac{\sigma}{h\rho(Q/h^2)^2}$$

and

$$\pi_4 = \frac{h^5 g}{Q^2} = \frac{gh}{(Q/h^2)^2}$$

Thus

$$\phi\left[\frac{B}{h}, \frac{\mu}{h\rho(Q/h^2)}, \frac{\sigma}{h\rho(Q/h^2)^2}, \frac{gh}{(Q/h^2)^2}\right] = 0$$

or

$$\frac{Q}{g^{\frac{1}{2}} h^{\frac{5}{2}}} = \phi_1\left(\frac{B}{h}, R, W\right) \text{ since } \frac{Q}{h^2} \text{ has units of velocity}$$

or

$$\frac{Q}{g^{\frac{1}{2}} h^{\frac{5}{2}}} = c_{\text{d}} \tan\frac{\theta}{2}$$

where again the coefficient of discharge is shown to be a function of Reynolds and Weber numbers.

EXAMPLE 5.8

When liquid flows down a vertical pipe and displaces a gas which moves up the same pipe, as shown in the figure, the gas moves up the pipe as bullet-shaped, nearly identical bubbles with a nearly constant spacing between adjacent bubbles. The upward velocity of the gas bubbles is also nearly constant along the length of the pipe.

(1) Make a dimensional analysis for the system shown in the figure and derive a complete set of independent dimensionless parameters. Assume the pipe is long enough that end effects are negligible and that the liquid level in the upper tank is constant and much smaller than the pipe length. Conditions are steady.

(2) If an arrangement similar to the one in the figure is proposed for the study of the influence of air bubbles on the coefficient of discharge c_{d}, suggest a plot of the experimental results. (U.S.T., Part III Mech., 1967.)

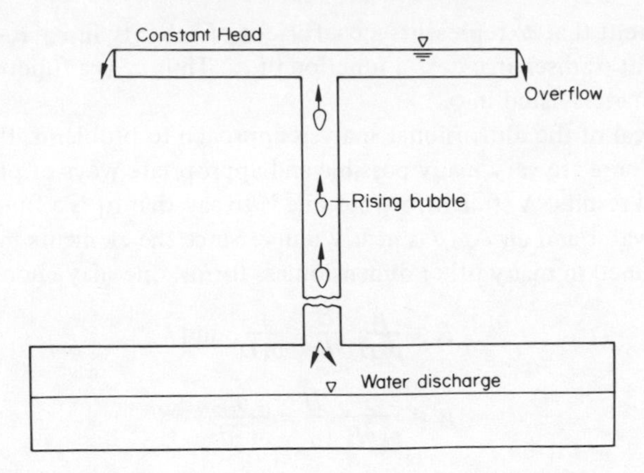

SOLUTION

The factors which might affect flow conditions are: pipe length, $H(L)$; pipe diameter, $D(L)$; size of bubble, $d(L)$, spacing of bubbles, $s(L)$; density of liquid, $\rho(M/L^3)$; density difference between liquid and gas, $\Delta\rho(M/L^3)$; surface tension $\sigma(M/t^2)$; viscosity of liquid, $\mu(M/Lt)$; velocity of liquid flow, $v(L/t)$; velocity of bubble, $v_b(L/t)$ and gravity, $g(L/t^2)$.

Thus

$$f(H, D, d, s, \rho, \Delta\rho, \sigma, \mu, v, v_b, g) = 0$$

or

$$\phi(\pi_1, \pi_2, \pi_3, \ldots \pi_8) = 0$$

There are eleven q's and three fundamental dimensions. Choose the k quantities, ρ, v and H

$$\pi_1 = \rho^a v^b H^c D = D/H$$
$$\pi_2 = \rho^a v^b H^c d = d/H$$
$$\pi_3 = \rho^a v^b H^c s = s/H$$
$$\pi_4 = \rho^a v^b H^c \Delta\rho = \Delta\rho/\rho$$
$$\pi_5 = \rho^a v^b H^c \sigma = \frac{\sigma}{\rho v^2 H} \equiv \text{Weber number}, W$$
$$\pi_6 = \rho^a v^b H^c \mu = \frac{\mu}{\rho v H} \equiv \text{Reynolds number}, R$$
$$\pi_7 = \rho^a v^b H^c v_b = v_b/v$$
$$\pi_8 = \rho^a v^b H^c g = \frac{gH}{v^2} \equiv \text{Froude number}, F$$

$\therefore$

$$\phi(D/H, d/H, s/H, \Delta\rho/\rho, W, v_b/v, R, F) = 0$$

or

$$\frac{v^2}{gH} = \phi'(D/H, d/H, s/H, \Delta\rho/\rho, v_b/v, W, R)$$

It is apparent that ϕ' represents a coefficient of velocity in $v = c_v\sqrt{(gH)}$. But the coefficient of discharge c_d is a function of c_v. Thus c_d is a function of the various parameters listed in ϕ'.

As is typical of the dimensional analysis approach to problems, the function is undefined. There are very many possible and appropriate ways of plotting the experimental results. A straightforward one is to say that c_d is a function of R and W. For water and air $\Delta\rho/\rho$ is nearly unity. Since the elements in the function can be combined in many other dimensionless forms, one may choose

$$R = \frac{\mu}{\rho v H}\,\frac{H}{D} = \frac{\mu}{\rho v D} \quad \text{and}$$

$$W = \frac{\sigma}{\rho v^2 H}\,\frac{H}{d} = \frac{\sigma}{\rho v^2 d}$$

Assuming s/H is very small and v_b/v has a minor effect only,

$$c_d = \phi''\left(\frac{\mu}{\rho v D}, \frac{\sigma}{\rho v^2 d}\right)$$

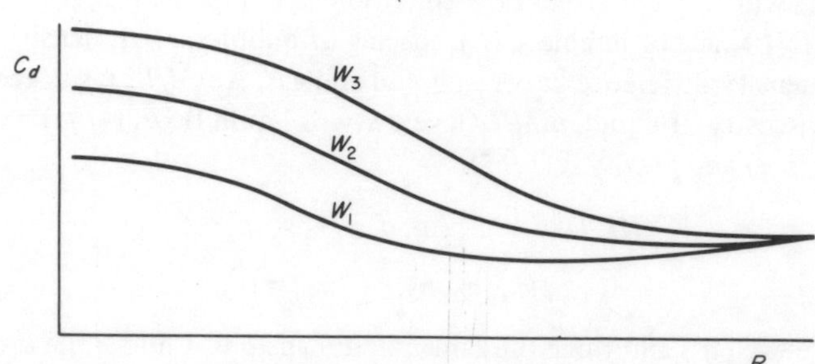

A suggested plot is shown in the sketch.

(Note. See Appendix on p. 401 for notes on flow measurement)

FURTHER READING

Allen, J., *Scale Models in Hydraulic Engineering,* Longmans.

Buckingham, E., 'Model Experiments and the Form of Empirical Equations'; *Trans, A.S.M.E.,* Vol. 37, 1915.

The Committee of the Hydraulics Division on Hydraulic Research, 1942, A.S.C.E. *Hydraulic Models.*

Daily, J.W. and Harleman, D.R.F., *Fluid Dynamics,* Addison–Wesley, U.S.A.

Dake, J.M.K. and Harleman, D.R.F., *An Analytical and Experimental Investigation of Thermal Stratification in Lakes and Ponds,* M.I.T. Hydro. Lab. Report No. 99, 1966.

Francis, J.R.D., *A Textbook of Fluid Mechanics for Engineering Students,* Edward Arnold.

Rouse, H. (ed.), *Engineering Hydraulics,* John Wiley and Sons, New York, London.

6 Water Pumps and Turbines

6.1 Introduction

A device which brings about an exchange of energy between a mechanical system and a fluid medium is a fluid machine. If the machine is driven mechanically to work on the fluid system and thereby transform mechanical energy into 'fluid' energy it is referred to as a *pump.* However, the word *pump* is generally applied to machines working on liquids only. Pumps which drive gases at low pressures are usually called *fans* and those which drive gases at high pressures are called *compressors.* The reverse of a pump, a device extracting energy from a fluid system and converting it into mechanical energy, is called a *turbine.* Like a pump, a turbine may be driven by liquid energy or gas energy giving a hydraulic turbine or a gas turbine.

The mechanical power for driving a pump may come from electrical power through a motor, a diesel motor, a gas or steam motor or any other source. Similarly, the mechanical energy extracted from the fluid by a turbine may subsequently be converted into electrical energy through a generator or used directly to drive another machine or do some other form of work.

The components of fluid machines and their arrangement vary greatly. Accordingly there is a large number of types of fluid machines. They may broadly be categorized as positive displacement machines and turbo or rotodynamic machines (see Fig. 6.1). In a positive displacement machine, energy is transmitted by virtue of the work done as a fluid under pressure positively displaces or is displaced by an element moving in a fixed and closely fitting case. Typical examples are the reciprocating pump and rotary machines.

Turbo or rotodynamic machines have elements known as blades, vanes or buckets rotating around an axis which is fixed in space. At some stage in the movement of a fluid particle through a rotodynamic machine it has a substantial velocity component tangential to a circle in a plane normal to the axis of the shaft of the rotor (the moving blades) of the machine. In pumps, fans and compressors the moving element forces the fluid to acquire the tangential or

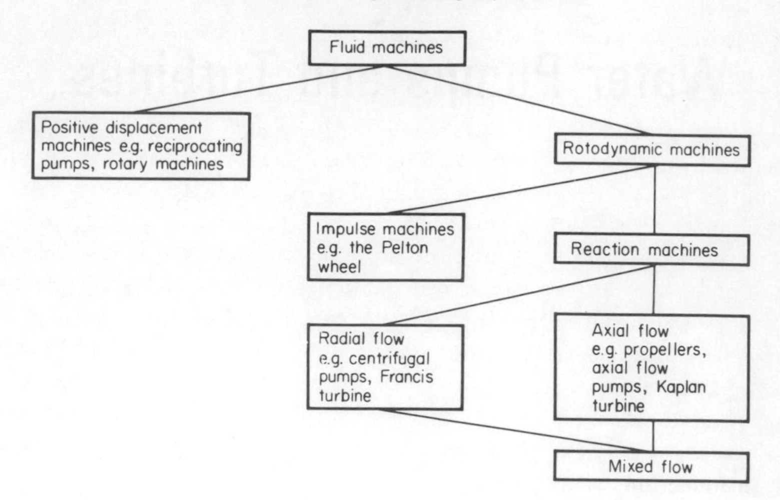

Fig. 6.1 Classification of fluid machines

whirl component of motion whereas the whirl of the fluid mass in a turbine forces its rotor to rotate and to continue to rotate. The whirl component of motion of the fluid, therefore, is related to the torque on the shaft of the machine.

There are many different arrangements for effecting the exchange of whirl motion in a rotodynamic machine. An *impulse machine*, for example the Pelton wheel, brings about the exchange through an impulsive action. A jet (or jets) of fluid is made to impinge on the blades (buckets) of the rotor and thereby force the rotor to move by an impulsive reaction. In other types of rotodynamic machines, for example the centrifugal pump, the Francis turbine or the Kaplan turbine, the transfer is effected by means of a 'smooth' interaction between the fluid and the blades or impeller of the machine. These are known as *reaction machines.* If the machine is designed such that the energy transfer takes place while the fluid moves principally radially through the impeller, it is referred to as a *radial flow* reaction machine. On the other hand, if the energy transfer takes place while the fluid flows principally in the axial direction, it is referred to as an *axial flow* reaction machine. The centrifugal pump and the Francis turbine are examples of radial flow reaction machines and propellers, the Kaplan turbine or axial flow pumps are axial flow reaction machines. *Mixed flow* machines combine features of both types and provide energy exchange while the fluid flows partially radially and partially axially. This results in the partial achievement of combined characteristics of radial flow (high head) and axial flow (high flow) machines.

One of the most interesting features of hydrodynamic machines is that their role is generally reversible, that is, many pumps can be used as turbines and vice versa. Pumps are designed so as to interrupt whirl imposed on the moving fluid after it leaves the impeller by a stationary part of the machine so that the resulting deceleration leads to an increased fluid pressure. In a turbine, a stationary part of the machine is designed to convert the fluid pressure of the oncoming fluid into a whirl motion before it enters the rotor. The rotor then removes the whirl and consequently acquires a torque which makes it rotate. Thus the design of pumps and motors does not lead easily to a reversal of roles and the efficiency of the reversed machine is generally very low.

Because of the identity of the basic principles on which the design and operation of pumps and turbines are based their treatment in the following pages will be as unified as possible. The characteristic performance of these machines will be explained and the considerations which lead to their selection, installation and use emphasized. Details for design and construction are omitted. The chapter is concentrated mainly on water pumps and turbines of the rotodynamic type, but the reciprocating pump is covered superficially in Section 6.6 which deals with pumping from wells.

6.2 The Pelton Wheel Turbine

The Pelton wheel, an impulsive turbine, is the simplest type of rotodynamic machine. It consists principally of a series of buckets mounted uniformly around a rigid circular frame on a rotating shaft (see Fig. 6.2). A jet (or jets) issues out of a nozzle (fed from a high head of water) tangentially to the mean circle of the runner and impinges on the system of buckets. The buckets are so designed that the jet of water splits into two parts and leaves the bucket deflected through an almost 180° angle. The impulsive reaction between the impinging jet and the

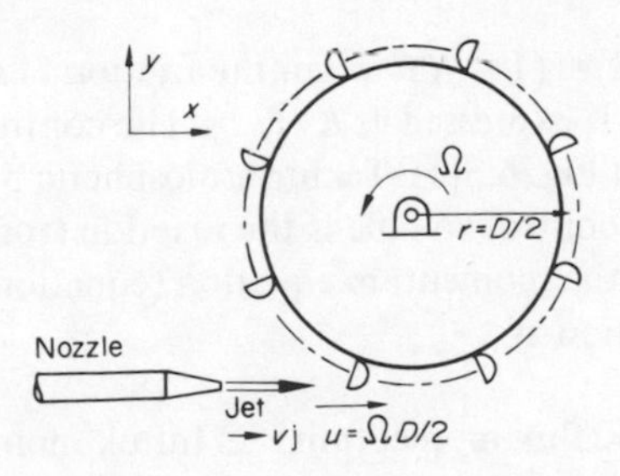

Fig. 6.2

bucket produces a torque on the shaft of the runner which makes the runner rotate and continue to rotate so long as there is a continuous jet stream coming and striking the buckets.

Let the velocity of the jet as it approaches the bucket be v_j ($= C_v \sqrt{(2gH)}$), where H is the head behind the nozzle, see equation (3.27). A stationary observer outside would see the fluid system as it strikes the moving bucket and is divided and deflected as in Fig. 6.3(a). In analysis, however, interest is in the reaction between the jet and the moving bucket. Therefore, the relative motion between the two as depicted in Fig. 6.3(b) is more relevant. The approach velocity of the jet relative to the bucket is $v_1 = v_j - u$, where u is the tangential bucket speed. The absolute velocity of the outgoing jet is indicated in the velocity diagram of Fig. 6.3 in which v_2 is the outgoing velocity relative to the bucket.

Assuming atmospheric pressure inside the turbine casing and neglecting gravity effects, Bernoulli's equation can be used to show that the outgoing jet

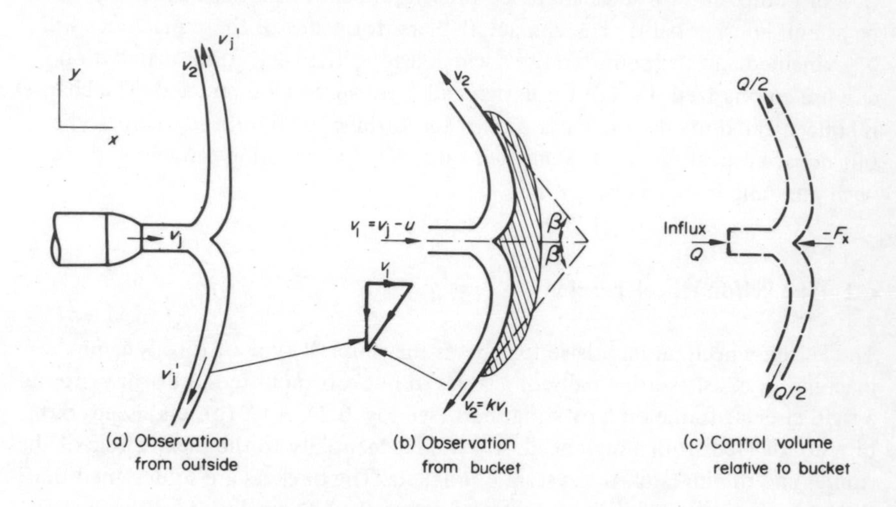

Fig. 6.3

velocity $v_2 = kv_1$ where $k = (1 - K)^{\frac{1}{2}}$ when the friction head loss between the bucket and the water jet is expressed as $Kv_1^2/2g$. The control volume relative to the bucket is as shown in Fig. 6.3(c). Taking atmospheric pressure as datum, the only force acting on the control volume is the reaction from the bucket $-F_x$. If the oncoming flow is Q, the momentum equation (equation 2.10) can readily be applied in the x-direction as

$$-F_x = \Sigma \text{ Efflux momentum} - \Sigma \text{ Influx momentum}$$

Thus

$$-F_x = -2\left(\rho \frac{Q}{2} v_2 \cos\beta\right) - \rho Q v_1 \tag{6.1}$$

where the angle β is shown in the figure and ρ is the fluid density.

Substituting for v_2 and v_1 the active force exerted on the bucket by the impinging jet

$$F_x = \rho Q(v_j - u)(1 + k\cos\beta) \tag{6.2}$$

The torque on the shaft is given by

$$T = \frac{1}{2}\rho QD(v_j - u)(1 + k\cos\beta) \tag{6.3}$$

and the power

$$P = T\Omega = \frac{1}{2}\rho QD\,\Omega\,(v_j - u)(1 + k\cos\beta) \tag{6.4}$$

where D is the mean diameter of the wheel and Ω is the speed of rotation with $u = \Omega D/2$.

Substituting $a_j v_j$ for the discharge Q (a_j is the cross sectional area of the on-coming jet), equation (6.4) can be transformed into

$$P = \rho\, a_j v_j^3 \frac{u}{v_j}\left(1 - \frac{u}{v_j}\right)(1 + k\cos\beta) \tag{6.5}$$

The corresponding hydraulic efficiency is given by

$$\eta = \frac{P}{\rho\, v_j^3 a_j/2} = 2\frac{u}{v_j}\left(1 - \frac{u}{v_j}\right)(1 + k\cos\beta) \tag{6.6}$$

Both equations (6.5) and (6.6) are parabolic for a fixed discharge from the nozzle. They predict that the maximum power transfer between the jet and the machine and the maximum efficiency occur when the bucket speed is half the jet speed ($u/v_j = 1/2$). The predicted maximum efficiency is $1/2\,(1 + k\cos\beta)$. However, due to mechanical losses at the shaft bearings and due to rotating parts (windage) and shock losses, the output power from the turbine is lower than that predicted by equation (6.5) and the maximum occurs when u/v_j is between 0·46 and 0·48. For the same reasons the actual power developed by a Pelton wheel and the corresponding efficiency are generally not a true parabolic curve (see Fig. 6.4).

The ideal deflection angle for a Pelton wheel bucket is 180° (i.e. $\beta = 0$) but for practical reasons the bucket angle generally chosen is 165° ($\beta = 15°$). The value of k must be as near unity as possible, a high polish of the bucket surface is essential, and the bucket is made as small as possible to reduce resistance to flow. The number of buckets must be such as to ensure a smooth interception of the jet for all points on the wheel.

It is possible to have more than one jet operating on a Pelton wheel and two jets are quite common. Ideally the power developed is directly proportional to the number of jets. However, because the water from one jet tends to interfere with the water from another, resulting in reduced efficiency, we now steer away from multi-jet Pelton wheels.

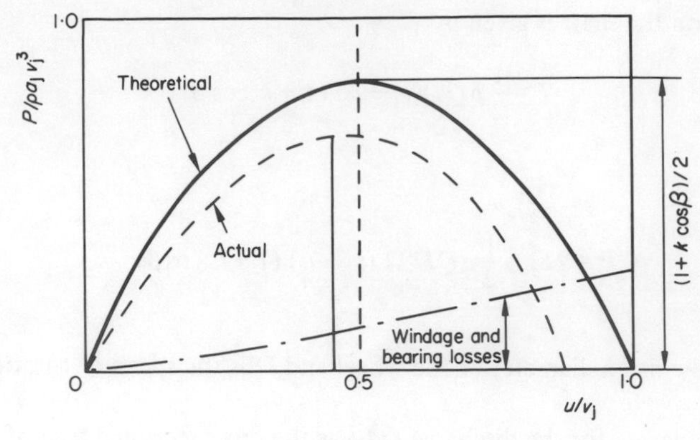

Fig. 6.4

The mechanics of converting a high head of water into a high speed jet were discussed in Chapter 3. It was demonstrated that maximum power transmission through a nozzle occurs when the total pipe loss equals a third of the available head. The jet diameter d_j which gives the maximum power efflux from the nozzle is determined from equation (3.44) as

$$\frac{d_j}{D_p} = \left(\frac{1 + K_n}{1 + 8fL/D_p}\right)^{\frac{1}{4}}$$

where D_p is the pipe (penstock) diameter.

From power consideration ($P \propto QH$) it is obvious that a low H necessitates large Q for any significant power to be developed by a Pelton wheel. However, if the available head is small the efflux velocity from the nozzle will also be small and a large nozzle area is required to obtain a significant discharge. This places practical limitations on Pelton wheels which are therefore used only when very high heads are available.

6.3 Reaction Machines

6.3.1 Radial Flow and Axial Flow Types

Figure 6.5 shows sectional drawings of two typical reaction turbines (a) a Francis turbine, which is a radial flow type and (b) a Kaplan turbine which is an axial flow type. Water flows through the spiral and reducing conduit (known as the scroll case) onto the guide vanes which are adjusted to direct flow onto the runner blades at an angle which gives the optimum swirl and reduces separation and shock. As the water flows over the runner blades the swirl is gradually

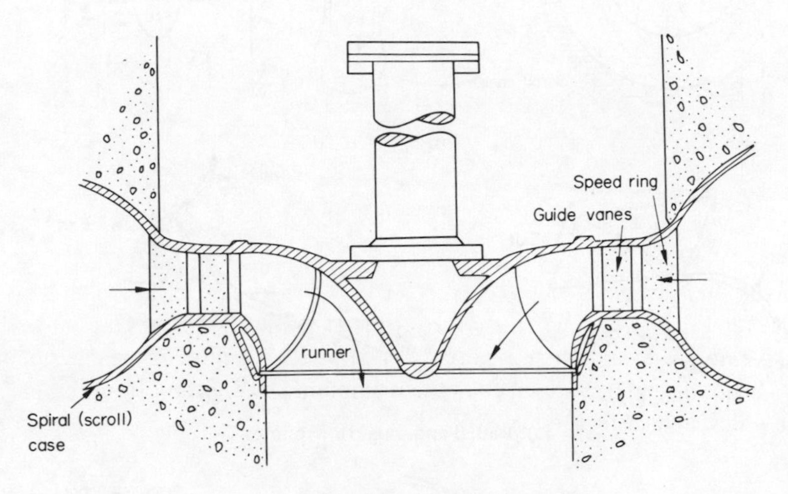

Fig. 6.5(a) Francis turbine $N_s \simeq 129$

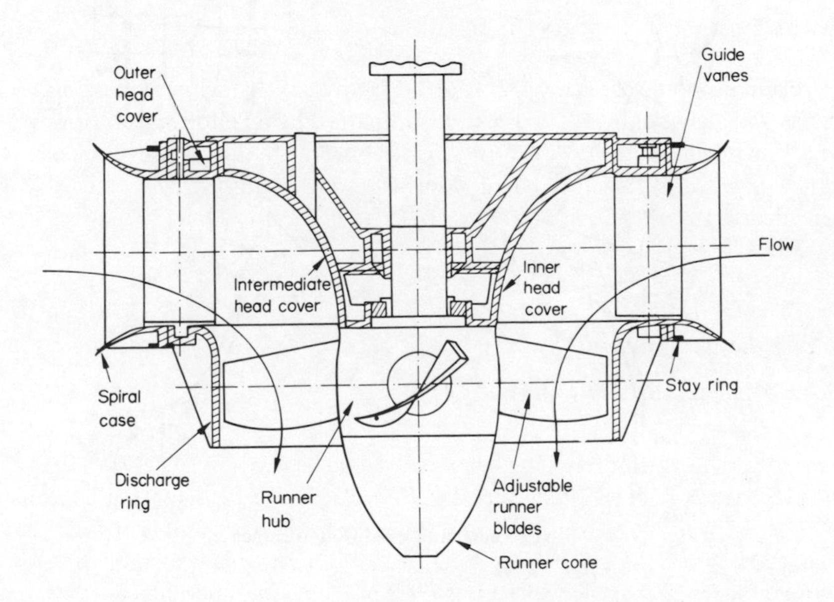

Fig. 6.5(b) Kaplan turbine $N_s \simeq 570$

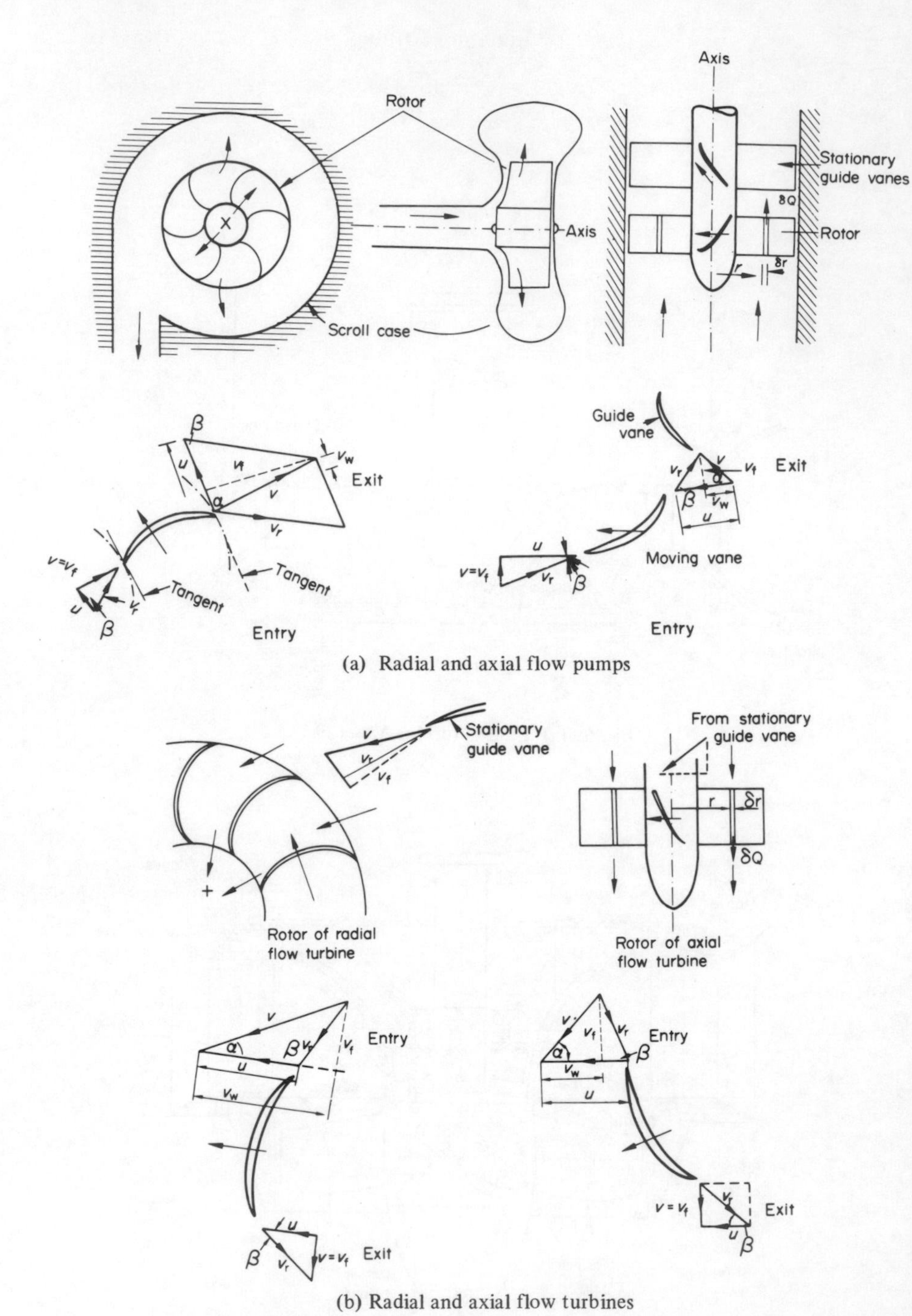

(a) Radial and axial flow pumps

(b) Radial and axial flow turbines

Fig 6.6

reduced until it exists virtually without a swirl. The curved and twisted nature of the runner blades facilities this conversion action.

The reverse action takes place in the corresponding centrifugal and axial flow pumps. The fluid enters in the direction opposite to that shown for the turbine, flows over the runner blades and exits into the scroll case and into a discharge pipe connected to the scroll case. In flowing over the blades, the mechanical effort of the rotor churns the water and makes it acquire a whirl motion. The whirl velocity is subsequently converted into pressure energy in the scroll case. Many modern radial flow pumps have the flow guide action incorporated into the runner blade design and therefore do not have separate guide vanes. Fig. 6.6 shows the diagrammatic arrangement for the impeller for radial and axial flow turbines and pumps and their corresponding velocity diagrams.

6.3.2 Performance Characteristics of Reaction Machines

With reference to Fig. 6.6 the dynamic equations for reaction machines can easily be derived. The water enters or leaves the machine across the inner and outer circles of mounting of the impeller blades in the case of a radial flow machine. The absolute velocity with which the fluid particle leaves or enters the tip of the impeller of a radial flow machine can be compounded from the velocity v_r of the particle relative to the tip of the blade and the velocity u of the tip of the blade (tangential to the peripheral circle). The absolute velocity can also be resolved radially and tangentially. The radial component v_f gives the flow velocity and the tangential component v_w the whirl velocity. In the case of the axial flow machine the water flows axially in the space between the hub on which the blades are mounted and the outside casing (the periphery of the blade tips). Nevertheless the absolute velocity of the fluid particle at the entry and exit ends of blade is made up of its relative velocity to the blade tip plus the speed due to angular rotation of the blade tip. In resolving the absolute velocity into flow and whirl components, however, it is resolved axially and tangentially to the circle in a plane normal to the axis of rotation.

The dynamic behaviour using a rotating finite control volume is governed by equation (2.21)

$$\underset{\text{External torque}}{T_e} = \underset{\text{Efflux angular momentum}}{\Sigma E_{MM}} - \underset{\text{Influx angular momentum}}{\Sigma I_{MM}} \tag{6.7}$$

In a radial flow machine the appropriate control volume is bounded by the inner and outer runner circles and the walls defining the width of the blades (normal to the plane of paper). The discharge Q is the same through both circles. Denoting efflux conditions by 2 and influx conditions by 1,

$$\Sigma E_{MM} = \rho Q v_{w2} r_2 \quad \text{and} \quad \Sigma I_{MM} = \rho Q v_{w1} r_1$$

where r is radius of the circle. If the absolute velocity makes an angle α with the tangent to the circle, $v_{w1} = v_1 \cos \alpha_1$, and $v_{w2} = v_2 \cos \alpha_2$. Substituting into equation (6.7), the external torque on the fluid system is given by

$$T_e = \rho Q\,(r_2 v_2 \cos\alpha_2 - r_1 v_1 \cos\alpha_1) \tag{6.8}$$

or

$$T_e = \frac{\rho Q}{\Omega}\,(u_2 v_2 \cos\alpha_2 - u_1 v_1 \cos\alpha_1) \tag{6.9}$$

where Ω is the angular speed of the rotor and $u = \Omega r$.

In an axial flow machine, the control volume is defined by the hub of the runner, the cylindrical space confining the extremities of the blades and the end planes defining the width of the blades. Taking an elemental lamina of width δr at a radial distance r from the axis and through which discharge δQ passes (see Fig. 6.6)

$$\Sigma E_{MM} = \int \rho dQ\, v_{w2}\, r_2 = \rho \int r_2 v_2 \cos\alpha_2\, dQ$$

and

$$\Sigma I_{MM} = \int \rho dQ\, v_{w1}\, r_1 = \rho \int r_1 v_1 \cos\alpha_1\, dQ$$

It is generally assumed that an irrotational flow condition exists in the impeller space and therefore it can be taken that the total head at each section (i.e. on a plane normal to the axis of rotation) is the same for all streamlines. The product of whirl velocity and radius may be assumed constant i.e. $v_w r$ = constant. Thus

$$\Sigma E_{MM} = \rho Q v_{w2}\, r_2 \text{ and } I_{MM} = \rho Q v_{w1}\, r_1$$

Substitution into equation (6.7) yields the same results for axial flow machines as are obtained for radial flow machines.

In practice as shown in Fig. 6.6 the machines are designed such that at the blade entry of a pump $\alpha_1 = 90°$ and correspondingly $\alpha_2 = 90°$ at the blade exit for a turbine. Equation (6.9) therefore reduces to

$$T_e = \rho \frac{Q}{\Omega}\, u_2 v_2 \cos\alpha_2, \text{ for a pump} \tag{6.10a}$$

and

$$T_e = -\frac{\rho Q}{\Omega}\, u_1 v_1 \cos\alpha_1, \text{ for a turbine} \tag{6.10b}$$

The signs show that work is done on the fluid in a pump while work is extracted from the fluid in a turbine. The power exchange between the fluid and the machine shaft in each case is given by (dropping the subscripts)

$$P_r = T_e\,\Omega = \pm\,\rho Q u v \cos\alpha \tag{6.11}$$

Using $P_r = \rho g Q H_r$, where H_r is the theoretical head developed by a pump or consumed by a turbine

$$H_r = \pm\, uv \cos \alpha/g \tag{6.12}$$

The theory presented above was first introduced by Euler after whom it is commonly named.

For pumps, the actual output head is lower than the theoretical head given by equation (6.12) because some energy is lost through friction, flow separation, shock, leakage and other sources. The actual head output from the pump is thus $H = H_r - H_L$ or $\eta_h H_r$ where η_h is known as the hydraulic efficiency and is given by

$$\eta_h = (1 - H_L/H_r)$$

where H_L is total head loss through flow conditions. On the contrary, a turbine consumes more water energy than it transmits to its shaft because of various hydraulic losses. The head absorbed $H = H_r + H_L$ or H_r/η_h where the hydraulic efficiency $\eta_h = 1/(1 + H_L/H_r)$. Bearing friction, poor transmission and moving parts contribute further to energy degradation known as mechanical losses. Thus a pump can develop a useful head H_p only if an equivalent energy of H_p/η is supplied by its motor. Similary if water head H is available a turbine can only supply a useful head H_T to its shaft given by ηH. In both cases, the total efficiency $\eta = \eta_h \times \eta_m$ where η_m is the mechanical efficiency.

In terms of motor power

$$P = \frac{\rho g Q H p}{\eta} \tag{6.13}$$

for a pump delivering a flow Q against a head H_p.

And

$$P = \eta \rho g Q H \tag{6.14}$$

for a turbine putting out P with a flow of Q under a pressure head H.

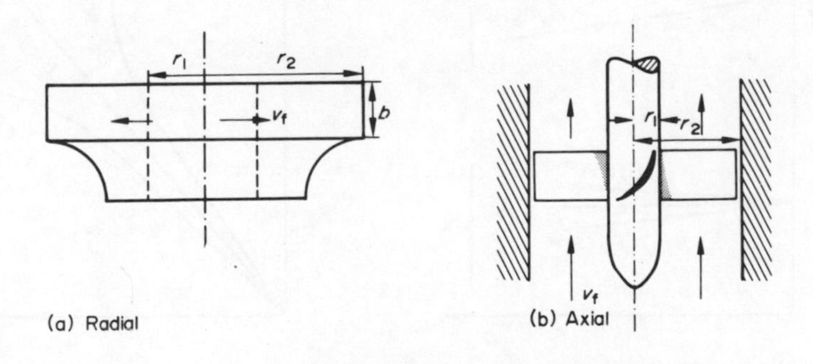

Fig. 6.7 Impellers for radial flow and axial flow machines

The discharge Q through a machine is given by the appropriate flow velocity v_f multiplied by the area of flow. The area of flow for a radial flow machine is $k\ 2\pi rb$, where b is the width of the impeller blades (see Fig. 6.7(a)). The area of flow for an axial flow machine is $k\pi(r_2^2 - r_1^2)$ where r_1 and r_2 are the radii of the hub and the impeller respectively (see Fig. 6.7(b)). In each case the factor k, less than unity is included to account for the thickness of the impeller blades.

Thus
$$Q = kv_{f1}\ 2\pi r_1 b = kv_{f2}\ 2\pi r_2 b \tag{6.15a}$$
for a radial flow machine

or
$$Q = k\pi(r_2^2 - r_1^2)\ v_f \tag{6.15b}$$
for an axial flow machine

In Fig. 6.6(a) the blade tip makes an angle β_2 with the tangent to the circle at exit Thus the relative velocity is inclined at angle β_2 to the tangential velocity u_2. From geometrical considerations,

$$v_2 \cos \alpha_2 = v_{w2} = u_2 - v_{f2} \cot \beta_2$$

Substituting into equation (6.12) gives the pump head as

$$H_r = \frac{u_2^2}{g} - \frac{u_2}{g} v_{f2} \cot \beta_2 = \frac{u_2^2}{g} - \left(\frac{u_2}{g} \cot \beta_2\right) \frac{Q}{\text{constant}} \tag{6.16}$$

where the constant is given by equation (6.15 a or b). The angle β_2 is fixed by design. Thus for a pump operating at a constant speed of rotation N the head developed must vary linearly with the discharge according to equation (6.16). The slope of the line depends on the value of β_2. This is illustrated in Fig. 6.8(a). Different sets of lines are obtained for different speeds of rotation. The corresponding power varies parabolically with discharge (see Fig. 6.8(b)) since

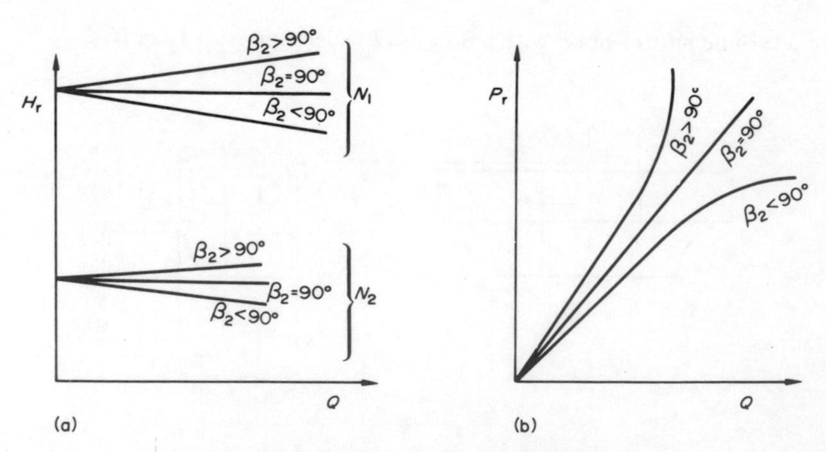

Fig. 6.8 Ideal performance curves for reaction pumps

$P_r \propto QH_r$. In practice in order to limit power demand by a pump β_2 is made less than 90°. The various losses referred to above modify these curves considerably as illustrated by the curves in Fig. 6.9 which are typical for centrifugal pumps.

Our principal interest is in the power developed by a turbine. The characteristics of turbines are conventionally presented as power versus speed for a fixed head or discharge. From Fig. 6.6(b),

$$v_1 \cos \alpha_1 = v_{w1} = u_1 - v_{f1} \cot \beta_1$$

Substituting into equation (6.11) gives

$$P_r = \rho Q (u_1 v_{f1} \cot \beta_1 - u_1^2) \tag{6.17}$$

or

$$P_r = C_1 N - C_2 N^2 \tag{6.18}$$

where the constants C_1 and C_2 depend on β_1, the size of impeller and the discharge Q. C_1 must always be positive for the machine to produce useful

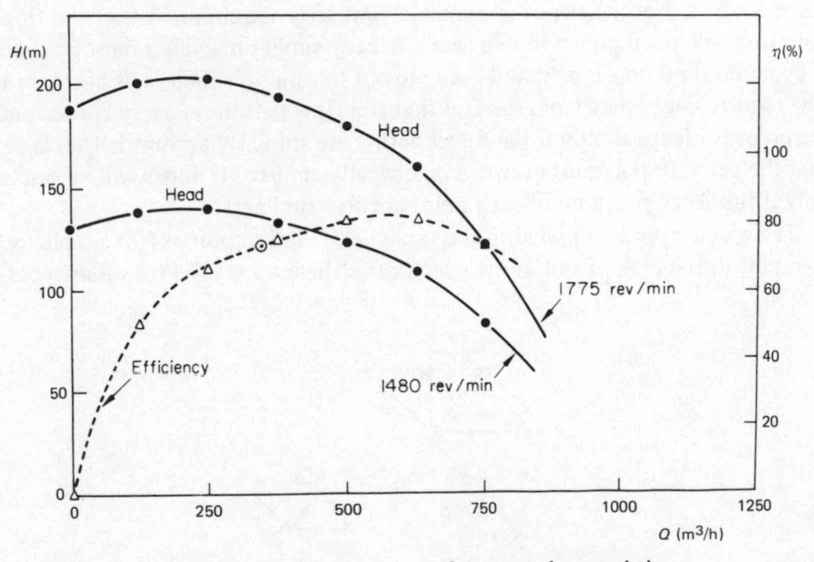

Fig. 6.9 Centrifugal pump performance characteristics

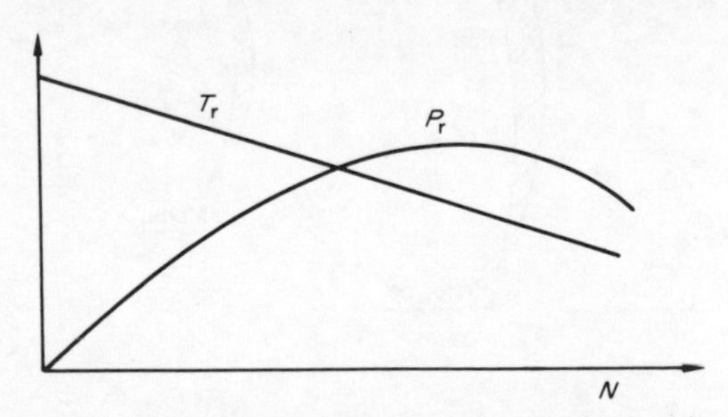

Fig. 6.10 Ideal performance curves for reaction turbines

power, so $\beta_1 < 90°$ in all practical reaction turbines. From $T_r = P/\Omega$, the torque equation is linear

$$T_r = C_1^1 - C_2^1 N \tag{6.19}$$

The shapes of the ideal performance curves for a reaction turbine as predicted by the above equations are shown in Fig. 6.10.

6.3.3 Comparative performance of rotodynamic machines

The basic principles of dynamic similitude discussed in Chapter 5 apply equally well to rotodynamic machines. Because they are invariably closed flows gravity effects are insignificant and viscosity (Reynolds) effects are generally also ignored because of the very turbulent nature of the flow. The Navier–Stokes equations cannot readily be used to derive the similitude laws for rotodynamic machines because of the large pressure differences and the strong vortex action involved in the machines. Nevertheless the dynamic similarity requirement that the flow pattern at identical points in two geometrically similar machines must be similar if dynamic similarity is achieved is employed to establish their laws of similitude. The requirement arises from the fact that the flow pattern at every corresponding point is identical only if the forces acting are similarly arranged. That is to say that the velocity polygons in two geometrically similar machines will be similar only if the force polygons at each point are also similar.

These concepts are applied to the runner of reaction pumps (for simplicity) to establish the laws of similitude which nevertheless are valid for other types of

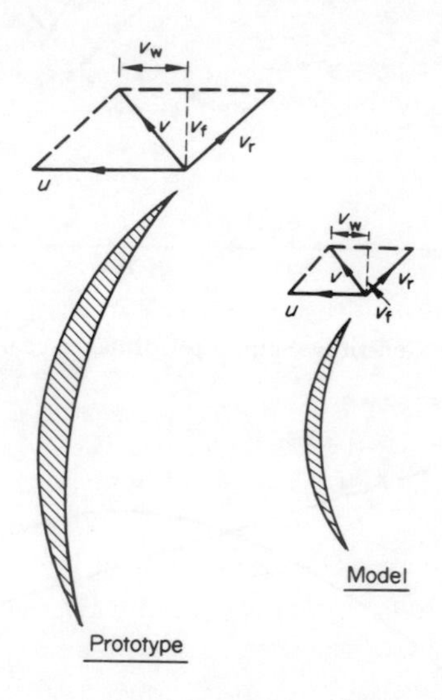

Fig. 6.11

rotodynamic machines. With reference to the exit end of the runner blade (Fig. 6.11) if the velocity triangles are similar for a prototype machine and its model, the ratio of corresponding sides of their velocity triangles will be equal.

Thus $$\left(\frac{v_f}{u}\right)_p = \left(\frac{v_f}{u}\right)_m = \text{constant}$$

But $$v_f = \frac{Q}{\text{area}} \propto \frac{Q}{D^2}$$

and $$u \propto ND$$

Thus $$\frac{Q}{ND^3} = \text{constant} \tag{6.20}$$

The forces acting are related to the pressure difference across the impeller. If these are assumed proportional to the head H under which the machine acts, the Euler equation can be used instead of actual force polygons to establish conditions for dynamic similarity

$$H = \eta\, H_r = \eta\, uv \cos \alpha/g$$

(for a turbine the equivalent equation is $H = -[uv \cos \alpha]/\eta g$). If the efficiency is constant,

$$H \propto u^2$$

since v/u = constant from similar velocity triangles

Thus $$\frac{H}{N^2 D^2} = \text{constant} \tag{6.21}$$

Equation (6.20) can be combined with equation (6.21) using $P \propto QH$ to obtain

$$\frac{P}{N^3 D^5} = \text{constant} \tag{6.22}$$

The three equations (6.20), (6.21) and (6.22) provide a powerful instrument for comparing the performance of pumps and turbines under changing conditions of speed, size and shape.

GEOMETRICALLY SIMILAR MACHINES

1. *Same size.* Equation (6.20) gives $Q \propto N$ which may be used to eliminate N from equation (6.21) to give $H \propto Q^2 \propto N^2$. Thus operating the same pump or a geometrically similar pump of the same size at different speeds gives discharges which are proportional to the pump speeds, but the corresponding heads are proportional to the square of the speeds, provided the efficiencies are the same. The locus of the operating points corresponding to a fixed efficiency (as the speed of a pump is varied) on an $H - Q$ plot is therefore parabolic as shown in Fig. 6.12.

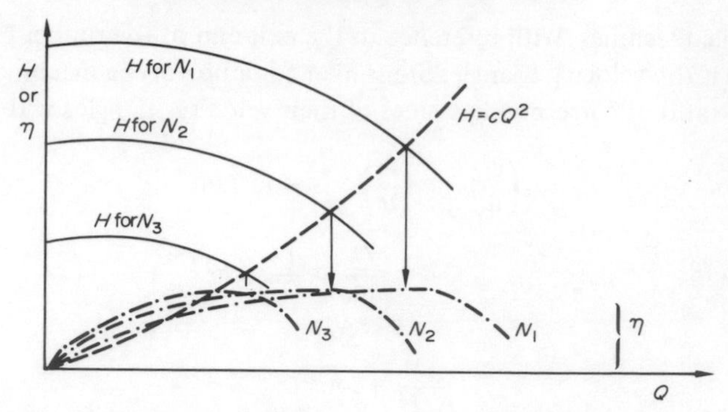

Fig. 6.12 Comparison of performance curves for similar pumps of same size

The terms *unit discharge* $Q_u = Q/H^{\frac{1}{2}}$; *unit speed* $N_u = N/H^{\frac{1}{2}}$; and *unit power* $P_u = P/H^{\frac{3}{2}}$ which define the characteristics when the operating head is unity are commonly used to characterize the performance of similar rotodynamic machines of the same size. With a known unit quantity, the quantity corresponding to a new head but with the same efficiency is obtained by simply multiplying the unit value with the appropriate power of the head.

2. *Different sizes*
From equation (6.21),

$$D = \text{constant } (H/N^2)^{\frac{1}{2}} \tag{6.23}$$

Substitute into (6.20),

$$N^2Q/H^{\frac{3}{2}} = \text{constant} = n_s^2 \tag{6.24a}$$

or into (6.22),

$$N^2P/H^{\frac{5}{2}} = \text{constant} = N_s^2 \tag{6.24b}$$

From equations (6.24, a and b)

$$n_s = N\sqrt{Q}/H^{\frac{3}{4}} \text{ (pumps)} \tag{6.25a}$$

$$N_s = N\sqrt{P}/H^{\frac{5}{4}} \text{ (turbines)} \tag{6.25b}$$

Both n_s and N_s of equations (6.25a and b) are constants for constant efficiencies and are designated specific speeds. By convention equation (6.25a) is used for pumps and (6.25b) for turbines. The convention is explained by the fact that the discharge is of principal interest in pumps while power is for turbines. It is also conventional to characterize rotodynamic machines by their specific speeds at maximum efficiency. The specific speed is, in fact, a summary of the optimum hydrodynamic behaviour of the machine and it varies from machine to machine. Two geometrically similar rotodynamic machines have the same specific speed only under dynamically similar operating conditions. Therefore the specific speed can be viewed as the speed of a model which delivers unit discharge or power

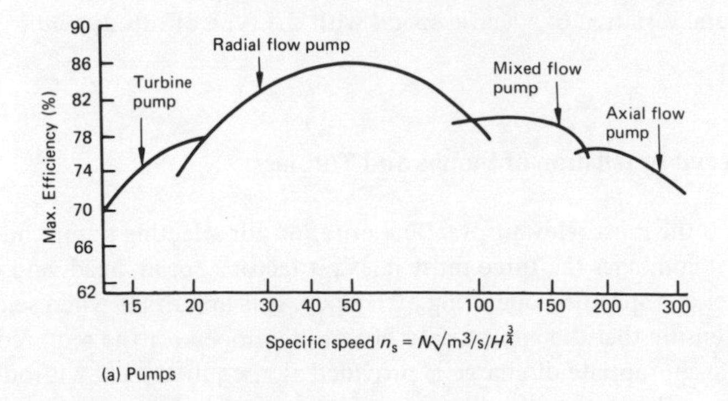

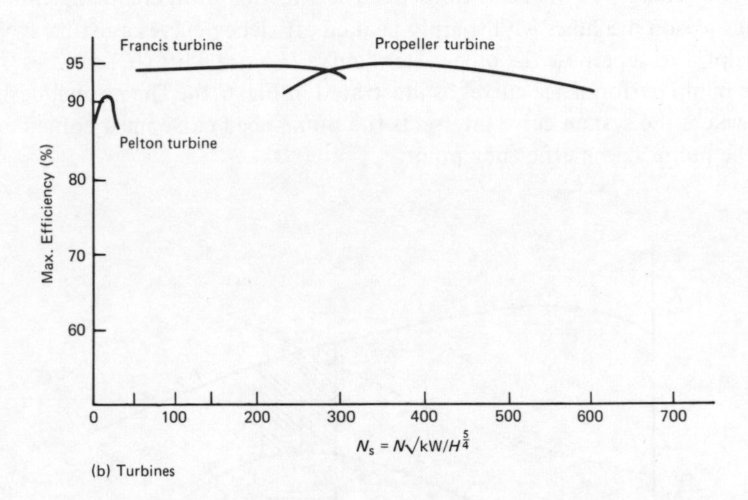

Fig. 6.13 Specific speed–efficiency graphs for water machines

against a unit head. In the SI system of units, N is expressed in rev/min, Q in m^3/s, H in metres, and P in kilowatts. Thus n_s (SI) = 0·0212 n_s (Imperial) and N_s (SI) = 3·8 N_s (Imperial). The units in the Imperial system are: N (rev/min), Q (gal/min), H (ft) and P (horse power). The dimensionless specific speed for pumps is given by $K_p = N\sqrt{Q}/(gH)^{\frac{3}{4}}$ and for turbines $K_t = N\sqrt{P/\rho}/(gH)^{\frac{5}{4}}$. These are arrived at by retaining the identities of g and ρ in equations (6.25a) and (6.25b).

DIFFERENT TYPES OF MACHINES

The specific speed is also related to the type and class of machine in view of the fact that each class of machine has its own peculiar dynamic conditions which limit the scope of its operation. For instance axial flow pumps which deliver large discharges against low heads have large specific speeds compared with centrifugal pumps which deliver smaller discharges to higher heads. Fig. 6.13

shows the general variation of specific speed with the type of rotodynamic machine.

6.4 Selection and Installation of Pumps and Turbines

Specific speed is the most relevant practical criterion for selecting pumps and turbines since it combines the three most relevant factors: speed, head, and discharge or power at a specified operating efficiency. It is important when selecting a pump to ensure that the appropriate head is developed for the required discharge or the appropriate discharge is provided at a required head without loss of efficiency. The range of head values to be covered and the corresponding range of discharge required also influence the selection of hydraulic machines. For this reason machines with sharply peaked efficiency curves must be avoided. It is helpful to superpose the pipe system curve H = static lift (H_s) + losses (h_f) on the pump performance curves as illustrated in Fig. 6.14. The normal design point where the system curve intersects the pump head curve must coincide with the pump's peak efficiency point.

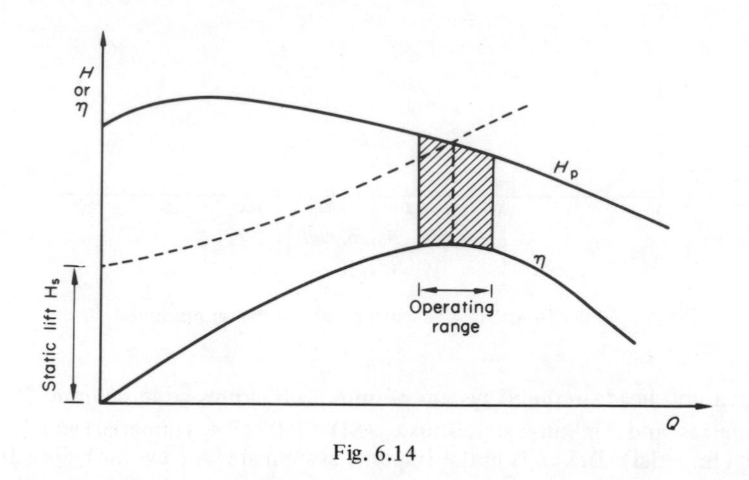

Fig. 6.14

Fluctuations brought about by variations in either the static lift or demand flow or both (shaded area) will not bring about a serious departure from peak efficiency so long as the efficiency curve is flattened about its peak value.

Figure 6.15(a) is a chart which indicates discharge and head values over which the various classes of pumps are suitable. A similar chart based on head and specific speed is shown for turbines in Fig. 6.15(b). Table 6.1 gives an approximate classification of turbines according to head and specific speed. In selecting a turbine runner an effort should be made to use a type which gives the required

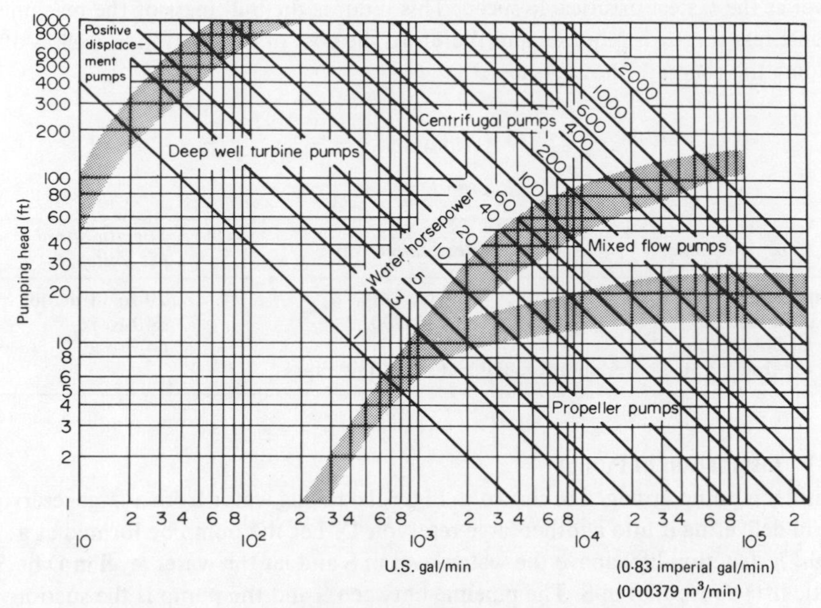

Fig. 6.15(a) Chart for selection of type of pump–Imperial units. (Fairbanks, Morse & Co.)

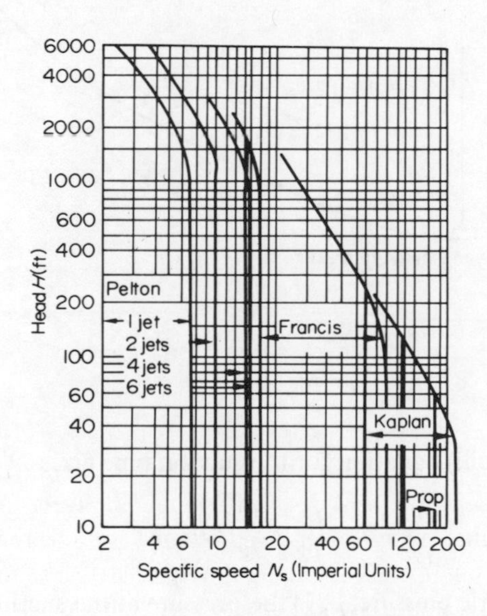

Fig. 6.15(b) Chart for turbine selection–Imperial units. (From Kempe's *Engineer's Year Book*)

power at the fastest practicable speed. This reduces the bulkiness of the machine because less torque is required and therefore the cost of installation as well as of building the generator house is less.

Table 6.1

Range of turbines

Type of turbine	*Head range* (m)	*Specific speed (SI units)*
Impulse turbine	above 240	20 to 11 per jet
Francis turbine	15 to 300	340 to 76
Axial flow turbine	3 to 24	646 to 323
Mixed flow turbine – overlapping radial and axial flow types.		

6.4.1 Installation of Pumps

Consider a pump arranged as shown in Fig. 6.16 taking water from a large reservoir S. and delivering it into another large reservoir D. Let the pump be located at a height h_s (suction lift) above the water level in S and let the water level in D be H_s (static lift) above that in S. The pipeline between S and the pump is the suction pipe and that between the pump and D is the delivery pipe.

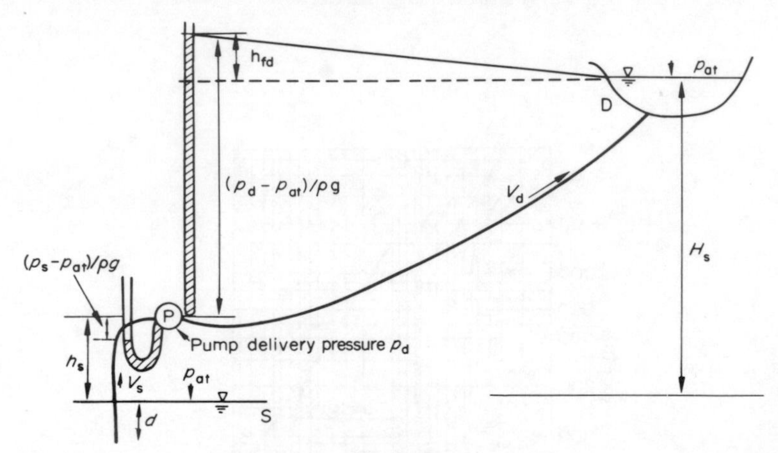

Fig. 6.16

The energy (Bernoulli) equation for the suction pipe gives

$$\frac{p_{at}}{\rho g} = h_s + \frac{p_s}{\rho g} + \frac{v_s^2}{2g} + h_{fs} \tag{6.26}$$

where p_{at} is atmospheric pressure, p_s is the pressure at the suction end of the pump, v_s is the flow velocity in the suction pipe and h_{fs} is the head loss in the suction pipe. Rearranging equation (6.26), the suction head is given by

$$\frac{p_s - p_{at}}{\rho g} = -h_s - h_{fs} - \frac{v_s^2}{2g} \tag{6.27}$$

or

$$h_s = \frac{p_{at} - p_s}{\rho g} - h_{fs} - \frac{v_s^2}{2g} \tag{6.28}$$

Equation (6.27) shows that the gauge pressure at the suction end of the pump is always negative unless the pump is located sufficiently far below the water level in the suction reservoir (thus providing a negative suction lift $(-h_s)$).

Equation (6.28) also demonstrates the importance of exposing suction reservoirs to atmospheric pressure. (Vents are provided for covered clear-water reservoirs for domestic water supply in order to achieve this.) The use of atmospheric pressure helps the suction pipe to lift water. The height at which the pump may be located above the reservoir level is limited by the vapour pressure p_v of water. The suction pressure p_s must be above the vapour pressure p_v of water or else the water will cavitate (see next section). The higher h_s is the higher the possibility of cavitation since an increase in h_s is invariably followed by a decrease in $p_s/\rho g$. A sufficient depth of immersion d of the suction pipe is necessary to account for water level fluctions and to prevent the pump from sucking in air and forming air pockets inside the casing - a situation which lowers the efficiency of operation and may cause damage to the pump impeller.

The pressure at the delivery end of the pump must be sufficient to overcome the static lift and friction losses in the delivery pipe. Its value is obtained by applying the energy equation between the delivery end of the pump and the delivery reservoir level as

$$\frac{p_d - p_{at}}{\rho g} = (H_s - h_s) + h_{fd} \tag{6.29}$$

The energy equation applied between S and D readily shows that the head H_p which must be supplied by the pump is given by

$$\frac{p_{at}}{\rho g} + H_p = \frac{p_{at}}{\rho g} + H_s + \frac{v_d^2}{2g} + h_{fs} + h_{fd}$$

or

$$H_p = \eta_h H_r = H_s + (h_{fs} + h_{fd}) + \frac{v_d^2}{2g} \tag{6.30}$$

The rejected kinetic energy term $v_d^2/2g$ is usually very small compared with the others and it has been neglected in applying the equation in the example of Fig. 6.14.

It may be necessary in practice to use two or more pumps together in series or in parallel. If the required head is more than can be provided by one pump, the pumps are connected in series as a booster arrangement (Fig. 6.17(a)). The same discharge passes through both pumps but the head developed by one augments the other. The total head developed is obtained by adding together the value of the head of each pump corresponding to the relevant discharge (see Fig. 6.17(b)).

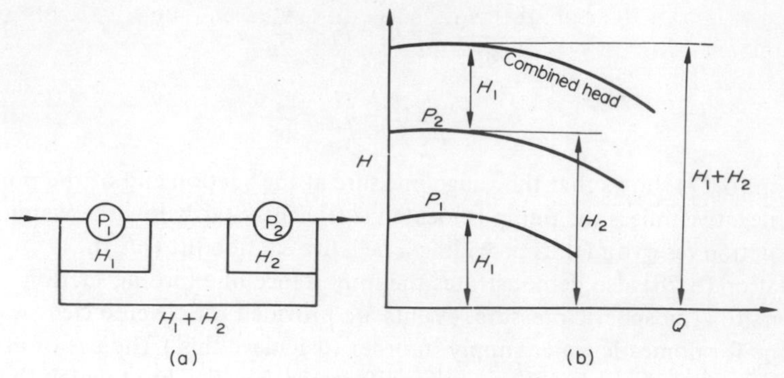

Fig. 6.17 Pumps in series

If the efficiency of pump P_1 is η_1 and of pump P_2 is η_2 the efficiency η of the two pumps in series is obtained from

$$\rho g Q(H_1 + H_2)/\eta = \rho g Q H_1/\eta_1 + \rho g Q H_2/\eta_2$$

$$\therefore \qquad \frac{1}{\eta} = \frac{H_1/\eta_1 + H_2/\eta_2}{H_1 + H_2}$$

$$\therefore \qquad \eta = \Sigma H/\Sigma H/\eta \tag{6.31}$$

Pumps are connected in parallel if the head developed by each pump is sufficient but the discharge is not. Thus the head across both pumps (Fig. 6.18) is the same but the discharges are different. The total discharge corresponding to

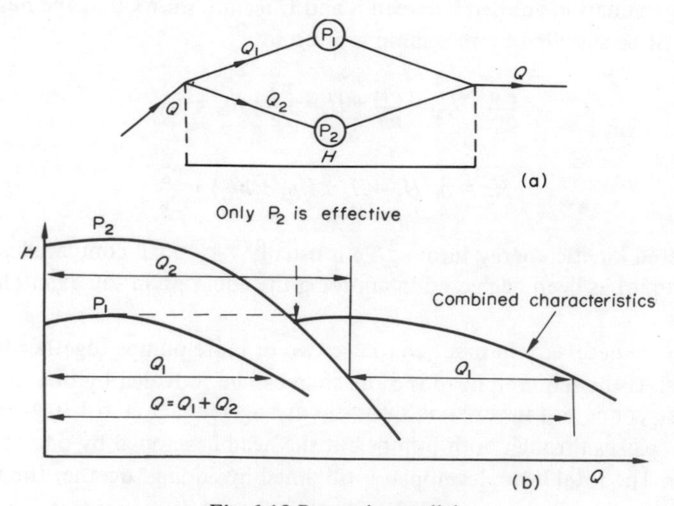

Fig. 6.18 Pumps in parallel

a particular head is the sum of the discharge through each pump at that head. The combined efficiency of the system is given by

$$\rho g (Q_1 + Q_2) H/\eta = \rho g Q_1 H/\eta_1 + \rho g Q_2 H/\eta_2$$

$$\therefore \qquad \eta = \Sigma Q / \Sigma Q/\eta \qquad (6.32)$$

A pump must never be started without ensuring that it is filled with water since the heat developed through friction may be sufficient to damage packing glands and bearings and cause metal parts which rub on each other to seal together, The process of filling up a pump casing with water is known as *priming*. Priming may be effected manually, by direct connection to water supply lines or by automatic devices. A check (foot) valve is normally provided at the suction end of the suction pipe to facilitate priming. The valve allows water to flow normally during pumping but prevents flow back into the sump when the pump stops or is being primed.

6.4.2 Installation of Reaction Turbines

Requirements for setting up a reaction turbine are generally the same as those for pumps. A draft tube is installed between the turbine and the downstream tail water as a counterpart to the suction pipe of a pump (Fig. 6.19). The draft tube expands gradually in cross section from the turbine to the tail water and therefore allows conversion of a large portion of the kinetic energy output from the turbine into pressure energy. This arrangement, apart from reducing the erosion potential of the flow in the tail race, permits the maximum possible pressure difference to be developed across the machine. Without the draft tube the pressure at the exit

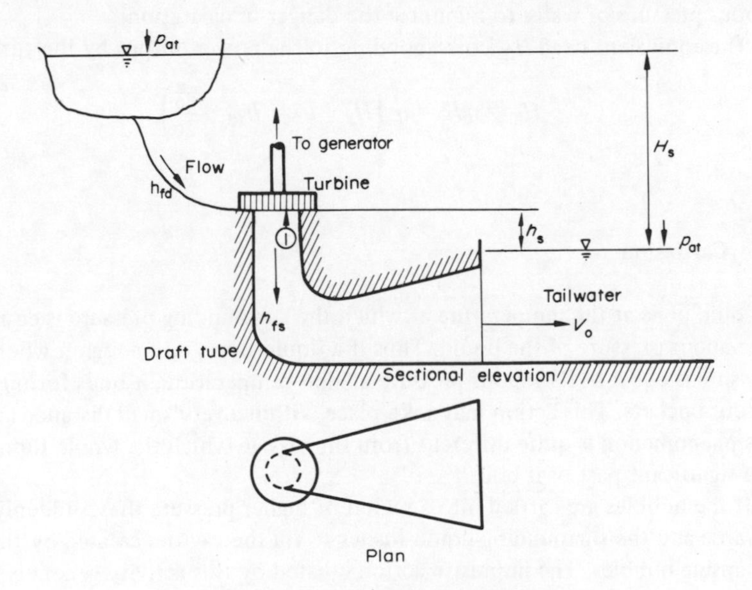

Fig. 6.19 Installation of a reaction turbine

end of the machine would be atmospheric. The draft tube makes it possible for pressures below atmospheric level to develop and therefore for more energy to be consumed by the machine. The draft tube also enables the turbine to be set above tail water level without loss of head, a condition which is often necessary in order to avoid flooding of the machine.

With reference to Fig. 6.19, the energy equation may be applied between outlet of the machine and the tailwater level

$$\frac{p_1}{\rho g} + \frac{v_1^2}{2g} + h_s = \frac{p_{at}}{\rho g} + h_{fs} + \frac{v_0^2}{2g} \tag{6.33}$$

where (1) refers to the outlet of the machine which is h_s above the tailwater level $v_0^2/2g$ is the rejected kinetic energy from the draft tube and h_{fs} is the friction head in the draft tube. From equation (6.33)

$$\frac{p_1 - p_{at}}{\rho g} = -h_s + h_{fs} - \left(\frac{v_1^2}{2g} - \frac{v_0^2}{2g}\right) \tag{6.34}$$

or

$$h_s = \frac{p_{at} - p_s}{\rho g} + h_{fs} - \left(\frac{v_1^2}{2g} - \frac{v_0^2}{2g}\right) \tag{6.35}$$

Equations (6.34) and (6.35) are similar to (6.27) and (6.28) especially if the relatively very small kinetic energy heads are neglected. The sign difference for the friction head terms is due to the fact that flow is toward a pump but away from a turbine. The negative gauge pressure at the exit of the turbine provides extra energy consumption across the machine but places a limitation on its height h_s above the tailwater level since the pressure must always be higher than vapour pressure of water to minimize the danger of cavitation.

The equivalent head H_T corresponding to the power output by the turbine is

$$H_T = \eta H_r = \eta \left(H_s - h_{fs} - h_{fd} - \frac{v_0^2}{2g}\right) \tag{6.36}$$

6.5 Cavitation

A liquid boils at the temperature at which the surrounding pressure is equal to the vapour pressure of the liquid. Thus if a liquid flows into a region where the pressure is equal to its vapour pressure at that temperature, it boils forming vapour pockets. This action may take place within a very small distance and this phenomenon is quite different from the way in which the whole fluid system or a significant part of it boils.

If the bubbles are carried into a region of higher pressure they suddenly collapse and the surrounding liquid rushes to fill the cavities created by the collapsing bubbles. The impulsive action created by this activity generally results in very high localized pressures which cause pitting and damage to solid surfaces over which the flow takes place. The phenomenon is known as *cavitation.*

Effects of cavitation are most noticeable in regions of high localized velocities which according to Bernoulli's equation tend to have low pressures. The inlet end of pump blades and the outlet end of turbine runners, especially on the curved side; the spillway of high dams especially in areas of small radii of curvature; and other structures with large local curvature are all susceptible to cavitation. The centrifugal acceleration (v^2/r) in a region becomes infinite when the radius of curvature is very small and the local pressure may be sufficiently low to produce cavitation.

A modified form of the Euler number, called the cavitation number, is generally used to measure the possibility or degree of cavitation. It is written as

$$\sigma = (p - p_v)/(\rho v^2/2) \tag{6.37}$$

where p is the local pressure, p_v the vapour pressure, ρ the density and v the velocity of the flowing fluid. Cavitation is less likely to occur if $p \gg p_v$ than if $p \sim p_v$ (or $\sigma \sim 0$). Two geometrically similar fluid systems would be equally likely to produce cavitation or would produce the same degree of cavitation if they had the same value of σ (see subsection 5.3.3).

The minimum pressure, particularly in turbo machines occurs along the convex side of the vanes especially near the suction side of a pump impeller or the exit end of a turbine impeller. Assuming that the suction pressure p_s of Fig. 6.16 or the pressure p_1 at the entrance to the draft tube of Fig. 6.19 corresponds to the pressure p_c at the critical point of the machine, equation (6.27) or (6.34) would give

$$\frac{v_c^2}{2g} = \frac{p_{at} - p_c}{\rho g} - h_s \pm h_{fs} \tag{6.38}$$

where v_c is the appropriate suction or exit velocity and the relatively small rejected kinetic energy from the draft tube is neglected. The positive sign is relevant to turbines and the negative sign to pumps. Putting $h_s' = h_s \mp h_{fs}$,

$$\sigma = \frac{v_c^2}{2gH} = \left(\frac{p_{at}}{\rho g} - \frac{p_c}{\rho g} - h_s'\right)/H \tag{6.39}$$

where σ is a form of cavitation number first proposed by D. Thoma and generally named after him and H is the total head produced by a pump or absorbed by a turbine. It is clear from the first part of equation (6.39) that σ is really a measure of the portion of the energy available in the machine which is kinetic at the critical location.

When $p_c \rightarrow p_v$ cavitation becomes imminent at the critical location. The term 'net positive suction head' (NPSH) is sometimes used to describe the difference between the static pump inlet head and the head corresponding to the vapour pressure of the liquid. This is given by

$$H_{sv} = p_{at}/(\rho g) - h_s' - p_v/(\rho g) = h_a - h_s' - h_v \tag{6.40}$$

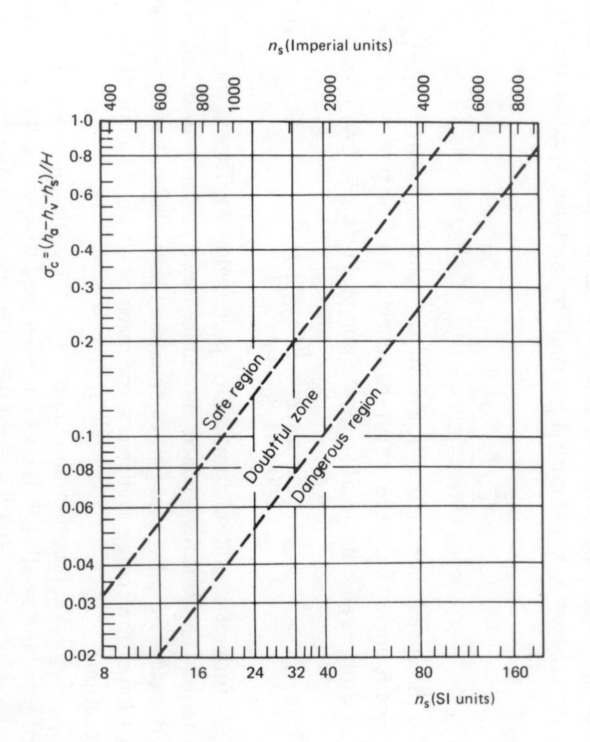

Fig. 6.20 Cavitation chart for centrifugal pumps

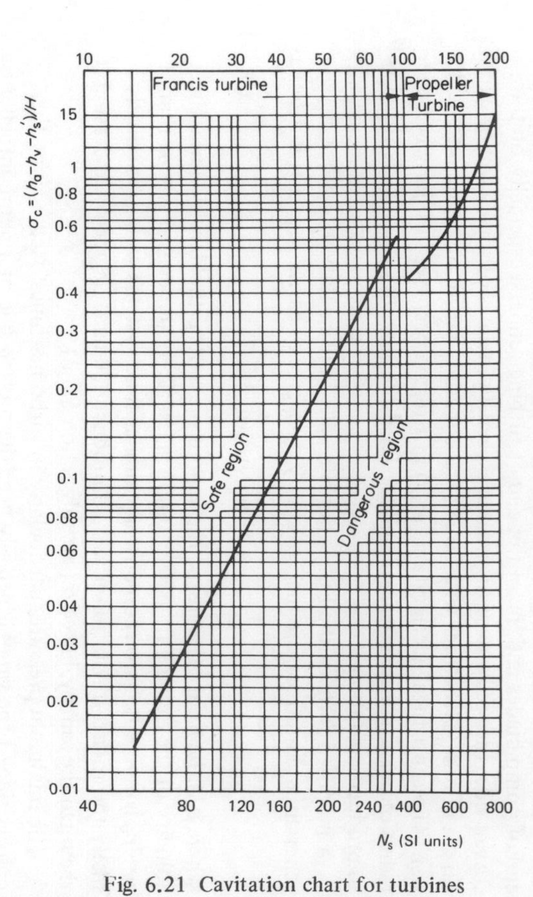

Fig. 6.21 Cavitation chart for turbines

The onset of cavitation is indicated by the critical cavitation number

$$\sigma_c = H_{sv}/H = (h_a - h_s' - h_v)/H \qquad (6.41)$$

A reaction machine is less likely to produce cavitation if the cavitation number σ given by equation (6.39) is very much greater than the critical cavitation number. The critical cavitation number is usually determined from model tests, and two geometrically similar machines would have the same cavitation potential if their cavitation numbers are equal. Figs 6.20 and 6.21 which are based on experimental results may be used to estimate the cavitation potential for single stage centrifugal pumps and turbines. Having determined the critical cavitation number σ_c, equation (6.41) can then be used to determine the maximum setting height which would avoid cavitation.

EXAMPLE 6.1

Pump P takes water from a stream whose level fluctuates between elevations 1.5 m and 4.6 m (see figure below). Two pumps (characteristic performance curves at 1500 rev/min are given below) are available. The delivery reservoir level is 5.2 m (the same as that of the pump axis elevation) and head loss against friction is given by

$$h_f \text{ (m)} = 3{\cdot}7 \times Q^2 \text{ (}Q \text{ in m}^3\text{/min)}$$

(1) Choose between pumps A and B showing clearly the reasons for your choice. (2) If the vapor pressure of water is 0.3 m and $f = 0{\cdot}006$ for the suction pipe, determine the minimum suction pipe diameter to avoid cavitation.
(3) Estimate the bending moment at the support shown when the stream level is at 4.6 m. The 10 cm design diameter pipe weighs 62 N/m. (4) What is the role of the foot valve?

Q (m^3/min)	0	0·45	0·90	1·35	1·57	1·80	2·05	2·24
Pump A; H (m)	15·2	16·5	16·5	14·6	12·8	10·3	7·6	4·6
η (%)	0	30	61	81	85	80	65	47
Pump B; H (m)	18·3	19·8	19·8	17·7	15·0	12·2	8·5	4·3
η (%)	0	34	74	89	84	68	41	20

(U.S.T., Part III, 1969.)

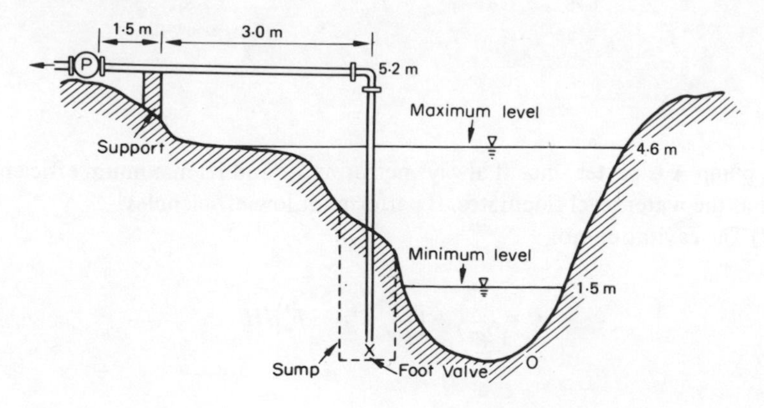

SOLUTION

(1) Required pump head, $H = H_s + h_f$

Discharge Q (m^3/min)	0	0·45	0·90	1·35	1·57	1·80
Friction head h_f (m)	0	0·74	3·0	6·8	9·1	12·0
Required pump head						
H_1 (H_s = 0·6) (m)	0·6	1·34	3·6	7·4	9·7	12·6
H_2 (H_s = 3·7) (m)	3·7	4·44	6·7	10·5	10·8	15·7

Plot H_1 and H_2 against Q on pump characteristic curves.

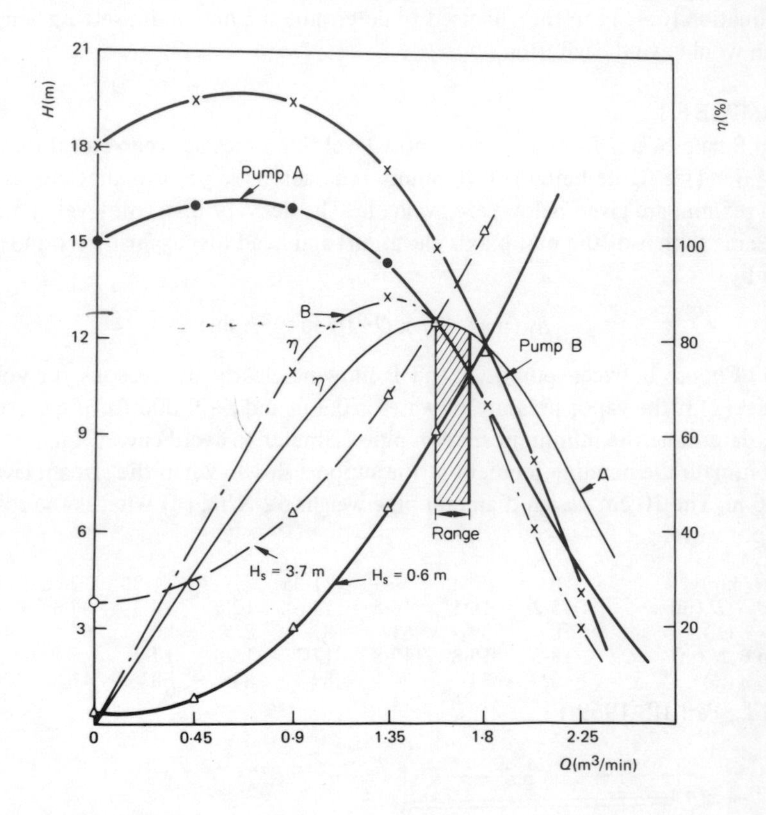

Thus pump A is better since it always performs around its maximum efficiency point as the water level fluctuates. B performs at low efficiencies.

(2) The cavitation no.

$$\sigma_c = \frac{v_s^2}{2gH} = \left(\frac{p_a}{\rho g} - \frac{p_v}{\rho g} - h_s'\right)/H$$

The situation is most critical when h_s is a maximum, i.e. 3·7 m. The total length of the suction pipe is 9·7 m. Thus

$$h_{fs} = \frac{0{\cdot}23 v_s^2}{D_p\, 2g}$$

$$\therefore \quad \frac{v_s^2}{2gH} = \frac{(10{\cdot}3 - 0{\cdot}3 - 3{\cdot}7)}{H} - \frac{0{\cdot}23 v_s^2}{D_p\, 2gH}$$

or

$$\frac{v_s^2}{2g}\left(1 + \frac{0{\cdot}23}{D_p}\right) = 6{\cdot}3$$

From the graph,

$$Q = 1{\cdot}5 \text{ m}^3/\text{min} = 0{\cdot}0262 \text{ m}^3/\text{s}$$

$$\therefore \quad \frac{(4 \times 0{\cdot}0262)^2}{\pi^2 D_p^4}\left(1 + \frac{0{\cdot}23}{D_p}\right) = 2g \times 6{\cdot}3 = 123$$

i.e.

$$\frac{D_p^4}{1 + 0{\cdot}23/D_p} = 0{\cdot}89 \times 10^{-3}$$

$$\therefore \quad D_p = \underline{0{\cdot}078 \text{ m}}$$

(3) The column of water in CD balances the hydrostatic force if we neglect the thickness of the pipe walls.

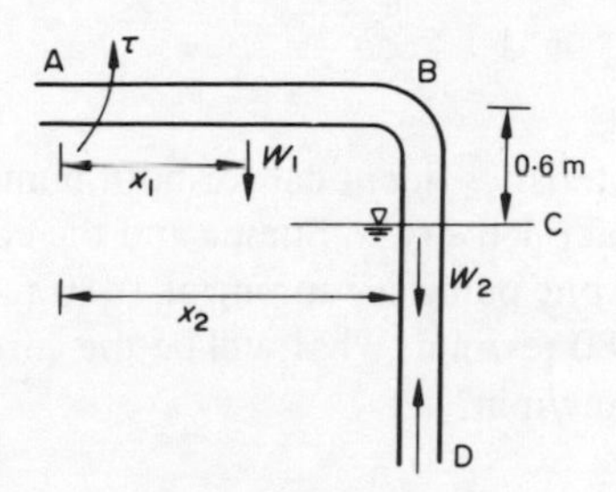

$$W_1 = \text{weight of pipe AB and water in AB}$$

$$= 62 \times 3 + 3\,\frac{\pi D_p^2}{4} \times 9{\cdot}81 \times 10^3 = 186 + 231$$

$$= 417\text{N acting through } x_1 = 1{\cdot}5 \text{ m from A}$$

$$W_2 = \text{weight of pipe BD + weight of water in BC}$$

$$= 62 \times 5{\cdot}2 + 231 \times 0{\cdot}6/3 = 322 + 46{\cdot}2$$

$$= 368{\cdot}2\text{N acting through } x_2 = 3 \text{ m from A}$$

Momentum flux at D

$$= \rho Q v = \frac{\rho Q^2}{a_p} = \frac{10^3 \times (0{\cdot}0262)^2 \times 4}{\pi \times 10^{-2}}$$

$$= 87{\cdot}4 \text{ N}$$

Torque $\qquad \tau = W_1 x_1 + W_2 x_2 - 87{\cdot}4\, x_2$

$$= 417 \times 1{\cdot}5 + 368{\cdot}2 \times 3 - 87{\cdot}4 \times 3$$

$$= \underline{1469{\cdot}4 \text{ N m anticlockwise}}$$

Bending moment at support $\qquad = \underline{1469 \text{ N m clockwise}}$

(4) Foot valve - see p. 185

EXAMPLE 6.2 (Imperial units)

The figure below shows a diagrammatic arrangement for water transmission from Owabi to Suame (Kumasi). The diameter, length and the Hazen–Williams C are given for each pipe on the figure. The Suame reservoir is 280 ft above that at

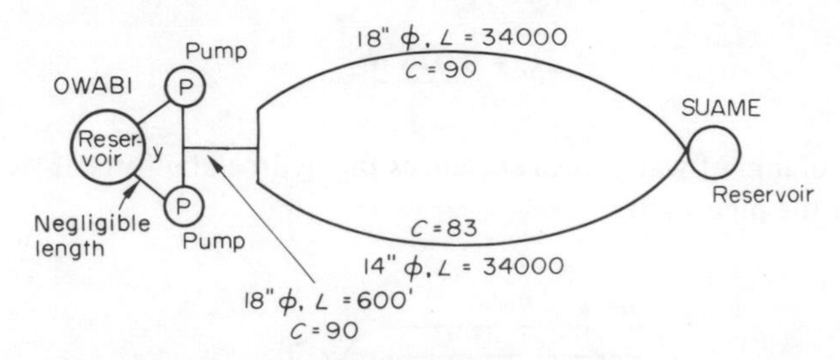

Owabi. The pump characteristics (identical for both pumps) are given below. Determine the rate of water delivery to Suame and the corresponding pump efficiency when (1) only one pump is running at 1480 rev/min and (2) both pumps are running at 1480 rev/min. What will be the corresponding values for pump speed (s) of 1775 rev/min?

(Hazen–Williams formula:

$$h_f = \frac{15 Q^{1{\cdot}85} L}{C^{1{\cdot}85} d^{4{\cdot}87}}$$

with h_f and L in ft, Q in gal/min and d in inches)

Pump characteristics ($N = 1480$ rev/min)

Q (gal/min)	0	500	1000	1500	2000	2500	3000
H (ft)	391	418	420	400	372	335	256
η (%)	0	51	66·6	75·3	80·5	82	73·5

(U.S.T. Part III, 1969.)

SOLUTION

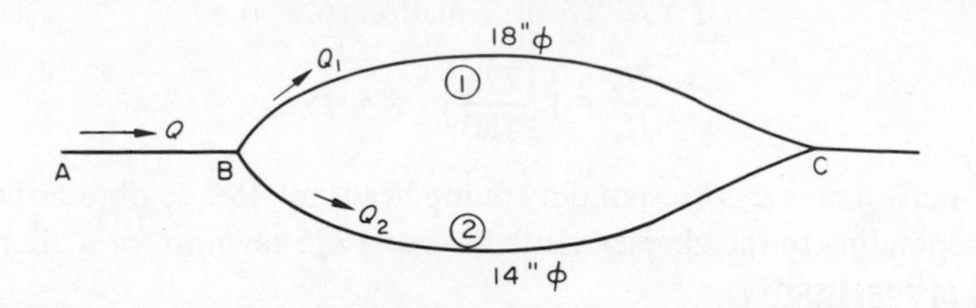

From continuity:

$$Q = Q_1 + Q_2 \rightarrow Q/Q_2 = 1 + Q_1/Q_2$$

From energy conservation:

$$h_{f1} = h_{f2}$$

i.e.

$$\frac{Q_1^{1 \cdot 85}}{C_1^{1 \cdot 85}\, d_1^{4 \cdot 87}} = \frac{Q_2^{1 \cdot 85}}{C_2^{1 \cdot 85}\, d_2^{4 \cdot 87}}$$

since the lengths are equal

$$\therefore \qquad \frac{Q_1}{Q_2} = \frac{C_1}{C_2}\left(\frac{d_1}{d_2}\right)^{2 \cdot 63}$$

Thus

$$\frac{Q}{Q_2} = 1 + \left(\frac{C_1}{C_2}\right)\left(\frac{d_1}{d_2}\right)^{2 \cdot 63} = 1 + \left(\frac{90}{83}\right)\left(\frac{18}{14}\right)^{2 \cdot 63}$$

$$\therefore \qquad Q_2 = 0 \cdot 323\, Q$$

and

$$Q_1 = 0 \cdot 677\, Q$$

Total head loss

$$h_f = h_{f(A \rightarrow B)} + h_{f1} = h_{f(A \rightarrow B)} + h_{f2}$$

$$= \frac{15Q^{1 \cdot 85} \times 600}{(90)^{1 \cdot 85}\,(18)^{4 \cdot 67}} + \frac{15Q_1{}^{1 \cdot 85} \times 34\,000}{(90)^{1 \cdot 85}\,(18)^{4 \cdot 87}}$$

Substitute for Q_1

$$h_f = 49 \cdot 7 \times 10^{-6}\, Q^{1 \cdot 85}$$

$\therefore$ Total head required from pump $H = 280 + h_f$

Q (gal/min)	1000	2000	3000	4000	5000
h_f (ft)	17·5	65	133	227	344
H (ft)	298	345	413	507	624

But $Q \propto N$ (from equation (6.20))

and $H \propto N^2$ (from equation (6.21))

$$\therefore \quad \frac{H_2}{H_1} = \left(\frac{1775}{1480}\right)^2 = 1{\cdot}45$$

Thus we must increase corresponding pump heads by 45% to obtain the value of H_2 corresponding to the characteristic head at 1775 rev/min for a discharge Q_2 given by (1775/1480) Q.

Q_2 (gal/min)	0	600	1200	1800	2400	3000	3600
H_2 (ft)	567	605	610	580	540	482	370
η (%)	0	51	66·6	75·3	80·5	82·0	73·5

We now plot the pump characteristics for single pumping and for double pumping in parallel and superimpose on them the pipe system curve.

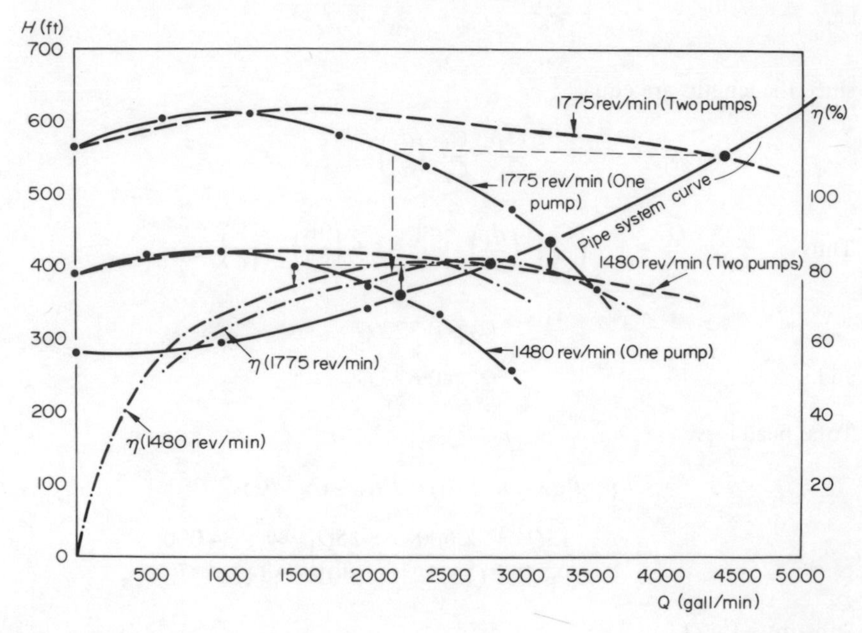

From the graphs

N = 1480 rev/min

With one pump $Q = 2220$ gal/min; $\eta = 82\%$

With two pumps $Q = 2850$ gal/min; $\eta = 75\%$

N = 1775 rev/min

With one pump $Q = 3300$ gal/min; $\eta = 79\%$

With two pumps $Q = 4450$ gal/min; $\eta = 79\%$

6.6 Pumping from Wells

Because a good proportion of the world's available water and other fluids are stored underground, deep well pumping deserves special attention, particularly in arid and semi-arid regions of the world. Wells, by their nature require pumps which can support a high lift (suction) or which can easily be set up underneath the ground surface in water or oil. Well pumps may generally be classified as reciprocating, jet, revolving vertical shaft or turbine, and air-lift.

6.6.1 Reciprocating Pumps

A one-stroke reciprocating pump is illustrated in Fig. 6.22(a). It comprises a power head (above ground level) which drives a crank shaft connected to a prime mover by gears or a belt drive, a cylinder and drop pipe and connecting rods. When the piston moves down, the foot valve closes while the piston valve opens so that the cylinder is filled with water. During an upward movement, the piston valve closes and water in the cylinder is forced up into the discharge pipe. The reduced pressure below the piston makes the foot valve open and water enters the cylinder from the well.

A reciprocating pump may also have a two-stroke arrangement (Fig. 6.22(b)) in which the rods C and D move in opposite directions. As D moves up, valve B closes and water in chamber E is emptied through valve A into chamber F and the delivery pipe. As rod C moves up valve A closes and the plunger forces the

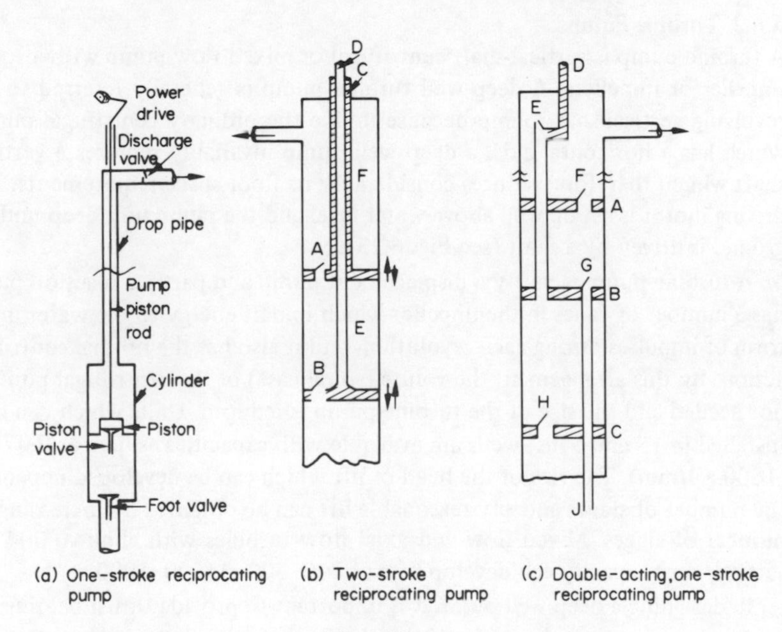

(a) One-stroke reciprocating pump (b) Two-stroke reciprocating pump (c) Double-acting, one-stroke reciprocating pump

Fig. 6.22

remainder of the water in F into the delivery pipe whilst valve B opens and water fills chamber E again from the well. The two-stroke pump provides a steadier flow rate than the one-stroke differential plunger arrangement of Fig. 6.22(a).

A double-acting, one-stroke reciprocating pump is illustrated in Fig. 6.22(c). The pump cylinder is divided into three chambers by the partitions A and C in which are located check valves F and H. As the piston B moves down, valves E and G open while F and H close. Water enters the uppermost chamber and the delivery pipe through E and the chamber AB is filled from the well. As B moves up, E and G close while F and H open. Water enters the uppermost chamber and the delivery pipe from AB and BC is filled through H from the well. Thus a double-action reciprocating pump provides a continuous stream in the delivery pipe. Its main disadvantage is that the pump requires long rods and is heavy.

Like a reaction pump a reciprocating pump must be primed before it operates. However, unlike a reaction pump whose discharge valve is normally closed before the pump is started, the delivery valve of a reciprocating pump should be opened before the pump is started to prevent high pressures which may damage the cylinder or the delivery pipe.

With their bulkiness, large power demands and the generally pulsating flow which they give, reciprocating pumps are gradually being replaced at large water works by rotary and turbine pumps. However, their high efficiencies and long life make them still attractive for oil wells and at small waterworks.

6.6.2 Turbine Pumps

A turbine pump is vertical-shaft centrifugal or mixed flow pump with a rotary impeller or impellers. A deep well turbine pump is generally referred to as a revolving vertical-shaft pump because unlike the ordinary centrifugal pump which has a horizontal axis, a deep well pump invariably requires a vertical shaft which, therefore, reduces considerably its floor space requirements. The driving motor is set up well above water level and the pump unit deep underground is driven by a shaft (see Fig. 6.23).

A turbine pump is partly a displacement pump and partly a reaction pump. It has a number of vanes in the impeller which impart energy to the water in the form of impulses during each revolution, and it also has the normal centrifugal action. By this arrangement the volute (spiral case) of the centrifugal pump is not needed and the size of the turbine pump is reduced. Units which can be installed in 15 cm (6 ins) wells are available with capacities as high as 0·075 m^3/s (1000 gal/min). The size of the head or lift which can be developed depends on the number of stages and any reasonable lift can be obtained by increasing the number of stages. Mixed flow and axial flow turbines with about 0·004 m^3/s (50 gal/min) capacity can develop lifts of 200–300 m (600–1000 ft).

In designing a deep well pump it is important to provide thrust bearings which can carry the full weight of the pump, shaft and the unbalanced weight

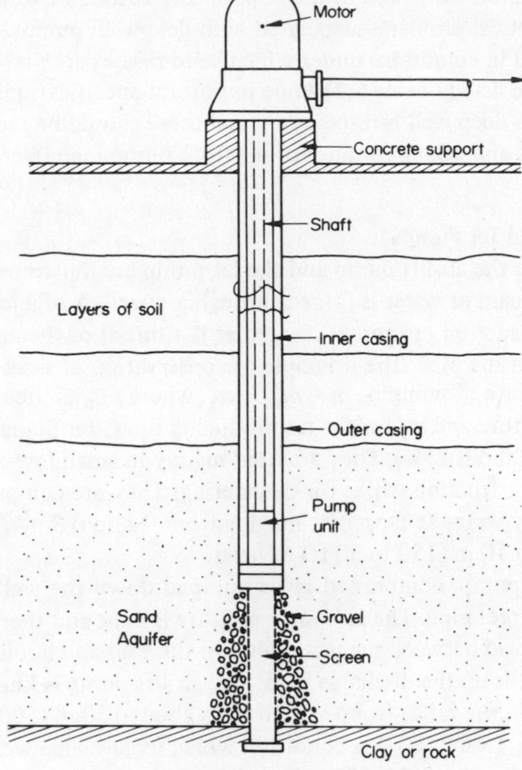

Fig. 6.23 Installation of turbine pump

Table 6.2

Turbine pump operating conditions

1. Pump capacity
2. Maximum speed
3. Elevation at site
4. Pump level below ground level at rated capacity, Z_p
5. Pump head above ground level including delivery pipe friction, H
6. Rated total head of pump, $Z_p + H$
7. Operating range: Maximum total head
   Minimum total head
8. Pump setting
9. Minimum inside diameter of well or casing
10. Total depth of well
11. Well straight to depth of Z
12. Static water level below ground surface
13. Draw-down? at flow?
14. Electric power available; maximum h.p., voltage, phase cycle, etc.
15. Other driver available; maximum h.p., engine type.

of the flowing water. Corrosion of metal parts and rotting of wooden parts are some of the practical problems associated with deep well pumps. Turbine pumps must not be used in conditions under which solid masses are present in the fluid. Babbit *et al.* have listed (Table 6.2) some important specifications required for the selection of a deep well turbine pump and these should be requested when making enquiries and asking for quotations from pump manufacturers.

6.6.3 Air-lift and Jet Pumps

The principles of the air-lift pump and the jet pump are illustrated in Fig. 6.24 and 6.25. Air, steam or water is forced through a nozzle A of a jet pump at a high velocity. A suction pressure is created at the throat of the nozzle which lifts water up through the pipe. The principle of conservation of mass may be applied to calculate the rate of pumping $m = m_m - m_j$, where m_m is the mass rate of flow for the mixture and m_j for the jet producing fluid. Jet pumps have very low efficiencies (about 25%). They are used mainly in small low-capacity installations and in construction works for dewatering. They are compact, light and can pump muddy water at capacities ranging from 0·1 to 0·3 m³/min (20 to 70 gal/min) at 50 to 10 m (150 to 30 ft) of head.

In an air-lift pump, compressed air is pumped down the well and up the eduction (discharge) pipe. The air-water mixture is light and therefore provides a difference of head between the water level in the well and liquid in the pipe which forces water up the discharge pipe. The air-lift pump is known to operate best if the ratio h_p/h_s varies from 2 when h_p is about 150 m (500 ft) to 0.5 when h_p is about 15 m (50 ft), a condition which makes deep wells necessary. Despite its low efficiency (25–50%) an air-lift pump can deliver large flows from small wells but it can raise water normally only to about ground level. Air-lift

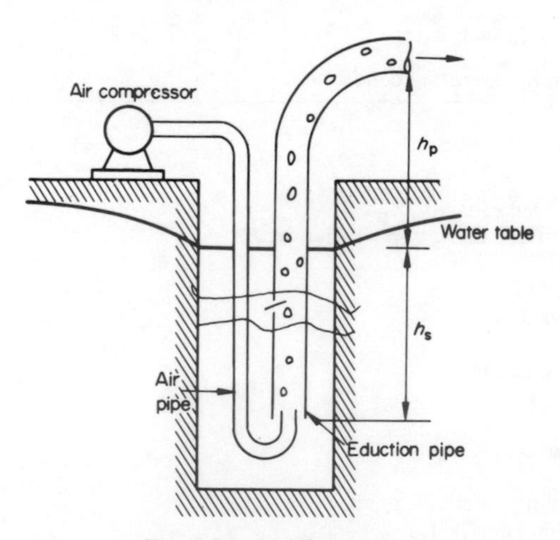

Fig. 6.24 Air-lift pump

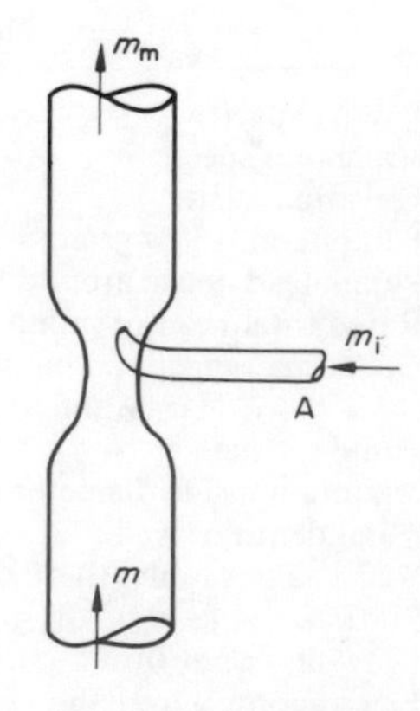

Fig. 6.25 Jet pump

pumps have a tendency to gush intermittently due to air bubbles coalescing and forming larger bubbles which move more rapidly than the water–air mixture.

The efficiency of an air-lift pump is related to its depth of submergence h_s. In Table 6.3 the recommended submergence ratios $h_s/(h_s + h_p)$ are given for optimum performance of an air-lift pump.

Table 6.3

Lift (m)	Up to 15	15–30	30–60	60–90	90–120	120–150
$h_s/(h_s + h_p)$	0·7–0·66	0·66–0·55	0·55–0·50	0·50–0·43	0·43–0·40	0·40–0·33

FURTHER READING

Babbit, H. E., Donald, J. J., and Cleasby, J. L., *Water Supply Engineering,* Chapters 10 and 11 McGraw-Hill, New York and London.

Barna, P. S., *Fluid Mechanics for Engineers,* Chapters 11, 12 and 13, Butterworths, London.

Daily, J. W., 'Hydraulic Machinery' in *Engineering Hydraulics* ed. Rouse, Wiley, New York.

Francis, J. R. D., *A Text-book of Fluid Mechanics*, Chapter 15, Edward Arnold, London.

Linsley, R. K., and Franzini, J. B., *Water Resources Engineering,* Chapter 12, McGraw-Hill, New York and London.

Stepanoff, A. J., *Centrifugal and Axial Flow Pumps,* Wiley, New York.

PART TWO:

Specialized Topics in Civil Engineering

7 Flow in Erodible Open Channels

7.1 Properties of Sediments

Erosion, transportation and deposition of soil are important in the control and utilization of the surface waters of the earth. Most river channels, earth canals and other unlined water courses undergo a continuous change due to the mobility of soil particles under the influence of hydrodynamic forces. The extent and pattern of movement of the soil, therefore, depend on the dynamic conditions of the stream as well as on the properties of the soil.

Soils can broadly be grouped into those that are sticky (cohesive) such as clays and those that are non-sticky (non-cohesive) such as sands. Normally, cohesive soils are fine grained and once the particles are agitated into suspension in a liquid they remain suspended and can only settle in still waters where conditions are favourable for them to join together in bigger units through coagulation or flocculation. Coagulation is an electrochemical process which is greatly influenced by the salinity of the water. Cohesive materials often go in open channels as wash loads (completely mixed up with the fluid and behaving like fluid particles). Deposition of these fine materials occurs mainly in reservoirs and estuaries where there are low velocities and the salinity is high.

Practically all non-cohesive material forming in the beds of alluvial streams is quartz sand with a specific gravity in the range 2·64 to 2·67 (average 2·65). Because of the larger particle size and the smaller tendency to flocculate these sands exhibit much more individuality in their movement in water. The movement of the sand particles is quite distinct from that of the water causing their movement. This chapter deals mainly with the bed movement of alluvial streams and we make only passing references to sedimentation in reservoirs.

7.1.2 Particle Size and Shape

As far as its motion is concerned the most important characteristics of a sediment particle are its size and shape. The size and shape of particles making up a stream bed vary widely. In dealing with the movement of the bed of a stream, therefore, statistical estimates of the size and shape of the particle become necessary. Sieve analysis is usually used to determine the gradation of sedimentary materials forming an alluvial river channel. Table 7.1 indicates a general description of the various sizes of sediments found in practice.

The size distribution of a sediment is often adequately described by the median size (D_{50}) and the geometric standard deviation (σ_g). The median size corresponds to the sieve diameter which retains 50% by weight of a representative sample of the sediment. The geometric standard deviation is the ratio of the sieve diameter which passes or retains 50% (by weight) of the sample to that which passes 15·9%, or the ratio of the sieve diameter which passes 84·1% to that which passes or retains 50%, i.e.

$$\sigma_g = D_{50}/D_{15\cdot 9} = D_{84\cdot 1}/D_{50}$$

(D_N represents the percentage N, by weight, of the sample which passes through a sieve with hole diameter D.) It is often more appropriate to define σ_g as the average of the two ratios

i.e. $$\sigma_g = \frac{1}{2}\left(\frac{D_{50}}{D_{15\cdot 9}} + \frac{D_{84\cdot 1}}{D_{50}}\right)$$

Apart from the fact that the shape of a sediment particle influences its passage through a sieve hole, its influence on the movement of the particle in water is significant. Geologists and engineers therefore give attention to the shape and roundness of sediment particles. The former refers to the form of the particle and the latter to the sharpness or otherwise of the edges. The shape is ideally defined in terms of a true sphere by taking the ratio of the surface area of the sphere with the same volume to the surface area of the particle. In practice it is more convenient to describe the shape in terms of a dimensionless shape factor (S.F.) using the lengths a, b and c of the longest, the intermediate and the shortest mutually perpendicular axes of the particle. Thus

$$\text{S.F.} = c/(ab)^{\frac{1}{2}}$$

The nearer the value of S.F. is to unity the more spherical the particle is. The concept of nominal diameter is sometimes used to refer the particle to (the diameter of) a sphere of equal volume.

Another important physical description of a sediment sample is the geometric mean size d_g or d. G. H. Otto has shown that this can be determined from a size-frequency graph plotted on logarithmic probability paper instead of the normal semi-logarithmic paper. The geometric mean size is read at the intersection of the 50% line and the straight line joining the cumulative 15·9% point to the 84·1% point. However, unless explicitly stated in the problems which follow, the median size D_{50} may be assumed equal to the geometric mean size.

Table 7.1

Sediment grade scale

Class name	Size range (mm)	Size range (μm)	Size range (in)	Approximate sieve mesh opening per inch: Tyler	Approximate sieve mesh opening per inch: United States standard
Very large boulders	4096–2048		160–80		
Large boulders	2048–1024		80–40		
Medium boulders	1024–512		40–20		
Small boulders	512–256		20–10		
Large cobbles	256–128		10–5		
Small cobbles	128–64		5–2·5		
Very coarse gravel	64–32		2·5 –1·3		
Coarse gravel	32–16		1·3 –0·6		
Medium gravel	16–8		0·6 –0·3	$2\frac{1}{2}$	
Fine gravel	8–4		0·3 –0·16	5	5
Very fine gravel	4–2		0·16–0·08	9	10
Very coarse sand	2·000 –1·000	2000–1000		16	18
Coarse sand	1·000 –0·500	1000–500		32	35
Medium sand	0·500 –0·250	500–250		60	60
Fine sand	0·250 –0·125	250–125		115	120
Very fine sand	0·125 –0·062	125–62		250	230
Coarse silt	0·062 –0·031	62–31			
Medium silt	0·031 –0·016	31–16			
Fine silt	0·016 –0·008	16–8			
Very fine silt	0·008 –0·004	8–4			
Coarse clay	0·004 –0·0020	4–2			
Medium clay	0·0020–0·0010	2–1			
Fine clay	0·0010–0·0005	1–0·5			
Very fine clay	0·0005–0·00024	0·5–0·24			

7.1.3 Fall Velocity

The drag D on a spherical object immersed in a fluid is determined by Stokes' Law (equation 7.1) when the Reynolds number ($R = \rho vd/\mu$) is less than 0·1 and by Newton's Law (equation 7.2) for larger Reynolds numbers.

$$D = 3\pi\mu dv \text{ (Stokes' formula)} \tag{7.1}$$

and

$$D = \pi/8(C_D \rho d^2 v^2)\text{(Newton's formula)} \tag{7.2}$$

where μ and ρ are the dynamic viscosity and density respectively of the fluid, d is the diameter of the sphere, v is the velocity of the fluid relative to the sphere and C_D is the coefficient of drag.

When a spherical particle falls freely through a fluid at a constant terminal velocity w, the drag force on the sphere balances its submerged weight. Thus equations (7.1) and (7.2) give respectively,

$$w = \frac{\gamma}{18\mu} d^2 \left(\frac{\gamma_s - \gamma}{\gamma}\right) \quad (7.3)$$

and

$$w^2 = \frac{4}{3} \frac{g}{C_D} d \left(\frac{\gamma_s - \gamma}{\gamma}\right) \quad (7.4)$$

where γ and γ_s are the specific weights of the fluid and the falling sphere respectively.

The coefficient of drag C_D is a function of Reynolds number and equation (7.4) reduces to (7.3) for $C_D = 24/R$. The general relationship as determined from experiments by various people is shown in Fig. 7.1. The auxiliary scale of $F/\rho v^2$

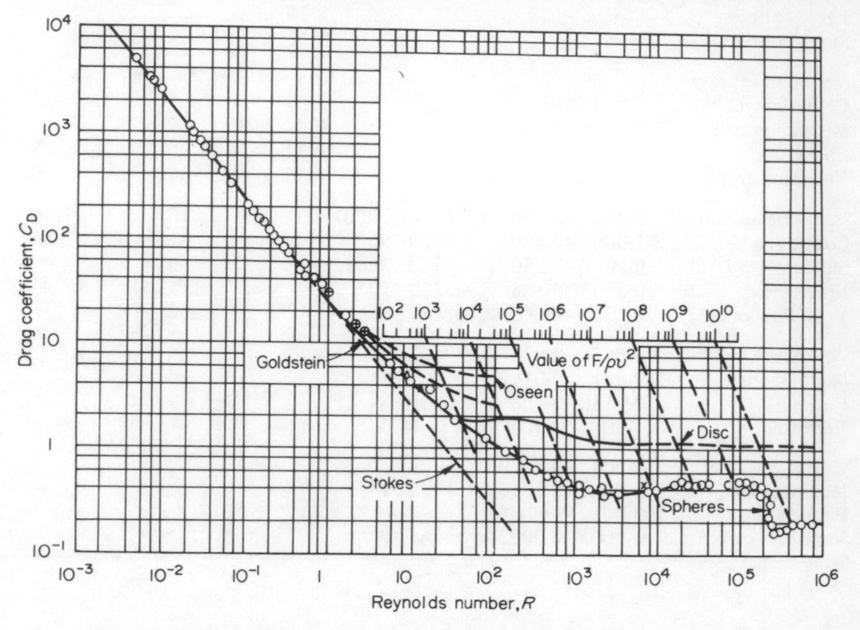

Fig. 7.1 Drag coefficient of spheres as a function of Reynolds number

helps in determining w once d, γ_s, γ and μ are known. F is the submerged weight, $\pi d^3(\gamma_s - \gamma)/6$. The value of R corresponding to a particular F is read from Fig. 7.1 and w is calculated from $R = \rho w d/\mu$ and checked against the value calculated using equation (7.4). For spheres of specific gravity of 2·65 settling in air or water Fig. 7.2 may be used directly to determine the fall velocity.

Theoretical and experimental attempts have been made in recent years to account for the non-spherical form of natural particles in determining their settling characteristics. Although the sedimentation diameter (defined as the diameter of a sphere of the same specific weight and the same terminal settling velocity as the given particle in the same sedimentation fluid) can be determined from a measured fall velocity using Fig. 7.1 and equation (7.4), its value depends on the temperature of the settling fluid. It is therefore necessary to standardize the determination by measuring a *standard fall velocity* in quiescent distilled water

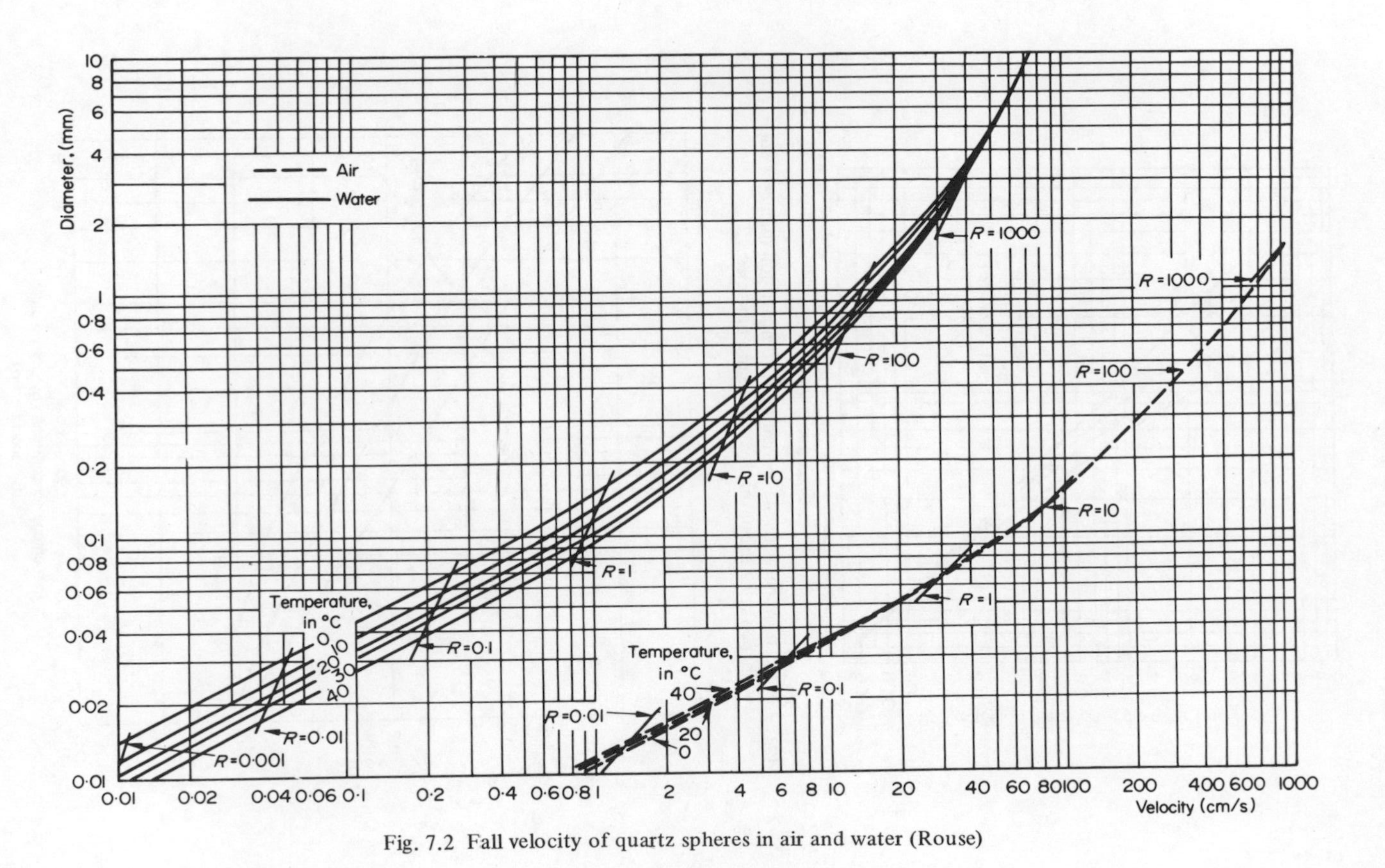

Fig. 7.2 Fall velocity of quartz spheres in air and water (Rouse)

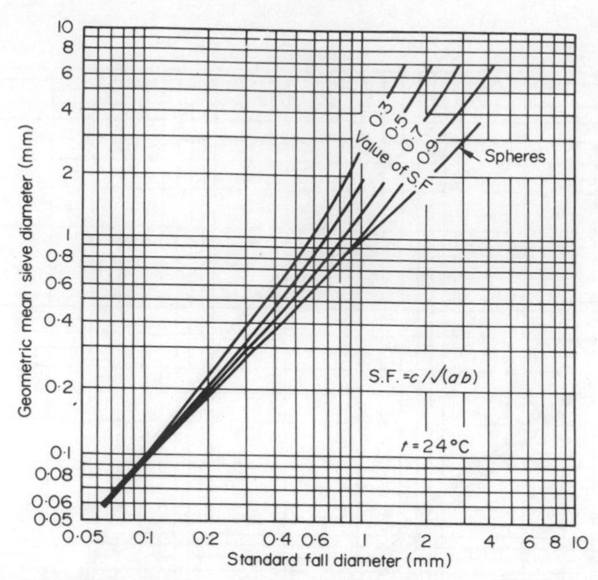

Fig. 7.3 Relation between sieve diameter, standard fall diameter and shape factor for naturally worn sand particles

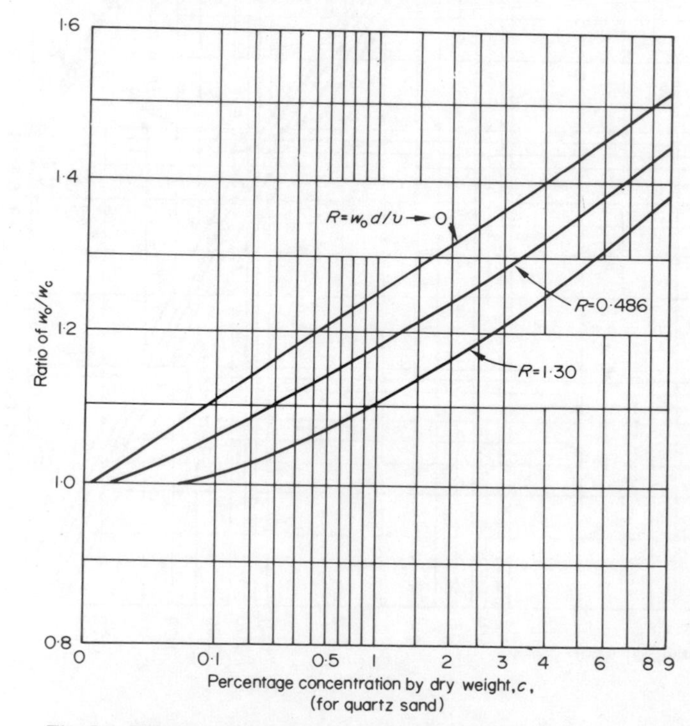

Fig. 7.4 Effect on concentration of settling velocity of uniform quartz spheres. (After McNown and Lin)

(of infinite extent) at a temperature of 24°C. Figure 7.3 shows the relation between the geometric sieve diameter and the standard fall diameter as a function of shape factor for naturally worn sand particles. In order to calculate fall velocity, the sedimentation diameter can be considered constant for temperature variations.

Sediment concentration naturally influences the settlement process. Very closely spaced particles tend to fall in a group with a velocity higher than that of a particle falling alone. McNown and Pin Nam Lin's theoretical prediction of the influence of concentration on settling velocity of sand is shown in Fig. 7.4. w_0 is the settling velocity in clear fluid and w_c is the velocity in a suspension of concentration c by dry weight, $R = w_0 d/v$. The curves are not applicable in cases when coagulation results from electrochemical processes.

7.2 Mechanics of Sediment Transport

7.2.1 Flow in Alluvial Channels

It has been well established through laboratory and field observations that changes in the bed configuration of alluvial streams occur according to a pattern governed by changes in flow rate. The following excerpt from the ASCE Sedimentation Committee report No. 31942 gives a vivid description of the progressive influence of flow rate on the movement of bed particles.

> Starting with an artificially flattened bed, at very low velocities, no sediment will move, but at some higher velocity individual grains will roll and slide along the bed. At a still higher velocity, some grains will make short jumps, leaving the bed for short instants of time and returning either to come to rest or to continue in motion on the bed or by executing further jumps. The material moved in this manner is said to saltate and is called the saltation load of the stream. If now the flow velocity is raised gradually, the jumps executed by the grains will occur more frequently and some of the grains will be swept into the main body of the flow by the upward components of the turbulence and kept in suspension for appreciable lengths of time. Sediment that is carried in suspension in this manner is known as the suspended load of a stream.

The net effect of the individual particle movements brings about a continuous change of the stream bed. The induced bed surface features translate and change with changing flow parameters. Four bed features may be developed as the flow velocity increases beyond the critical or threshold velocity.

(1) Ripples which are three-dimensional regular wavy features (but irregular in plan) between 10 cm and 30 cm long and up to about 7 cm high. They have a flat upstream face and a steep downstream slope.

(2) Dunes which are essentially enlarged ripples with long straight crests. Small ripples may be observed superposed on dunes. Ripples begin to appear on the surface of the stream and 'boils' caused by eddies downstream of the dune crests begin to appear as dunes form. A dune may be from 60 cm to several metres long and from 15 cm to a metre or so high.

(3) Further increase in flow velocity wipes out the dunes, resulting in flat or slightly wavy beds.

(4) Excessive velocities produce anti-dunes which are reversed dunes in appearance translating upstream. The water surface shows definite wavy patterns with the formation of anti-dunes.

In addition to a continuously changing bed configuration, alluvial streams exhibit a continuous change of longitudinal alignment. Local variations of the numerous parameters such as roughness, depth, soil gradation, channel width, etc., produce local hydrodynamic effects which cause the stream to deviate from a straight path. The initial tendency to deviate causes further imbalance with the

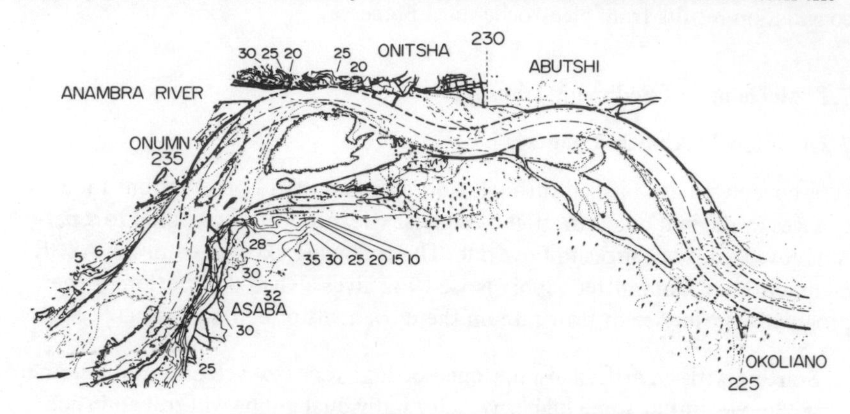

Fig. 7.5 River regulation project, lower Niger (Nigeria)
(Proposed by NEDECO)
Note that the training walls are provided at the outside of bends where erosion is most critical

result that the stream becomes snake-like and winding and is said to meander. Material is constantly eroded from the outside (concave bank) of bends and deposited along the inside (convex bank) of the bend (Fig. 7.5).

The processes responsible for the meandering of rivers are not fully understood. Experiments have however established that the phenomenon depends on the discharge and hydraulic properties of the channel, the sediment load and the relative erodibility of the channel bed and banks. The tendency of streams to erode material from the outside of bends and deposit it on the inside can partially be explained although mathematical analysis has so far been unsatisfactory.

Centrifugal forces give rise to raised water surface levels at the outside of bends over those at the inside. Ippen and Drinker concluded from experimental observations that the total water surface rise as well as the surface profile at the upstream portion of the bend conformed quite closely to the trend predicted for free vortices. They further concluded that at decreased curvatures, the irrotational pattern of flow was suppressed in the downstream half of the bend. The implication therefore is that high velocities are transferred to the outer bank only at the beginning of the bend. The pressure imbalance created by the raised water level along the outer banks produces fairly strong under currents in the downstream half of the bend. The result is a clockwise spiral motion (facing downstream) on a left bend and counter-clockwise on a right bend. The phenomenon is further complicated by the existence of bed shear and the attendant velocity distribution with higher centrifugal accelerations for fluid particles nearer the surface. The general picture is therefore of strong surface currents moving toward the outer bend and strong undercurrents moving away from the outer bend, as demanded by continuity. Alluvial material is consequently removed from the outer bend toward the inner bend and, under favourable conditions, deposited.

It is important to appreciate, in view of the ever-changing hydrological conditions, that an alluvial river can hardly attain an equilibrium position. The complexity of the problem is emphasized by the fact that 'the current which erodes the channel is, in turn, controlled by the ever-changing shape of the channel it has eroded.'

7.2.2 Threshold of Particle Movement

The point at which a particle on a stream bed begins to move is the most significant factor in the mechanics of sediment transport. Attempts at prediction follow two lines: the so-called rational approach which seeks to express analytically the resultant force acting on a typical particle, and the analysis based on physical and dimensional consideration of the various unknown factors. The success of the rational approach has been limited by the complexities of the interaction of particles with each other on the stream bed, the irregular shapes of the particles and the nature and intensity of fluid turbulence near the bed. Until these factors and their influence are sufficiently well understood, the semi-empirical dimensional approach will continue to be more convenient.

The parameters defining the channel geometry, flow, fluid and sediment properties are numerous but the most significant ones may be listed as:

y_0, depth of flow (L)
B, average width of channel (L)
v, average velocity of flow (L/t)
S_f, slope of the energy grade line (L/L)
G, rate of sediment transport (weight per unit time) (ML/t^3)
d_g or d, average (geometric mean) size of sediment particle (L)
ρ, density of the flowing liquid (M/L^3)
μ, viscosity of the flowing liquid (M/Lt)
ρ_s, density of the bed material (M/L^3)
τ_0, bed shear stress (M/Lt^2)
and g, gravitational acceleration (L/t^2)

Combinations of the various parameters which affect the sediment transport can be obtained using dimensional analysis. The form of the relationship can, however, only be determined experimentally. Taking buoyancy into account the effective density of the solid particle is given by $\Delta\rho = \rho_s - \rho$. The parameters may be combined with v, d and ρ (see Section 5.5) to give a function f as

$$f(y_0/d, B/d, S_f, G/(\rho v^3 d), \mu/(\rho v d), \Delta\rho/\rho, \tau_0/(\rho v^2), gd/v^2) = 0 \tag{7.5}$$

The dimensionless parameters of equation (7.5) may be combined in many ways to give different dimensionless groups. For example, the last three can be replaced by one by dividing the seventh term by the sixth and eighth to give

$$\frac{\tau_0}{\rho}\Bigg/\left(g\,\frac{\Delta\rho}{\rho}\,d\right)$$

The fifth term can also be modified by dividing it by the square root of the seventh to

$$\mu/[\rho d\sqrt{(\tau_0/\rho)}]$$

These new numbers are forms of the Froude number and the reciprocal of Reynolds number respectively with the shear velocity defined

$$v' = \sqrt{(\tau_0/\rho)} = \sqrt{(g y_0 S_f)}$$

(see equation (4.10)). Equation (7.5) may thus be replaced by a new function ϕ

$$\phi\left(\frac{y_0}{d}, \frac{B}{d}, S_f, \frac{G}{\rho v^3 d}, R', \frac{F'}{S_s - 1}\right) = 0 \tag{7.6}$$

or

$$R' = \phi_1\left(\frac{y_0}{d}, \frac{B}{d}, S_f, \frac{G}{\rho v^3 d}, \frac{F'}{S_s - 1}\right) \tag{7.7}$$

where

$$R' = \frac{\rho v' d}{\mu}, \quad F' = \frac{v'^2}{gd}$$

and S_s is the specific gravity of the solid particle (ρ_s/ρ) referred to the fluid density.

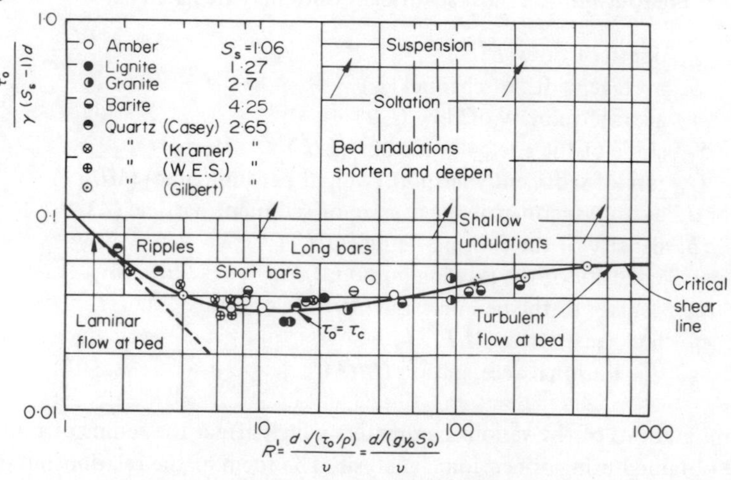

Fig. 7.6 Analysis by Shields (sediment transport)

Shields performed an experiment to establish the functional relationship for parameters similar to those in equation (7.7). His results (Fig. 7.6) have not clearly established the influence of flow depth, channel width, S_f, and the amount of sediment transport. They do however give a clear indication of threshold conditions particularly the bed shear stress at which the particle begins to move. Shields' critical stress values agree quite well for reasonably flat beds but not for already duned or rippled beds.

7.2.3 Sediment Load

When it has started moving the bed particle may be transported by the stream in a number of different ways including suspension, saltation, sliding, rolling and combinations of these. Apart from extremely fine grained particles which stay permanently in suspension as wash load, sediment load may broadly be classified as bed load or suspended load. Bed load materials do not leave the bed for long and suspended load materials travel predominantly in suspension.

Professor J. R. D. Francis described the activity below the water surface of a moveable bed river in these words:

> As the water speed increases, more and more sand moves, sliding over the remainder of the particles which remain nearly still. At higher speeds the particles are more agitated, clearly coming more under the influence of the fluid turbulence, and being suspended in the body of the stream for appreciable periods. Two mechanisms can be seen at work, one close to the bed where the particles collide so often that there is a dispersive pressure acting upward and so lifting them; and the other further from the bed where fluid turbulence is at work. It should not be assumed that the latter mechanism only affects small and claylike particles, for in flood (when rivers change their beds more rapidly) large stones can be brought near the water surface and carried downstream large distances.

By bed load in this context is meant materials that slide, roll and saltate. The deciding factor is that bed load material is of such a size and weight that it cannot stay in suspension for long. Smaller particles are easier to keep in suspension by turbulence and follow a distribution similar to that shown in Fig. 7.9. Typical sediment distribution in the Niger and Benue are illustrated by measurements indicated in Table 7.2.

One of the challenging problems connected with erodible bed streams is the prediction and measurement of the rate of transport of the sediment load. The total load transport is made up of the bed load and the suspended load. Since each type differs significantly with respect to characteristic movement it is obvious that their determination would also be significantly different.

BED LOAD

One of the earliest published attempts to predict sediment transport rate of rivers

Table 7.2

Yearly transports of sediment in the Niger and Benue (Nigeria)

(After NEDECO)

Station	Year	Bed load		Suspended load		Wash load		Total
		$10^6 m^3$	% of total	$10^6 m^3$	% of total	$10^6 m^3$	% of total	$10^6 m^3$
Upper Niger								
Baro	1956	0·29	8	0·29	8	3	84	3·6
	1957	0·44	7	0·82	13	5	80	6·3
	average 1934–57	0·31	6·5	0·55	11	4	82·5	4·9
Benue								
Yola	1956	0·20	6	0·17	5	3·0	89	3·3
	1957	0·28	6	0·24	5	4·2	89	4·7
	average 1934–57	0·20	6	0·17	5	3·0	89	3·3
Makurdi	1956	0·57	5	0·87	8	9·5	87	11
	1957	0·86	5·5	1·3	9	13	85·5	15
	average 1932–57	0·61	5	0·95	8	10	87	12
Lower Niger								
Shintaku	1956	0·72	5	1·0	7	13	88	15
	1957	1·2	5·5	1·8	9	18	85·5	21
	average 1915–57	0·88	5	1·3	8	15	87	17
Onitsha	1956	0·76	5	1·2	8	14	87	16
	1957	1·3	6	2·2	10	19	84	22
	average 1925–57	1·0	5·5	1·7	9	16	85·5	19

was by Du Boys in 1879. Du Boys hypothesized that the bed of a stream could be considered as a series of superposed layers with a velocity varying linearly from zero below the surface to a maximum value at the interface between the fluid and the solid bed. This tractive force approach led him to develop a formula for predicting the rate of transport of bed material as

$$q_b = C_s \tau_0(\tau_0 - \tau_c) \qquad (7.8)$$

where q_b is the rate of material transport per unit width of stream (L^2/t); C_s is a sediment parameter; τ_0 is bed stress and τ_c may be considered the critical value of the shear stress. Values of C_s for different quartz sand sizes and the values of τ_c were established by Straub, and are shown in the table below.

d (mm)	0·125	0·25	0·50	1	2	4
C_s (10^{-5} m^6/N^{2}s)	3·29	1·95	1·18	0·69	0·41	0·24
τ_c (N/m^2)	0·76	0·81	1·05	1·53	2·44	4·30

The structure of most of the later empirical formulae (proposed mainly in the

1930's) follow that of Du Boys which can also be written using Manning's formula (uniform flow) as

$$q_b/q = C_s n^{1\cdot 2} \gamma^2 (q^{0\cdot 2} - q_c^{0\cdot 6}/q^{0\cdot 4}) \tag{7.9}$$

where S_0 is the bed slope; q the fluid discharge per unit width and q_c the fluid discharge per unit width corresponding to the critical shear stress τ_c.

Some of the more recent formulae are:

Shields $$q_b = 10qS_0 \frac{\tau_0 - \tau_c}{\gamma (S_s - 1)^2 d} \tag{7.10}$$

Chang $q_b = An\, \tau_0(\tau_0 - \tau_c)$ for uniform sand

U.S. Waterways Experiment Station $q_b = A\, (\tau_0 - \tau_c)^m/n$ for graded sand

Meyer-Peter $q_b = (Aq^{\frac{2}{3}} S_0 - Bd_{90})^{\frac{3}{2}}$

Schoklitsch $q_b = AS_0^{\frac{3}{2}}\, (q - q_c)/\sqrt{d}$ for uniform sand

McDougall $q_b = A\, S_0^m\, (q - q_c)$ for graded sand

In the above formulae, A, B and m are constants and S_0 is the average bed slope.

In the 1940s a new batch of research was carried out, notably by H. A. Einstein and Kalinske who apparently for the first time attempted to predict bed sediment transport taking simultaneously into account fluid turbulence, the boundary layer and the statistical distribution of particle size and velocity. Later theoretical attempts by Einstein, Babarosa and Bagnold also included studies of the influence of bed configuration based on the assumption that the shear stress could be split up into two components due to grain roughness and bed form (i.e. skin friction and form drag (see Section 1.7)).

The divergence of the various theories as illustrated by Fig. 7.7 is clear enough evidence of the divergence of opinion on the mechanics of sediment transport. The problem is further complicated by the difficulties of field measurement. The general principle of the measuring devices is the trapping and screening of the moving material. None of the trapping devices so far developed has been found entirely satisfactory for determining bed load, suspended load or the total load. This is generally responsible for the scatter of measured data which sometimes diverge as much as the theories which seek to predict them.

SUSPENDED LOAD

Fine grained particles are kept in suspension by turbulent agitation. The vertical components of turbulent motion carry fluid particles and with them sediment particles up and down. The net rate of flow through a unit area in the horizontal

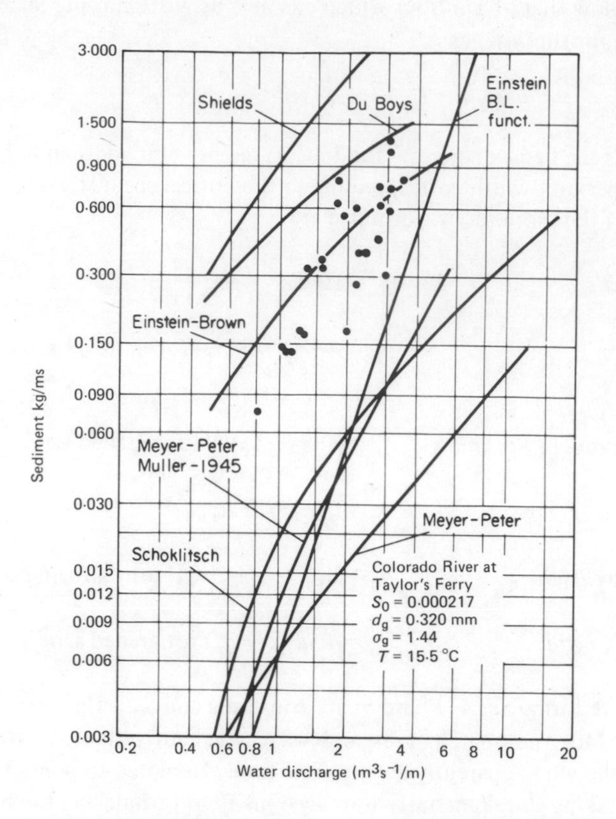

Fig. 7.7 Theoretical relationships between sediment transport and water discharge in the Colorado River, with some observed points added. Notice the large range of predictions given by the various theories, and the scatter of the observations. (After Francis)

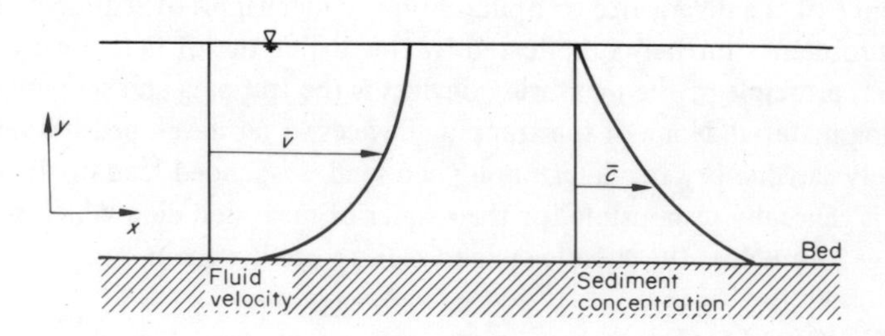

Fig. 7.8

plane is however zero under steady two-dimensional conditions. The vertical motion of sediment particles can thus be viewed as the resultant of a net turbulent mixing (diffusion) process determined by the intensity of turbulence and the concentration of sediments on the one hand and a downward settling movement governed by the bouyant weight of the particle on the other. Assuming an average sediment concentration $\bar{c}$ in the vertical y direction (Fig. 7.8) the flux of sediment due to turbulent diffusion is given according to equation (1.10b) by

$$w' = -E_m \frac{\partial \bar{c}}{\partial y}$$

The flux due to settlement with a velocity w is given by $\bar{c}w$. Under steady state conditions the net flux at any point is zero, and the settling velocity w may be assumed equal to the terminal value. Replacing Eddy diffusivity E_m by the extensively used symbol ϵ_s the suspended load distribution is therefore given by

$$\bar{c}w + \epsilon_s \frac{d\bar{c}}{dy} = 0 \tag{7.11}$$

or

$$\log_e \bar{c}/\bar{c}_a = -w \int_a^y \frac{dy}{\epsilon_s} \tag{7.12}$$

where the sediment concentration is $\bar{c}_a$ at $y = a$. Equation (7.11) is sometimes referred to as Schmidt's equation.

The main problem in the solution of (7.12) concerns the distribution of ϵ_s. The simplest approach relates it to momentum transfer. Assuming that it is proportional to Eddy viscosity ϵ it can be related to shear stress (see equation (4.1)) by

$$\tau = \frac{\rho \epsilon_s}{\beta} \frac{d\bar{v}}{dy}$$

where β is a constant of proportionality such that $\epsilon_s = \beta\epsilon$.

If it is further assumed that the shear stress is linearly distributed such that it is a maximum τ_0 on the bed and is zero at the water surface

$$\epsilon_s = \beta \frac{\tau_0}{\rho} \frac{(1 - y/y_0)}{d\bar{v}/dy} \tag{7.13}$$

where y_0 is the depth of flow.

Furthermore from the Prandtl–Karman hypothesis (equation (4.2)),

$$\frac{d\bar{v}}{dy} = \frac{K}{y} = \frac{2{\cdot}5\sqrt{(\tau_0/\rho)}}{y} \tag{7.14}$$

From (7.13) and (7.14)

$$\epsilon_s = 0{\cdot}4\, \beta v'\, y(1 - y/y_0) \tag{7.15}$$

Substitution of (7.15) into (7.12) yields

$$\frac{\bar{c}}{\bar{c}_a} = \left[\frac{y_0 - y}{y} \frac{a}{y_0 - a}\right]^Z \tag{7.16}$$

where

$$Z = 2{\cdot}5w/(\beta v') \tag{7.17}$$

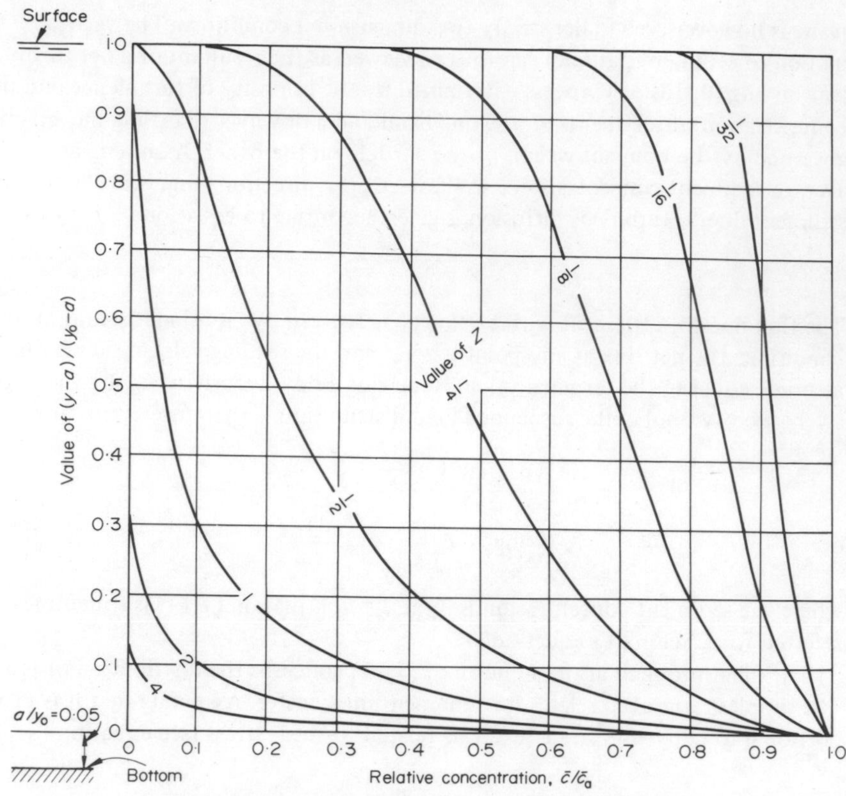

Fig. 7.9 Graph of suspended load distribution equation for $a/y_0 = 0{\cdot}05$ and several values of Z

A plot of equation (7.16) for different values of Z is given in Fig. 7.9. It is observed that for low values of Z (small fall velocity and/or large turbulence) the suspended sediment concentration tends to be uniform over the depth whereas for large values of Z (large fall velocity and/or small turbulence) the load tends to be concentrated near the bed. Experimental measurements and data collected from rivers have been reported which confirm the shape of distribution predicted by equation (7.16). The major problem has been how to determine Z and to a lesser extent a. The determination largely depends on the best fit approach. It has also been observed that the Karman constant (equation 7.14) can be as high as 5 for flows with high concentrations as opposed to the clear water value of 2·5 used above. A refinement of the above theory which takes into account the space occupied by the sediment particles and uses von Karman's definition of mixing length

$$\left(l = k\,\frac{\mathrm{d}\bar{v}}{\mathrm{d}y}\bigg/\frac{\mathrm{d}^2\bar{v}}{\mathrm{d}y^2}\right)$$

has been proposed by J. N. Hunt. A summary of Hunt's analysis is in Raudkivi's *Loose Boundary Hydraulics.*

The transport rate of suspended load may be obtained by integrating the product of stream velocity $\bar{v}$ and concentration over the depth $y = a$ to $y = y_0$. The transport rate per unit width of channel is:

$$q_s = \int_a^{y_0} c\bar{v}\, dy \tag{7.18}$$

Adopting a velocity distribution attributed to Keulegan as (compare with equation (4.3)),

$$v = 2{\cdot}5\, v' \log_e 30{\cdot}2y/\Delta \tag{7.19}$$

where $\Delta = k_s/\chi$ (k_s is grain roughness ($= D_{65}$) and χ is a roughness correction factor which is equal to unity for uniform grain sizes), H. A. Einstein expressed equation (7.18) using (7.16) as:

$$q_s = \int_a^{y_0} \left\{ \bar{c}_a \left[\frac{y_0 - y}{y} \cdot \frac{a}{y_0 - a} \right]^Z 2{\cdot}5\, v' \log_e \frac{30{\cdot}2y}{\Delta} \right\} dy \tag{7.20}$$

Defining dimensionless numbers $A = a/y_0$, $\eta = y/y_0$

$$q_s = 2{\cdot}5 v' \bar{c}_a y_0 \left(\frac{A}{1-A} \right)^Z \int_A^1 \left\{ \left(\frac{1-\eta}{\eta} \right)^Z \log_e \frac{30{\cdot}2 y_0 \eta}{\Delta} \right\} d\eta \tag{7.21}$$

After expansion, equation (7.21) becomes

$$q_s = 2{\cdot}5 v' \bar{c}_a y_0 \left(\frac{A}{1-A} \right)^Z \left[J_1 \log_e \frac{30{\cdot}2 y_0}{\Delta} + J_2 \right] \tag{7.22}$$

where

$$J_1 = \int_A^1 \left(\frac{1-\eta}{\eta} \right)^Z d\eta \tag{7.22a}$$

and

$$J_2 = \int_A^1 \left\{ \left(\frac{1-\eta}{\eta} \right)^Z \log_e \eta \right\} d\eta \tag{7.22b}$$

7.3 Design of Stable Alluvial Channels

7.3.1 Objectives

Channels should be designed so that neither deposition nor erosion of the channel material will occur. If, therefore, a channel carries sediment laden water or is made through an erodible material the flow velocity must be high enough to avoid sedimentation and at the same time low enough so as not to initiate scouring. In other words, if the water entering a reach of a channel brings with it an appreciable sediment load, the design must be such that the sediment transport capacity of the flow is equal to the rate of sediment supply.

The way a channel is to be used naturally also influences its design. For example, a canal intended to transport water from a higher to a lower level over a short distance would have a high slope whereas a canal requiring maximum conservation of potential head would have as gentle a slope as possible. In all cases, economic considerations demand the minimum amount of excavation and the bank slopes must be stable (see subsections 4.1.4 and 4.1.5).

The basic criteria for a successful design of alluvial channels may be summarized as:

(1) Keeping the Froude number (v^2/gy_0) low (less than about 0·1).

(2) Keeping the sediment discharge concentration, $\frac{q_b + q_s}{q}$, low (less than about 500 ppm).

(3) Avoiding long straight reaches in a windy country. This avoids large wind-generated waves which complicate the sediment transport properties of the channel (see Chapter 10).

(4) Ensuring that the sediment transport capacity of the channel is equal to the quantity of sediment diverted into the channel.

The design of alluvial channels is approached in two different ways; the empirical approach based on the so-called 'regime' concept and formalized methods based on tractive force considerations. Both approaches are summarized in the next two subsections.

7.3.2 The Empirical Approach (Regime Method)

A channel is said to be in regime if its net erosion or deposition is zero over a hydrological cycle (of high and low discharges). One of the earlier attempts to summarize observations of such a river in the form of a design formula was due to Kennedy (1895). He suggested that the velocity v is related to the depth y_0 by

$$v = 0{\cdot}84\, y_0^{0{\cdot}64} \tag{7.23}$$

The deficiencies of Kennedy's formula which does not take into account the physical properties of the channel (slope, size and bed materials) are obvious. Lacey attempted to remove this deficiency by modifying the formula to

$$v = 1{\cdot}17\, (f_s m)^{\frac{1}{2}} \tag{7.24}$$

where m is the hydraulic mean depth and f_s is a form of friction factor since designated as Lacey's silt factor. Lacey related f_s to Manning's n by

$$n = 0{\cdot}022 f_s^{\frac{1}{5}} \tag{7.25}$$

and to particle size by

$$d = f_s^2/64 \tag{7.26}$$

where d is in inches.

Equations (7.24) and (7.25) (with some simplifying assumptions) can be combined with Manning's formula to determine the desired slope for an alluvial channel in regime carrying a discharge Q as:

$$S_0 = 0{\cdot}00039 f_s^{\frac{3}{2}}/Q^{\frac{1}{9}} \tag{7.27}$$

which does not indicate the influence of the size and shape of the channel section. Table 7.3 was given by Lacey as typical values of f_s for various materials.

Table 7.3

Lacey's silt factor

	f_s
Large stones	38·60
Large boulders, shingle and heavy sand	20·90
Medium boulders, shingle and heavy sand	9·75
Small boulders, shingle and heavy sand	6·12
Large pebbles and coarse gravel	4·68
Heavy sand	2·00
Coarse sand	1·56–1·44
Medium sand	1·31
Standard Kennedy silt (Upper Bari Doals)	1·00

The above summary of Lacey's regime equations is principally meant to indicate his general line of thinking which originated from the general form of Kennedy's formula $v = ay_0^b$ where a and b are constants. Lacey's subsequent modifications yielded the following empirical formulae for design purposes:

wetted perimeter $$P = 2{\cdot}67Q^{\frac{1}{2}} \quad (7.28)$$

cross-sectional area $$A = 1{\cdot}26Q^{\frac{5}{6}}/f_s^{\frac{1}{3}} \quad (7.29)$$

average velocity $$v = 0{\cdot}794Q^{\frac{1}{6}}/f_s^{\frac{1}{3}} \quad (7.30)$$

bed slope $$S_0 = 0{\cdot}000\,55 f_s^{\frac{5}{3}}/Q^{\frac{1}{6}} \quad (7.31)$$

The slope given by equation (7.31) is obtained by defining n in the Manning formula in terms of the Lacey's so-called absolute roughness factor n_a as:

$$n_a = 0{\cdot}939 m^{\frac{1}{12}} n = 0{\cdot}0225 f_s^{\frac{1}{4}} \quad (7.32)$$

which implies that $v \propto m^{\frac{3}{4}} S_0^{\frac{1}{2}}$

According to Lacey's 'theory', therefore, once the discharge and silt factor are known a stable channel can be designed. According to equation (7.28), the wetted perimeter for a stable channel is constant for a given discharge irrespective of the fineness or otherwise of the bed material. Furthermore, the shape of the channel cross-section defined by P/m can be shown from equations (7.28) and (7.29) to be dependent on the velocity only. But field observations do not support these inferences.

Raudkivi on the other hand has used Shields' bed-load equation (7.10) to show that Lacey's silt factor is dependent on many important variables such as sediment load, discharge, particle size, depth of flow and channel slope. Thus its value may differ appreciably from river to river and from the values given in Table 7.3. Furthermore, the narrow range of silt sizes over which the results of Lacey's observations were made calls for extreme caution in their application.

Blench, following a similar approach to Lacey, introduced factors for the bed and sides which he claimed represented the mean tractive force intensity. The depth of flow y_0 and average width

$$B = \frac{\text{wetted cross-sectional area}}{y_0}$$

were given in terms of the bed factor F_b and side factor F_s as

$$y_0 = (F_s Q/F_b^2)^{\frac{1}{3}} \tag{7.33a}$$

$$B = (F_b Q/F_s)^{\frac{1}{2}} \tag{7.33b}$$

The combination of (7.33a) and (7.33b) gives, for rectangular channels,

$$F_b = v^2/y_0 \quad \text{and} \quad F_s = v^3/B$$

The tractive force claim was based on the deduction that F_s multiplied by $\rho\mu$ (a constant) gives the dimensions of the square of stress. This claim has, however, been questioned (see Raudkivi) on the grounds that B is a lateral dimension while a longitudinal dimension for a similar drag equation derived from laminar boundary layer theory is $\rho\mu v^3/l$.

Blench recommended the values of 0·3 (highly cohesive banks), 0·2 (moderately cohesive banks) and 0·1 (slightly cohesive banks) for the side factor, F_s. The bed factor F_b was postulated to depend on the median sediment size D_{50} (mm) and on the bed load concentration c for values higher than 20 ppm. The expressions given were as follows:
For $c < 20$ ppm,

$$F_b = 1{\cdot}9\sqrt{D_{50}} \tag{7.34}$$

and

$$S_0 = \frac{F_b^{\frac{5}{6}} F_s^{\frac{1}{12}} \upsilon^{\frac{1}{4}}}{3{\cdot}63 g\, Q^{\frac{1}{6}}} \tag{7.35}$$

and for $c > 20$ ppm (for non-uniform sands)

$$F_b = 1{\cdot}9\sqrt{D_{50}}\,(1 + 0{\cdot}012c) \tag{7.36}$$

and

$$S_0 = \frac{F_b^{\frac{5}{6}} F_s^{\frac{1}{12}} \upsilon^{\frac{1}{4}}}{3{\cdot}63 g(1 + 0{\cdot}000429c) Q^{\frac{1}{6}}} \tag{7.37}$$

where

$$\upsilon = \mu/\rho$$

The results of Blench's regime equations (7.33) together with (7.34) and (7.35) or (7.36) and (7.37) offer an improved alternative to Lacey's equations for the design of an alluvial channel. The difficulties with regard to the choice of the bed and side factors, especially the latter, nevertheless persist. It is important to realize the local nature of the regime equations which have been based on very limited number of physical variables. It is also important to point out that the constants given in the above equations are based on the engineers' (ft, lbf, s) system of units. A summary of the existing regime theory equations can be found in the American Society of Civil Engineers' Task Force report on friction factors in open channels (see Further Reading).

7.3.3 Tractive Force Methods

A tractive force method of alluvial channel design has as its premise the stability or otherwise of a bed or bank particle or layer under the influence of fluid shear, solid friction and gravity. Three physical phenomena are of major importance in trying to rationalize the equilibrium of channel materials: namely the fact that the fluid shear or drag intensity is not constant over the wetted boundaries, that the resistance to displacement of a particle on the side slopes is different from

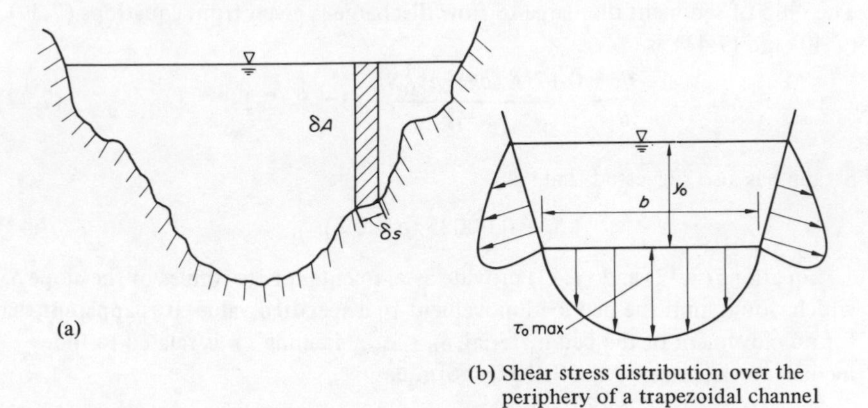

(a)

(b) Shear stress distribution over the periphery of a trapezoidal channel

Fig. 7.10

that of a particle on the flat part of the bed and that the shear strength (capacity to withstand drag forces) of soils varies widely with soil type. For flow under normal conditions the average shear stress τ_0 on a segment of the boundary δs (see Fig. 7.10(a)) is given according to equation (4.10) by

$$\tau_0 = \rho g S_0 \frac{\delta A}{\delta s} \tag{7.38}$$

A typical shear stress distribution on the boundaries of a trapezoidal channel is illustrated in Fig. 7.10(b). The maximum shear stress $\tau_{0\ max}$ is estimated to be about 140% of the average for the whole section $\rho g m S_0$, and the maximum value on the sides about 108%, occurring at 0·1 to 0·2 of the depth (see Lane, 1953). The rational (tractive force) method of design seeks to specify the critical shear stress τ_c and provide a design which guarantees that the shear stress is everywhere less than the critical value.

The earliest published rational analysis of the sediment problem is believed due to Du Boys (equation 7.8). Using $\tau_c = \rho g y_0 S_c$, the equation is

$$q_b = C_s(\rho g)^2 y_0^2 S_0^2 (1 - S_c/S_0) \quad (7.39)$$

where S_c is the slope that would sustain critical conditions. From Manning's equation, the discharge per unit width for a wide channel is

$$q = \frac{1{\cdot}0}{n} y_0^{\frac{5}{3}} S_0^{\frac{1}{2}} \quad (7.40)$$

Straub has suggested that the sediment parameter C_s may be expressed in terms of the average sediment size d in millimetres as

$$C_s = 0{\cdot}17/d^{\frac{3}{4}} \quad (7.41)$$

The ratio of sediment discharge to flow discharge is given from equations (7.39), (7.40) and (7.41) as

$$\frac{q_b}{q} = \frac{0{\cdot}17/\text{A}\,(\rho g)^2 n y_0^{\frac{1}{3}} S_0^{\frac{3}{2}}}{d^{\frac{3}{4}}} (1 - S_c/S_0) \quad (7.42)$$

Straub has also suggested that

$$S_c = 0{\cdot}00025\,(d + 0{\cdot}8)/y_0 \quad (7.43)$$

Equations (7.42) and (7.43) provide assessments for the values of the slope S_0 which would limit the bed load movement to a specified value. It is apparent that for no movement of the bed material, $S_0 < S_c$, Manning's n is related to the median particle size by the Strickler formula

$$n = D_{50}^{\frac{1}{6}}/31{\cdot}3 \text{ (when } D_{50} \text{ is in inches)} \quad (7.44)$$

$$= D_{50}^{\frac{1}{6}}/53{\cdot}8 \text{ (when } D_{50} \text{ is in mm)}$$

Raudkivi has indicated that other bed-load functions can be treated in a fashion similar to that illustrated above using empirical coefficients. He has used Shields' equation, for example, to obtain

$$q_b/q = \frac{10 S_0}{(S_s - 1)} \left[\frac{y_0 S_0}{(S_s - 1)d} - 0{\cdot}056 \right] \quad (7.45)$$

This is based on the relation

$$\tau_c/(\rho g(S_s - 1)d) = 0{\cdot}056$$

for turbulent flows (see Fig. 7.6). For sands ($S_s = 2{\cdot}6$), this corresponds to

$$d = \frac{\tau_c}{0{\cdot}09 \rho g} = 11 y_0 S_c \text{ approximately} \quad (7.46)$$

Using the approximation in equation (7.46), Raudkivi has derived approximate expressions for the average velocity, slope and discharge as

$$v = 28m^{\frac{1}{2}}S_0^{\frac{1}{3}} \tag{7.47}$$

$$Q = 28Pm^{\frac{3}{2}}S_0^{\frac{1}{3}} \tag{7.48}$$

and

$$S_0 = 0{\cdot}45d^{1{\cdot}15}\,Q^{-0{\cdot}46} \tag{7.49}$$

More recent approaches initiated by Lane have used more sophisticated methods for determining the equilibrium of a particle on the side slopes. Shear forces as well as particle and fluid weights are resolved in the longitudinal (flow) as well as down-slope directions. With the assumption that the shear stress on an element of the boundary is balanced by the component of weight of water vertically above it and that the maximum shear stress ($\tau_{0\,max} = \tau_c$) occurs at

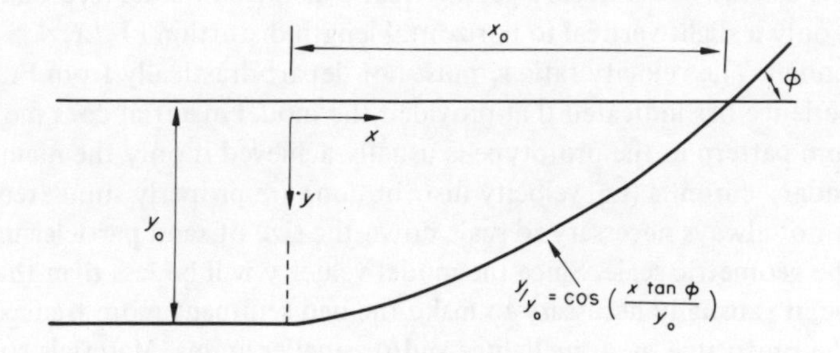

Fig. 7.11 The stable alluvial section

maximum depth (Fig. 7.11) the U.S. Bureau of Reclamation has derived the following results for a stable cross section (see Raudkivi or Lane for details)

$$\text{cross-sectional area}\quad A = 2y_0^2 \cot\phi \tag{7.50}$$

$$\text{wetted perimeter}\quad P = \frac{2y_0}{\sin\phi}E \tag{7.51}$$

where ϕ is the angle of repose, y_0 is the maximum depth and

$$E = \int_0^{\pi/2} [1 - \sin^2\phi \sin^2\theta]^{\frac{1}{2}}\, d\theta \tag{7.52}$$

with

$$\theta = \frac{x \tan\phi}{y}$$

Their results show that the stable cross-sectional shape is a cosine curve, the half-width of which is

$$x_0 = \frac{\pi}{2}y_0 \cot\phi \tag{7.53}$$

and the hydraulic mean depth

$$m = \frac{y_0 \cos \phi}{E} \tag{7.54}$$

We conclude by emphasizing that the above predictions have been based on the conditions for the inception of particle movement. The equations can therefore not be applied to beds rippled or covered with dunes, a sure indication that the bed is already moving. An interesting comparison between the regime and tractive force methods of stable channel design is given in Raudkivi's *Loose Boundary Hydraulics*.

7.4 Moveable Bed Models

The principal requirement in a moveable bed model is that the bed material must move. This usually necessitates a geometrically distorted model (see Chapter 5) although only a slight vertical to horizontal length distortion ($Y_r/X_r \leqslant 5$) is recommended. The velocity ratio v_r must not depart drastically from Froude's law. Experience has indicated that provided the model material does move the equilibrium pattern in the prototype is usually achieved if only the main flow and secondary currents (i.e. velocity distribution) are properly simulated. It is therefore not always necessary to scale down the size of sand particles in proportion to the geometric scale. Since the model velocity will be less than that of the prototype, it is usually necessary to make the bed sediment more transportable than in the prototype by using lighter and/or smaller grains. Materials commonly used are processed coal, pumice, perspex, plastics and crushed fruit stones.

Model verification is essential in moveable bed studies. Since the hydraulic resistance of a sand bed is variable and hard to predict, the velocity ratio must be established in the model as that which gives correct energy slope and water

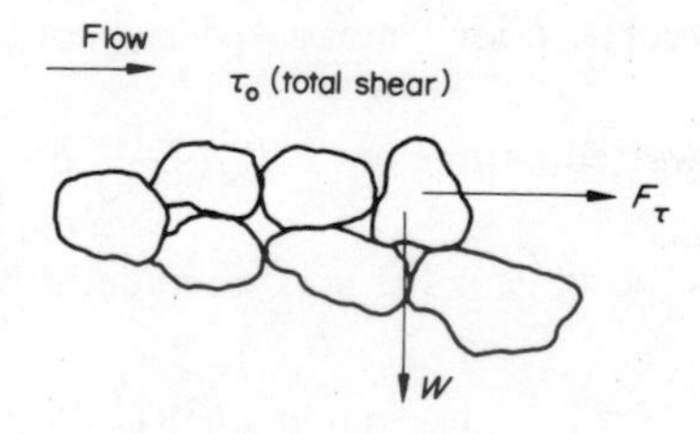

Fig. 7.12

surface. A model is verified if, after being adjusted to reproduce accurately known events in the prototype, it is able to indicate future occurrences as well.

Two distinct forces act on a bed-load particle; the buoyant weight of the particle W and the shear force F_τ (Fig. 7.12). The movement of the particle is dependent on the relative magnitude of these forces. If a leading dimension of the particle is its mean diameter d the two forces are given by

$$W \propto (\gamma_s - \gamma_f)d^3$$

and

$$F_\tau \propto \tau_0' d^2$$

where γ_s and γ_f are the unit weights of the solid particle and fluid respectively and τ_0' is the component of the shear stress due to the grain roughness (skin friction). This plus the drag τ_0'' due to the form of the bed make up the total shear stress τ_0, i.e.

$$\tau_0 = \tau_0' + \tau_0'' \tag{7.55}$$

$$\frac{F_\tau}{W} = \text{constant} \times \frac{\tau_0'}{(\gamma_s - \gamma_f)d} \tag{7.56}$$

For the particle to move at all the shear stress must exceed a critical value which may be determined from Shields' diagram of Fig. 7.6. It is apparent from equation (7.56) that dynamic similarity of the grain movement will be achieved when

$$\left[\frac{\tau_0'}{(\gamma_s - \gamma_f)d}\right]_m = \left[\frac{\tau_0'}{(\gamma_s - \gamma_f)d}\right]_p \tag{7.57}$$

Similarity for the settlement of the suspended particles is achieved when the ratio of the fall velocity w to the shear velocity v' is the same in the model and prototype.

i.e.

$$(w/v')_m = (w/v')_p \tag{7.58}$$

The shear velocity $v'(=\sqrt{(\tau_0/\rho)})$ is related to the average velocity v of the stream by

$$(v'/v)^2 = f/2 \tag{7.59}$$

where the friction factor is defined in accordance with equation (7.55) as

$$f = f' + f'' \tag{7.60}$$

where f is the total friction factor corresponding to τ_0
f' is the friction factor due to grain roughness corresponding to τ_0'
and f'' is the friction factor due to bed form corresponding to τ_0''

The settling velocity in water may be obtained from Fig. 7.2 if the particle is sand (specific gravity = 2·65) and f' may be determined from a pipe friction diagram.

As the result of observations by various workers the following criteria have been established as a guide to the classification of the problems involved.

If $w/v' < 0{\cdot}1$, wash load dominates and the problem has to be re-examined.

If $0{\cdot}1 \leqslant w/v' \leqslant 0{\cdot}6$, suspended load dominates and modelling should be according to equation (7.58).

If $w/v' > 0{\cdot}6$, bed load dominates and modelling should be according to equation (7.57).

In establishing the above criteria the interaction of particles has been ignored and the problem therefore oversimplified. It is also impossible in practice to have only suspended load movement or only bed-load movement. Consequently the criteria can only be used as an aid to choosing the initial model and cannot be a substitute for proper model verification. The solution involves trial and error as demonstrated in the example below.

EXAMPLE 7.1

We want to build a model to investigate a reach of a sand bed stream with the following properties:

$$\text{slope,}\quad S_0 = 0{\cdot}0036$$
$$\text{average velocity,}\quad v = 2{\cdot}1\text{m/s}$$
$$\text{depth,}\quad y_0 = 3.7\text{ m}$$

The mean diameter of quartz bed materials (specific gravity = 2·65) is 1·20 mm. The prototype channel is very wide and the water temperature is 20°C for both prototype and model. If the horizontal scale is $X_r = 1/150$ and the vertical scale is $Y_r = 1/50$, investigate the suitability of quartz as model sand.

Prototype

$$R_p = \frac{4y_{0p}v_p}{\upsilon} = \frac{4 \times 3{\cdot}7 \times 2{\cdot}1}{1{\cdot}04 \times 10^{-6}} = 2{\cdot}97 \times 10^7$$

∴ The flow is fully turbulent

From Fig. 7.2 the settling velocity w for a sand grain of average diameter = 1·20 mm at 20°C is 20 cm/s approximately.

$$\text{Shear velocity}\quad v' = \sqrt{(\tau_0/\rho)} = \sqrt{(gy_0S_0)}$$
$$= \sqrt{(9{\cdot}81 \times 3{\cdot}7 \times 0{\cdot}00036)}$$
$$= 0{\cdot}113\text{ m/s}$$

$$(w/v')_p = \frac{20}{11{\cdot}3} = 1{\cdot}77 >> 0{\cdot}6$$

∴ Sediment transport in the stream is predominantly by bed-load movement.

Shear Parameters

The equivalent of the relative roughness k_s/D on a pipe friction diagram is

$$d_p/4y_{0p} = \frac{1{\cdot}2}{4 \times 3700} = 0{\cdot}000081$$

From the pipe friction diagram (Fig. 3.5) and with $R_p = 3 \times 10^7$ the friction factor due to grain roughness $f' = 0{\cdot}003$.

From equation (7.59),

$$f = 2(v'/v)^2 = 2(0{\cdot}113/2{\cdot}1)^2$$
$$= {\cdot}0057$$

$\therefore$ Bed-form friction factor $\quad f_p'' = f_p - f_p' = {\cdot}0027$

Check if the bed material does in fact move.

$$R' = \frac{dv'}{\upsilon} = \frac{0{\cdot}0012 \times 0.113}{1{\cdot}04 \times 10^{-6}} = 132$$

From Shields' diagram (Fig. 7.6) the critical shear stress for movement of bed materials is

$$\frac{\tau_c}{(\gamma_s - \gamma_f)d} = 0{\cdot}048$$

This is less than

$$\frac{\tau_0'}{(\gamma_s - \gamma_f)d} = \left(\frac{\rho v^2 f'}{2}\right)\frac{1}{(\gamma_s - \gamma_f)d}$$
$$= \frac{10^3 \times (2{\cdot}1)^2 \times 0{\cdot}003}{2 \times 9{\cdot}81 \times 10^3 \times 1{\cdot}65 \times 1{\cdot}2 \times 10^3}$$
$$= 0{\cdot}35$$

$\therefore$ The prototype bed material definitely moves and it moves predominantly as bed load.

Model

Model must satisfy

$$\left[\frac{\tau_0'}{(\gamma_s - \gamma_f)d}\right]_m = \left[\frac{\tau_0'}{(\gamma_s - \gamma_f)d}\right]_p = 0{\cdot}35 \qquad (7.61)$$

From Chezy's law, velocity ratio, v_r

$$= \sqrt{\left(\frac{Y_r}{f_r}\,\frac{Y_r}{X_r}\right)}$$
$$= \sqrt{\left[\left(\frac{1}{50}\right)^2 \cdot \frac{150}{f_r}\right]} = \frac{0{\cdot}246}{f_r^{\frac{1}{2}}} \qquad (7.62)$$

Friction factor ratio, $\quad f_r = \dfrac{f_m' + f_m''}{f_p' + f_p''}$

Assuming bed forms are similar and therefore

$$f_m'' = f_p'' = 0{\cdot}0027$$

Then $\quad f_r = (f_m' + 0{\cdot}0027)/0{\cdot}0057 \qquad (7.63)$

Trial and error solution

Assume a value for $f'_m = 0{\cdot}005$. Corresponding to this friction factor, from Fig. 3.5, the relative bed roughness $d_m/4y_{0m}$ should lie between zero when $R = 6 \times 10^4$ and about 0·001 when $R \geqslant 1{\cdot}4 \times 10^6$. In practice however the model should operate in the fully turbulent region. From (7.63)

$$f_r = \frac{0{\cdot}0077}{0{\cdot}0057} = 1{\cdot}35$$

∴ from (7.62)

$$v_m = \frac{0{\cdot}246}{f_r^{\frac{1}{2}}} v_p = 0{\cdot}44 \text{ m/s}$$

From (7.61)

$$0{\cdot}35\, d_m = \frac{\tau'_0}{(\gamma_s - \gamma_f)} = \frac{\rho v_m^2 f'_m}{2(\gamma_s - \gamma_f)}$$

∴

$$d_m = \frac{0{\cdot}44^2 \times 0{\cdot}005}{0{\cdot}70 \times 9{\cdot}84 \times 1{\cdot}65}$$

$$= 0{\cdot}085 \text{ mm}$$

Relative roughness $d_m/4y_{0m} = 0{\cdot}0003$ since $y_{0m} = y_{0p}/50 = 0{\cdot}074$ m. According to this relative roughness and the corresponding friction factor, the model will operate quite close to a smooth bed at a Reynolds number of about 8×10^4. This model is therefore unsuitable in practice. Using this relative roughness for a new estimate, $f'_m = 0{\cdot}0038$ if the model operates near the fully turbulent zone at $R = 5{\cdot}5 \times 10^6$.

New: $f_r = 1{\cdot}12$, $v_m = 0{\cdot}48$ m/s

$$d_m = \frac{(0{\cdot}48)^2 \times 0{\cdot}0038}{0{\cdot}70 \times 16{\cdot}2} = 0{\cdot}078 \text{ mm}$$

Verifying: $d_m/4y_{0m} = 0{\cdot}00026$

This is quite close. Take $d_m = 0{\cdot}08$ mm

Check

Froude law $v_r = \sqrt{Y_r} = 1/7{\cdot}1$

From (7.62) $v_r = 1/4{\cdot}7$

The departure is reasonable. Will the model bed move?

$$R' = \frac{d_m v'}{\upsilon_m} = \frac{0{\cdot}08 \times 10^{-3}}{1{\cdot}04 \times 10^{-6}} \times v_m \sqrt{(f'_m/2)}$$

$$= 1{\cdot}64$$

And thus from Fig. 7.6

$$\frac{\tau_c}{(\gamma_s - \gamma_f)d_m} = 0{\cdot}06$$

which is less than the operative model value of 0·35. Thus the model particles will move.

Settling velocity w = 0·6 cm/s (from Fig. 7.2)

$$(w/v')_m = \frac{0{\cdot}006}{0{\cdot}021} = 0{\cdot}29 < 0{\cdot}60$$

Thus the model particles will move predominantly as suspended load. The model as planned is therefore unsuitable. Lighter but bigger model sediment materials are required.

FURTHER READING

American Society of Civil Engineers, *Progress Report of the Task Committee on Preparation of Sedimentation Manual*.

(i) 'Introduction and Properties of Sediment', *Proc. ASCE, Hydraulics Div.*, Vol. 88, No. HY4, 1962.

(ii) 'Erosion of Sediment', *Proc. ASCE, Hydraulic Div.*, Vol. 88, No. HY4, 1962.

(iii) 'Suspension of Sediment', *Proc. ASCE, Hydraulic Div.*, Vol. 89, No. HY5, 1963.

(iv) 'Density Currents', *Proc. ASCE, Hydraulics Div.*, Vol. 89, No. HY5, 1963.

Blench, T., *Regime Behaviour of Canals and Rivers*, Butterworths, London, 1957.

Brown, C. B. 'Sediment Transportation', *Engineering Hydraulics* (ed. H. Rouse), Chapter XII, Wiley, 1950.

Francis, J. R. D., Inaugural Lecture as Professor of Hydraulics, Imperial College of Science and Technology, March, 1967.

Ippen, A. T. and Drinker, P. A., 'Boundary shear stress in curved trapezoidal channels', *Proc. ASCE, Hydraulics Div.*, Vol. 88, No. HY5, 1962.

Lane, E. W., 'Progress Report on Studies on the design of stable channels for the (U.S.) Bureau of Reclamation', *Proc. ASCE*, No. 180, 1953.

Leliavsky, S., *An Introduction of Fluvial Hydraulics,* Constable, London, 1955.

'Progress Report of the Task Force on Friction Factors in Open Channels', *Proc. ASCE, Hydraulics Div.*, Vol. 89, No. HY2, 1963.

Raudkivi, A. J., *Loose Boundary Hydraulics*, Pergamon London, 1967.

8 Physical Hydrology and Water Storage

8.1 Introduction - The Hydrologic Cycle

Hydrology deals with the occurrence, circulation and distribution of the waters of the earth. It is also concerned with their chemical and physical properties and their reaction with their environment, including their relation to living things. Hydrology, which is, in fact, a branch of physical geography, is one of the oldest branches of science. Like many others, it started with the ancient philosophers, who considered water as a basic element along with the earth, air and fire.

The earliest application of hydrological concepts was naturally in the Mediterranean and the middle eastern areas, because of their precarious water situation. Greek and Roman philosophers were among the first to conceive theories about the occurrence and movement of water and Egyptian engineers were among the first to gather hydrological data to plan and to build structures to forecast or control water flow.

The basic concept of modern hydrology is the principle of the conservation of mass which relates water storage to evaporation, infiltration and precipitation through what is known as the hydrologic cycle. The concept of the hydrologic cycle is believed to have been initiated by da Vinci. He is frequently quoted as saying that 'water goes from the rivers to the sea and from the sea to the rivers' and that 'the saltness of the sea must proceed from the many springs of water which as they penetrate the earth, find mines of salt, and these they dissolve in part and carry with them to the ocean and other seas, whence the clouds, the begetters of rivers, never carry it up'.

The scope of hydrology now embraces the movement of water over and through the earth's crust, its use by plant life and its redistribution through the combined activity of the oceans and atmosphere. The interaction of these components forms a chain of events called the hydrologic cycle. This cycle is depicted

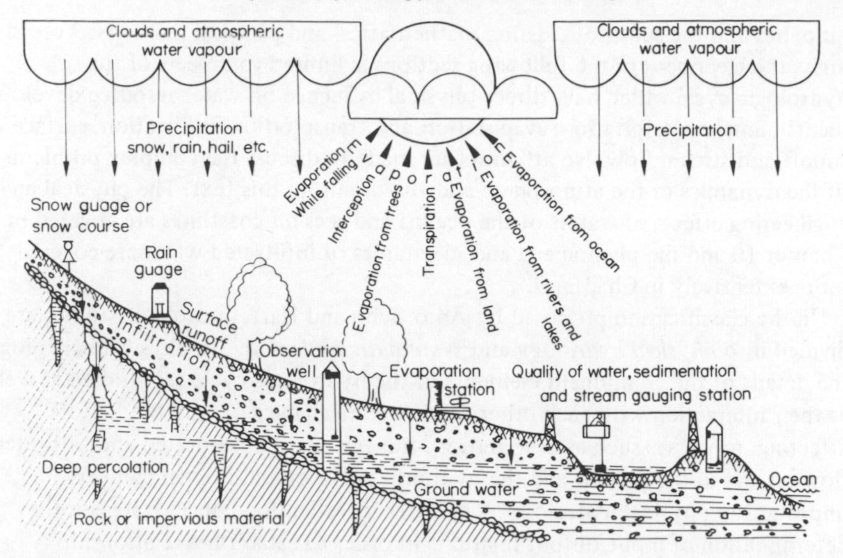

Fig. 8.1 The hydrologic cycle (From *ASCE Hydrology Handbook*, 1949, by UCOWR)

schematically in Fig. 8.1 following the *American Society of Civil Engineers Hydrology Handbook*.

Waters from the oceans and lakes evaporate into the atmosphere. The rising vapours are cooled, condensed and fall back to the earth in the form of rain, snow, hail, etc. Some of this precipitation is intercepted by vegetation and some of it falls directly onto the oceans and lakes and is re-evaporated into the atmosphere. The precipitation which reaches the ground either flows into streams and rivers and ultimately into the oceans or infiltrates below the earth's surface. Some of the subsurface water returns to the atmosphere through evaporation or transpiration by plants and some returns to the surface as streams or into the ocean, and the remainder is stored as groundwater.

A wide understanding of the fields of oceanography, meteorology, geology (earth sciences), biology (life sciences) and fluid mechanics is necessary for a thorough study of hydrology. The engineering hydrologist, however, is primarily concerned with the occurrence of water on the land surface as snow or ice or as liquid water that flows or is impounded, as well as with the occurrence of water below the surface in the interstices of the soil and the underlying rocks. He is normally only interested in atmospheric water in so far as precipitation and evaporation affect the surface and subsurface water.

8.1.1 Scope

A comprehensive study of hydrology also involves qualitative and quantitative identification of all relevant factors and the simultaneous solution of a set of equations governing all the physical processes of the hydrologic cycle. This is quite beyond the scope of existing mathematical and physical methods. Accordingly, the treatment in the following sections is limited to aspects of the hydrologic cycle which have direct physical influence on water resource development: namely precipitation, evaporation and transpiration, infiltration, surface runoff and stream flow. No attempts are made to discuss the complex problems of the dynamics of the atmosphere and the oceans in this text. The physical and engineering effects of waters of the oceans and seas on coastlines are covered in Chapter 10 and the phenomena and mechanics of infiltrated water are covered more extensively in Chapter 9.

In the classification proposed by Amorocho and Hart, hydrologic studies are divided into *physical hydrology* and *systematic hydrology*. In physical hydrology the details of the component elements in the hydrologic cycle are studied as well as their interaction with each other. Thus the mechanics of, and the factors affecting, processes such as evaporation, transpiration, unsaturated and saturated flow on and in the ground and the hydraulics of overland flow are all very important in physical hydrology. A systems investigation concentrates on the determination of input/output relations in order to establish a relationship suitable for prediction purposes.

A systems approach to a hydrologic basin may take one of two forms: (1) a deterministic or parametric method in which the system input functions (e.g. precipitation) are given explicit algebraic functional relations and (2) a probabilistic or stochastic method in which no specific functional relations are ascribed. A deterministic approach may assume lumped parameters in which physical relations at each point of the system are not of interest (a black box system) or distributed parameters showing variations (linear or non-linear) from point to point. A stochastic system may be purely probabilistic in which case the natural order of events is not considered or it may be purely stochastic demanding serious consideration of the natural sequence of events.

It is apparent from these remarks that no serious systems approach to a hydrologic basin is possible without a good knowledge of the nature of the physical elements of the basin. Success of the systems approach also depends on the quality and length of historical records. This book deals mainly with physical hydrology, the understanding of which is necessary for system hydrological studies at postgraduate level. However an introduction to modern trends of thinking in relation to systematic hydrology is given in Section 8.6.

8.2 Precipitation

8.2.1 Types of Precipitation

The term precipitation refers to all forms of water which falls from the atmosphere to the surface of the earth. It is generally estimated that only about a fourth of the total amount of precipitation which falls on continental land masses is returned to the seas by direct runoff and underground flow, the remainder is stored underground or returned to the atmosphere through evaporation and transpiration.

Several factors, including the amount of moisture evaporated into the atmosphere, land relief and meteorogical conditions, influence the amount of precipitation which falls in an area. All other things being equal, the amount of precipitation in an area varies directly with the amount of moisture available in the atmosphere above it. One must understand clearly the nature of temperature variations in the atmosphere to appreciate the influence of relief on condensation of moisture and the resultant precipitation.

From the first law of thermodynamics the adiabatic expansion of ideal gases (dry air) can be shown to follow the law

$$\frac{dT}{T} = \frac{R}{JC_p} \frac{dp}{p} \tag{8.1}$$

where T is the absolute temperature, p the absolute pressure, R the universal gas constant, C_p is the specific heat at constant pressure and J is the mechanical equivalent of heat.

Taking z in the upward vertical direction and assuming a static pressure distribution,

$$\frac{dp}{dz} = -\rho g \tag{8.2}$$

where ρ is the density of atmospheric air at temperature T and g is the gravitational constant. Substituting for $\rho = p/RT$

$$\frac{dp}{p} = -\frac{g}{RT}\, dz \tag{8.3}$$

Equation (8.3) may be substituted into (8.1) and the relationship $R/J = C_p - C_v$ used to obtain

$$\frac{dT}{dz} = -g\,\frac{(1-n)}{R} \tag{8.4}$$

where $n = C_v/C_p$ and C_v is the specific heat at constant volume.

The factor $\Gamma = -g(1 - n)/R$ is known as the dry adiabatic lapse rate and indicates the rate at which unsaturated air will cool as it is forced up. Its value for dry air is approximately -9·8°C/km (-5·5°F/1000 ft). The ambient atmospheric temperature, however, generally falls at a rate α = -6°C/km (-3·3°F/1000 ft) approximately. Thus if a 'packet' of dry air is forced up it cools faster than the surrounding air (since $|\Gamma| > |\alpha|$). Assuming the same pressure, the air inside the packet will thus become denser than that of the surrounding air. If the lifting force is removed the packet will fall back to its original equilibrium position. The condition is therefore said to be stable. If, however, the numerical value of the dry adiabatic lapse rate of a packet of gas is less than that for the atmosphere, the packet will continue to rise even if the lifting force is removed since it will be less dense than the ambient air. The condition is therefore unstable.

A similar unstable condition exists when the lapse rate of the atmosphere is greater than the rate of decrease in temperature of an ascending packet of air cooling by adiabatic expansion. As a moist unsaturated air packet is forced up, it cools and its relative humidity increases until at some elevation a condition of saturation is reached. Its adiabatic lapse rate is consequently reduced to a wet lapse rate of about 6°C/km (3°F/1000 ft) and the condition becomes unstable. Further cooling results in condensation of moisture and the latent heat acquired during evaporation is released as latent heat of condensation. The moisture is released from the packet and the released energy tends to warm up the air mass left in the parcel. The rate of release of energy together with other atmospheric conditions influence the form (liquid or solid) in which the precipitation reaches ground level. Liquid precipitation falls as drizzle or rain and solid precipitation as freezing rain, snow, hail or sleet.

There are three main causes of the initial lifting of air masses.

1. *Cyclones* The main horizontal air circulation in the earth's atmosphere is known to be concentrated about cells of high and low pressures. Cyclonic motions are produced when air currents rush into a low pressure cell from high pressure zones and anticyclones are produced when the reverse occurs. In the case of cyclones, the warmer and therefore lighter air is lifted over the colder, denser air. This results in a large turbulence and vortex motion. The rising motion is potentially unstable when the rising air is moist. Large atmospheric disturbances, clouds and precipitation are thus generally associated with cyclones. The cyclone is termed frontal if it is produced by the meeting of extensive fronts of air currents coming from cold and warm belts. It is non-frontal if the lifting is caused by horizontal convergence on a low pressure zone.

2. *Convection* Solar and other forms of heating of an unsaturated air mass near the ground cause it to expand and rise by convection. The convectional process is believed to follow the dry adiabatic pattern until the dew-point temperature is reached after which the lapse rate rapidly falls to a wet value rendering atmospheric conditions unstable. Condensation occurs and clouds form. A critical temperature is soon attained after further ascent and precipitation occurs. The convection process is responsible for most thunderstorms especially in the tropics. Convectional thunderstorms tend to be localized, brisk and short-lived as compared with cyclonic storms which are regional and continue for many hours or days.

3. *Geographical relief* Mountain barriers generally force up air masses in winds. Moisture-laden winds blowing from the oceans to the land encounter high land masses and go through the processes of lifting and condensation and give rise to relief or orographic precipitation. Orographic precipitation is normally heavier on the windward side of mountain barriers and it forms rain shadows on the leeward side.

8.2.2 Measurement of Rainfall

Rainfall is measured by a rain gauge which, in its simplest form, is an open container with vertical sides. Modern rain gauges are non-recording or self-recording. The non-recording rain gauge consists of a cylindrical receiver which passes the rainfall into a cylindrical measuring jar. Rainfall is measured on the basis of a vertical depth of water which would accumulate on a plane horizontal surface if all of it remained there. The measuring jar may be pre-calibrated in which case the depth of rainfall is read directly or a calibrated rod is used to measure the amount collected in the measuring receiver. Recording gauges give a continuous record of rainfall as a pen trace on a clock driven chart. Thus they give not only the amount of rainfall but also its intensity, and are particularly useful in remote areas which are not easily accessible. Two types, weighing or volumetric, are generally in use. One sophisticated design consists of a series of buckets which are filled through a funnel in turn. The bucket is so designed that when a fixed quantity of rainfall collects in it, it tips and empties the water into a storage can and at the same time brings another bucket under the funnel.

8.2.3 Area Variation of Precipitation and Methods of Averaging

Most hydrologic problems involve large river basins over which precipitation can hardly be uniform. The uniformity of rainfall over an area is believed to be dependent on the type, quantity and duration of storm and on the season as well as on the size and nature of the area. In hydrologic studies it is convenient to use an average precipitation and various methods of determining its value have been proposed. The error between the average value of rainfall over a basin and the value measured at a single station increases with a decrease in total rainfall over

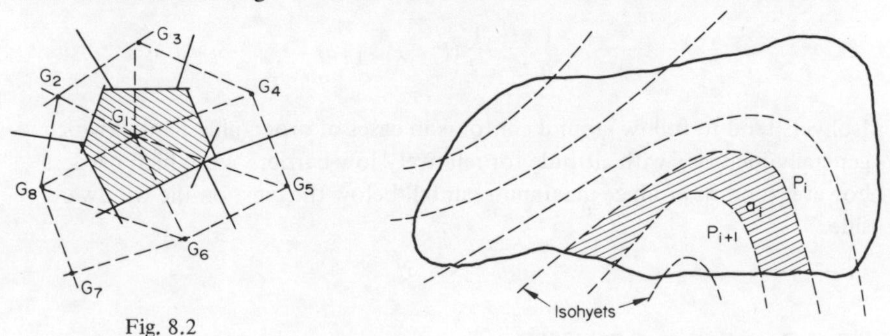

Fig. 8.2

Fig. 8.3 Isohyetal map

the basin and with increasing basin area and decreases with increasing duration of rainfall. It is also believed to be greater for convective and orographic rainfalls than it is for cyclonic rainfalls.

The simplest method of determining average rainfall over a basin is to average arithmetically the amounts measured by gauges located on the basin. This approach is satisfactory only when a large number of gauges are distributed uniformly over the area. It is important when adopting this type of averaging to consider orographic influences on precipitation in the selection of gauge sites in mountainous regions. The large cost involved in providing dense networks of gauging stations is usually prohibitive. The *Thiessen method* of averaging is more extensively used than arithmetic averaging. It allows for irregularities in rain gauge spacing by weighting the record of each gauge in proportion to the area which it is supposed to cover. The area includes all points which are closer to a particular gauge than to any other. This is found by drawing the perpendicular bisector to the lines joining adjacent stations (see Fig. 8.2). The resulting polygon surrounding a particular station identifies its effective area. Let the area of the polygon surrounding the ith station which records rainfall P_i be a_i. The weighted average rainfall for the basin is given by

$$\frac{1}{A} \sum_{i=1}^{N} a_i P_i$$

where A is the total area of the basin and N is the number of stations. The Thiessen method is more accurate than the arithmetic mean but the method is very inflexible since addition or removal of a new station necessitates readjustment of polygons and of weightings.

A more general method which is also considered to be more accurate than the Thiessen method is the *isohyetal method.* Contours of equal precipitation (isohyets) are drawn from the station records (see Fig. 8.3). The average precipitation of a basin is obtained by weighting the average precipitation between two contours with the area between them, totalling for the whole basin and dividing by its area. For example, if the area between isohyets P_i and P_{i+1} is a_i, the average rainfall for the basin of area A with N contour spacings is given by

$$\frac{1}{A} \sum_{i=1}^{N} \frac{1}{2} (P_i + P_{i+1})\, a_i$$

Isohyets tend to follow ground contours in cases of orographic rainfall which generally increases with altitude for relatively low barriers. Very high barriers however tend to produce maximum rainfall below the crest on the windward side.

8.2.4 Presentation of Rainfall Data

Many water resource development problems, e.g. storage for hydro power, require knowledge only of the total annual precipitation and its seasonal variations. For such projects rainfall data are usually required as annual (sometimes monthly) total precipitation over a number of years. Flood control problems, on the other hand, require a more thorough knowledge of changes in rainfall. Thus daily and hourly records become necessary. In designing drainage facilities for an area, the engineer needs a more precise knowledge of individual rainfalls; their amount, duration and intensity.

The average rainfall intensity is obtained by dividing the amount (depth) by the duration. Sometimes a mass curve of the rain storm (i.e. the cumulative depth versus time) is plotted and the slope of the curve at a point gives the intensity at the corresponding time. Self-recording rain gauges are useful in the determination of rainfall intensities.

A formula commonly used in estimating the *average* intensity of rainfall storms of duration less than 2 hours is

$$i_{av} = \frac{R}{t_r + C} \tag{8.5}$$

where i_{av} is the average intensity, t_r is the duration of rainfall, R is a rainfall coefficient and C is a constant. Both R and C vary from one locality to another and R also varies with the amount of rainfall. Intensities of rainfalls of duration greater than 2 hours are frequently expressed in the form

$$i_{av} = a/t_r{}^n \tag{8.6}$$

where a and n are constants dependent on locality and amount of rainfall.

8.3 Evaporation and Transpiration

Precipitation reaching the earth's surface is transmitted to surface and subsurface reservoirs by means of surface and/or subsurface flow. The water stored in the reservoirs is subsequently transferred back into the atmosphere by the processes of evaporation and transpiration. Evaporation is the process by which a liquid changes directly into vapour and transpiration is the transformation from liquid into vapour through plant metabolism.

While the former generally involves surface stored water, the latter generally involves subsurface water. The exceptions are evaporation of soil moisture and transpiration of surface water by some water plants. The total amount of water taken up by vegetation for tissue building and transpiration plus the evaporation of soil moisture, when added to water intercepted by plants during precipitation makes up what is known as *consumptive use* which receives considerable attention in irrigation engineering. All water losses in an area through evaporation of surface water, snow, ice and intercepted water plus transpiration by plants constitute *evapo-transpiration.*

8.3.1 Evaporation

Evaporation involves the transfer of fluid masses from a fluid surface into the atmosphere and accordingly would be expected to follow the mass diffusion law discussed in Section 1.5. The basic equation would thus be expected to be of the form

$$E = -k \frac{de}{dz} \tag{8.7}$$

where E is rate of evaporation, e is vapour pressure (indicating the concentration of fluid mass in the air), z is vertical distance and k is a transfer coefficient. Except the rare case of very stable atmospheric conditions under which turbulence does not exist, the transfer coefficient depends on atmospheric conditions such as wind speed, pressure, energy from the sun, effectiveness with which the water is heated, etc. Vapour pressure is dependent on temperature, relative humidity and salinity. The simplest form of equation (8.7) which is commonly used is Dalton's law

$$E = k\,\frac{(e_w - e_a)}{\Delta z} \tag{8.8}$$

where e_w is the saturated vapour pressure corresponding to the temperature of the water surface, e_a is the vapour pressure of the air above the water surface and Δz is the thickness of a thin film at the surface over which the vapour pressure is supposed to have dropped from e_w to e_a. Δz is often absorbed into the transfer coefficient to give

$$E = b(e_w - e_a) \tag{8.9}$$

The practical difficulty lies in determining the factor b. Controlled experiments (models) using standard evaporation pans are usually performed to establish equation (8.9) in terms of atmospheric conditions. The pan filled with water is installed on land or on the surface of a reservoir and the level changes in the pan are measured regularly together with wind speed, atmospheric temperature and water temperature. Modified forms of some of the results obtained from pan tests are given in the list below.

(1) Proposed by Morton

$$E = 42{\cdot}4\,(0{\cdot}6 + 0{\cdot}1u)\,\frac{e_w - e_a}{p}$$

(2) Proposed by Rohwer

$$E = 0{\cdot}0771\,(1{\cdot}465 - 0{\cdot}000733p)\,(0{\cdot}44 + 0{\cdot}118u)\,(e_w - e_a)$$

(3) Proposed by Horton

$$E = 0{\cdot}04\,[\{2 - \exp(0{\cdot}2u)\}\,e_w - e_a]$$

(4) Other formulae (Penman)

$$E = 0{\cdot}035\,(1 + 0{\cdot}24u)\,(e_a - e_d)\ \text{(grassland)}$$

and

$$E = 0{\cdot}050\,(1 + 0{\cdot}24u)\,(e_a - e_d)\ \text{(from water surface)}$$

In all expressions, E is measured in cm per day, u is the wind speed in mph at pan rim level, p is atmospheric pressure head in mm of mercury, e_w, e_a are vapour pressures at surface and air temperature respectively in mm of mercury and e_d is the dew-point vapour pressure also in mm of mercury. e_a in Penman's formulae is the saturation vapour pressure corresponding to the air temperature.

Reliance on pan evaporation formulae to predict evaporation from large natural bodies of water is limited by many factors among which are: (1) the fact that heat transfer from a small volume of water in the pan is certain to be different from that of a large body of water (because penetrating radiation warms up the pan water significantly more than a field reservoir with the result that the rate of evaporation is higher from the pan). A 'pan coefficient' (about 0·7 for a land pan and 0·8 for a floating pan) is usually introduced when pan formulae are applied to moderate and large bodies of water; (2) the nature and size of the exposed surface which have a significant effect on evaporation rates. Evaporation rates may not be proportional to the pan area because of the influence of side walls, vegetation, etc.; (3) the effect of waves, ripples and other disturbances which affect thermal stratification and density instabilities; and (4) differences in the level at which wind velocity, temperature and other atmospheric quantities are measured.

8.3.2 Transpiration and Potential Evapo-transpiration

The role of the atmosphere in the transpiration process is the same as it is in evaporation. It takes up the moisture and it exerts influence on the process through climatic and atmospheric conditions. The soil acts as the reservoir from which transpiration originates except in the case of acquatic plants. Two principal forces are believed to provide the mechanism for the transpiration process – osmotic force and capillary force. The osmotic force acts as a pump, sucking in water from the interstices of the soil. With the combined action of the osmotic and capillary forces the water is conducted through the low resistance conduit of the plant's xylem. However, H. L. Penman, one of the foremost modern authorities on evapo-transpiration feels that the role of osmosis might have been exaggerated in evaluation of transpiration. In transit, the water provides the plant with its nutrients and assists in the general metabolic process. The leaves of the plant provide spongy mesophyll cells around stomatal cavities from the surface of which the water escapes to the atmosphere through stomatal openings. The stomatal openings are self regulating and the opening depends on the quantity of water arriving in the cavities. Under unfavourable conditions, certain plants are known to dispose of over-stored water by guttation. This is believed to be through special organs called hydathodes which cause water to collect at the edges and tips of leaves by means of root pressure originating from the root cells. Because transpiration rates are extremely difficult to measure it is usually combined with evaporation and dealt with as *potential evapo-transpiration*.

Potential evapo-transpiration is defined as the maximum evapo-transpiration which can take place from an area. Potential evapo-transpiration is relevant if more water arrives at the leaves of the plant than is transpired. The rate of transpiration Q_T is estimated assuming that the vaporization process is due entirely to solar power P.

Thus

$$Q_{\mathrm{T}} = \frac{P}{\gamma H_{\mathrm{v}}} \tag{8.10}$$

where H_v is the mechanical equivalent of the latent heat of vaporization and γ is specific weight of water.

A plant eventually withers if Q_T is greater than the incoming flow into its leaves and would gutter if the reverse is true. One of the latest and reasonably accurate expressions for estimating potential evapo-transpiration (P.E.) is Hammon's equation

$$\text{P.E. (mm/yr)} = 0{\cdot}14D^2 \, \rho_s \tag{8.11}$$

where D is the number of possible hours of sunshine in units of 12 h and the saturated vapour density ρ_s (= 0·622 (e_w/RT)) is expressed in g/m^3. The saturated vapour pressure e_w is that corresponding to the mean daily surface temperature T (absolute) and R is the universal gas constant (dry air).

8.4 Infiltration

Infiltration is the process by which part of precipitation (rain water) enters the subsurface. During a rain storm, water particles enter the voids in the soil and fill them to saturation under sufficient rainy conditions and the water particles move down freely to join the underground reservoir. The rate at which water percolates

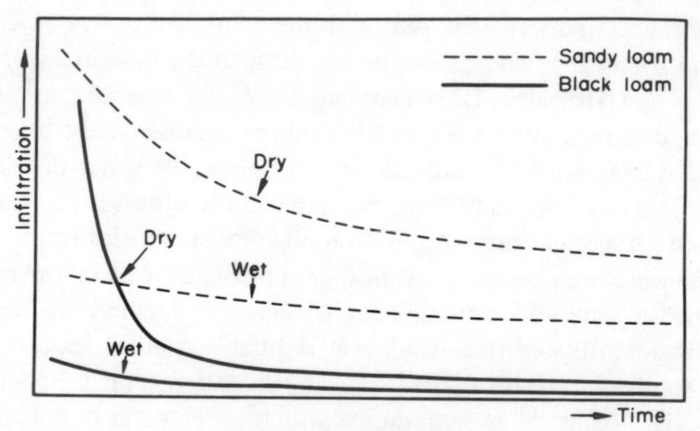

Fig. 8.4 Infiltration rates

into the ground is known as the infiltration rate or f-rate. Typical variations of infiltration rate with time during a storm are illustrated in Fig. 8.4. It is clear from the figure that the rate at which the soil absorbs moisture decreases with time and with the degree of saturation.

The maximum rate at which a particular soil absorbs moisture gives its infiltration capacity or f-capacity (f_p).

The infiltration capacity of a soil depends on the availability of water, the porosity, the vegetation cover as well as on the intensity of rainfall. A loose, permeable soil has a higher infiltration capacity than a compacted clayey soil. A high intensity of rainfall tends to clog the soil interstices with finer particles through impact action and therefore tends to reduce the rate of infiltration. Grass and other vegetation cover reduce the impact of rain drops on the soil and therefore enhance its infiltration. Fig. 8.5 illustrates the influence of soil moisture

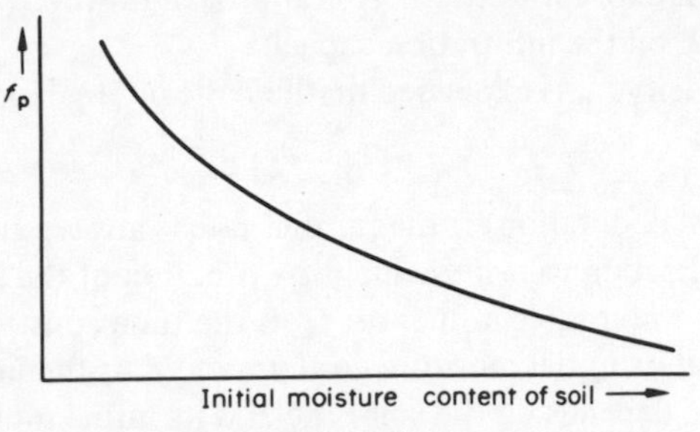

Fig. 8.5

content on infiltration capacity. Since the water content of a soil changes as it absorbs water, the infiltration capacity of the ground changes continuously during a rain storm until saturation is reached. Before saturation the infiltration rate is normally equal to the intensity of rainfall, both being less than the infiltration capacity.

However as saturation is approached the infiltration rate approaches the reduced infiltration capacity. The excess of rainfall over the infiltrating water accumulates and begins to flow as surface run-off. Fig. 8.6 illustrates the situation

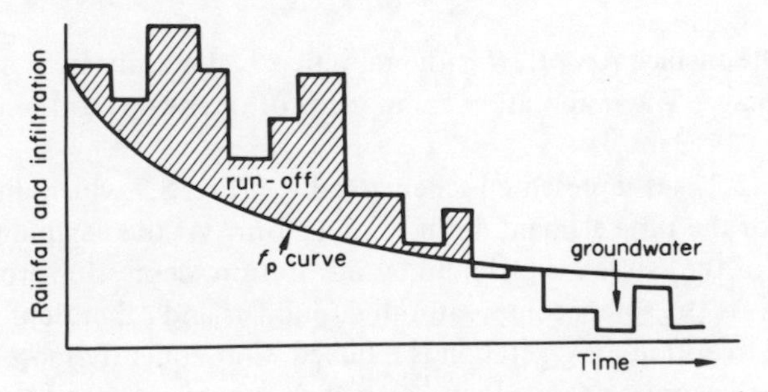

Fig. 8.6

for cleared and compacted areas (e.g. airports, roads, built-up areas) with homogeneous infiltration characteristics. The rainfall rate is generally higher than the infiltration capacity and the residual which runs off must be properly collected and drained away.

Measurement of infiltration rates can be direct or indirect. Using the indirect approach, the quantity of water penetrating the soil surface is measured by assuming it to be equal to the difference between rainfall and run-off. Such measurements may be broadly classified into laboratory experiments using aritificial rainfall, field experiments using artificial rainfall and field determinations on isolated plots or small drainage basins based on natural rainfall and run-off. It is difficult to determine reliably the variation of infiltration rate with time and with soil water content using this method. With the direct approach, small tubes partially imbedded in the soil are filled with water. The rate of fall of level in a tube indicates the rate of infiltration. This approach however, ignores the influence of rainfall on the infiltration capacity.

Infiltration capacity, f_p is expressed mathematically by Horton's formula as

$$f_p = f_\infty + (f_0 - f_\infty)\, e^{-t/\tau_w} \tag{8.12}$$

for $0 \leqslant t \leqslant t_r$ (rainfall duration) if the rainfall rate is always greater than f_p. f_∞ is the ultimate infiltration capacity which is a function of the soil type only, under insignificant water table influence. τ_w is the time constant which depends on the soil type and its initial moisture content and f_0 is the initial infiltration capacity which also depends on the soil type and its initial moisture content.

8.5 Surface Run-off (Overland Flow)

Theoretically the run-off resulting from a given storm is obtained by subtracting the amount of infiltration and surface retention from the rainfall. Surface retention includes depression storage (interception by vegetation, and water stored in puddles, ditches, etc.) and evaporation during rainfall. Thus mathematically, surface run-off is given by

$$Q = R - I - S \tag{8.13}$$

where Q is the surface run-off, R is the rainfall, I is the infiltration and S is the retention storage. The evaporation component of retention storage is considered to be small.

Equation (8.13) is represented schematically in Fig. 8.7 which illustrates the significance of the time element during a rain storm. At the beginning of the storm, most of the rainfall is taken up by infiltration, depression storage and interception. As the soil becomes saturated, puddles and other depressions become filled and interception by vegetation is reduced. Consequently more and more of the rain water becomes detained temporarily as a sheet over the basin. This is known as *surface detention* and is distinct from depression storage. The detention depth provides the necessary hydraulic gradient for flow and as the depth increases the flow increases until equilibrium is attained between run-off and rainfall.

Consider drainage from a plane, wide, paved area (Fig. 8.8), on which rain is falling at a uniform intensity i_0. The discharge from the area measured at the outlet B varies from zero at the time the rain starts, to a maximum at t_c. At t_c

conditions are uniform, and it is the time when the first particles from A, the farthest point from B, arrives at B provided the rain has not stopped before then. At time $t \geqslant t_c$, until the rain stops, as much water flows out of the basin as falls on it. Thus the discharge per unit width of a rectangular plot is $q = i_0 L$, where L is the distance between A and B. The uniform flow discharge may be assumed to obey Chezy's law. Thus

$$q = i_0 L = C d^{\frac{3}{2}} (\sin\theta)^{\frac{1}{2}} \tag{8.14}$$

since for small θ, $\sin\theta \simeq \tan\theta$. C is the Chezy constant, d is the detention depth and θ is the angle of slope of the line AB. The time of concentration t_c has been

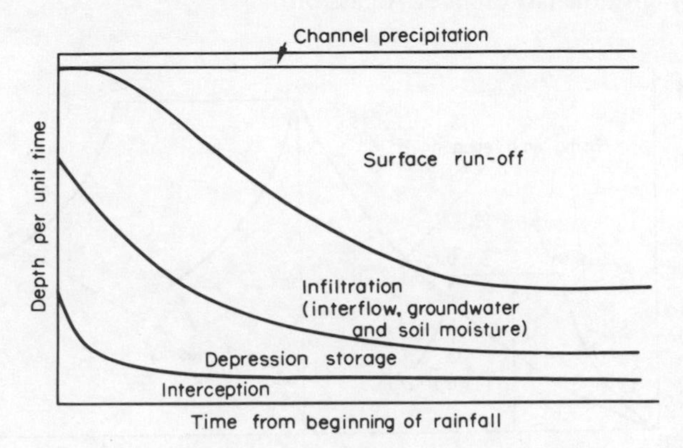

Fig. 8.7 Disposal of rain storm

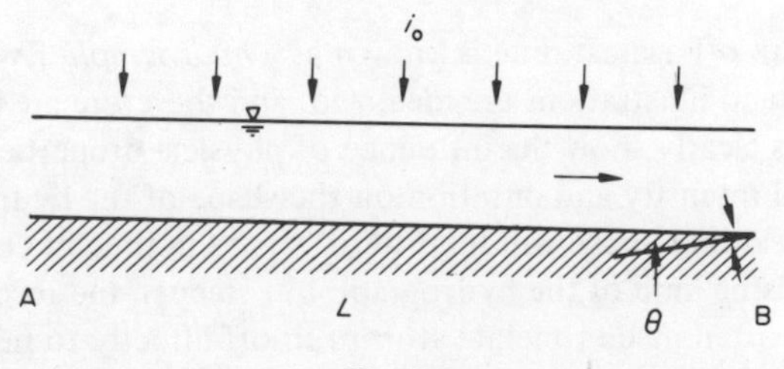

Fig. 8.8

shown by various investigators to be a function of the rainfall intensity and the properties of the drainage area. It is given by

$$t_c = \left[\frac{L}{C(i_0 \sin\theta)^{\frac{1}{2}}}\right]^{\frac{2}{3}} \tag{8.15}$$

Various researchers have used the method of characteristics in solving basic hydrodynamic and continuity equations to predict that the rising flow during $0 \leqslant t \leqslant t_c$ is given by

$$q = C(\sin\theta)^{\frac{1}{2}} i_0^{\frac{3}{2}} t^{\frac{3}{2}} \tag{8.16}$$

Three cases are illustrated in Fig. 8.9. In the first case the rainfall duration t_r is less than the time of concentration. The rising limb Oa_1 follows equation (8.16) until the rain stops. A constant discharge is maintained corresponding to the maximum detention depth d_{max} ($q_{max} = C(\sin\theta)^{\frac{1}{2}} d_{max}^{\frac{3}{2}} < i_0 L$) until the last rain drops at A arrive at B at $t = t_p$. The discharge then falls rapidly to zero along $b_1 c_1$. In the second case $t_r = t_c$. The discharge rises according to equation (8.16) along Oa_2 to a value given by equation (8.14) at $t = t_c$ and thereafter falls progressively along $a_2 c_2$ until the last rain drops from A arrive at B. The third case in which $t_r > t_c$ is represented by $Oa_2 b_2$. The discharge is maintained at the constant value $i_0 L$ along $a_2 b_2$ until the rain stops and it falls progressively along $b_2 c_3$ until the last drops at A pass out.

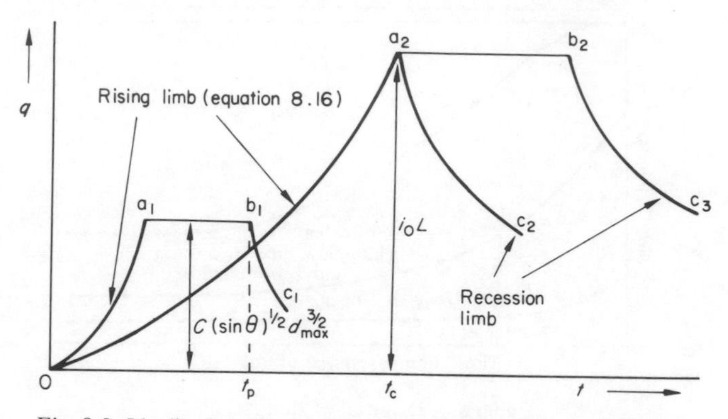

Fig. 8.9 Idealized outflow hydrographs (Adapted from Eagleson)

The plot of run-off against time is known as a *hydrograph*. Even though the above equations and illustrations are idealized, and therefore are limited in scope, they nevertheless clearly show the influence of physical properties of the drainage basin and rainfall intensity and duration on the shape of the hydrograph. Infiltration and depression storage will tend to increase the time of concentration and flatten the rising limb of the hydrograph but steepen the recession limb.

Attempts are often made to relate storm run-off directly to rainfall. Infiltration indices are sometimes used as a means for estimating run-off volumes from large areas. Two of the commonest infiltration indices are the W index and the ϕ index. The W index is defined by the equation

$$W = \frac{R - Q - S}{t_r} \tag{8.17}$$

where R is the amount of rainfall, Q is the surface run-off due to the storm, S is the effective surface storage (interception and depression storage) and t_r is the duration of the rainfall. The ϕ index is defined as that rate of rainfall above which the rainfall volume equals the run-off volume. This is illustrated in Fig. 8.10. Both indices vary from storm to storm depending on the previous conditions.

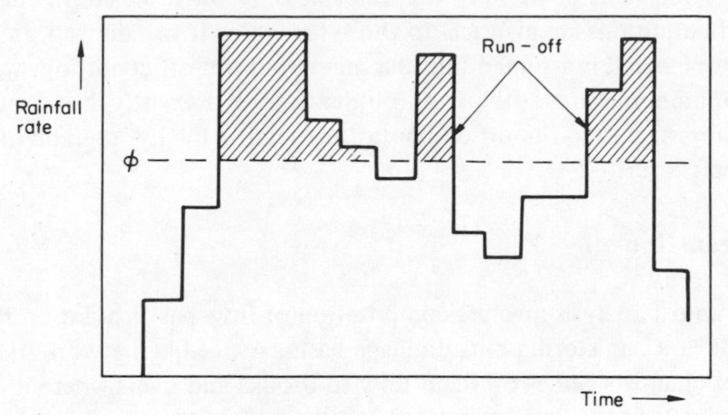

Fig. 8.10 Definition of the ϕ-index

Fig. 8.11 Pra/Ankobra drainage basins

Thus a derived value of an index is applicable only when the storm characteristics and conditions are identical to those for which it was derived. In many design problems, it is assumed that the maximum run-off conditions will prevail and the minimum value of the W or ϕ index is used. Recently, however, more rational approaches to run-off computations such as the hydrograph method have come into favour.

8.6 Stream Run-off

Stream run-off analysis involves computation of flow past a point on the stream as a result of a rain storm on its drainage basin. A tree-like network of rivulets and other channels guides overland flow to brooks and rivers where it is joined by groundwater contributions and eventually finds its way to the sea – probably after serving various needs of man. Two typical Ghanaian coastal drainage basins (Pra and Ankobra) are shown in Fig. 8.11.

The problem of estimating run-off of rivers is complex. It involves consideration of the area and time variations of the input (rainfall), the geological structure of the basin, variations in climatic conditions, vegetation cover, water use, etc. The time variation of precipitation includes differences with respect to an individual rainfall as well as daily, monthly and seasonal changes. In dealing with large basins, therefore, daily, monthly or yearly averages of rainfall are used. In working with time averages, the length of time of records becomes important. The longer the period of observation the more valid average computations are. The basic requirements of run-off relations as applied to design and operation is of prime importance in streamflow predictions. In design, it is immaterial whether or not streamflow and its related parameters (rainfall, infiltration etc.) are considered concurrently. In operation, on the other hand, run-off must be predicted and therefore its time relation to the input is of supreme importance. This means that the seasonal or annual volume of streamflow must not only be related to rainfall, but also to factors which can be measured in advance.

8.6.1 Time Distribution of Run-off (Hydrograph Analysis)

A plot of the time distribution of discharge through a stream gives the hydrograph for the basin above the point of gauging. A hydrograph analysis represents a study of the run-off characteristics of a basin. These characteristics depend on the characteristics of precipitation as well as on the physical and meteorogical properties of the basin. The waters of a rain storm reach a stream through four main routes.

(1) Direct precipitation falling on the water surface of the stream. This is usually very small since the water surface hardly makes up 5% of the basin area. The direct precipitation is usually included in overland flow.

(2) Overland flow which carries water on the surface of the soil to the stream. It is the residue after infiltration, interception and depression storage. Overland

drainage is the principal factor in the creation of flood peaks through streams because of the concentrated nature of its flow.

(3) Interflow or subsurface flow whereby infiltrated water comes back to the surface and finally reaches a stream as surface flow. Water penetrating the soil surface spreads sideways and may reappear on the ground surface on side slopes. Interflow is governed by the soil structure and is given by total infiltration minus soil moisture recharge and deep percolation. Since it is difficult to measure it is usually combined with overland flow and direct precipitation to give what is called direct *run-off*.

(4) Groundwater flow. Rainwater percolating saturated soils flows by gravity to join the groundwater. As groundwater, it may eventually move into streams or springs. Discharge from groundwater storage into a stream may take place several months after the original rainfall – the time lag depending on the geology of the area. Water entering a stream from underground may also come from precipitation on another basin. Underground flow provides the most significant antecedent condition for a stream run-off.

The ratio of the total annual run-off at a point on a stream to the total annual rainfall on the basin above that point is known as the *run-off coefficient*. It is governed by geological, meteorological and other physical conditions of the drainage basin. In Ghana, for instance, most of the rivers and streams have run-

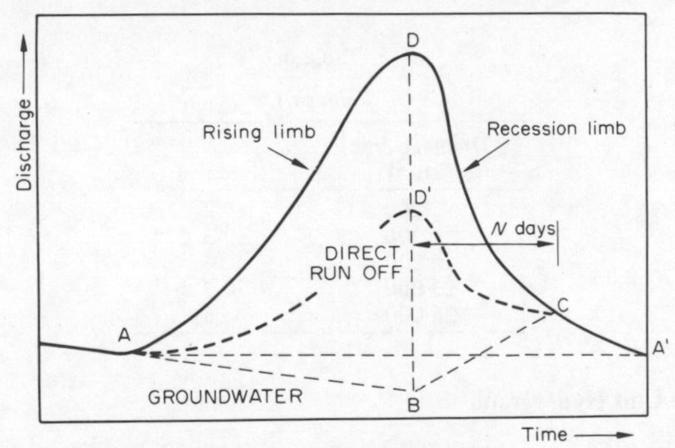

Fig. 8.12 A flood hydrograph

off coefficients of only about 10%, most of the annual rainfall being lost through evaporation and transpiration. All other things being equal, steeply sloping streams have higher coefficients than gently sloping, slow flowing ones.

The main features of a hydrograph resulting from a single rain storm on a basin are (Fig. 8.12) a rising limb, a peak and a recession limb. Multiple peaks caused by the geography of the basin or more often resulting from rain storms separated by periods of little or no rain or by areal variation in storms are sometimes noticed. The recession represents the case of depletion of water temporarily stored in the basin.

The hydrograph is used in computing storage and outflow from reservoirs. In order to predict movement of flood waves due to a particular storm it is necessary to be able to separate the discharge of the stream due to direct run-off, which is the principal cause of quick flooding, from the groundwater or base flow, which moves much more slowly. Numerous methods of separating the hydrograph are used. Three of these are indicated in Fig. 8.12. It is important to emphasize the arbitrariness of all the methods. No reliable method is available because stream gauges simply cannot be used to indicate the constituent parts. The simplest method of separation is to draw a horizontal line AA′, A being the smallest discharge point before the rainstorm inflow. This method yields an extremely long time base for the direct run-off hydrograph which depends on the storm and the flow at the time of rise. To reduce the time base an alternative method is to draw a straight line AC, where C is on the recession limb N days after the peak.

Suggested values of N based on studies of different sizes of drainage basins located in the U.S.A. are shown in Table 8.1. These values do not show the effect of the physical shape and geography of the basin and their use is limited. Using the third method, which is generally considered more satisfactory than the preceding two, the lines ABC are drawn, the point B being the continuation of the antecedent line to a point under the peak of the hydrograph.

Table 8.1

Values of N

Drainage area (km^2)	N (days)
260	2
1 300	3
5 200	4
13 000	5
26 000	6

8.6.2 The Unit Hydrograph

The direct run-off from a specified rainstorm is identified in Fig. 8.12 as the area ABCD. If the rainfall had been half the value of that producing the run-off ABCD, provided the ratio of infiltration to amount of rainfall is the same for both storms, the area representing the direct run-off of the new storm will be half that of the first. If the patterns of areal variation of the two storms are identical and provided the base conditions ABC are the same, their direct run-off hydrographs ABCD and ABCD′ will also be similar (for the same duration of storm) and their ordinates will be proportional to their areas (volumes of run-off). The hydrograph due to a storm whose direct run-off volume (represented by ABCD) is unity is known as the *unit hydrograph*. Hence the unit hydrograph due to any storm may be obtained by dividing the ordinates of the direct run-off hydrograph by the direct run-off volume.

The unit hydrograph, in effect, represents one of the earliest attempts at systems hydrology. It seeks to predict an input–output relationship for a basin whose characteristics are known. It is based on the blackbox concept.

The unit hydrograph is used as a 'rational' basis for predicting the hydrographs for storms whose duration and areal pattern are the same as that for which the unit hydrograph has been prepared. The application is based on the hypothesis that identical storms with the same antecedent conditions produce identical hydrographs and that all hydrographs resulting from rainfalls of a given duration have the same time base. To obtain the hydrograph for a storm which satisfies the above conditions in relation to one for which a unit hydrograph has been obtained, the ordinate values of the unit hydrograph are multiplied by the volume of the direct run-off due to the storm.

The limitations of the application of the unit hydrograph are inherent in the assumptions made in its derivation. Similar areal and time distribution of storms for large areas is rare and the method is generally regarded as suitable for small areas (under 5000 km^2 or 2000 sq. miles) only. The shape and length of a basin, even though it is small, may also bring about significant variations in the rainfall pattern and therefore render the use of the unit hydrograph unsuitable. Identical storms with the same antecedent conditions are also rare and the assumption that the time bases of all floods caused by rainfalls of equal duration are the same has been known not to be true. The time required for flows to recede to some fixed value are observed to increase with the initial flow. Despite these limitations, unit hydrograph methods have been applied to many hydrological design problems with reasonable success.

EXAMPLE 8.1

Below are the observed flows from a storm of 4 hours duration at a gauging station where the drainage area is 1500 km^2.

Time (h)	0	3	6	9	12	15	18	21	24	27	30
Flow (m^3/s)	37·5	75	165	225	220	176	150	105	70·6	47·5	37·5

Derive the 4 hour unit hydrograph assuming a constant base flow of 37·5 m^3/s. Use the unit hydrograph to predict the hydrograph for a similar 4 hour rainfall whose direct run-off volume is 16·6 mm over the basin.

SOLUTION

Plot the hydrograph (marked I) as shown in the figure.

Base flow = area below line AB = 1125 cumec-h*

Direct run-off = area above line AB = 2700 cumec-h

$$\text{Direct run-off} = \frac{2700 \times 3600}{1500 \times 10^6}$$

$$= 6{\cdot}48 \text{ m}$$

* Cumec is sometimes used to denote m^3/s.

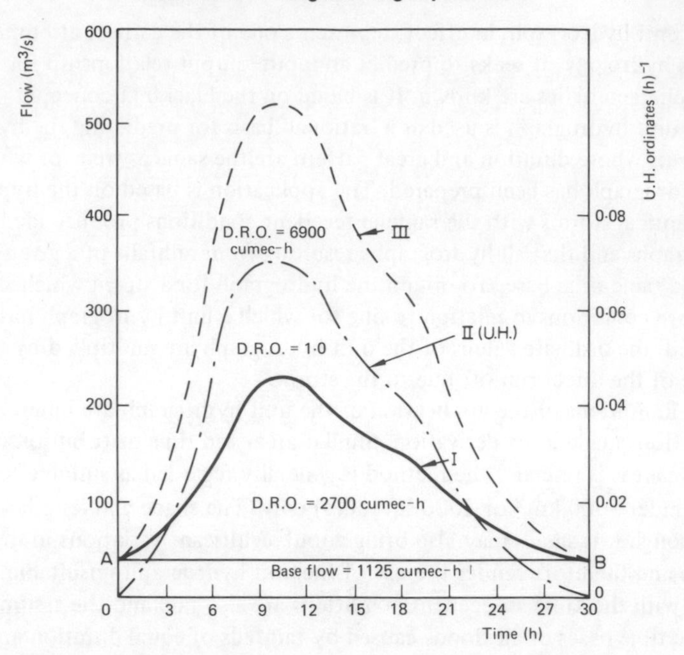

The table below shows the rest of the computations. Column (3) gives the ordinates for the direct run-off component (D.R.O.) of the original rainfall. Column (4) gives the normalized ordinates i.e. column (3)/2700 (h^{-1}). These are plotted and labelled curve II, the appropriate scale* being given on the right. The ratio of the direct run-off volume of the expected rainfall to that for which the unit hydrograph (U.H.) is drawn is16·6/6·48 = 2·56. Thus the volume = 2700 x 2·56 = 6900 cumec-h. The actual flow rates, column (6), are obtained by multiplying the ordinates of the unit hydrograph by the volume (6900 cumec-h) column (5), and adding the result to the base flow (37·5 m^3/s). The predicted run-off is marked III on the graph.

Time (h)	*Flow* (m^3/s)	*D.R.O.* (m^3/s)	*U.H. ordinates* (h^{-1})	*D.R.O.* (m^3/s)	*Flow* (m^3/s)
0	37·5	0	0	0	37.5
3	75	37·7	0·0139	96·0	133·5
6	165	127·5	0·0472	326	363·5
9	225	187·8	0·0695	479	576·5
12	220	182·9	0·0676	465	502·5
15	176	128·2	0·0509	351	388·5
18	150	112·7	0·0417	287	324·5
21	105	67·5	0·0250	172	209·5
24	70·6	33·1	0·0120	82·8	120·3
27	47·5	10·0	0·0037	25·5	63·0
30	37·5	0	0	0	37·5

* In practice it is common that the ordinates representing the direct run-off are divided by D.R.O. in depth (inches or mm). Consequently the resulting U.H. represents a volume of unit depth of direct run-off.

8.6.3 Modern Trends

It has probably been clear from the discussions above that the civil engineer's interest in hydrologic studies lies in forecasting the averages and extremes of surface water and groundwater flows at a given location on the earth's surface. His main concern is that given values of inputs f_i and the state of the hydrologic system S, he will be able to predict values of the output. This concept is represented diagrammatically in Fig. 8.13. The advent of the computer and recent

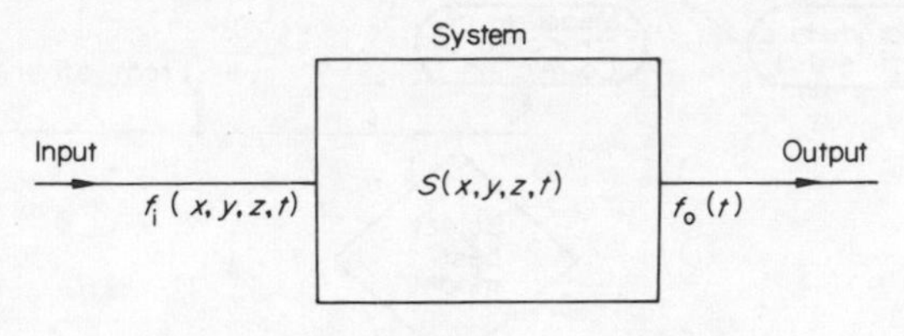

Fig. 8.13

advances in systems analysis using linear and dynamic programming, has brought hopes for the success of systemized approach to hydrologic problems. Two modern approaches using linear regression and correlation analysis and linear systems analysis are summarized in this subsection. The interested reader is referred to Eagleson's *Hydrology Systems** and the paper by Amorocho and Hart for details and other approaches to systematic hydrology.

The mathematical technique of regression and correlation analysis involves fitting assumed functions to observed dependent (output) and independent (input and system parameters) physical variables and then measuring the strength of the relationship between pairs of the variables. In hydrologic studies the dependent variable is the stream run-off and the independent variables are precipitation and the physical parameters of the drainage basin. To facilitate correlation analysis on a linear basis it is necessary to reduce the actual distributed system to a *lumped system* – i.e. the spatial variation is removed from the system. Thus the input variable $f_i\,(x, y, z, t)$ becomes a lumped input $X(t)$ and the drainage basin is represented by average parameters P_i (mean slope, shape roughness, etc.) and the dependent output variable $f_o(t)$ (stream run-off) is redesignated $Y(t)$. One of the principal difficulties concerns the limit of drainage area within which physical parameters can be lumped.

Using linear correlation analysis an attempt is made to relate the output variable linearly to the input variable and the system parameters through arbitrary functions ϕ (of the input variable) and ψ (of the system parameters). The linear correlation equation is generally of the form:

$$Y(t) = a_0 + \sum_i a_i\, \psi_i\, P_i + \sum_j a_j\, \phi_j\, X(t - \tau_j) \tag{8.18}$$

where a_0, a_i and a_j are constants and τ_j is a time lag which introduces the effect of antecedent precipitation. The technique of solving equation (8.18) is shown in the flow chart of Fig. 8.14.

* Published as part of *Dynamic Hydrology*, McGraw-Hill.

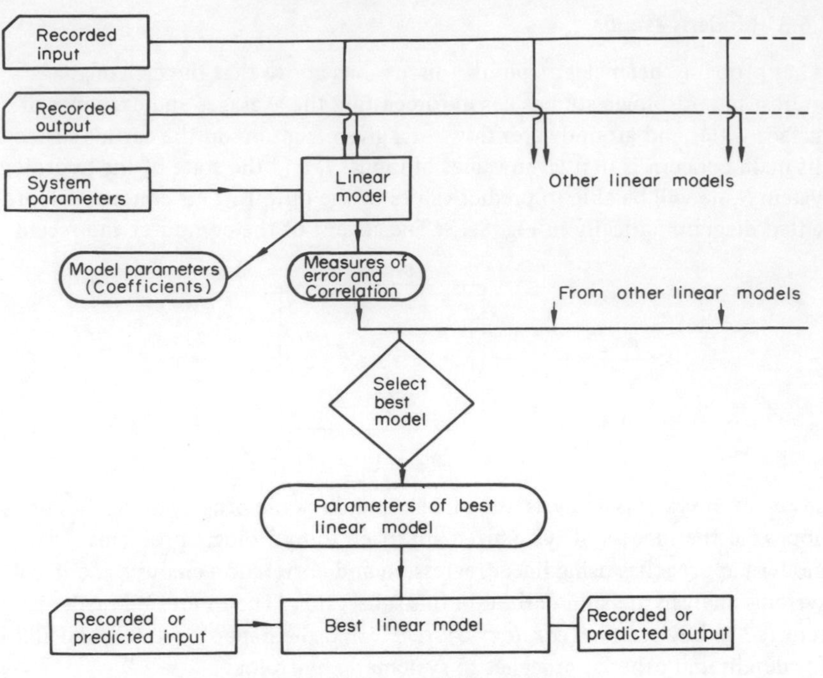

Fig. 8.14 Linear correlation analysis

A set of values is assumed for ϕ_j and ψ_i and with measured values of $X(t)$ and $Y(t)$, the coefficients a_0, a_i and a_j which provide the best fit under a chosen error criterion are determined by multiple linear regression (see books on methods of numerical analysis e.g. by K. L. Neilson). The assumed functional relationships are tested by calculating a set of error estimates and correlation coefficients. If the results are unsatisfactory a new set of values is assumed for the functions ϕ and ψ and the process is repeated until satisfactory results are obtained.

In systems analysis we seek to establish a *unique* relationship between an input and the corresponding output of a system by mathematical manipulations known as convolution. Using this approach we ignore the detailed mechanism of the system and need to know only whether the system is linear or non-linear and whether the input is stationary or time varying, distributed or lumped. The input may be continuous (i.e. can assume any value within a range at any time) or discrete and periodic (i.e. repeating itself exactly at regular intervals), aperiodic (transient only) or random. Almost all hydrologic inputs are random – i.e. no matter how much data is available, the variable cannot be predicted precisely at any time in the future.

Mathematical methods are used to generate input–output relations of proved uniqueness using measured inputs of periodic, aperiodic or random nature. The basis of the convolution concept is illustrated in Fig. 8.15.

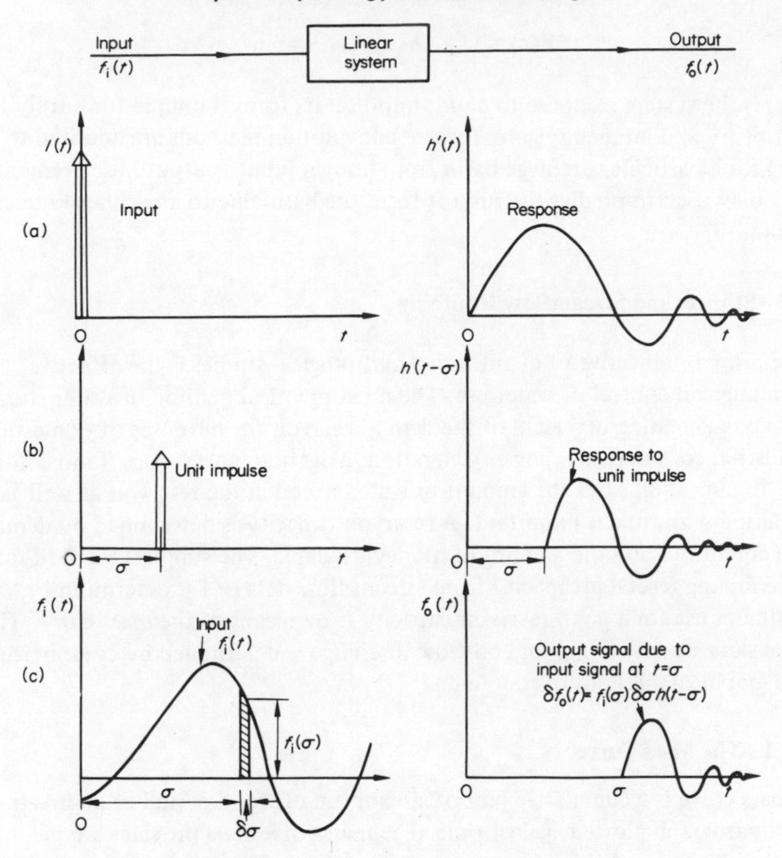

Fig. 8.15 Superposition of impulse responses

Assume that an impulse $I(t)$ into a system produces a response $h'(t)$ (Fig. 8.15(a)). For a stationary linear system a time shift σ of the input produces a similar time shift in the response (Fig. 8.15(b)). Thus if an impulse of unit strength located at $t = \sigma$ produces a response of $h(t-\sigma)$, an impulse of strength $I(\sigma)$ at $t = \sigma$ will produce a response $h'(t-\sigma) = I(\sigma)\,h(t-\sigma)$ for a linear system. Thus a small signal at $t = \sigma$ of width $\delta\sigma$ of an extended input function $f_i(t)$ will give an output signal

$$\delta f_0(t) = f_i(\sigma)\,\delta\sigma\,h(t-\sigma)$$

The total response to an input function $X(t)$ for a finite time interval $0 \leqslant t \leqslant T$ therefore results in the convolution or superposition integral:

$$Y(t) = \int_0^T h(t-\sigma)\,X(\sigma)\,\mathrm{d}\sigma$$

or
$$Y(t) = \int_{t-T}^{t} h(\tau)\, X(t-\tau)\, \mathrm{d}\tau \qquad (8.19)$$

$h(\tau)$ is the system response to a unit impulse. Its form is unique for a truly linear, stationary and lumped system. Inverse convolution methods are adopted to find $h(\tau)$ for a particular drainage basin from known input–output measurements. It is then used to predict the run-off from the basin due to any other known rainfall.

8.7 Storage and Streamflow Routeing

The primary objective of quantitative hydrological studies is the effective planning and control of water use. The most practical method of water storage is damming, creating upstream of the dam a reservoir for purposes of domestic and industrial consumption, power generation, irrigation, navigation, flood control, etc. In almost all cases the amount of water stored in the reservoir as well as its variation with time is important. A reservoir capacity is determined by demand, site conditions and the amount of money available. The simplest method for determining reservoir capacity from stream flow data or for determining optimum use for a given reservoir capacity is by means of the *mass curve*. The time variation of storage and outflow discharge is determined by computational and graphical methods of routeing.

8.7.1 The Mass Curve

A mass curve is a cumulative plot of stream run-off for a period or an integral of the hydrograph plotted against time. The mass curve thus provides a quick graphical picture of the flow capacity of a stream over an extended period and enables a quick inspection of records over that period to be made. The slope of the curve at a point gives the net rate of flow at the particular point, accounting for evaporation, leakage and other losses. It therefore provides a simple means of estimating the storage capacity of reservoirs.

For example, the mass curve in Fig. 8.16 can be used to estimate the usable storage capacity of a proposed dam at Weija on the Densu River to supply a required demand. The figure is derived from the synthesized (derived) flow record of Table 8.2 (p. 268).

Evaporation and other expected losses were taken into account in obtaining the monthly flows indicated in the table. The slope of the straight line joining any two points on the mass curve represents a uniform flow that would have given the increment in the total flow of water past the gauging station in the same period.

In Fig. 8.16 the difference in the ordinates of A and B indicates a total flow volume of 40 000 million gallons (Mgal) (180 million m^3) between the end of December, 1945 and the end of August, 1947 (20 months). This would have been given by a uniform flow of 2000 million gallons (9 millon m^3) per month which

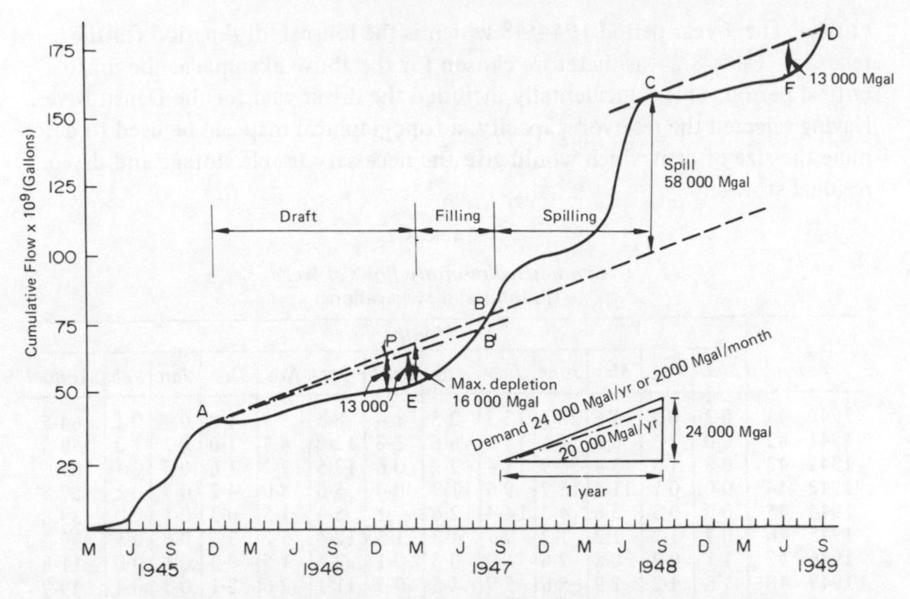

Fig. 8.16 Determination of reservoir capacity from a mass curve

is the slope of the straight line AB. A reservoir which is full at the end of December 1945 and from which water is being withdrawn at a steady rate of 2000 Mgal (9×10^6 m^3) month will therefore not be full again until the middle of September 1947, since more water would be withdrawn per month than the average monthly inflow for any period between the two dates. At any time during this period the length of the ordinate between the straight line and the mass curve measures the draft from the reservoir, i.e. the amount by which the reservoir falls short of being full. The maximum ordinate (16 000 Mgal = 70×10^6 m^3) therefore measures the total volume of storage required to guarantee a steady withdrawal of 2000 Mgal (9×10^6 m^3)/month throughout the period.

Beyond B the difference in the ordinates of the mass curve and the straight line measures the spill from the reservoir. Hence until the next 'dry' season starting at the end of August, 1948 (point C), the total spill would be 58 000 Mgal (260×10^6 m^3). There would be a total depletion of 13 000 Mgal (60×10^6 m^3) before the reservoir fills up again at D if water is continuously withdrawn at the rate of 2000 Mgal (9×10^6 m^3)/month. Since the latter storage volume cannot satisfy the demand conditions AB, the reservoir must be designed to store a minimum usable capacity of 16 000 Mgal (70×10^6 m^3).

In practice, the approach involves superposing the demand line at the peak points on the mass curve, drawn for the critical dry periods in the gauging record, and determining the maximum depletion and therefore the minimum usable storage. It must be emphasized that the entire record has to be examined to determine the most critical period. The critical low flow year does not necessarily determine the selection of a reservoir capacity required to store the necessary volume to supply a demand. A series of moderately dry years may be more critical. The 4-year period 1944/48 which is the longest 'dry' period for the record in Table 8.2 was therefore chosen for the above example as the most critical period. This coincidentally included the driest year for the Densu River. Having selected the reservoir capacity, a topographical map can be used to determine the size of dam which would give the necessary usable storage and dry or residual storage.

Table 8.2

Synthesized monthly flows at Weija
(in thousand Megagallons)

Year	*Month*												
	Mar.	*Apr.*	*May*	*June*	*July*	*Aug.*	*Sept.*	*Oct.*	*Nov.*	*Dec.*	*Jan.*	*Feb.*	*Annual*
1940–41	0·2	0·3	1·6	28·5	13·3	2·5	1·6	8·5	5·7	1·8	0·4	0·1	64·5
1941–42	0·0	3·2	6·0	4·6	14·1	6·6	5·3	12·8	4·7	1·0	0·2	1·3	58·8
1942–43	0·8	1·5	13·4	34·9	11·4	2·2	0·6	13·5	7·2	1·6	0·3	0·1	87·6
1943–44	0·0	0·1	11·4	15·7	9·6	1·9	0·4	5·5	8·0	4·2	0·7	0·1	57·8
1944–45	0·0	0·3	3·8	4·3	14·4	2·6	7·1	5·4	1·5	0·3	0·1	0·0	39·8
1945–46	0·7	1·2	1·2	1·2	7·7	4·3	1·5	12·4	5·2	3·1	0·6	0·1	39·2
1946–47	1·1	0·9	0·6	2·6	1·9	0·3	0·1	2·4	1·5	0·3	0·1	0·0	11·8
1947–48	1·6	4·2	1·9	5·6	5·9	4·5	9·7	11·1	2·6	2·1	0·3	0·4	49·9
1948–49	5·5	4·5	6·1	27·0	12·4	2·0	0·4	0·1	2·8	1·1	0·4	0·1	62·4
1949–50	0·9	0·2	0·6	3·9	25·4	10·6	10·1	10·0	1·8	2·1	4·2	0·8	70·6
1950–51	1·2	4·8	5·7	3·7	2·9	0·6	0·1	0·5	5·5	1·0	0·2	0·0	27·2
1951–52	1·1	1·6	3·0	8·0	12·8	3·4	0·7	20·8	24·8	5·7	1·1	0·2	83·2
1952–53	3·5	4·5	3·9	21·6	14·9	3·8	5·9	23·0	11·3	2·1	0·4	0·1	95·0
1953–54	0·2	0·1	2·4	18·6	14·3	4·1	0·8	11·3	3·5	0·7	0·1	2·1	58·2
1954–55	1·4	2·9	3·4	5·4	10·9	1·7	1·8	7·2	2·8	0·6	0·1	0·0	38·2
1955–56	1·8	1·8	0·4	7·1	17·0	4·8	1·0	7·4	23·9	3·4	0·7	0·1	69·4
1956–57	0·0	4·5	7·0	11·5	4·6	0·9	0·2	0·1	0·3	0·3	0·1	0·0	29·5
1957–58	0·0	2·2	5·6	31·7	57·8	12·0	2·3	2·7	0·6	0·1	0·0	0·0	115·0
1958–59	0·8	0·9	13·8	37·3	9·1	1·7	0·3	0·8	0·5	0·1	0·0	0·0	55·2
1959–60	1·9	2·3	11·9	10·3	21·1	6·2	1·2	5·8	7·1	1·5	0·3	0·1	69·7
1960–61	0·4	6·7	1·9	13·5	17·0	2·5	1·3	11·4	2·3	0·7	0·3	0·1	58·1
1961–62	0·0	0·3	0·4	13·9	37·6	11·8	2·3	2·4	1·1	0·2	0·1	0·0	70·1
1962–63	0·6	0·5	1·9	28·9	37·9	11·4	2·2	0·0	10·1	11·8	1·8	0·3	111·4
1963–64	1·0	3·6	2·1	8·3	41·2	32·6	52·0	81·5	14·6	2·6	0·5	0·1	240·1
1964–65	1·4	7·5	12·4	13·6	7·2	1·2	0·3	0·2	0·1	0·0	0·0	0·6	44·5
1965–66	0·5	3·1	4·4	27·4	48·4	13·7	6·0	30·2	4·7	0·9	0·2	0·1	139·6
1966–67	3·6	0·5	2·7	7·7	16·2	6·9	5·4	9·4	9·5	1·5	0·3	0·1	63·8
Mean	1·1	2·3	4·8	15·1	18·0	5·8	4·5	11·1	6·1	1·9	0·5	0·3	71·5

The mass curve can also be used to determine the maximum withdrawal from a reservoir whose storage capacity is fixed. In Fig. 8.16, if the reservoir's usable capacity had been limited to 13 000 Mgal (60×10^6 m^3), erecting ordinates equivalent to that value at the trough points E and F would give straight lines AB′ and CD from the 'peak' points. The slope of AB′ is 20 500 Mgal (90×10^6 m^3) /yr or 1710 Mgal ($7{\cdot}5 \times 10^6$ m^3)/month. This is the maximum demand that can be satisfied during the 1946/47 dry season with the 13 000 Mgal (60×10^6 m^3) storage available. A higher withdrawal rate say AB would deplete all the useful storage at P (end of January 1947).

8.7.2 Stream Flow Routeing

Stream flow routeing involves solving the continuity equation to ascertain the rate of storage of water in a reservoir.

Rate of storage = rate of inflow – rate of outflow

or
$$\frac{dS}{dt} = Q_i - Q_o \tag{8.20}$$

where S is the reservoir volume, t is time and Q_i and Q_o are the inflow and outflow run-offs respectively.

Written in finite difference form using average flow values

$$\bar{Q}_i = \frac{1}{2}(Q_{i1} + Q_{i2}) \text{ and } \bar{Q}_o = \frac{1}{2}(Q_{o1} + Q_{o2})$$

for a small time interval, equation (8.20) is

$$S_2 - S_1 = \frac{1}{2}(Q_{i1} + Q_{i2})\,\Delta t - \frac{1}{2}(Q_{o1} + Q_{o2})\,\Delta t \tag{8.21}$$

where 1 and 2 refer to the beginning and end of the time interval. The equation may be rearranged as

$$(Q_{i1} + Q_{i2}) + \left(\frac{2S_1}{\Delta t} - Q_{o1}\right) = \left(\frac{2S_2}{\Delta t} + Q_{o2}\right) \tag{8.22}$$

The inflow quantities are generally provided in the form of a hydrograph or a table. Thus if the storage and the outflow at the beginning of a time interval are known the right hand side of equation (8.22) can be determined. From a predetermined graph of $([2S/\Delta t] + Q_o)$ against Q_o the value of outflow at the end of a time interval Δt can be read and the corresponding storage and elevation determined. Normally the storage S is obtained as a function of elevation from a contour map of the reservoir area and the outflow (spill) characteristics are also given in relation to elevation, e.g. Q_o = constant x $H^{\frac{3}{2}}$ for an overflow spillway where H is the reservoir level above the spillway crest. For large reservoirs, velocities in the reservoir can be considered as small and therefore the water level horizontal throughout the reservoir.

EXAMPLE 8.2

The following example illustrates the procedure for an uncontrolled spill (i.e. gate and valve positions are maintained throughout the period under consideration). Fig. 8.17 gives a plot of storage and outflow against elevation (Table 8.3) from which are determined $([2S/\Delta t] + Q_o)$ and $([2S/\Delta t] - Q_o)$ choosing Δt = 2 h. These are plotted in Fig. 8.18. Fig. 8.19 (Table 8.4, column (2) gives the inflow hydrograph for a flood wave entering the reservoir). With reference to Table 8.4 the semi-graphical procedure may be summarized as follows:

Table 8.3
Storage and spill characteristics

Water level above crest of dam (m)	*S (storage)* ($\times 10^6$ m³)	Q_o *(outflow)* ($\times 10^3$ m³/s)	$\frac{2S}{\Delta t} + Q_o$ ($\times 10^3$ m³/s)	$\frac{2S}{\Delta t} - Q_o$ ($\times 10^3$ m³/s)
0·25	0·003	0·014	0·015	−0·013
0·30	0·013	0·061	0·065	−0·057
0·40	0·047	0·106	0·119	−0·093
0·60	0·268	0·161	0.235	−0·087
0·80	0·749	0·202	0·410	+0·006
1·00	1·545	0·236	0·665	+0·193
1·20	2·848	0·265	1·054	+0·524
1·40	4·780	0·293	1·621	+1·025
1·60	7·360	0·317	2·361	+1·727
1·80	10·500	0·340	3·257	+2·577
2·00	14·110	0·361	4·280	+3·558
3·50	28·22	0·40	5·734	+4·931

Table 8.4
Routeing for storage

(1)	(2)	(3)	(4)	(5)	(6)	(7)
Time (h)	Q_i (m³/s)	$Q_{i1} + Q_{i2}$ (m³/s)	$\frac{2S}{\Delta t} + Q_o$ (m³/s)	Q_o (m³/s)	$\frac{2S}{\Delta t} - Q_o$ (m³/s)	*Reservoir elevation* (m)
00	21	–	–	14	−13	0·25
02	60	81	68	80	−90	0·34
04	215	275	185	140	−90	0·51
06	345	560	470	210	+50	0·85
08	380	725	775	245	270	1·06
10	430	810	1080	266	540	1·20
12	475	905	1445	290	860	1·37
14	525	1000	1860	300	1270	1·45
16	605	1130	2400	320	1750	1·61
18	665	1270	4150	360	3450	1·98
20	730	1395	4845	377	4100	2·12
22	560	1290	5390	391	4600	2·31
24	310	870	5470	393	4700	2·32
26	200	510	5210	387	4500	2·25
28	160	360	4860	378	4080	2·16
30	130	290	4370	365	3650	2·03
32	100	230	3880	352	3160	1·90

1. Add algebraically the first value of $([2S/\Delta t] - Q_o)$ in column (6) to the first sum of flow $(Q_{i1} + Q_{i2})$ in column (3) to obtain the new value of $([2S/\Delta t] + Q_o)$ in column (4), i.e. 81 − 13 = 68 m³/s.

2. From Fig. 8.18, obtain the new outflow (80 m³/s) and the new value of $([2S/\Delta t] - Q_o)$ corresponding to $([2S/\Delta t] + Q_o) = 68$ m³/s.

3. From Fig. 8.17, read the new reservoir elevation, column (7) (0·34 m) and the new storage (not shown) which corresponds to conditions at $t = 2$ h.

4. With the new value of $([2S/\Delta t] - Q_o)$ and the next value of $(Q_{i1} + Q_{i2})$ repeat the procedure to obtain new values for columns (4), (5) and (7) and the new storage at $t = 4$ h. Continue until the routeing is completed.

The resulting outflow hydrograph and reservoir elevation are plotted in Fig. 8.19. The difference between the occurrence of the peak inflow and the peak outflow is known as the *lag effect*. In this example the maximum outflow of 394 m^3/s occurs 3·2 h after the maximum inflow of 740 m^3/s. There is net storage when the inflow exceeds the outflow and depletion when the outflow exceeds the inflow. It may be observed that column (6) can easily be obtained from column (4) by subtracting $2Q_o$, column (5) from the latter. The little discrepancies in the values of column (6) from those obtained this way are due to inexactness of graph readings. These discrepancies, however, make very little difference in estimates for storage.

When routeing for storage in the case where gate operations take place during the period of interest the differences in outflow due to gate adjustments must be included. The procedure is the same as outlined above except that the Q_o, $([2S/\Delta t] + Q_o)$ and $([2S/\Delta t] - Q_o)$ curves of Figs 8.17 and 8.18 are in families instead of being single. The particular curve to be used depends on the gate conditions at the particular time in the routeing process.

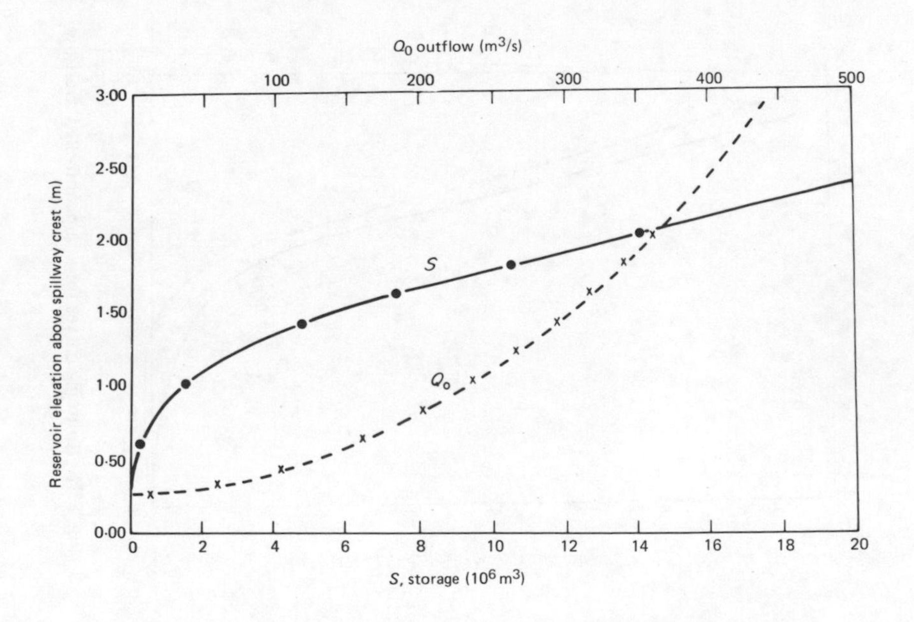

Fig. 8.17 Reservoir storage and spill characteristics

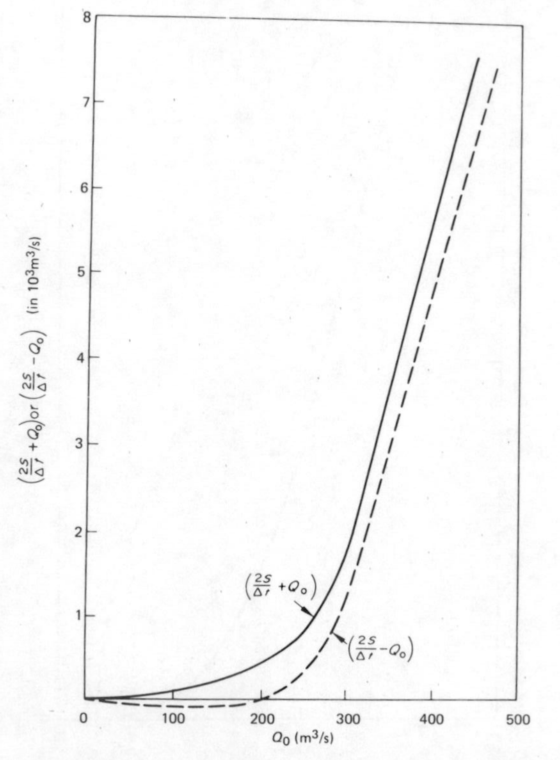

Fig. 8.18 Routeing curves for an uncontrolled overflow

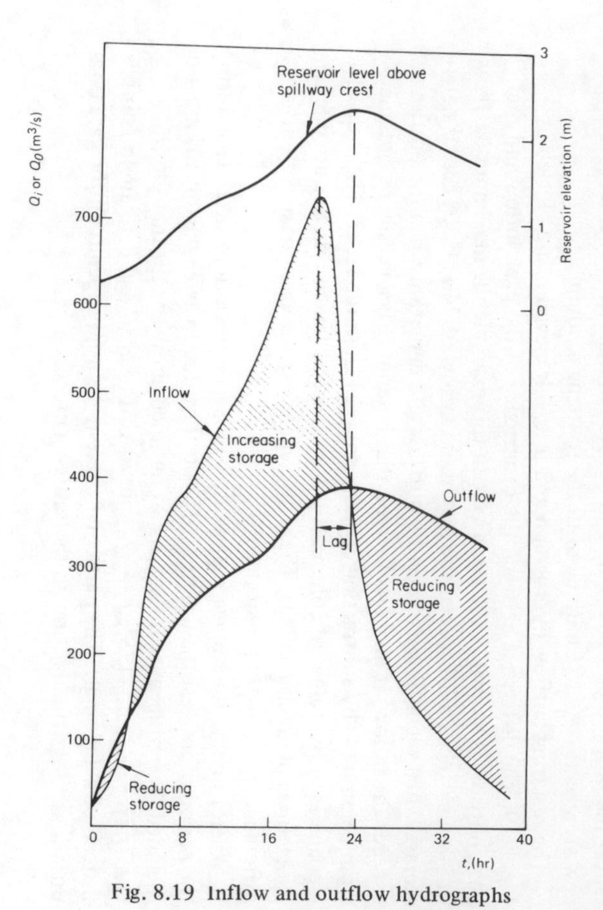

Fig. 8.19 Inflow and outflow hydrographs

8.7.3 Routeing of Floods in River Channels

Consider a flood wave entering a reach of a river at section A and coming out at section B across the valley. The flood spreads within the reach and consequently parts of the entering flow are stored temporarily in the channel valley between A and B. However, unlike the case of reservoirs, there is no simple relationship between outflow and storage. It is recognized that both inflow and outflow have profound influence on the valley storage. The relationship is generally expressed by the empirical relationship:

$$S = \frac{b}{a}\left[xQ_i^{m/n} + (1-x)\,Q_o^{m/n}\right] \tag{8.23}$$

where S is storage (L^3); Q_i is inflow at A (L^3/t) and Q_o is outflow at B (L^3/t). The constants a and n are believed to be measures of the stage-discharge relations at the two ends of the channel reach while b and m relate to the stage-storage relationship. The dimensionless constant x is a weighting factor indicating the relative influence of Q_i and Q_o on the channel storage capacity.

In a large number of practical applications the ratio m/n is assumed as unity and $K = b/a$, which has the dimension of time (t), is taken as the wave travel time through the reach. Thus equation (8.23) can be simplified to the linear relationship:

$$S = K\left[xQ_i + (1-x)\,Q_o\right] \tag{8.24}$$

Equation (8.24) when substituted into equation (8.20) gives the so-called Muskingum routeing equation as:

$$Q_{o2} = c_0\,Q_{i2} + c_1 Q_{i1} + c_2 Q_{o1} \tag{8.25}$$

where

$$c_0 = -(Kx - 0{\cdot}5\Delta t)/(K - Kx + 0{\cdot}5\Delta t) \tag{8.26a}$$

$$c_1 = (Kx + 0{\cdot}5\Delta t)/(K - Kx + 0{\cdot}5\Delta t) \tag{8.26b}$$

$$c_2 = (K - Kx - 0{\cdot}5\Delta t)/(K - Kx + 0{\cdot}5\Delta t) \tag{8.26c}$$

and

$$c_0 + c_1 + c_2 = 1{\cdot}0 \tag{8.26d}$$

Equation (8.26d) provides a check on calculations of c_0, c_1, and c_2. It is also important that K and Δt should be in the same units (usually hours or days).

The empirical factors K and x are usually determined by trial and error. Values are assumed for x (usually varying between 0·1 and 0·3 for natural channels) until the loop formed by a plot of $[xQ_i + (1-x)Q_o]$ versus storage S approximately conforms to a straight line (see Fig. 8.20). The slope of the straight line gives the value of K as can be seen from equation (8.24).

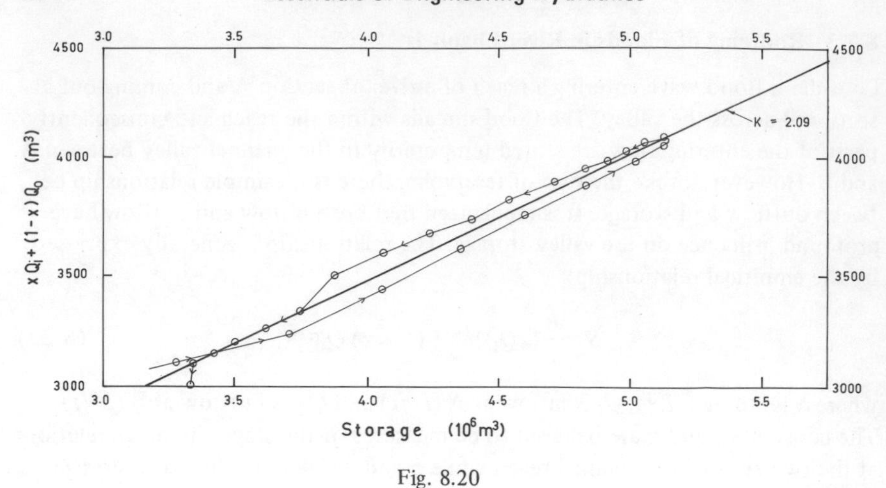

Fig. 8.20

Alternatively, the following theoretical/graphical method may be used. From equation (8.24):

$$\frac{\mathrm{d}S}{\mathrm{d}t} = Kx\,\frac{\mathrm{d}Q_\mathrm{i}}{\mathrm{d}t} + K\,(1-x)\,\frac{\mathrm{d}Q_\mathrm{o}}{\mathrm{d}t} \tag{8.27}$$

At the part where the inflow and outflow hydrographs intersect, Q_i is equal to Q_o and storage within the valley attains its maximum value, thus $\mathrm{d}S/\mathrm{d}t = 0$. Consequently the ratio of the tangents to both hydrographs at the point of their intersection,

$$\frac{\mathrm{d}Q_\mathrm{i}}{\mathrm{d}t} \Big/ \frac{\mathrm{d}Q_\mathrm{o}}{\mathrm{d}t} = (x-1)/x \tag{8.28}$$

After obtaining x according to equation (8.28), normally a first approximation, K is determined as illustrated in Fig. 8.20.

EXAMPLE 8.3

An agricultural land area in a valley is to be protected against floods. The hydrograph for a possible dam site located 12 hours flood crest travel time upstream is shown in Fig. 8.21. If the required flood crest reduction at the farm land is 40%, determine the minimum storage capacity due to the dam. Neglect local inflow and assume Muskingum $x = 0{\cdot}1$.

SOLUTION

We need to determine the consequent flood at the farm land without the upstream dam. This is obtained by routeing the inflow using equation (8.25). Choosing

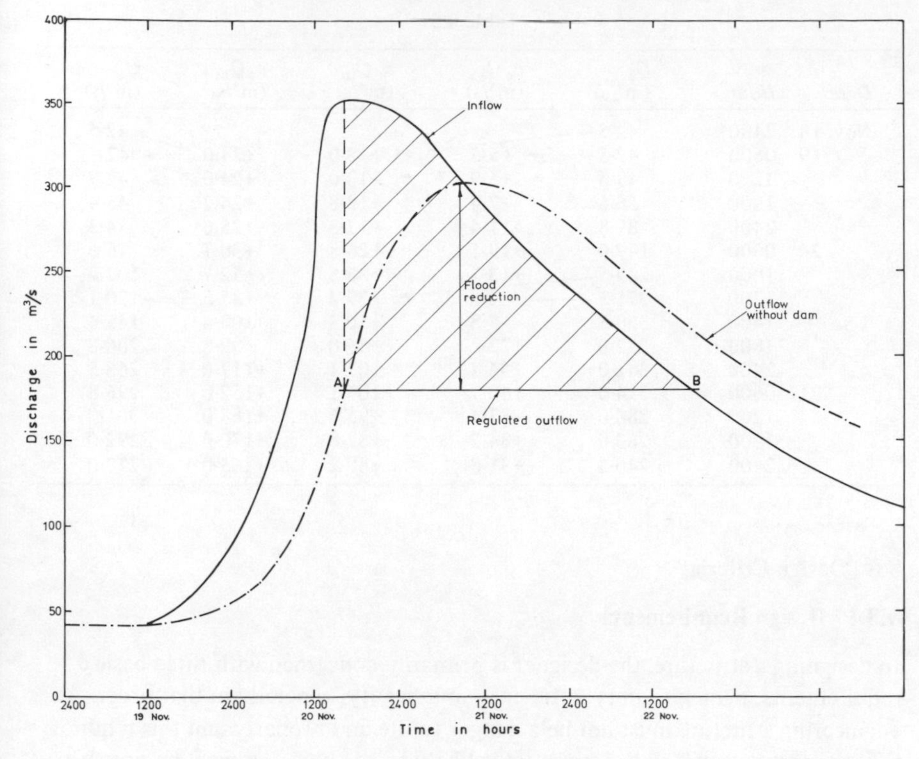

Fig. 8.21

time intervals Δt = 2, 6 and 12 h, the following parameters are calculated (equations (8.26)).

Δt	2 h	4 h	6 h	12 h
c_0	−0·017	0·063	0·130	0·286
c_1	0·187	0·250	0·305	0·429
c_2	0·831	0·687	0·565	0·286
$c_0 + c_1 + c_2$	1·001	1·000	1·000	1·001

Application of the Muskingum method is illustrated in Table 8.5. The resultant flow at the farm land, without the dam, is represented by the last column of the table. This is plotted in Fig. 8.21. Consequently, without the dam, the peak flood of the farm land is 302 m^3/s. This must be reduced by 40% (i.e. to 181·2 m^3/s) by regulation of the dam. Outflow from the dam would be regulated along line AB. Thus the excess flow represented by the shaded area must be stored. The required reservoir capacity is approximated 16 x 10^6 m^3.

Table 8.5

Date	*Hour*	Q_i (m³/s)	c_0Q_{i2} (m³/s)	c_1Q_{i1} (m³/s)	c_2Q_{o1} (m³/s)	Q_o (m³/s)
Nov. 18	2400	42·5	–	–	–	42·5
19	0600	42·5	+5·5	+13·0	+24·0	42·5
	1200	45·3	+5·9	+13·0	+24·0	42·9
	1800	56·6	+7·4	+13·8	+24·2	45·4
	2400	87·8	+11·4	+17·3	+25·6	54·3
20	0600	147·0	+19·1	+26·8	+30·7	76·6
	1000	209·5	+13·2	+36·8	+52·7	102·7
	1200	271·5	−4·6	+39·2	+85·5	120·1
	1400	339·5	−5·8	+50·8	+100·4	145·4
	1800	352·0	+22·2	+85·0	+99·6	206·8
	2400	342·0	+44·4	+107·1	+117·0	268·5
21	0600	314·0	+40·7	+104·1	+152·0	296·8
	1200	288·0	+37·4	+95·6	+168·0	301·0
	1800	263·0	+34·2	+87·8	+170·0	292·0
	2400	240·5	+31·8	+80·2	+165·0	277·0

8.8 Design Criteria

8.8.1 Design Requirements

In designing a structure, the designer is primarily concerned with three basic requirements, namely: safety, efficiency and beauty, generally in that order. An engineering structure must not be a danger to life and property and this requires that the forces to which the structure is likely to be subjected must be properly ascertained. In the absence of a definite knowledge of these forces, the designer estimates the magnitude of the forces and allows for eventualities by including a reasonably large factor of safety in his calculations. This factor of safety is based on past experiences of similar structures and on economic considerations. Thus he decides that, beyond a certain limit, he is prepared to take the risks of the structure failing since in his judgment it would be less expensive to accept its failure than to spend extra money making the structure safe beyond this limit. After all, he reasons, the structure may not be stretched beyond this limit. This is referred to in engineering design as *calculated risk*.

Economic considerations are particularly important in water resources development planning. The optimum design is the best compromise between efficient and safe functioning of the structure and its usefulness as well as monetary investment and expected benefits. Economic considerations which lead to optimum design and use of water resource development projects are discussed in Chapter 11.

Structural beauty is principally an architectural concern and demands that the structure must not be an eyesore or a nuisance to the general public or impair the beauty of the surroundings.

As already mentioned, the main difficulty confronting the designer is reliable prediction of the forces to which his structure will be subjected. The complexity of the analysis of hydrologic events is obvious from earlier discussions. Failure of hydraulic structures such as dams, levees, by-passes, etc., often results in major losses of life and property. Unfortunately, however, present knowledge is inadequate and models cannot be relied upon to predict the occurrence of hydrologic events. The hydrologist, therefore, depends on past records to predict future events as best as he can. Some of the methods he adopts are presented in this section.

8.8.2 Probability Concepts

A hydraulic structure is designed to cope with the severest flood which can be expected within its lifetime. For instance, a dam which is expected to last for 50 years would be designed to withstand the severest flood which would occur once in 50 years. The implication in this concept is that the designer and the owner accept the risk of the dam's failure should a flood bigger than the 50-year flood occur since they do not consider the extra investment involved in catering for bigger floods worth it. Probability concepts are applied to records of past flow conditions to determine flood frequencies or recurrence intervals.

Theoretically the recurrence interval of flood of a specified magnitude is the time interval which would be expected to elapse between two consecutive occurrences of such floods. Conceptually, it is equivalent to the period of occurrence of a harmonic event. In practice, however, the theoretical interval is hardly ever the same as the observed interval. Thus despite their importance in hydrologic data analysis, flood frequencies or recurrence intervals must be understood to provide design criteria with only small physical significance in relation to the actual occurrence of floods. A 50-year flood may occur or be exceeded the year following its occurrence or it may not occur within the following 50 years or more.

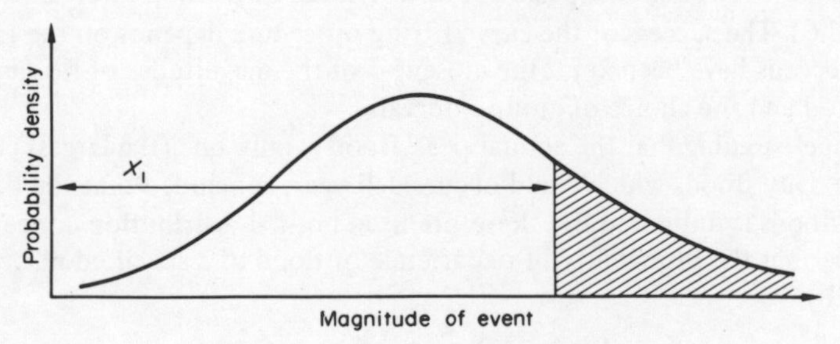

Fig. 8.22 Ideal frequency distribution

Ideally, any statistical data corresponding to a continuously (densely) distributed event conform to the normal distribution curve when the data are plotted as probability density (representing the number of events in a group in relation to the total number) against the magnitude of the event, as in Fig. 8.22. The area under the curve (shaded area) above any event of magnitude X_1 is the probability that the event will be equalled or exceeded. The area under the whole curve, therefore, represents the probability that any event in the series will occur at all – its value is obviously unity.

Attempts are frequently made to reduce river flow records to a form of the normal, Gaussian or other distribution curves so that the known properties of the standard curve can be used to predict the occurrence of specified flood magnitudes for the river. These efforts generally make use of the peak (maximum) flow in a year for a series of years for which records are available.

The frequency histogram of Fig. 8.23, based on estimated annual peak flows at Weija (Densu River) vividly illustrates the difficulties in seeking a smooth curve

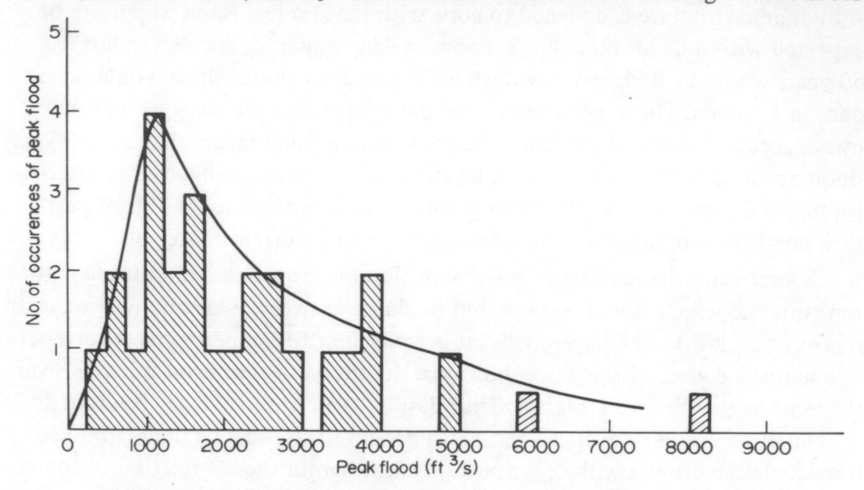

Fig. 8.23 Frequency histogram of flood peaks (Densu River at Weija)

to describe hydrological records covering a small number of years. The figure gives the number of annual peak floods which fall into group intervals of 250 ft³/s (Table 8.6). The success of the curve fitting procedure depends on the time for which records have been kept, the closeness of the magnitudes of floods within the period and the choice of group intervals.

Gumbel, arguing that the annual peak flood is only one (the largest) of 365 possible daily floods which could occur each year, concluded that the distribution of floods is unlimited. He therefore used normal distribution concepts to conclude that the probability of occurrence of flood of a specified magnitude X is given by

$$P = 1 - e^{-e^{-b}} \tag{8.29}$$

where

$$b = \frac{1}{0{\cdot}7797\,\sigma}\,(X - \bar{X} + 0{\cdot}45\,\sigma) \tag{8.30}$$

and $\bar{X}$ is the mean of all floods in the series of N items, the standard deviation σ being given by

$$\sigma = \sqrt{[\Sigma(x-\bar{x})^2/(N-1)]} \tag{8.31}$$

and e is the exponential constant (= 2·71828).

Table 8.6
Flood data at Weija on Densu River

Year	*Peak flow X* (ft³/s)	*Order of magnitude* *m*	$X - \bar{X}$	$(X - \bar{X})^2$	t_p(yr) (*see equation* (8.33))	t_p (yr) (*Gumbel*)
1963	8 150	1	5 744	32 993 536	28·0	110
1957	5 780	2	3 374	11 383 876	14·0	22·2
1965	4 840	3	2 434	5 924 356	9·33	11·2
1962	3 790	4	1 384	1 915 456	7·00	5·48
1961	3 760	5	1 354	1 833 316	5·60	5·38
1958	3 730	6	1 324	1 752 976	4·67	5·26
1942	3 490	7	1 084	1 175 056	4·00	4·55
1940	2 850	8	444	197 136	3·50	2·99
1948	2 700	9	294	86 436	3·11	2·76
1949	2 540	10	134	17956	2·80	2·50
1955	2 390	11	−16	256	2·44	2·32
1952	2 300	12	−106	11 236	2·33	2·20
1951	2 080	13	−326	106 276	2·15	1·96
1953	1 860	14	−546	298 116	2·00	1·76
1960	1 700	15	−706	498 436	1·87	1·63
1966	1 620	16	−786	617 796	1·75	1·58
1943	1 570	17	−836	698 896	1·64	1·54
1964	1 360	18	−1 046	1 094 116	1·55	1·42
1941	1 280	19	−1 126	1 267 876	1·47	1·38
1945	1 240	20	−1 166	1 359 556	1·40	1·36
1959	1 190	21	−1 216	1 478 656	1·33	1·33
1956	1 150	22	−1 256	1 577 536	1·27	1·32
1954	1 090	23	−1 316	1 731 856	1·22	1·29
1947	970	24	−1 436	2 062 096	1·17	1·24
1944	710	25	−1 696	2 876 416	1·12	1·16
1950	570	26	−1 836	3 370 896	1·07	1·12
1946	260	27	−2 146	4 605 316	1·03	1·06
SUM	64 970			80 935 432		
MEAN	2 406			σ = 1730		

Gumbel's theoretical conception is one of the most recent in hydrologic frequency analysis and equation (8.29) has been found to match satisfactorily data from gauging stations with *long* periods of record. The *recurrence interval*, i.e. the average interval between the occurrence of a flood of magnitude equal to or greater than that for which P is calculated is given by

$$t_p \text{ (in years)} = 1/P \tag{8.32}$$

Recurrence intervals calculated from equations (8.29) and (8.32) when plotted against annual peak flow X on Gumbel's paper gives a straight line. The paper is constructed by laying out on a linear scale of b (equation (8.30)), corresponding values of $t_p = 1/P$. The annual peak flow X is plotted on a linear scale. The plot thus looks like a semi-logarithmic plot (see Fig. 8.24).

Earlier flood frequency analyses made use of the fact that the m th largest flood in a data series of annual peak floods over N years, would have been equalled or exceeded m times in the period of record. Many different formulae,

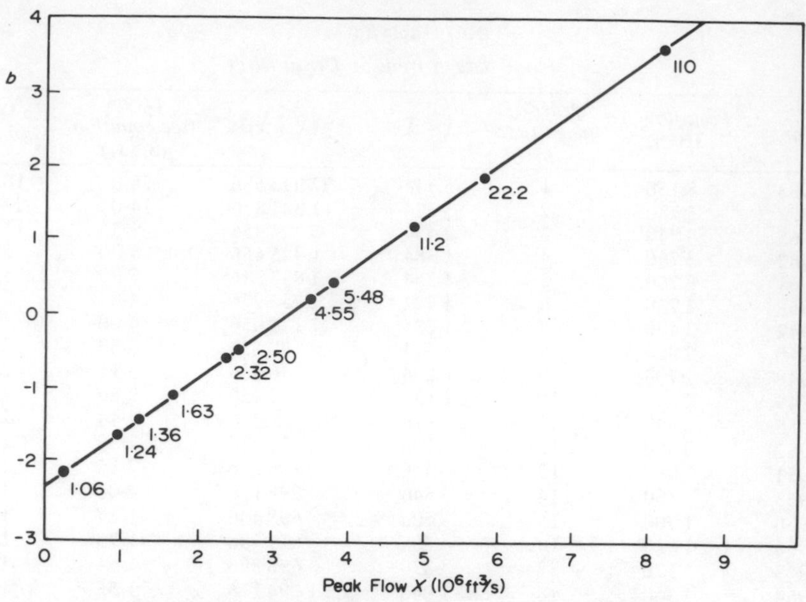

Fig. 8.24 Gumbel frequency plot (Densu River at Weija)
Figures on graph are t_p values.

making use of this concept, are still in use. Logically, it would appear that an estimate of the recurrence interval of the m th largest flood would be given by (N/m) years. However, to correct for the fact that the average probability of occurrence is based on a finite number of (and therefore insufficient) data the recurrence interval is usually calculated from

$$t_p = (N + 1)/m \tag{8.33}$$

Similarly the exceedence interval defined as the average number of years between the occurrence of a flood and the next greater flood is given by

$$t_e = \frac{N + 1}{m - 1} \tag{8.34}$$

It may be observed that the addition of unity to N in equation (8.33) and (8.34) is insignificant if N is large. In Table 8.6 recurrence intervals calculated using equation (8.33) are compared with those predicted by Gumbel's formula. A 27-year record would normally be considered short. Lack of long periods of data is one of the major difficulties in the developing countries. Empirical formulae and other methods (for example, the method of synthesis used in estimating the flows in Table 8.2) have been very helpful in the design of many projects. However, it must be emphasized that the application of these formulae and methods must take into account the basis on which they were developed and appropriate allowances must be made to avoid disaster or excessive costs.

FURTHER READING

Amorocho, J. and Hart, W. E., A Critique of Current Methods in Hydrologic Systems Investigation, *Trans. American Geophysical Union,* Vol. 45, No. 2, June 1964.

Eagleson, P. S., *Hydrologic Systems – Their Analysis and Synthesis*, M.I.T. Course Notes, 1964, published as part of *Dynamic Hydrology*, McGraw-Hill, 1970.

Linsley, R. K., Kohler, M. A., and Paulhus, J. L. H., *Applied Hydrology*, McGraw-Hill, New York, 1949.

Linsley, R. K., and Franzini, J. B., *Water Resources Engineering,* Chapters 2, 3 and 5, McGraw-Hill, 1964.

Linsley, R. K., The Relation Between Rainfall and Runoff, *Journal of Hydrology,* Vol. V, No. 4, October, 1967.

9 Groundwater and Seepage

9.1 Introduction

The three sources of the world's fluid supply – the atmosphere, the earth's surface, and the subsurface (underground) – are of major importance in physical hydrology. Underground is the most important reservoir for some of the most valued fluids of the earth – water, crude oil, natural gas and others. In this chapter we discuss the occurrence of water underground and the basic principles governing its movement with a view to understanding the problem of extraction and recharge of groundwater and seepage through embankments. The discussions will be limited to the zone in which the water occupies all the voids within a geologic stratum. As this is a basic undergraduate introductory course theoretical considerations are also limited mainly to steady flow situations.

9.1.1 Occurrence of Ground Water

Apart from connate water entrapped in the interstices of sedimentary rocks at the time of deposition and water of magmatic, volcanic or cosmic origin, groundwater forms an essential part of the hydrologic cycle. A simplified form of the hydrologic cycle as presented by Linsley and Franzini is shown in Fig. 9.1

Precipitated moisture may penetrate the soil directly to form part of the groundwater or may enter surface streams or reservoirs and subsequently percolate underground. Only after interception by leaves, depression storage and soil moisture interception can precipitated moisture reach the saturated zone of underground water (the water table). Thus water is stored underground only after prolonged precipitation of sufficient intensity.

Soil pores near the earth's surface known as the zone of aeration generally contain air and water in varying amounts. Farther down in the earth's crust is the zone of saturation where all the interstices are filled with water. This zone may be bounded at the top by either a limiting surface of saturation (phreatic or piezometric surface) or an impermeable stratum and at the bottom by an impermeable stratum of clay or rock. The water contained within this saturated zone is differentiated in this text from the general ground water by the use of the one word *groundwater*. The free surface ground water is said to be unconfined while confined groundwater is bounded at the top by an impermeable stratum.

Geologic formations which have structures which permit appreciable storage of water and its movement through them under ordinary field conditions are

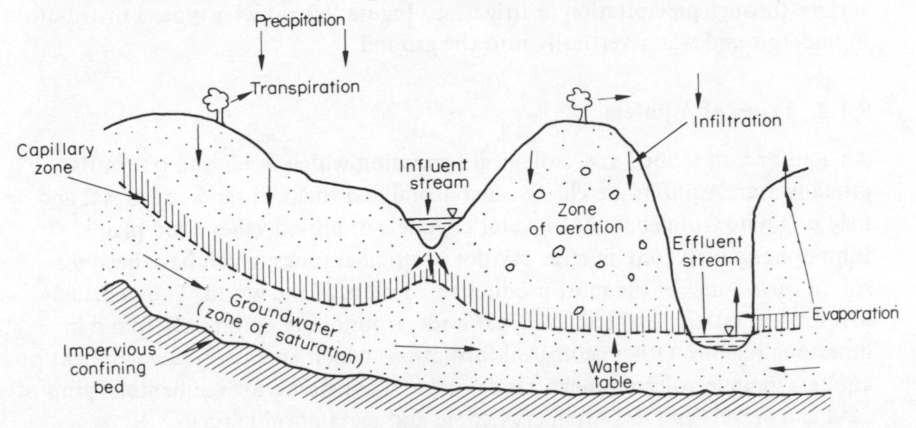

Fig. 9.1

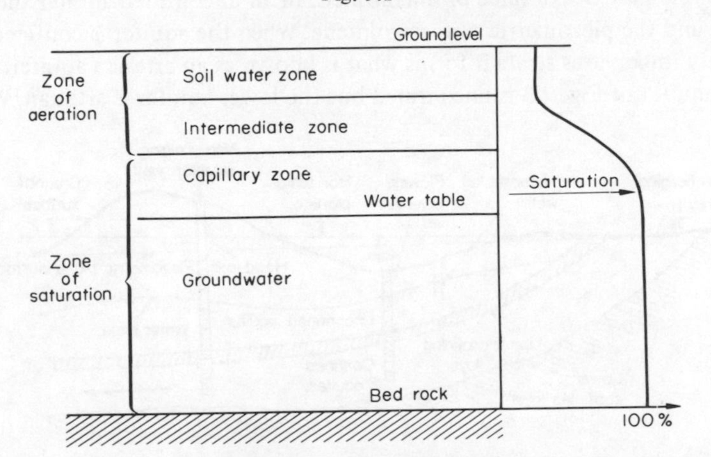

Fig. 9.2 Ground water distribution

known as aquifers. Water may enter these formations from the ground or from water bodies at the ground surface, travel for long or short distances and reappear at the surface as a natural flow or by action of man or plants. The amount of water entering the groundwater zone from influent streams (i.e. streams whose levels are higher than the groundwater table and which therefore seep into the ground) is limited by the characteristics of the underlying material particularly its extent, permeability and water content. Groundwater may reappear as an effluent stream if the water table is higher than the level of the stream.

Just above the groundwater table there may be a thin zone of saturation due to capillarity. The moisture is held in the soil interstices against gravity by the force of surface tension. The extent of the capillary zone depends on the nature of soil, being higher in finer grained soils than in coarser ones. An intermediate zone joins the capillary zone to the soil water zone in which the water normally exists at less than saturation except when excessive water reaches the ground surface through precipitation or irrigation. Figure 9.2 shows a typical distribution of underground water vertically into the ground

9.1.2 Types of Aquifers

An aquifer is described as a geological formation which holds and transmits groundwater. Aquifers are chiefly unconsolidated rocks of sand and gravel and may originate from abandoned water channels or buried valleys and plains. Limestones which contain large cavities of original rocks which have been dissolved form some of the most productive aquifers of the world. The so-called lost streams (streams which disappear underground) are commonly found in limestone regions. Other geological formations which encourage development of aquifers include volcanic rocks, sandstone and conglomerate (cemented forms of sand and gravel) and less often crystalline and metamorphic rocks.

Aquifers may be confined or unconfined. In an unconfined aquifer the phreatic surface and the piezometric surface coincide. When the aquifer is confined by relatively impervious strata it forms what is known as an artesian aquifer. The upper aquifer of Fig. 9.3 is unconfined but the lower aquifer is artesian. Water

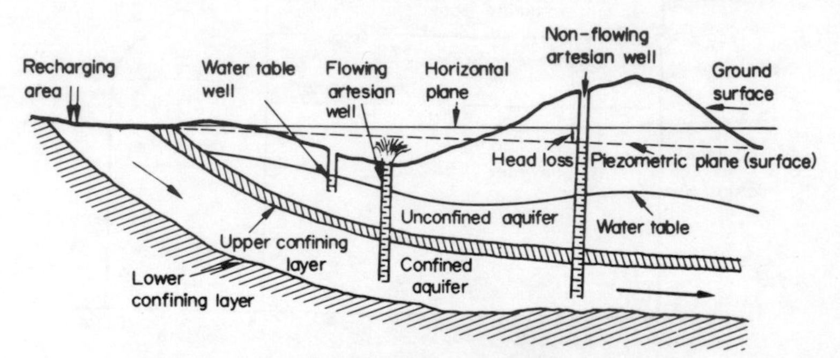

Fig. 9.3 Aquifers (adapted from Linsley and Franzini)

enters a confined aquifer at a location where the confining bed rises to the surface. Water movement through a confined aquifer is similar to flow through a pipe. The hydraulic head at a point within the aquifer is equivalent to the elevation of the water table in the recharge area less the head loss up to the point in question. The water level rises above the confining bed as shown in Fig. 9.3 to the level of the local static head. Just as water levels in piezometers in a pipe determine the piezometric surface, so do the water levels in a group of artesian wells penetrating a confined aquifer. The artesian piezometric surface is equivalent to its water table. If the piezometric surface lies above ground level, a flowing well occurs and water extraction can take place without pumping.

The yield from artesian aquifers is usually small compared with unconfined aquifers because of their relatively small recharge areas. They are however generally more economical because of the fact that they transmit water for large distances and an artesian well can be operated with little pumping cost.

Perched aquifers (Fig. 9.4) occur when percolating water is intercepted by an impervious stratum of limited size. The main water table may be far below the

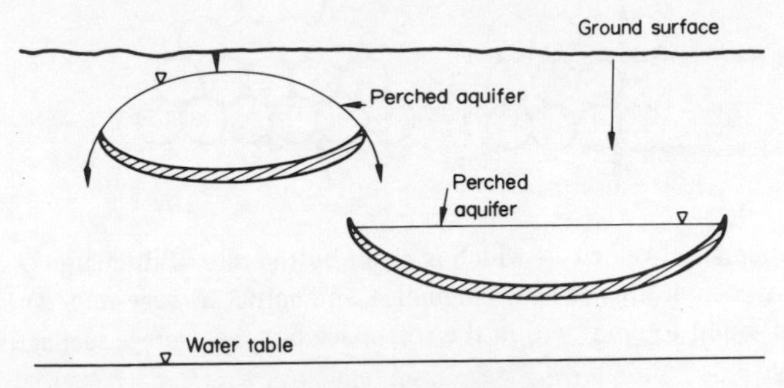

Fig. 9.4 Perched aquifers

perched groundwater. The perched aquifer is a special case of the unconfined aquifer. The yield from a perched aquifer is usually small and temporary.

The yield of an aquifer is determined by its size and porosity. The total amount of water that can be withdrawn assuming no further recharge is always less than the total pore space of the aquifer because some of the water is invariably retained by capillary forces. The specific yield, which is defined as the ratio of the volume of water which will drain freely from the aquifer to the total volume of the aquifer, is accordingly always less than the average porosity of the aquifer. Specific yield tends to be higher for coarse grained aquifers than for fine grained ones since the former experience less capillary action.

9.2 Fundamentals of Groundwater Hydraulics

9.2.1 Basic Soil Characteristics

Figures 9.5(a) and (b) show the extreme arrangements of perfect spherical objects of identical size which have been piled up. Case (a) represents the unstable cubical array of the spheres and case (b) the stable rhombohedral arrangement. For the cubical arrangement of spheres, the porosity n defined by the ratio of volume of voids V_V to the total volume can be calculated to be 0·476 and for the rhombohedral packing the porosity is 0·26. Assuming water flows through any porous medium constituted as shown in Fig. 9.5, it is quite obvious that the liquid will not pass through regular passages but through pore spaces of cavernous cells interconnected by narrower channels.

The flow path in natural soils is still more complicated since natural soil particles deviate considerably from ideal spheres and are not uniform. In dealing with groundwater flow, however, it is not necessary to consider flow through individual channels. It is sufficient to consider the porous medium as a continuous medium of many interconnected openings which serve as the fluid carrier and satisfactory results are obtained by considering macroscopic flow across a section of many pore channels. Thus in seepage problems distinction must be made

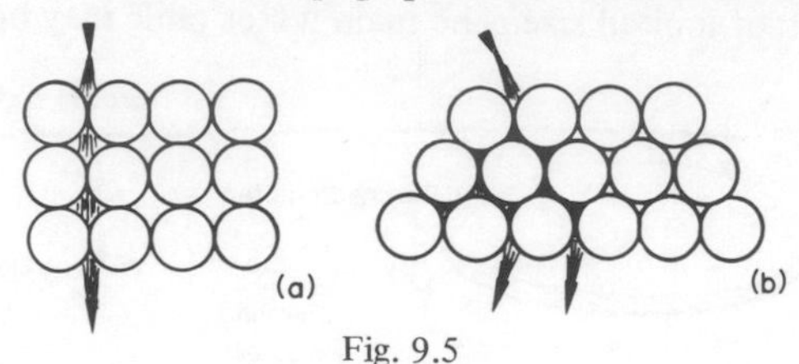

Fig. 9.5

between superficial velocity v which is given by the rate of discharge Q divided by the macroscopic area of flow channel A and actual *seepage velocity*. The latter varies from point to point within the soil space but the average seepage velocity v_s is given by

$$v_s = \frac{Q}{\text{pore space area}} = \frac{Q}{nA} = \frac{v}{n}$$

Unless otherwise specified, the velocity referred to throughout this text is the superficial velocity v.

9.2.2 Darcy's Law

In 1856, Henry Darcy published the results of experiments he had performed in Paris and proposed a law which is presently acknowledged as the basic principle

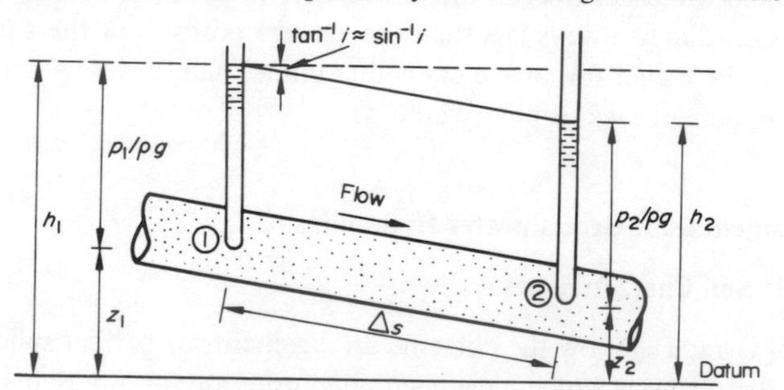

Fig. 9.6 Flow through sand filter

of flow through porous media. He used an experimental set-up similar to that shown in Fig. 9.6 to study the flow of water through a sand filter bed. By varying the length Δs of sand column and the head of water on the sand, he discovered that the flow rate was proportional to the head drop across the bed $(h_1 - h_2)$ and inversely proportional to Δs for a particular type of sand having the same cross sectional area, A.

The results may be summarized in mathematical language as

$$Q = k \frac{h_1 - h_2}{\Delta s} A \tag{9.1}$$

where the coefficient of proportionality k is now known in soil mechanics as the *coefficient of permeability*. It has the units of velocity and by analogy to conductance in electricity may be referred to as *hydraulic* conductivity. In the limit as Δs approaches zero, Darcy's law may be written as

$$v = Q/A = ki \tag{9.2}$$

where the *hydraulic gradient*

$$i = -\lim_{\Delta s \to 0} \Delta h/\Delta s = -\mathrm{d}h/\mathrm{d}s$$

The negative sign emphasizes the point that head loss is in the direction of flow. The piezometric head h is made up as usual of the elevation head z and the pressure head of the flowing liquid ($h = p/\rho g + z$).

Darcy's law is valid only for very slow flows through soils. Various investigators have given the limit of validity based on the Reynolds number ranging from 1 to 12. It is therefore safe to assume that the limit of validity is a Reynolds number of unity or

$$R = vd/\upsilon = 1 \tag{9.3}$$

where d is the average diameter of the soil particles (see Chapter 7) and υ is the coefficient of kinematic viscosity of the flowing fluid.

The law has however been satisfactorily applied to situations in which the Reynolds number does not exceed 10.

9.2.3 Coefficient of Permeability

Up to date information on the relationship between permeability and the geometric properties of soils and the physical properties of the seeping fluid is incomplete. However, in order that the student may appreciate more fully the influence of the various factors on permeability, the following model of flow through porous media as flow through an irregular pipe section is presented. This is a rather simplified approach to the physics of seepage but it clearly exhibits the significant influence of both soil and fluid properties on the coefficient of permeability.

The Hagen–Poiseuille equation for laminar discharge Q through a circular duct of diameter D (see equation (3.8)) is,

$$Q = -\frac{\pi \rho g}{128\mu} D^4 \frac{\mathrm{d}h}{\mathrm{d}s} \tag{9.4}$$

or

$$Q = -\frac{\rho g A' m^2}{2\mu} \frac{\mathrm{d}h}{\mathrm{d}s} \tag{9.5}$$

where $m = D/4$ is the hydraulic mean radius, A' is the cross-sectional area of the duct and μ is the coefficient of dynamic viscosity.

For an irregular duct (Fig. 9.7), equation (9.5) may be applied provided a

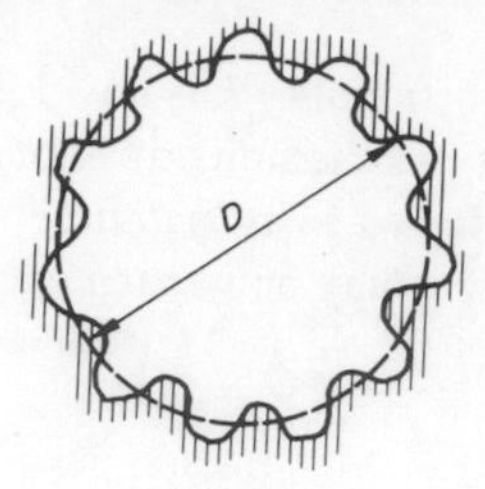

Fig. 9.7

shape factor C_s is applied to account for the irregular shape and the hydraulic radius is based on the mean diameter of the duct. Thus

$$Q = -C_s \frac{\rho g}{2\mu} A'm^2 \frac{dh}{ds} \tag{9.6}$$

For soil the combined microscopic flow channels may be considered to provide a macroscopic duct. The equivalent hydraulic mean radius may be assumed proportional to the ratio of volume of voids to the total surface area of the solids. Thus

$$m = K \frac{\text{volume of voids}}{\text{surface area of solids}} = K \frac{nV}{a_s} \tag{9.7}$$

where n is porosity, V is the volume of soil, a_s is the surface area of solids, and K is a constant of proportionality. In relation to the void ratio e,

$$m = K \frac{eV_s}{a_s} = K'ed \tag{9.8}$$

where V_s is the volume of solids, d is the average size of particles and K' is a new constant. Also $A' = n \times$ (macroscopic area of soil) $= nA$.

Thus from equation (9.6), absorbing K' into the shape factor,

$$Q = \left[\frac{C_s \rho g}{\mu} e^2 d^2 n\right] iA \tag{9.9}$$

or

$$Q = kiA$$

which is identical with Darcy's law, equation (9.2). The coefficient of permeability

$$k = \frac{C_s \rho g}{\mu} d^2 e^2 n = \frac{C_s \rho g}{\mu} d^2 \frac{e^3}{1+e} \tag{9.10}$$

It may therefore be inferred that the coefficient of permeability depends on both the soil characteristics and the physical properties of the fluid. Separating the two:

$$k = \frac{\rho g}{\mu} C_s d^2 \frac{e^3}{1+e} \tag{9.11}$$

or
$$k = \frac{\rho g}{\mu} k_0 \tag{9.12}$$

where
$$k_0 = C_s d^2 \frac{e^3}{1+e}$$

k is sometimes referred to as the specific or intrinsic permeability (dimension L/t) as distinct from k_0 (dimension L^2). Equation (9.12) must be used when dealing with a fluid flow with significant temperature variation. In the groundwater and seepage problems encountered in civil engineering practice, however, k is normally taken as constant for a particular soil type with a uniform degree of compaction. k_0 is measured in Darcy units which are equivalent to $9{\cdot}87 \times 10^{-13}$ m^2 or $10{\cdot}62 \times 10^{-12}$ ft^2.

Some typical values of the coefficient of permeability attributed to Silin-Bekchurin are quoted in Table 9.1 for guidance. The determination of permeability by laboratory and field methods receives considerable attention in soil mechanics. Pumping tests are used to determine permeability under field conditions and this will be referred to in subsection 9.3.3.

Table 9.1

Some typical values of coefficient of permeability (quoted from Harr)

Soil type	k (cm/s)
Clean gravel	1·0 and greater
Clean coarse sand	1·0–0·01
Mixed sand	0·01–0·005
Fine sand	0·05–0·001
Silty sand	0·002–0·0001
Silt	0·0005–0·00001
Clay	0·000001 and smaller

9.2.4 General Flow Equations and the Flownet

It was stressed in the introduction to Chapter 2 that the principles of conservation of mass, momentum and energy plus the equation of state form the basis of all fluid flow problems. This is also true of the hydraulics of groundwater. The basic equations of continuity and of motion in hydrodynamic theory are relevant and can readily be applied. Many groundwater and seepage problems make use only of the principle of conservation of mass and the basic principles involved are discussed in this section.

We will now derive the general equation for the conservation of mass in a differential control volume (Fig. 9.8). If the velocity components are u, v and w in the three principal cartesian directions x, y and z respectively, the net rate of storage of fluid in the control volume of sides δx, δy and δz due to the x component of velocity is

$$-\frac{\partial(\rho u)}{\partial x}\,\delta x \delta y \delta z$$

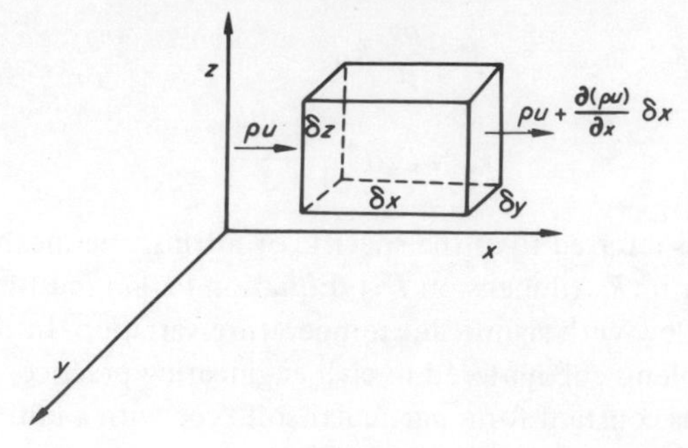

Fig. 9.8

and due to flow in the y and z directions are

$$-\frac{\partial(\rho v)}{\partial y}\,\delta x\delta y\delta z$$

and

$$-\frac{\partial(\rho w)}{\partial z}\,\delta x\delta y\delta z$$

respectively.

Thus the total net rate of storage of fluid mass in the control volume is given by

$$-\left[\frac{\partial(\rho u)}{\partial x}+\frac{\partial(\rho v)}{\partial y}+\frac{\partial(\rho w)}{\partial z}\right]\delta x\delta y\delta z$$

But the fluid mass inside the control volume at any instant of time is $\rho nS\delta x\delta y\delta z$, where n is the porosity of the soil mass and S is the degree of saturation. Thus

$$-\left[\frac{\partial(\rho u)}{\partial x}+\frac{\partial(\rho v)}{\partial y}+\frac{\partial(\rho w)}{\partial z}\right]\delta x\delta y\delta z=\frac{\partial(\rho nS\delta x\delta y\delta z)}{\partial t}\tag{9.13}$$

The control volume is independent of time, thus

$$-\left[\frac{\partial(\rho u)}{\partial x}+\frac{\partial(\rho v)}{\partial y}+\frac{\partial(\rho w)}{\partial z}\right]=\frac{\partial(\rho nS)}{\partial t}\tag{9.14}$$

For a homogeneous fluid with a constant density,

$$\frac{\partial u}{\partial x}+\frac{\partial v}{\partial y}+\frac{\partial w}{\partial z}=-\left[n\frac{\partial S}{\partial t}+S\frac{\partial n}{\partial t}\right]\tag{9.15}$$

The general form of Darcy's law is

$$q=-k\frac{\partial h}{\partial s}\tag{9.16}$$

where the absolute velocity q is in the direction of flow s. A porous medium may have a permeability which varies with direction, thus k is a tensor quantity which has components k_x, k_y and k_z in three principal cartesian directions. Thus

$$u = -k_x \frac{\partial h}{\partial x}, \quad v = -k_y \frac{\partial h}{\partial y}, \quad w = -k_z \frac{\partial h}{\partial z} \tag{9.17}$$

Substituting equation (9.17) into (9.15)

$$k_x \frac{\partial^2 h}{\partial x^2} + k_y \frac{\partial^2 h}{\partial y^2} + k_z \frac{\partial^2 h}{\partial z^2} = n \frac{\partial S}{\partial t} + S \frac{\partial n}{\partial t} \tag{9.18}$$

For an isotropic soil medium ($k_x = k_y = k_z = k$), permanently completely saturated ($S = 100\%$), equation (9.18) becomes

$$k \left[\frac{\partial^2 h}{\partial x^2} + \frac{\partial^2 h}{\partial y^2} + \frac{\partial^2 h}{\partial z^2} \right] = \frac{\partial n}{\partial t} \tag{9.19}$$

Equation (9.19) is the three-dimensional consolidation equation in soil mechanics. K. Terzaghi employed the special one-dimensional case as the basis for his well-known theory of consolidation, an excellent treatment of which is given in Taylor's *Fundamentals of Soil Mechanics*. Underground water extraction produces changes in pore water pressure and commonly results in settlement. The transient situation may be analysed using equation (9.18) or (9.19). Under steady state conditions, there are no changes in void ratio (porosity) and

$$\frac{\partial^2 h}{\partial x^2} + \frac{\partial^2 h}{\partial y^2} + \frac{\partial^2 h}{\partial z^2} = 0 \tag{9.20a}$$

Equation (9.20a) is a form of Laplace's equation. In fact, equation (9.16) corresponds to the definition of velocity potential ϕ given by

$$q = -\nabla \phi \tag{9.20b}$$

with kh being equivalent to ϕ. Just as the equipotential surface is everywhere normal (orthogonal) to streamlines, the surface represented by kh = constant in groundwater flow is everywhere normal to the macroscopic streamline. By analogy, there exists a stream function ψ for all two-dimensional and axisymmetrical steady, incompressible groundwater flows and the ψ = constant curve is

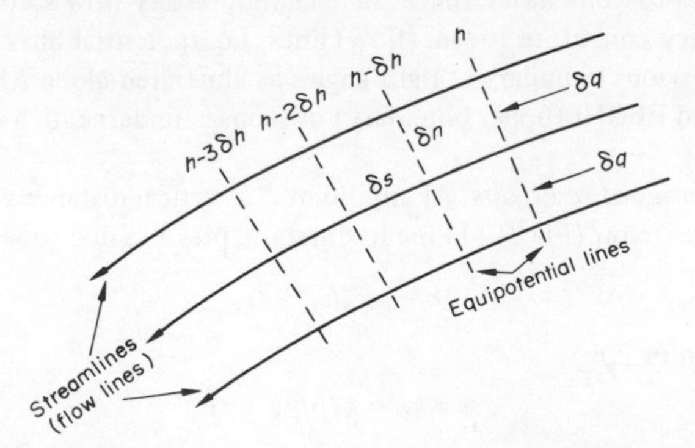

Fig. 9.9 Orthogonal flownet of streamlines and equipotential lines

everywhere orthogonal to the kh = constant curve. The entire flow regime can be imagined covered by a network of equipotential lines and streamlines as illustrated in Fig. 9.9.

Considering the portion of the flownet shown in Fig. 9.9 the hydraulic gradient i is given by

$$i = -\frac{dh}{ds} \tag{9.21}$$

and flow between two adjacent stream lines separated by distance δn per unit thickness normal to the plane of paper is given by

$$\delta q = ki\delta n = -k\frac{dh}{ds}\delta n \tag{9.22}$$

If the flownet is drawn as approximate squares, $\delta n \simeq \delta s$, equation (9.22) will reduce to

$$dq = -k\,dh$$

or

$$dq = k\,h/m \tag{9.23}$$

where h is the total head drop across m squares ($h = -\Sigma\Delta h$).

For N flow channels, the total flow per unit thickness is given by

$$q = kh\,N/m = S_F kh \tag{9.24}$$

where $S_F = N/m$ is known as the flownet shape factor.

9.2.5 Boundary Conditions

The use of a flownet in seepage problems is analogous to the graphical solution of the two-dimensional Laplace equation. Like any other form of solution of a differential equation the appropriate boundary conditions must be satisfied. Four boundary conditions are necessary (two-dimensional second-order partial differential equation) and these are prescribed at the impervious boundary (boundaries), the boundaries of the reservoirs in the case of dams, and the free surface in the case of unconfined aquifers. These are discussed briefly below.

(1) *Impervious boundaries*. Since there cannot be any flow across impervious boundaries they constitute stream (flow) lines. Equipotential lines must therefore meet an impervious boundary at right angles as illustrated along AB (bottom boundary) and EFGHI (upper boundary) of seepage underneath a concrete dam in Fig. 9.10.

(2) *Boundaries of reservoirs*. At any point P a vertical distance z from the bottom of a reservoir (Fig. 9.11) the hydrostatic pressure due to water is given by

$$p = \rho g\,(h_1 - z) \tag{9.25}$$

From equation (9.20b)

$$\phi = kh = k(p/\rho g + z)$$

$\therefore$ At the boundary

$$\phi = kh_1\text{, a constant} \tag{9.26}$$

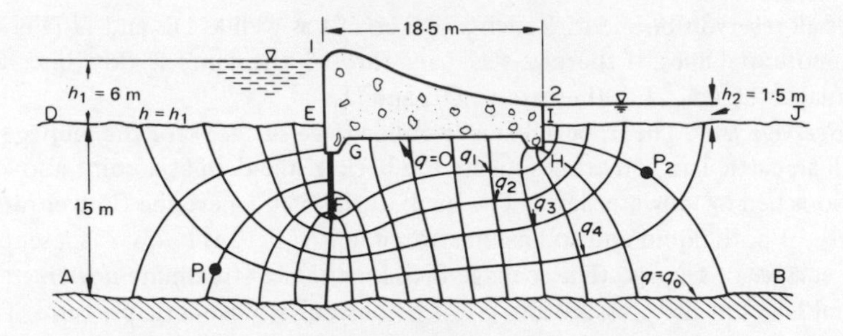

Fig. 9.10 Seepage underneath a dam

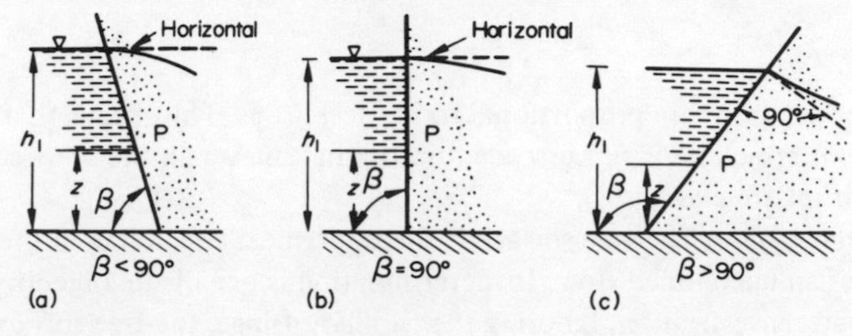

Fig. 9.11

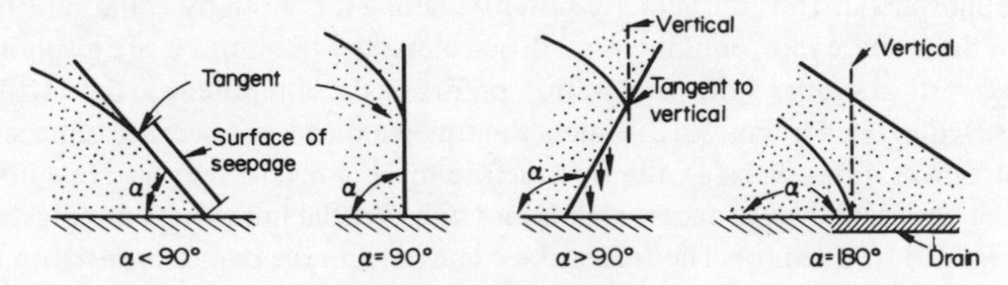

Fig. 9.12

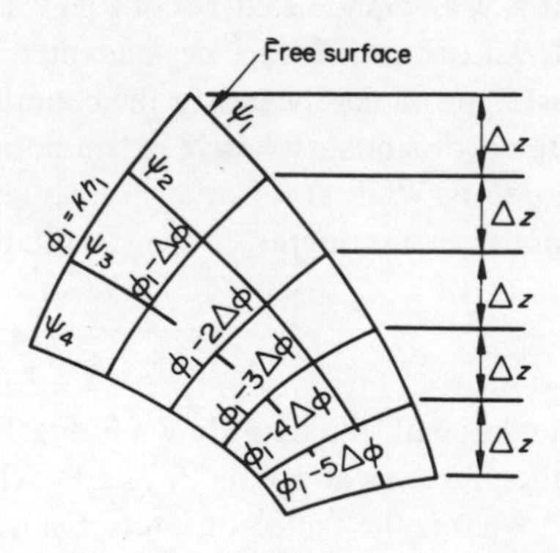

Fig. 9.13

Thus all reservoir boundaries such as 1E and 2I as well as DE and IJ (Fig. 9.10) are equipotential lines. If there was seepage through the dam the flow lines would be normal to 1E and 2I as they are to DE and IJ.

(3) *Seepage face*. The uppermost streamline (free surface) for the seepage through an earth dam sometimes meets the back of the dam at a point above an impervious bed or tailwater level. The back of the dam where the flow enters a zone free of both liquid and soil as illustrated in Fig. 9.12 is known as a seepage face or surface of seepage. The seepage face is neither a streamline nor an equipotential line, but the pressure along it is a constant atmospheric pressure. Thus along a seepage face, the variation of the potential ϕ given by

$$\phi = kh = k(p_{at}/\rho g + z) \tag{9.27a}$$

is

$$\Delta\phi = k\Delta z \tag{9.27b}$$

The potential drops are proportional to vertical drops. This means that at the points of intersection of successive equipotential lines with the seepage face, the vertical drops are equal.

(4) *Free surface*. The free surface is the uppermost streamline in the flow domain of an unconfined flow. Its determination is one of the objectives of groundwater investigation. Ignoring the capillary fringe, the free surface or line of seepage separates the saturated groundwater region from the unsaturated soil above it. The pressure along a free surface is constant and is normally assumed to be atmospheric. Thus the velocity potential along it is given by equation (9.27a). Like the seepage face, equipotential drops along the free surface are given by equal vertical drops of the intersection points of the equipotential lines with it. This is illustrated in Fig. 9.13. Unless the upstream end of a seepage zone slopes upstream as in Fig. 9.11(a), the free surface must enter the seepage zone normal to the upstream dam surface which is an equipotential line. These are illustrated in Figs. 9.11(b) and (c). The free surface leaves from the dam tangential to the back surface when the slope α is not greater than 90° and it is tangential to the vertical when α is greater than 90°. The proof of this statement is beyond the scope of this book, but A. Casagrande's curves of Fig. 9.12 are reproduced for guidance. Figures 9.11, 9.12 and 9.13 have been adapted from Harr.

A good flownet must approximately satisfy the conditions in (1) to (4) in addition to satisfying the orthogonality between equipotential lines and streamlines at all intersecting points. While it is not a necessary condition that the net must be made up of approximate squares, it does facilitate computations and it is very convenient.

EXAMPLE 9.1

Calculate the seepage underneath the dam shown in Fig. 9.10 with h_1 = 6 m and h_2 = 1·5 m. What are the pressures at points P_1 and P_2 which are 2 m and 12 m respectively above AB? What is the factor of safety against piping near I? k = 2·5 × 10^{-3} cm/s.

SOLUTION

From the flownet in Fig. 9.10 the number of flow channels $N = 5$, and the number of equipotential drops $m = 16$. The shape factor $S_F = 0{\cdot}312$.

The total head drop $h = h_1 - h_2 = 4{\cdot}5$ m.

The discharge per unit width of dam

$$q = 0{\cdot}312 \times 2{\cdot}5 \times 10^{-5} \times 4{\cdot}5 \times 24 \times 3600 = 3{\cdot}02 \text{ m}^2/\text{day}$$

The head drop per equipotential drop $= 4{\cdot}5/16 = 0{\cdot}28$ m (i.e. $h_1 - 2 \times 0{\cdot}28 = 20{\cdot}44$ ı

P_1 lies on the 20·44 m equipotential (relative to AB).

∴ The pressure at $P_1 = 20{\cdot}44 - 2 = \underline{18{\cdot}44 \text{ m or } 180 \text{ kN/m}^2}$

Similarly the pressure at $P_2 = \underline{4{\cdot}22 \text{ m or } 41{\cdot}2 \text{ kN/m}^2}$

The seepage gradient $i = -\mathrm{d}h/\mathrm{d}s$ near I is approximately equal to the potential drop $\Delta h = h/m$ divided by the average length of the side of the square adjoining I. Using the approximate linear scale for the drawing

$$i = 0{\cdot}28/1{\cdot}75 = 0{\cdot}16$$

According to soil mechanics theory, piping or boiling is about to occur when the seepage gradient equals the ratio of the buoyant unit weight of the soil, γ_b, to the unit weight of the flowing fluid, γ_w. It is approximately equal to unity for most sandy soils. Thus the factor of safety against piping is approximately

$$1/i = \underline{6{\cdot}25}$$

9.2.6 Non-homogeneous Porous Media

Due to different stages of deposition and consolidation, sedimentary soils are often naturally laminated. Such soil media are described as *non-homogeneous*

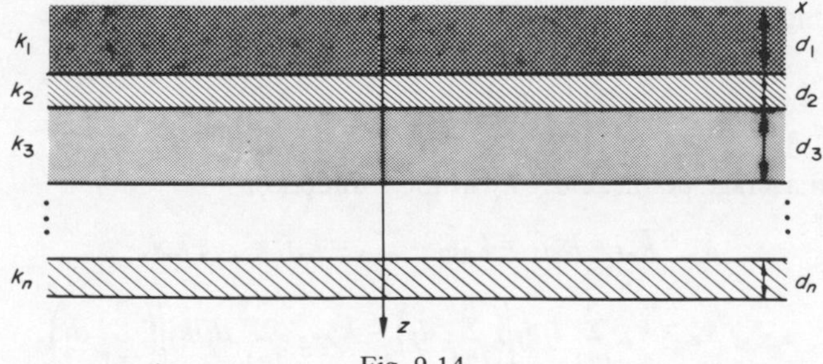

Fig. 9.14

and the different thin layers often have different coefficients of permeability. If, however, the coefficient of permeability is the same in all directions for a particular layer, the medium would be non-homogeneous but isotropic. Should the coefficient of permeability vary in different directions, it would be non-homogeneous and anisotropic. Even in most homogeneous soil deposits, because of natural consolidation, the coefficient of permeability in the vertical direction is generally less than the coefficient in the horizontal direction. The opposite is however known to be true in loesses. The detailed consideration of anisotropy is beyond the scope of this book and the interested reader is referred to Harr's *Groundwater and Seepage* and advanced soil mechanics textbooks. In this book we limit our discussion to two-dimensional parallel beddings with flow along the laminations or normal to the bedding planes.

Figure 9.14 depicts such a sedimentary bedding of n layers of thickness $d_1, d_2, \ldots d_n$ and permeability $k_1, k_2, \ldots k_n$. For flow along the bedding in the x direction the total flow per unit width of soil is given by the sum of flow in the n beddings,

$$q_x = \sum_{j=1}^{n} q_j$$

$$\therefore \qquad q_x = \sum_{j=1}^{n} k_j i d_j \tag{9.28}$$

where $j = 1, 2, \ldots n$.

The seepage gradient i is the same for all beds since the total head drop is the same over the same distance along the bedding. The average seepage velocity

$$\bar{v}_x = q_x \Bigg/ \left(\sum_{j=1}^{n} d_j \right) = i \sum_{j=1}^{n} k_j d_j \Bigg/ \left(\sum_{j=1}^{n} d_j \right) = \bar{k}_x i \tag{9.29}$$

where the average coefficient of permeability in the x direction

$$\bar{k}_x = \sum_{j=1}^{n} k_j d_j \Bigg/ \left(\sum_{j=1}^{n} d_j \right) \tag{9.30}$$

This is equivalent to the expression for average electrical conductance for conductors in parallel.

For flow normal to the bedding planes, the same volume of water passes through unit area at all levels in the z direction. If i is the average head loss per unit length in flowing across the beddings,

$$\text{Total head drop} = i \sum_{j=1}^{n} d_j = \sum_{j=1}^{n} i_j d_j \tag{9.31}$$

With an average permeability $\bar{k}_z$ in the z direction

$$v_z = \bar{k}_z i = k_1 i_1 = k_2 i_2 = \ldots = k_j i_j = \ldots k_n i_n \tag{9.32}$$

or

$$v_z = \bar{k}_z \sum_{j=1}^{n} i_j d_j \Bigg/ \left(\sum_{j=1}^{n} d_j \right) = \bar{k}_z v_z \sum_{j=1}^{n} d_j / k_j \Bigg/ \left(\sum_{j=1}^{n} d_j \right) \tag{9.33}$$

since from (9.32) $\qquad i_j = v_z / k_j$ and v_z is constant.

Thus

$$\bar{k}_z = \sum_{j=1}^{n} d_j \Bigg/ \left(\sum_{j=1}^{n} d_j / k_j \right) \tag{9.34}$$

Equation (9.34) is similar to the expression for the conductance of electrical conductors in series.

9.2.7 Dupuit's Approximation

In 1863, J. Dupuit proposed a theory, which now bears his name, for the analysis of certain types of unconfined flows. An unconfined flow is one which has a free surface as one of the boundaries of the flow domain as opposed to a confined

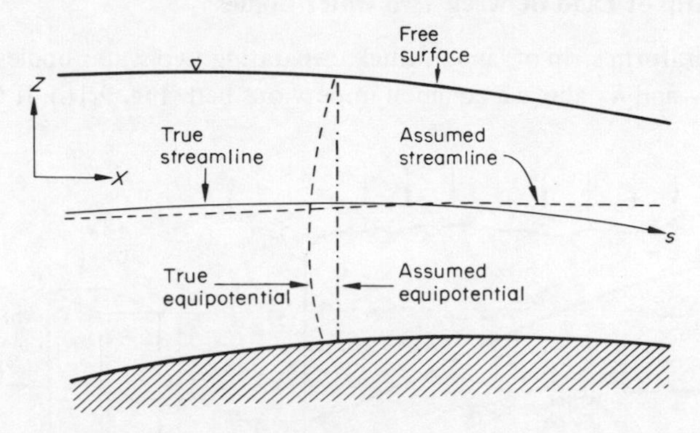

Fig. 9.15

flow whose boundaries are structurally fixed. Consider an unconfined two-dimensional flow domain bounded by an impervious 'straight' bottom and a free surface, as shown in Fig. 9.15. Taking any streamline in the general direction s, the velocity along the streamline is given by equation (9.16) as

$$v = -k\frac{\mathrm{d}h}{\mathrm{d}s}$$

Dupuit's theory is based on two assumptions: (1) that for small inclinations of the free surface, the streamlines may be taken as horizontal and thus the equipotential lines vertical. Thus

$$v = -k\frac{\mathrm{d}h}{\mathrm{d}s} \simeq -k\frac{\mathrm{d}h}{\mathrm{d}x} \tag{9.35}$$

and (2) that the hydraulic gradient is equal to the slope of the free surface and does not vary with depth. At the free surface,

$$\frac{\mathrm{d}h}{\mathrm{d}x} = \frac{\mathrm{d}(p_{\mathrm{at}}/\rho g + z)}{\mathrm{d}x} = \frac{\mathrm{d}z}{\mathrm{d}x}$$

since the pressure p_{at} at the free surface is constant. According to Dupuit's two assumptions,

$$v = -k\frac{\mathrm{d}z}{\mathrm{d}x}\bigg|_{\text{at the free surface}} \tag{9.36}$$

9.3 Some Practical Groundwater Flow Problems

This section is devoted to the application of the theories discussed in Section 9.2 to some simple practical groundwater and seepage problems. Complete and free saturation is assumed in all cases and flow is by gravity only, although in some special cases, such as in the well problems, pumping may be responsible for creating the imbalance necessitating groundwater (gravity) flow.

9.3.1 A Strip of Land Between Two Water Bodies

Imagine a uniform strip of land, L thick, separating two water bodies whose levels are h_1 and h_2 above a common impervious bed (Fig. 9.16). If there is a

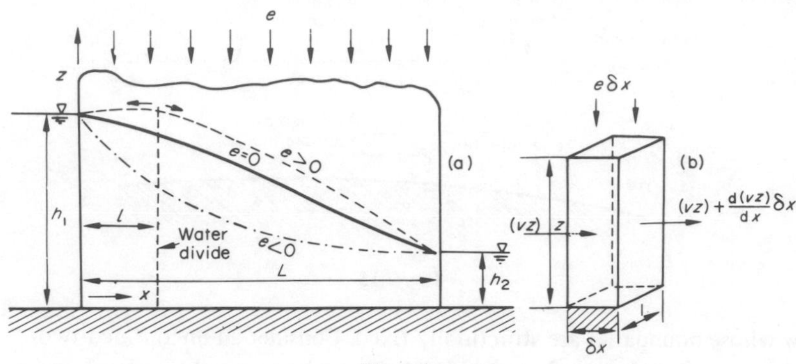

Fig. 9.16

uniform infiltration e per unit area entering the mound, the problem is to locate the water table and to determine the seepage rate through the mound.

Assuming the free surface to be a distance z above the impervious bed, and at a distance x from the water reservoir with the higher level, the continuity equation for the isolated element shown in Fig. 9.16(b) is

$$e - \frac{\mathrm{d}(vz)}{\mathrm{d}x} = 0 \tag{9.37}$$

Using Dupuit's approximation,

$$v = -k\,\frac{\mathrm{d}z}{\mathrm{d}x} \tag{9.38}$$

Combining (9.37) and (9.38)

$$e + \frac{\mathrm{d}(kz\mathrm{d}z/\mathrm{d}x)}{\mathrm{d}x} = 0 \tag{9.39}$$

Integration of (9.39) gives

$$\tfrac{1}{2}kz^2 + \tfrac{1}{2}ex^2 + C_1x + C_2 = 0 \tag{9.40}$$

Using the boundary conditions, $z = h_1$ when $x = 0$ and $z = h_2$ when $x = L$ the constants of integration C_1 and C_2 are determined and

$$z^2 = h_1^2 - (h_1^2 - h_2^2 - eL^2/k)\, x/L - ex^2/k \tag{9.41}$$

Equation (9.41) represents an ellipse for a positive value of e (uniform infiltration), a hyperbola for a negative e (uniform evaporation), and a parabola for e equal to zero.

In the case of uniform infiltration, the lengthwise rate of change of discharge per unit width through a vertical plane at x is given from equation (9.37) as

$$\frac{\mathrm{d}(vz)}{\mathrm{d}x} = \frac{\mathrm{d}q}{\mathrm{d}x} = e \tag{9.42}$$

Integrating with respect to x,

$$q = ex + q_0 \tag{9.43}$$

where q_0 is the seepage rate from the higher body of water ($x = 0$). From equation (9.38)

$$q = -kz\,\frac{\mathrm{d}z}{\mathrm{d}x} \tag{9.44}$$

Substituting (9.44) into (9.43) and integrating gives

$$-\frac{k}{2}z^2 = \frac{e}{2}x^2 + q_0 x + C \tag{9.45}$$

Using the boundary conditions, $z = h_1$ when $x = 0$ and $z = h_2$ when $x = L$, the constant of integration is determined and q_0 is obtained as

$$q_0 = k\,\frac{(h_1^2 - h_2^2)}{2L} - \frac{eL}{2} \tag{9.46}$$

Thus from equation (9.43) the discharge at any section is given by

$$q = \frac{k(h_1^2 - h_2^2)}{2L} - e\left(\frac{L}{2} - x\right) \tag{9.47}$$

Let $q = 0$ at $x = l$, and from (9.47)

$$l = \frac{L}{2} - \frac{k(h_1^2 - h_2^2)}{2eL} \tag{9.48}$$

$x = l$ is known as the water divide and locates the maximum level of the water table. It represents the plane which separates the zones in which water flows in opposite directions. Notice that equation (9.41) does not predict the existence of a seepage face. This is because of Dupuit's approximations and the simplified exit boundary conditions used.

If $e = 0$, the seepage gradient at the low water level end of the mound is obtained from (9.41) as

$$i = -\left.\frac{\mathrm{d}z}{\mathrm{d}x}\right|_{x=L} = \frac{1}{2L}\left(\frac{h_1^2}{h_2} - h_2\right) \tag{9.49}$$

The factor of safety against piping is approximately

$$\frac{2Lh_2}{h_1^2 - h_2^2}$$

The Keta sea erosion problem, in its simplified form, can be likened to the situation represented by Fig. 9.16. Keta is situated on a narrow sand bank separating the Keta lagoon on the one side from the sea on the other. In the dry season the lagoon level is normally lower than the mean sea level. Salt water therefore would seep from the sea toward the lagoon thereby increasing the salt content of soils inland. However in the wet season when the lagoon level is higher than the mean sea level the seepage direction is reversed. Depending on the head difference, the seepage gradient (equation (9.49)) on the sea side may be quite high. A high seepage gradient means a high seepage force which may loosen the natural state of the soil substantially and make it easier for sea waves to remove the soil. At Keta the wet season generally coincides with high seas and high wave energy (see Chapter 10). This probably explains why sea erosion at Keta, apart from flooding from the Keta lagoon, assumes critical dimensions between August and October of almost every year.

9.3.2 Unconfined Flow on an Inclined Impervious Bed

Consider the lower impervious boundary of an unconfined flow regime inclined at a slope S_0 as shown in Fig. 9.17. If the water table is at a distance z above an

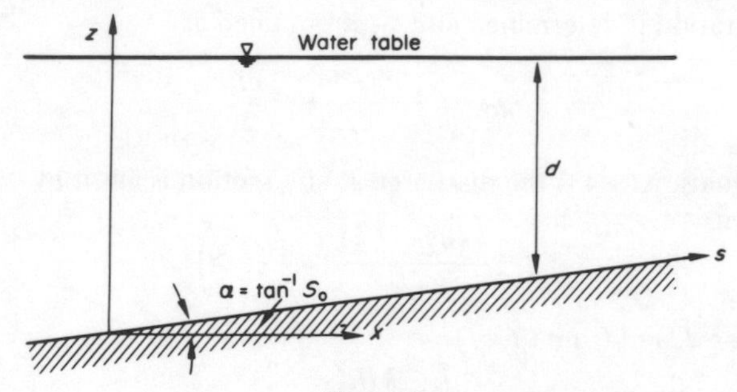

Fig. 9.17

arbitrary horizontal level, the rate of flow per unit width according to Dupuit's law is given by

$$q = -kd \frac{dz}{dx} \tag{9.50}$$

where d is the level of the water table above the confining bed at a distance x from an arbitrary origin.

For a mild slope, $S_0 = \tan \alpha \simeq \sin \alpha$

$$q = -k(z - S_0 x) \frac{dz}{dx} \tag{9.51}$$

or

$$\frac{dx}{dz} - k\frac{S_0 x}{q} = -\frac{kz}{q} \tag{9.52}$$

Solving equation (9.52) for a constant q (from continuity) gives

$$x = -C\exp(kS_0 z/q) + z/S_0 + q/(kS_0^2) \tag{9.53}$$

where the constant of integration C is to be determined from the boundary conditions. Suppose $C = 0$,

$$z - S_0 x = -q/(kS_0) = \text{constant}, z_0 \tag{9.54}$$

where $z_0 = z - S_0 x$ is the depth of flow.

The flow according to equation (9.54) is uniform, the water table is parallel to the impervious bed and the discharge

$$q = -kS_0 z_0 \tag{9.55}$$

From equations (9.53) and (9.54), putting $C_1 = S_0 C$,

$$z - (z_0 + S_0 x) = C_1 \exp(-z/z_0) \tag{9.56}$$

Let $z = z_1$ when $x = x_1$, and $z = z_2$ when $x = x_2$. Then

$$\frac{z_1 - z_0 - S_0 x_1}{z_2 - z_0 - S_0 x_2} = \exp[(z_2 - z_1)/z_0]$$

or

$$z_2 - z_1 = z_0 \log_e\left(\frac{z_1 - z_0 - S_0 x_1}{z_2 - z_0 - S_0 x_2}\right) \tag{9.57}$$

Thus knowledge of the free surface at any two locations enables the normal depth z_0 to be determined from equation (9.57) by trial and error or otherwise and subsequently the discharge q is obtained from equation (9.54). It should be observed that equation (9.54) is in fact asymptotic to equation (9.56) (see Fig. 9.18(a) and (b)); therefore the quantity of discharge is the same. This point will become clearer after the three possible conditions expressed in equation (9.56) have been examined below. C_1 may be determined (after obtaining z_0) from

$$C_1 = \frac{z_1 - z_0 - S_0 x_1}{\exp(-z_1/z_0)} = \frac{z_2 - z_0 - S_0 x_2}{\exp(-z_2/z_0)} \tag{9.58}$$

According to equation (9.56) the free surface will always lie above the normal depth line if $C_1 > 0$ and below it if $C_1 < 0$. These conditions are illustrated in Fig. 9.18 for different orientations of the impervious bottom bed and relative levels of the headwater and tailwater. For a low tailrace z_2 relative to the headwater z_1, the water table lies entirely below the normal depth line (Fig. 9.18(a)). As the tailrace level rises a normal depth condition may be encountered when $d_2 = d_1$. For $d_2 > d_1$ but $z_2 < z_1$, Fig. 9.18(b) is appropriate and the water table lies entirely above the normal depth line. As $z_2 > z_1$ the flow is reversed in the uphill direction and the sense of S_0 becomes negative. The equivalent of equation (9.56) for this case is

$$z + z_0 + S_0 x = C_1 \exp(z/z_0) \tag{9.59}$$

It may be observed that the general equation (9.59) is asymptotic to zero but only for $z \to -\infty$. The only physically meaningful solution in this case is the falling water table Fig. 9.18(c). The geometrical significance of equation (9.59) is illustrated in Fig. 9.18(d).

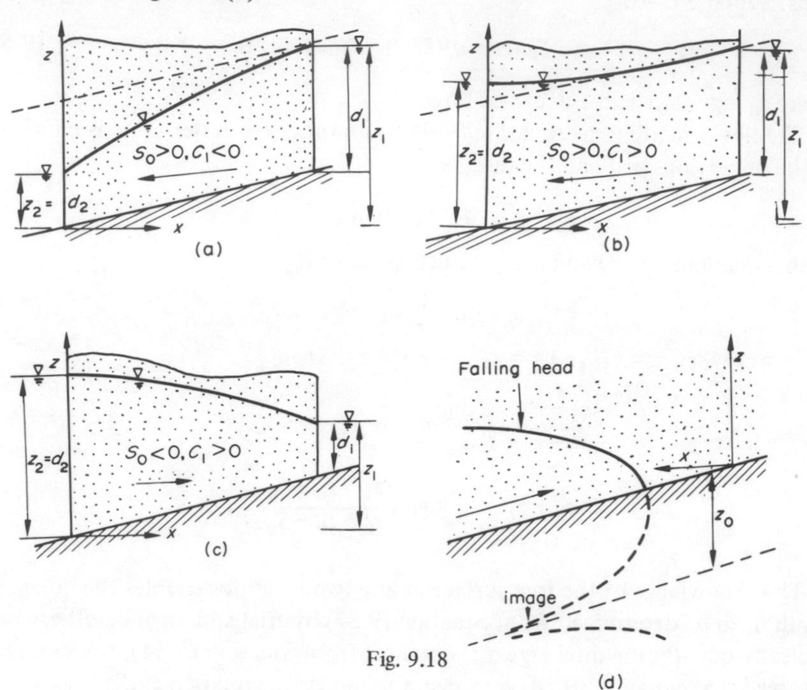

Fig. 9.18

The similarity between the curves in Fig. 9.18 and the so-called backwater curves in open-channel hydraulics must be noted. (See Fig. 4.19). The curve of Fig. 9.18(a) is equivalent to an M2 curve, of (b) to an M1 curve and of (c) to an A2 curve.) However, it must be emphasized here that unlike open-channel turbulent flow conditions which result in complex mathematical equations, the use of Darcy's law and Dupuit's approximations for the generally laminar groundwater flow has yielded simple mathematical solutions as indicated above. A more detailed discussion of the subject based on Pavlovsky's solutions is given in Harr's *Groundwater and Seepage*.

9.3.3 The Hydraulics of Wells

The importance of the hydraulics of wells cannot be over-emphasized since wells constitute the major mode of aquifer water extraction. Whenever water is pumped from a well the water table or piezometric surface in the surrounding aquifer is lowered. There are two main types of problems concerned with wells; problems concerned with confined aquifers and those with unconfined aquifers. The method of solution based on Dupuit's theory is identical in both cases, except that Dupuit's method is exact in the case of a confined aquifer of a uniform depth.

Figure 9.19 shows a well of radius r_w penetrating the whole depth b of a confined aquifer. If water is being pumped out of the well at a constant rate Q and

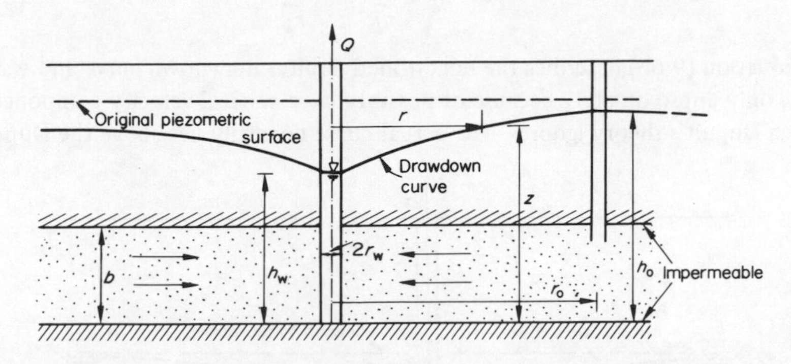

Fig. 9.19 Radial flow to a well penetrating an extensive confined aquifer

if it is assumed that water flows into the well from all radial directions,

$$Q = (2\pi rb)v_r \tag{9.60}$$

where v_r is the velocity at radius r from the centre of the well.

From equation (9.35)

$$v_r = k\frac{dz}{dr} \tag{9.61}$$

The positive sign shows that the velocity is in the negative r direction.

From (9.60) and (9.61)

$$dz = \frac{Q}{2\pi kb}\frac{dr}{r} \tag{9.62}$$

Integrating and using the boundary condition, $z = h_w$ at the well $r = r_w$

$$z - h_w = \frac{Q}{2\pi kb}\log_e r/r_w \tag{9.63}$$

The drawdown curve described by equation (9.63) is often used in practice to determine the field coefficient of permeability. Knowing the water level h_0 in a well at $r = r_0$ which reaches the aquifer but from which water is not being pumped gives

$$k = \frac{Q}{2\pi b(h_0 - h_w)}\log_e r_0/r_w \tag{9.64}$$

In the unconfined aquifer the piezometric surface coincides with the water table. If the well penetrates to the lower confining impermeable bed, Dupuit's approximation gives

$$Q = 2\pi rzk \frac{dz}{dr} \tag{9.65}$$

For a constant rate of pumping Q,

$$z^2 - h_w^2 = \frac{Q}{\pi k} \log_e \frac{r}{r_w} \tag{9.66}$$

Equation (9.66) describes the unconfined aquifer drawdown curve and water table only approximately because of possibly large vertical velocity components which Dupuit's theory ignores. The actual curve normally lies above the Dupuit

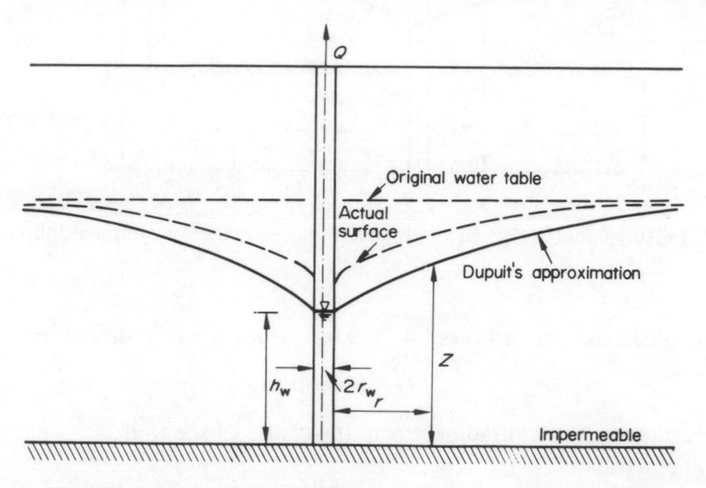

Fig. 9.20 Radial well penetrating an extensive unconfined aquifer

parabola (equation (9.66)) as indicated in Fig. 9.20. They both coincide at large radii at which the vertical velocity components become very small.

An interesting variation of the unconfined well problem arises when there is uniform recharge into the aquifer. This is illustrated in Fig. 9.21. The constant rate of recharge through rainfall or water spreading is e per unit area and water is being pumped from the well at a constant rate of Q_w.

Equation (9.65) still applies but Q is now not constant although Q_w is. The increase of flow rate is obtained by considering a control volume defined by the cylinder of thickness δr at radius r

$$-dQ = 2\pi er dr \tag{9.67}$$

Thus

$$Q - Q_w = -\pi er^2 \tag{9.68}$$

since at the well $r = r_w \rightarrow 0$ and $Q = Q_w$.

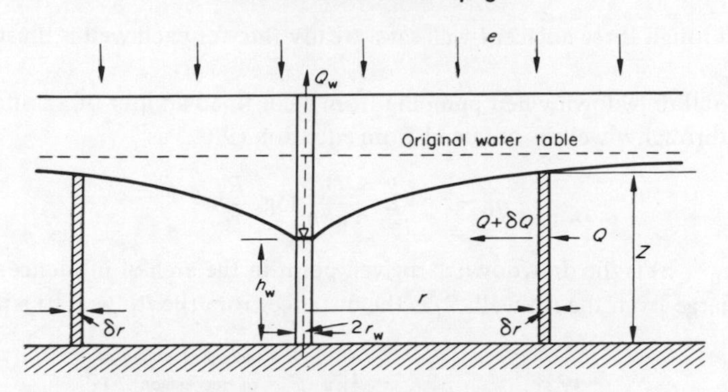

Fig. 9.21 Recharge of an unconfined aquifier

Substituting (9.68) into (9.65) gives

$$Q_w - \pi e r^2 = 2k\pi r z \frac{dz}{dr} \tag{9.69}$$

Integrating (9.69) and using the conditions $z = h_w$ at $r = r_w$ gives

$$z^2 - h_w^2 = \frac{e(r_w^2 - r^2)}{2k} + \frac{Q_w}{\pi k} \log_e \frac{r}{r_w} \tag{9.70}$$

which shows that recharge has the tendency to flatten the Dupuit parabola.

9.3.4 Multiple Well Systems

It is necessary for the sake of efficient planning and water utilization to determine the influence of adjacent wells on one another and of adjacent streams and rivers on wells and vice versa. Multiple well operations are also prominent in construction works where they are employed to lower the groundwater table to facilitate reasonably dry excavations.

The principle of superposition is employed to calculate the total drawdown due to a system of wells. This presupposes that the interaction of wells is linear. While this can be argued to be reasonably true for confined aquifers it is known not to be valid for unconfined aquifers. Nevertheless the concept is also generally applied to unconfined aquifers in which the drawdown is relatively small. The number and geometrical arrangement of wells are also important in the determination of drawdowns.

Generally if a system of n wells has independent drawdowns $(\Delta h)_1, (\Delta h)_2, \ldots, (\Delta h)_n$ the superposed drawdown is given by

$$(\Delta h) = \sum_{i=1}^{n} (\Delta h)_i \tag{9.71}$$

where $(\Delta h)_i$ may be positive or negative depending on whether there is withdrawal or recharge in the well i. A simple case of withdrawal from a confined aquifer through three adjacent wells at a steady rate for each well is illustrated in Fig. 9.22.

The total drawdown when pumping from a confined aquifer of a uniform depth b through n wells is obtained from equation (9.62) as

$$h_0 - z = \sum_{i=1}^{n} \frac{Q_i}{2\pi kb} \log_e \frac{R_i}{r_i} \tag{9.72}$$

where $(h_0 - z)$ is the drawdown at a given point in the area of influence, Q_i is the discharge from the ith well, R_i is the distance from the ith well to a point at

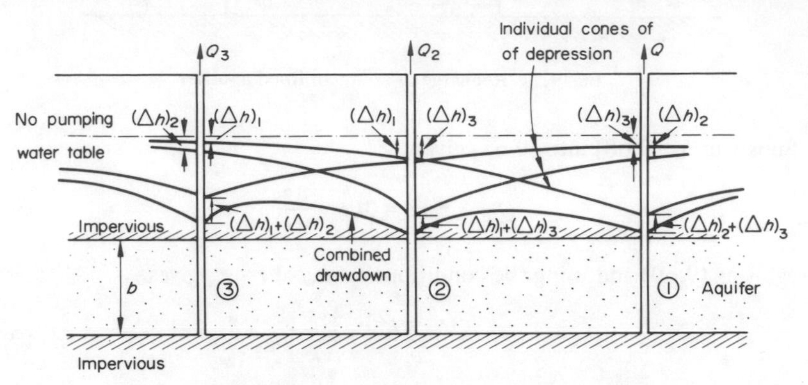

Fig. 9.22 Composite drawdown in a confined aquifer

which the drawdown is negligible and r_i is the distance from the ith well to the point of interest. The corresponding drawdown in an unconfined aquifer is derived from equation (9.65) for *relatively small* drawdowns as

$$h_0^2 - z^2 = \sum_{i=1}^{n} \frac{Q_i}{\pi k} \log_e \frac{R_i}{r_i} \tag{9.73}$$

Because of the non-linearity of the unconfined aquifer equation (9.65) and the inherent inexactness of Dupuit's approximations in relation to unconfined aquifers, equation (9.73) gives reasonably good results only when the drawdown is small and vertical components of velocity can be neglected. It must not be used for large rates of pumping.

Equation (9.71) may be applied to two wells located in an aquifer with one acting as an aquifer recharge well and the other as a normal supply well as shown in Fig. 9.23. A relatively high head must be available in the recharge well to provide the necessary hydraulic gradient which makes water seep into the aquifer. The resultant build-up curve is as shown. If the same flow is pumped down the recharge well as is extracted from the supply well a zero drawdown will be maintained midway between the two wells since the build-up value $(-\Delta h)$ of recharge balances the drawdown value $(+\Delta h)$ of withdrawal. Their combined effect is like a source (recharge) and a sink (withdrawal) if the wells are considered as having infinitesimally small diameters. The resulting flownet for a uniformly confined

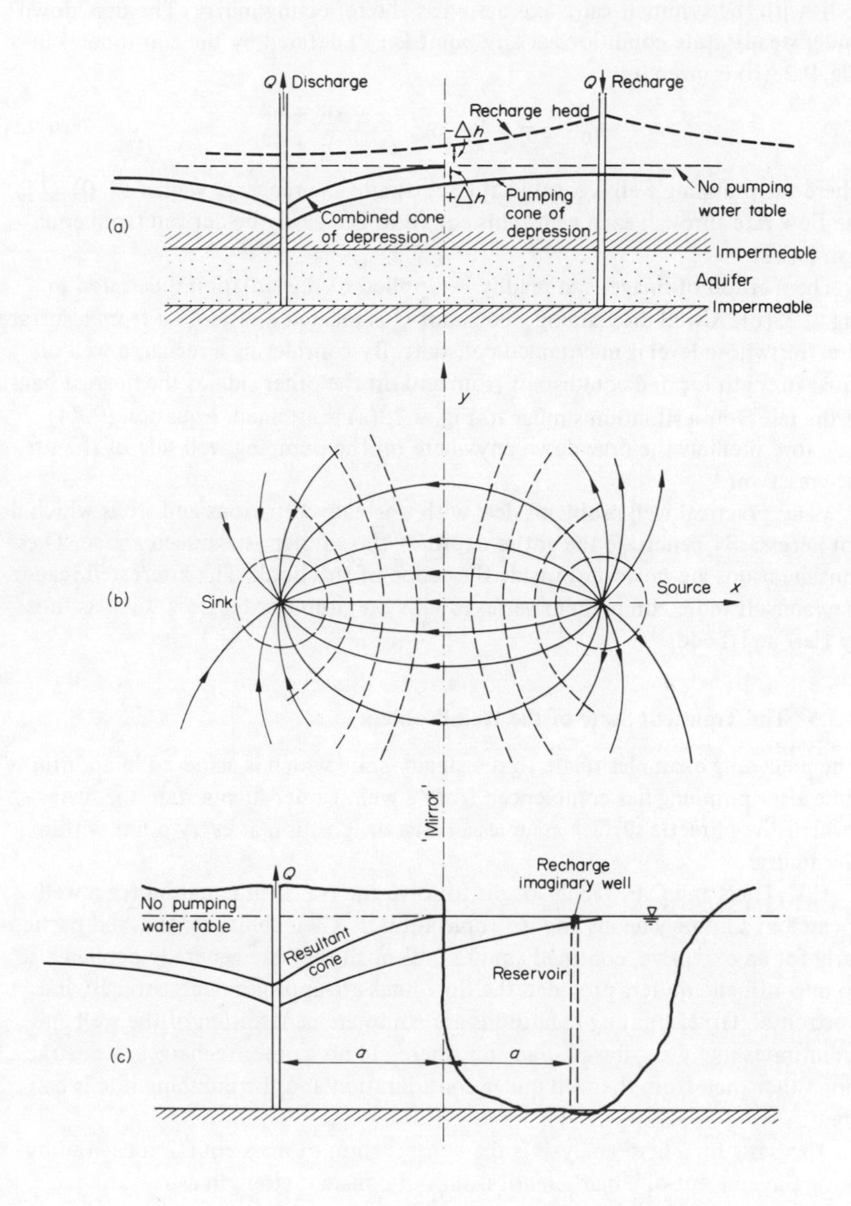
Q Discharge
Q Recharge
Recharge head
−Δh
+Δh
No pumping water table
Pumping cone of depression
Combined cone of depression
(a)
Impermeable
Aquifer
Impermeable
y
(b)
Sink
Source
x
'Mirror'
Q
Recharge imaginary well
No pumping water table
Resultant cone
Reservoir
(c)
a
a

Fig. 9.23

aquifer as predicted by classical hydrodynamic theory is shown in Fig. 9.23. The source (recharge well) may be viewed as a reflected image of the sink (supply well) with the symmetrical y axis acting as the reflecting mirror. The drawdown under steady state conditions at any point (x, y) defined by the coordinates in Fig. 9.23(b) is given by

$$h_0 - z = \frac{Q}{4\pi kb} \log_e \frac{(x-a)^2 + y^2}{(x+a)^2 + y^2} \tag{9.74}$$

where the pumping well is located at $(-a, 0)$ and the recharge well at $(a, 0)$. Q is the flow rate through each well. This equation can easily be derived from equation (9.72).

The method of images can readily be applied to the situation illustrated in Fig. 9.23(c). A well discharging a constant Q is located a distance a from a surface reservoir whose level is maintained constant. By considering a recharge well of equal strength located equidistant from and on the other side of the nearest bank of the reservoir a situation similar to Fig. 9.23(a) is attained. Equation (9.74) therefore predicts the drawdown anywhere on the pumping well side of the surface reservoir.

Many practical well problems deal with unsteady situations and wells which do not necessarily penetrate the entire depth of the aquifer as assumed above. These considerations are however outside the scope of this book. The interested reader may consult more comprehensive textbooks on groundwater flow, such as those by Harr and Todd.

9.3.5 The Transient State of the Well Problem

The preceding examples relate to the steady state which is achieved in an infinite time after pumping has commenced from a well. Under such a state the water level or the phraetic surface assumes a constant position at every point within the aquifer.

C.V. Theis and C.E. Jacob sought to solve the transient equation for a well located in a homogeneous and isotropic aquifer. Their solutions are valid particularly for an extensive, confined aquifer, although they are generally applied also to unconfined aquifers provided the flow lines are approximately straight and horizontal. Other limiting conditions are complete penetration of the well and an infinitesimally small well diameter; there are no aquifer recharges or abstraction other than from the well under consideration and the pumping rate is constant.

The basis for Theis' analysis is the conservation of mass equation derived by using the concept of a mathematical sink of constant strength as:

$$\frac{\partial^2 z}{\partial r^2} + \frac{1}{r}\frac{\partial z}{\partial r} = \frac{S}{T}\frac{\partial z}{\partial t} \tag{9.75}$$

where S = specific storage $(S_s) \times b$ and $T = kb$.

Specific storage S_s has the dimension $1/L$ and is interpreted as the amount of water in storage that is released from a unit volume of the aquifer when the piezometric head drops a unit. The dimensionless factor S is known as the *storage constant*, being the volume of water removed from a column of the aquifer of unit area when the piezometric surface is lowered by a unit. T, whose units are L^2/t, is known as *transmissibility*.

In accordance with equation (9.62) as $r \to 0$

$$r\frac{dz}{dr} = \frac{Q}{2\pi kb} = \frac{Q}{2\pi T}$$

Other conditions for the solution of equation (9.75) are:

$$z \to h_0 \text{ as } r \to \infty \text{ for } t > 0 \text{ and } z = h_0 \text{ for } t < 0$$

The general solution of equation (9.75) is

$$(h_0 - z) = \frac{Q}{4\pi T}\int_u^\infty \frac{e^{-u}}{u}\,du \tag{9.76}$$

where

$$u = \frac{r^2 S}{4Tt} \tag{9.76a}$$

The integral of equation (9.76) is known as the exponential integral, tabulated in mathematical tables commonly as $-Ei(-x)$. Wenzel, however, has calculated it as a well function $W(u)$. Thus the drawdown is given by:

$$d = (h_0 - z) = \frac{Q}{4\pi T}W(u) \tag{9.77}$$

From equations (9.76a) and (9.77)

$$\frac{d}{(r^2/t)} = \text{constant} \times \frac{W(u)}{u}$$

Thus graphs of d versus r^2/t and $W(u)$ versus u should be similar. This characteristic is used in calculating pumping test results. Field observations of r, t and d are plotted on a logarithmic paper (d vs. r^2/t) and superposed on the standard $W(u)$ vs. u curve drawn using the same scale and adjusted until there is good matching while the co-ordinate axes of both plots are parallel, (see Fig. 9.24). The co-ordinates of a common point are used in equations (9.76a) and (9.77) to determine storage constant and transmissibility for the aquifer.

EXAMPLE 9.2

The following readings were taken in an observation well 47 m from a well pumped at a constant rate of 1·3 m^3/min. Find the transmissibility and storage constant for the aquifer.

Time (h)			0·4	1·8	2·7	5·4	9·0	18·0	54·0
Drawdown (cm)			9·0	27·2	37·0	55·4	72·6	88·1	124·0
r^2/t (m^2/h)			5530	1225	818	409	246	123	41
u	0·01	0·02	0·05	0·10	0·2	0·50	1·0	2·0	5·0
$W(u)$*	4·04	3·35	2·47	1·82	1·22	0·56	0·22	·05	0·001

* see *Geohydrology*, by de Wiest (John Wiley) Appendix B, or *Water Resources Engineering* by Linsley and Franzini (McGraw-Hill) Table 4.2.

The d vs. r^2/t plot superposed on the $W(u)$ vs. u curve is shown in Fig. 9.24. The identified common point has the following co-ordinates:

$W(u) = 1{\cdot}92$, $u = 0{\cdot}09$ and $r^2/t = 409$ (m^2/h), $d = 0{\cdot}55$ (m)

From equation (9.76a), $0{\cdot}09 = \dfrac{409}{4}\dfrac{S}{T}$

From equation (9.77), $0{\cdot}55 = \dfrac{(1{\cdot}3 \times 60)}{4\pi}\dfrac{(1{\cdot}92)}{T}$

Thus $\underline{T = 2{\cdot}17 \text{ m}^2\text{/h} \text{ and } S = 0{\cdot}0019}$

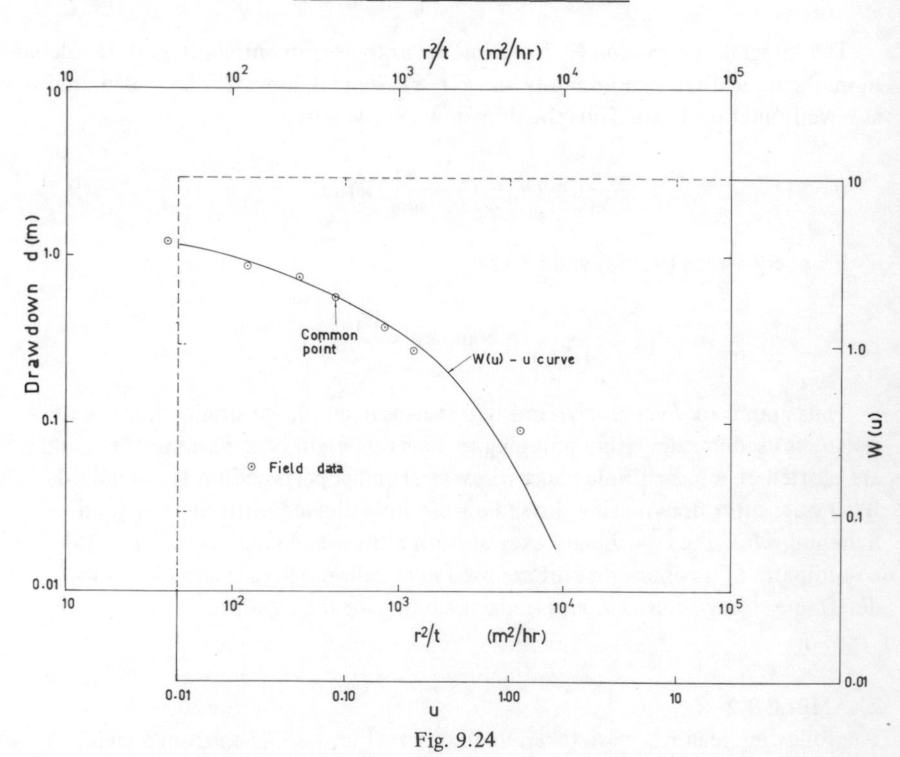

Fig. 9.24

C. E. Jacob simplified Theis' equations for values of u less than 0·01, i.e. for small r and/or large t. By neglecting u^2 and higher orders in the series expansion for $W(u)$, the drawdown equation (9.77) is simplified to

$$d = \frac{Q}{4\pi T}\left(\log_e \frac{1}{u} - \log_e 1{\cdot}78\right)$$

or

$$d = \frac{2{\cdot}30Q}{4\pi T} \log_{10} \frac{2{\cdot}25\ Tt}{r^2 S} \tag{9.78}$$

Equation 9.78 may be used under three conditions of field measurements:

(i) When measurements are done in one well at different times plot d against t on a semi-logarithmic paper. The slope of the resulting straight line is equal to $2{\cdot}30Q/(4\pi T)$ from which T can be calculated. S can then be determined using the co-ordinates at any point on the line and equation (9.78). Alternatively the intercept t_0 when d is equal to zero may be used.

From

$$\log \frac{2{\cdot}25Tt_0}{r^2 S} = 0, S = \frac{2{\cdot}25Tt_0}{r^2} \tag{9.79}$$

(ii) When measurements are done in different wells at the same time, plot d against r on a semi-logarithmic paper. The slope of the resulting straight line is equal to $-2{\cdot}30Q/(2\pi T)$ from which T can be calculated, S may be determined from the intercept r_0 when d is equal to zero.

$$S = \frac{2{\cdot}25Tt}{r_0^2} \tag{9.80}$$

(iii) When measurements are made in different wells at different times; plot d against t/r^2 on a semi-logarithmic paper. The slope of the straight line is equal to $2{\cdot}30Q/4\pi T$. If the intercept at d equal to zero is $(t/r^2)_0$,

$$S = 2{\cdot}25T/(t/r^2)_0 \tag{9.81}$$

9.3.6 Seepage through Earth Embankments

It is always desirable from the point of view of both water conservation and slope stability analysis in the design of earth dams to estimate as accurately as possible the seepage rate through the dam. The flownet method is often helpful but its reliability depends very much on proper determination of the free surface. The discussions in subsection 9.2.5 have stipulated certain entrance and exit conditions that must be satisfied by the free surface. Various investigators, notably Schaffernak and Van Iterson and L. Casagrande (brother of A. Casagrande), have proposed semi-analytical methods for a quick calculation of the seepage through a homogeneous earth embankment on an impervious base. A fuller treatment of their reasoning is given in soil mechanics and seepage textbooks. Here it will be sufficient to give their results and show how the important qualities may be determined.

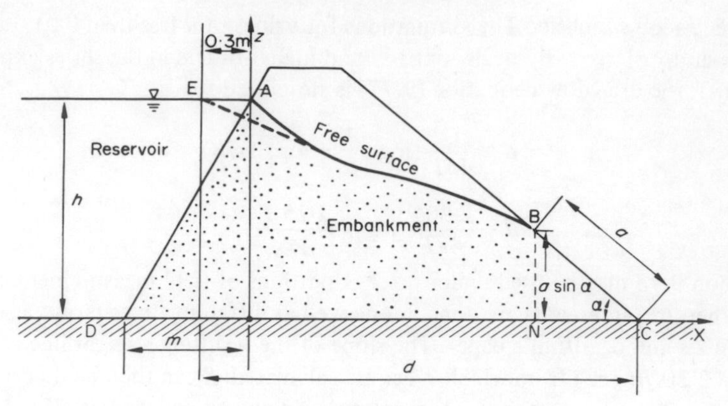

Fig. 9.25 Seepage through an earth dam

Consider a long earth embankment sloping uniformly on both sides, and on an impervious base DC as shown in Fig. 9.25. There is no tailwater and the free surface meets the back of the embankment tangentially at B such that the length of the seepage face BC is a. A. Casagrande had recommended that the free surface be imagined to start from E instead of A as shown in Fig. 9.25 (the normal entrance requirement is then sketched in as illustrated in the figure). Schaffernak and Van Iterson give an approximate value along Dupuit's line of the discharge per unit width as

$$q = -kz \left.\frac{\mathrm{d}z}{\mathrm{d}x}\right|_{\text{free surface}} \tag{9.82}$$

Thus at B,

$$q = -k\,(\text{length of BN})\,(-\tan\alpha) = ka\sin\alpha\tan\alpha \tag{9.83}$$

where α is as indicated in Fig. 9.25 and a is given by

$$a = \frac{d}{\cos\alpha} - \sqrt{\left(\frac{d^2}{\cos^2\alpha} - \frac{h^2}{\sin^2\alpha}\right)} \tag{9.84}$$

in which h is the reservoir depth and d is defined in Fig. 9.25. L. Casagrande employed the more exact seepage gradient at B to analyse the same problem. From

$$q = -kz \left.\frac{\mathrm{d}z}{\mathrm{d}s}\right|_{\text{free surface}} \tag{9.85}$$

where s is measured along the free surface, he obtained

$$q = ka\sin^2\alpha \tag{9.86}$$

and

$$a = \sqrt{(d^2 + h^2)} - \sqrt{(d^2 - h^2\cot^2\alpha)} \tag{9.87}$$

9.3.7 Salt Water Intrusion into Coastal Aquifers

Quite often wells located in coastal regions yield salt water even though the well is apparently well within a fresh groundwater aquifer. There is normally a delicate equilibrium between fresh water flow and sea water flow along coastlines. The specific gravity of sea water is about 1·03; thus under normal conditions 3 m of fresh water could exist above sea level for every 100 m below sea level where fresh water flow enters the sea. However perfect hydrostatic pressure does not exist because of the hydraulic gradient introduced by the sloping water tables. Fresh groundwater tends to flow on top of sea water toward the sea and salt sea water under the fresh water, landward. Equilibrium is normally attained with a fresh water–sea water interface as illustrated in Fig. 9.26(a).

Pumping lowers the groundwater table thus upsetting the delicate equilibrium. If the pumping rate exceeds a certain limit the situation indicated in Fig. 9.26(b)

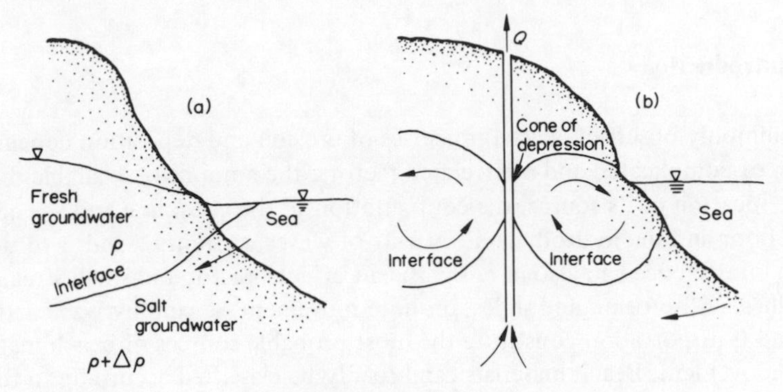

Fig. 9.26 Salt water intrusion

ensues and only salt water is extracted from the aquifer. The safest solution to this problem is that the well should not penetrate below sea level but this seriously limits the capacity of the well since such a well can only operate on a small drawdown.

FURTHER READING

De Wiest, R. J. M., *Geohydrology*, Wiley, New York.
Harr, M. E., *Groundwater and Seepage*, McGraw-Hill.
Linsley, R. K. and Franzini, J. B., *Water-resources Engineering*, McGraw-Hill.
Polubarinova-Kochina, P. Ya., *Theory of Ground Water Movement* (Translated from Russian by de Wiest), Princeton University Press.
Scott, R. F., *Principles of Soil Mechanics*, Addison-Wesley.
Taylor, D. W., *Fundamentals of Soil Mechanics*, Wiley, New York.
Todd, D. K., *Ground Water Hydrology*, Wiley, New York.

10 Sea Waves and Coastal Engineering

10.1 Introduction

The commonly observed coastal processes of erosion and deposition depend on a number of complicated and interrelated factors: the amount of available material and the location of its source, the configuration of the coast line and the adjoining sea floor and the hydrodynamic effects of waves, currents, winds and tides. Human interference can also have significant effects. Major and small streams and gullies, cliff erosion and slides, onshore movement of sand by wave action and wind transportation constitute the most probable sources of beach material and its movement. Beach materials can broadly be classified according to their mode of movement. Generally pebbles roll, coarse sand moves by a series of hops or leaps (i.e. saltation) and fine sand and silt move in suspension. Gravel, pebbles and coarse sand generally move as bed load while fine sand and silt make up the bulk of suspended load.

Bed load motion is due to intense near sea-bottom fluid velocities. Turbulent agitation near the bed resulting principally from the action of breaking waves may give rise to entrainment of large quantities of sediment. The bigger particles fall back to the sea floor because of gravity but turbulence keeps the finer ones in suspension until sufficiently calm conditions prevail. Transport of fine material as suspended load is similar to local fluid mass transport. This transport may be in the offshore direction due to a rip current and other return flows or in the onshore direction due to fluid mass movement. Longshore currents carry materials parallel to the shore line. Bed load movement is oscillatory with a net drift which may be onshore or offshore with a longshore component depending on the characteristics of the longshore currents.

Winds are also believed to play a significant role in the transportation of beach material. Apart from direct carriage of dry beach material from one place to another, winds influence the net transport of fluid mass. An onshore (from sea to shore) wind drags the surface layers of the water shoreward. To satisfy the continuity requirement of no net flow shoreward, there is a net offshore drift near the bed which gives rise to a net sedimentary material movement toward the sea because of non-uniform concentration of sediment. An offshore wind produces the opposite effect (see Fig. 10.1). No sea is however still and the wind effects illustrated in Fig. 10.1 should be considered as indicating only one

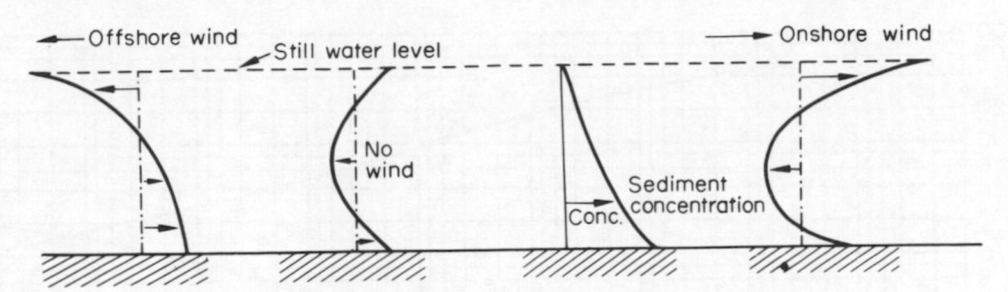

Fig. 10.1 Net drift of water in a still sea

element of the complex coastal process. With a breaking or spilling wave, the shoreward flow of water at the surface is compensated by a return flow seaward with a net seaward drift of water near the bed.

10.2 Wave Generation and Propagation

It is quite clear from the brief introduction above that no adequate insight into coastal processes is possible without a good knowledge of the mechanics of water waves. A water wave, as observed in the ocean, is the outward manifestation of the transmission of a series of disturbances caused by winds, earthquakes, volcanoes and landslides, etc. The waves reaching a coastline may have been created many thousands of miles away. Transfer of energy from winds to the sea is mainly responsible for the generation of ocean waves. It is known that the transfer is effected through shear action (momentum transfer) but the exact method by which this is achieved is not yet fully known quantitatively.

The characteristics of waves at the source of generation depend on the wind speed U, its duration t and the effective distance over which the wind blows, known as the fetch, F. Dimensionless plots of wave height, steepness and period (as determined by Bretschneider from a series of data collected by Sverdrup and Munk) against the reciprocal of a Froude number based on the wind speed and fetch are reproduced in Fig. 10.2. These curves are applicable to deep water wave generation only. The designations for wave height $H_{1/3}$ and $\overline{H}$ and for wave period $T_{1/3}$ and $\overline{T}$ must be properly noted. A wide spectrum of different sizes of waves is normally produced by a wind. This arises from the fact that no wind ever blows uniformly and at a constant rate and because of the continuously changing irregularity of the water surface. The weaker and smaller waves are more quickly damped out because of the viscosity of water than are the stronger and higher waves. Oceanographic considerations distinguish between the average wave height $\bar{H}$ (and the corresponding average wave period $\bar{T}$) and the significant wave height $H_{1/3}$ (and the corresponding significant wave period $T_{1/3}$). The significant wave height and period are the average height and period of the highest third of the waves generated. Also plotted on Fig. 10.2 is tU/F which gives the time required for the generation of maximum energy for a particular fetch and wind

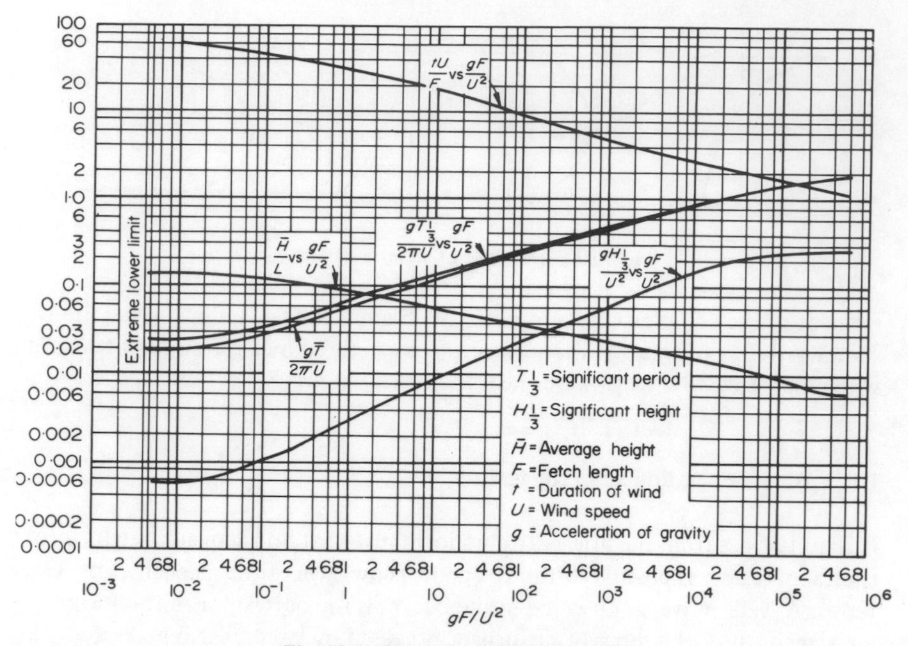

Fig. 10.2 Fetch graph for deep water

velocity. There cannot be any higher values of wave height and length for a specified wind blowing over a particular fetch for a duration longer than that indicated by the tU/F curve.

The waves of various sizes interact and as a result of viscous damping the waves which leave the generating area differ significantly from those originally generated. When the waves leave the generating area (or the generating wind dies down) they cease to grow and gradually attenuate with time and distance. In deep water the wave speed does not change but the wave height and apparent period vary. Many different empirical and analytical approaches have been suggested by various investigators for forecasting the attenuation of waves in deep water as they leave the fetch zone. None of these methods has so far been found to be universally applicable. It appears that local conditions play a significant role in the process. Rapid frictional damping of the small immature waves and the reduction of the energy of the larger ones as they leave the generating area tend to remove the extreme irregularities from the wave train and instead produce the smooth characteristic swells that are observed near the coast.

The four main processes by which the wave decays in deep water after leaving the generating area are:

(1) The process of selective attenuation whereby the wave energy becomes spread over an ever-increasing distance in the direction of the wave train. There is a spectrum composed of different sizes of waves and the longer wave elements tend to travel faster thus 'stretching' the distance into which the particular set of waves was accommodated in the generating zone.

(2) Lateral diffraction of the wave energy as a result of two-dimensional dispersion from the localized source. This is believed to be the major cause of attenuations immediately after the waves leave the generating zone.

(3) Air resistance, especially of winds blowing directly opposite to the direction of wave travel.

(4) Interference by currents. Waves are attenuated when they over-run currents flowing in the same direction. Opposing currents would tend to steepen the waves thus increasing their height and possibly initiating wave breaking. Changes in wave characteristics in shallow water will be discussed in Section 10.5.

The method of wave forecasting presented above is commonly known as the SMB method after Sverdrup, Munk and Bretschneider. Sverdrup and Munk originally initiated the method employing dimensional analysis but their results were later revised by Bretschneider into the form presented in Fig. 10.2. The SMB method is considered generally appropriate when the wind field is reasonably regular and the storm movement is slow. Under such conditions average wind speed, direction, fetch length and wind duration may be used. A graphical method proposed by Wilson is, however, considered more appropriate for ill-defined variables, especially for hurricanes. It must be remembered that winds within a hurricane are not constant in speed and are circular in direction instead of straight as assumed by Bretschneider. A full discussion of Wilson's method is beyond the scope of this text and the interested reader may consult the literature listed at the end of the chapter.

Another important contribution to wave forecasting was made by Darbyshire in England and Pierson, Neuman and James in the United States. Their contribution is based on the analyses of wave spectra. The method which is generally referred to as the PNJ method after the American workers can be used to predict the spectrum of waves from which one may obtain the significant wave height and the statistical distribution of the waves. The SMB and the PNJ methods are not really different. Both are based on analyses of field data utilizing Longuet-Higgins' theoretically derived distribution function. In fact, if both methods predict exactly the same significant wave height they will also give exactly the same wave spectrum. However, whereas SMB predicts a significant wave height proportional to U^2 for a fully developed sea, PNJ predicts it to be proportional to $U^{5/2}$. The two methods give approximately the same results when the wind speed lies between 55 and 65 km/h. The SMB method gives higher values for lower wind speeds and lower values for higher wind speeds. For comparison Table 10.1 lists the values for significant wave height, minimum wind duration and minimum fetch according to SMB and PNJ methods. According to Bretschneider when the sea is fully developed $gH_{1/3}/U^2 = 0{\cdot}282$, when it is 90% developed the value is 0·254 and when it is 80% developed it is 0·226.

Table 10.1
Significant wave height as a function of wind speed, minimum duration and minimum fetch (After Bretschneider)

U	t_{min} (h)		F_{min} (km)		$H_{1/3}$ (m)	
(km/h)	SMB	PNJ	SMB	PNJ	SMB	PNJ
18	102	2·4	1600	18	0·76	0·43
36	205	10·0	6450	140	3·05	2·41
54	307	23·0	14500	520	6·71	6·62
72	410	42·0	25900	1300	12·26	13·57
90	512	69·0	40000	2630	19·15	23·73
100	572	88·0	48600	3880	23·94	31·42

EXAMPLE 10.1

A steady 32 km/h wind blows in the region of Lake Victoria and effectively covers 32 km of the lake in the wind direction. What should be the minimum depth of the lake in the area before it could be considered deep in relation to the waves generated by the wind? What should the minimum duration of the wind be in order to develop the highest possible waves? Assuming both conditions are satisfied, determine the main characteristics of the resultant wave.

SOLUTION

Figure 10.2 may be used to forecast waves produced in deep water. In the problem, wind speed $U = 32$ km/h $= 8{\cdot}9$ m/s and fetch $F = 32$ km.

Thus

$$gF/U^2 = 3960$$

From the figure the corresponding $g\bar{T}/2\pi U = 0{\cdot}67$. Thus the average period generated by the wind in deep water would be $\bar{T} = 0{\cdot}67 \times 8{\cdot}9 \times 0{\cdot}64 = 3{\cdot}82$ s. The deep water wave length (see equation (10.7)).

$$L_0 = 1{\cdot}57\ \bar{T}^2 = 23{\cdot}1 \text{ m}$$

For deep water $$d/L_0 > 1/2$$

∴ Minimum depth required = 11·5 m

From Fig. 10.2 $tU/F = 3{\cdot}4$ from which it can be calculated that the minimum duration of the wind for maximum energy input into the lake is 3·4 h.

$\bar{H}/L = 0{\cdot}019$, thus the average wave height = 0·44 m

The significant period $T_{1/3}$ and wave height $H_{1/3}$ can also be determined from Fig. 10.2. Their respective values are 4·0 s and 0·89 m.

10.3 Small Amplitude Wave Theory

Eagleson and Dean described small amplitude wave theory as 'a first approximation to the complete theoretical description of wave behaviour. As is the case in many physical problems, this first approximation is very rewarding in that it yields a maximum of useful information for a minimum investment in mathematical endeavour'.

In applying small amplitude wave equations the limitations must be clearly appreciated. As the name implies, the use of small amplitude wave theory implies that the wave height is very small compared with the depth of the medium of transmission or the wavelength. Thus a wave which starts out at sea as a small amplitude wave will not be small as it approaches the shore.

10.3.1 Wave Velocity, Length and Period

Small amplitude waves are known to be sinusoidal in nature. Two-dimensional classical hydrodynamic theory can be used to show that the liquid surface elevation variation relative to the undisturbed surface when small amplitude gravity waves are transmitted follows the sinusoidal law

$$\eta = a \sin (kx - \sigma t) \tag{10.1}$$

where k (the wave number) is given by $2\pi/L$ and σ (the wave frequency) is given by $2\pi/T$, L and T being the wave length and wave period respectively. The wave length is the distance between two successive crests or troughs (Fig. 10.3) and the

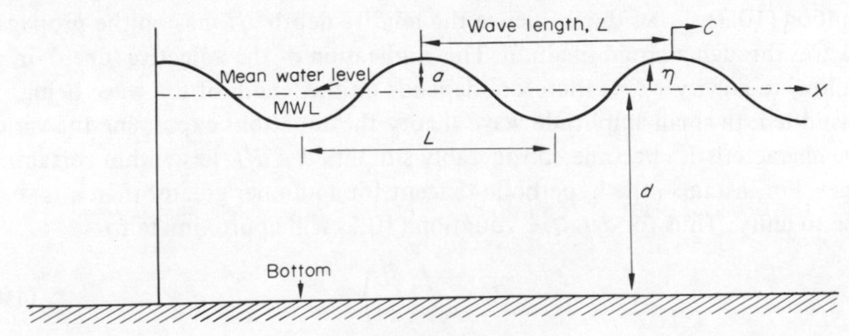

Fig. 10.3 Definitions

wave period is the time interval between the occurrence of two consecutive crests or troughs at a fixed location. a is known as the wave amplitude, x is distance away from the observer and t is time.

Equation (10.1) shows that small amplitude wave motion is periodic both in distance x and time t. If an observer moves along with the wave such that his position relative to the wave form remains fixed, he will see the same surface configuration at all times, and thus relative to him

$$kx - \sigma t = \text{constant}$$

The observer's speed of movement C which must equal the speed of wave travel is given by

$$C = \mathrm{d}x/\mathrm{d}t = \sigma/k = L/T \tag{10.2}$$

Potential flow theory can further be used to show that the speed of propagation is given by

$$C = \sqrt{\left[\frac{g}{k} \tanh (kd)\right]}$$

or

$$C = \sqrt{\left[\frac{gL}{2\pi} \tanh \left(\frac{2\pi d}{L}\right)\right]} \tag{10.3}$$

where d is the undisturbed depth of the liquid medium.

Since

$$C^2 = \frac{L^2}{T^2} = \frac{gL}{2\pi} \tanh \left(\frac{2\pi d}{L}\right)$$

$$L = \frac{gT^2}{2\pi} \tanh \left(\frac{2\pi d}{L}\right) \tag{10.4}$$

The wave length is thus entirely determined by the wave period and the depth of liquid. Figs. (10.4) and (10.5) give plots of equations (10.3) and (10.4).

10.3.2 Deep and Shallow Water Waves

Equation (10.3) shows the influence the relative depth d/L has on the propagation of waves through a liquid medium. The application of the adjective 'deep' or 'shallow' to gravity waves therefore depends on the length of the wave being transmitted. In small amplitude wave theory the equations expressing the various wave characteristics become considerably simplified if d/L lies within certain ranges. For instance, the hyperbolic tangent for a number greater than π is very close to unity. Thus for $d/L \geqslant \frac{1}{2}$ equation (10.3) will approximate to

$$C = \sqrt{\left(\frac{gL}{2\pi}\right)} \tag{10.5}$$

Waves which satisfy this condition are said to be *short* or are moving in *deep water*. At the other end, if d/L is equal to or less than 1/20, $\tanh (2\pi d/L)$ is very small and is very nearly equal to the value of $2\pi d/L$. Thus for *long* or *shallow water* waves,

$$C = \sqrt{(gd)} \tag{10.6}$$

The deep water wave length L_0 from equation (10.4) is

$$L_0 = \frac{gT^2}{2\pi}\ 1{\cdot}57T^2 \tag{10.7}$$

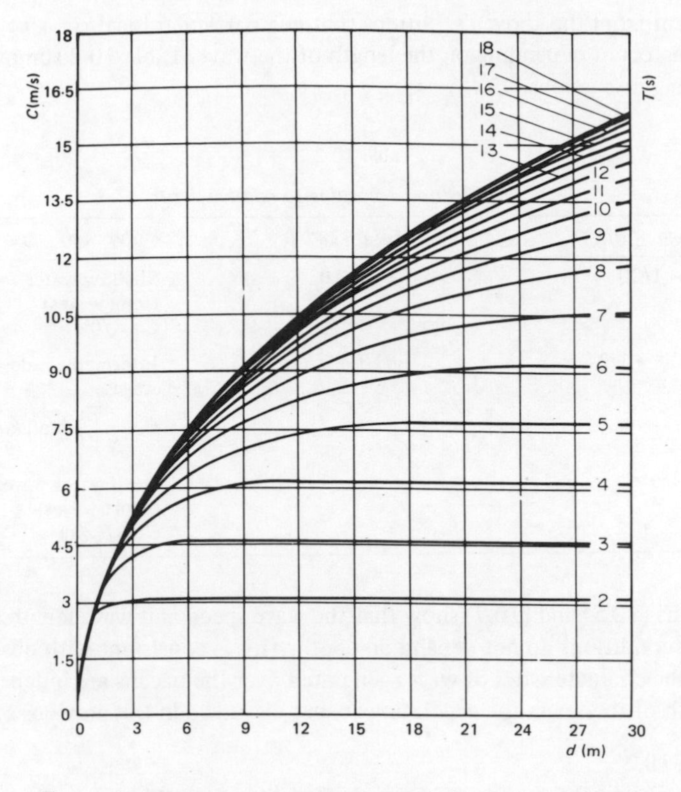

Fig. 10.4 Relationship between wave period, velocity, and depth. (Equations 10.3 and 10.4)

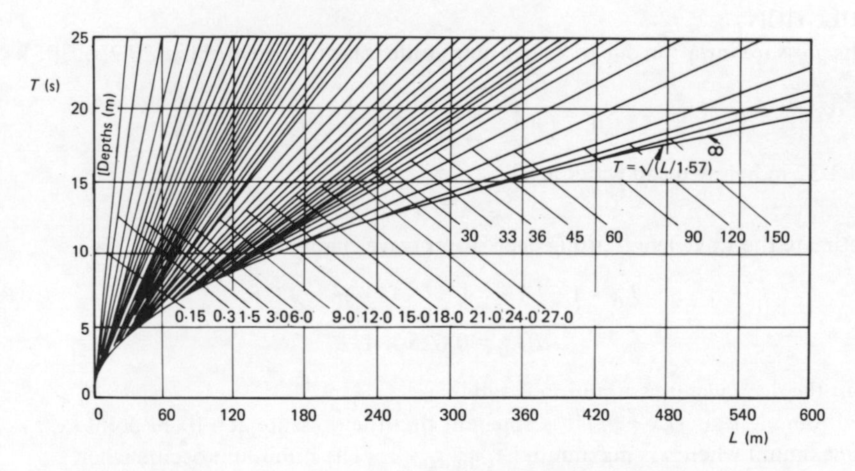

Fig. 10.5 Relationship between wave period, length, and depth (Equation 10.4)

It is apparent from the above definitions that in a particular location, all types of waves can occur depending on the length of the wave. Table 10.2 summarizes the foregoing discussions.

Table 10.2

Wave classification according to relative depth

Range of d/L	*Range of 2πd/L*	*Types of wave*
$0 \rightarrow 1/20$	$0 \rightarrow \pi/10$	Shallow water waves (long waves) $C = \sqrt{(gd)}$
$1/20 \rightarrow 1/2$	$\pi/10 \rightarrow \pi$	Intermediate depth waves $C = \sqrt{\left[\frac{g}{k} \tanh kd\right]}$
$1/2 \rightarrow \infty$	$\pi \rightarrow \infty$	Deep water waves (short waves) $C = \sqrt{(g/k)}$

Equations (10.5) and (10.7) show that the wave speed and wave length for deep water conditions do not depend on depth. This is consistent with observations that the characteristics of waves generated over the oceans are independent of the depth of the sea in the generating area as discussed in the previous section.

EXAMPLE 10.2

A small amplitude progressive wave travels through a 6·1 m deep lake. If the wave amplitude is 22·8 cm and the period is 2·5 s, calculate the maximum and minimum pressures 1·52 m below the mean water level.

SOLUTION

The pressure variation due to a small amplitude wave is. (See problem 10.2, p. 395).

$$p = \rho g \left(\frac{\eta \cosh k(d + z)}{\cosh kd} - z \right)$$

At 1·52 m below mean water level

$$z = -1{\cdot}52 \text{ m}$$

Estimate the wave length using deep water wave theory

$$L_0 = 1{\cdot}57\, T^2 = 1{\cdot}57 \times (2{\cdot}5)^2 = 9{\cdot}8 \text{ m}$$

$$d/L_0 = 0{\cdot}625 > 1/2$$

thus the deep water assumption is valid.

From $\eta = a \sin (kx - \sigma t)$ it is apparent that the pressure at a fixed point (x, z) is maximum when η is maximum i.e. $\eta_{max} = a$. (The minimum occurs when $\eta_{min} = -a$.)

$$p_{max} = \rho g \left[\frac{0{\cdot}228 \cosh 2\pi\,(6{\cdot}1 - 1{\cdot}52/9{\cdot}81)}{\cosh 2\pi(6{\cdot}1)/9{\cdot}81} + 1{\cdot}52 \right]$$

$$p_{max} = 1{\cdot}60\,\rho g$$

$$p_{min} = 1{\cdot}44\,\rho g$$

For fresh water

$$p_{max} = \underline{15{\cdot}7\ \text{kN/m}^2}$$

$$p_{min} = \underline{14{\cdot}1\ \text{kN/m}^2}$$

10.3.3 Orbital Motions

The passage of a wave through a liquid induces the fluid particles to move in an oscillatory manner. This particle motion is one of the most important factors which influence material transport. In referring to particle motion, the reader must be clear in his mind about the distinction between wave velocity and fluid particle velocity. The wave velocity is the speed at which the message of disturbance is transmitted. The transmission imparts momentum and energy to individual fluid particles which causes them to move with a speed much smaller in magnitude than that of the wave itself. The situation is analogous to that of a long and dense queue of football enthusiasts waiting at the entrance to a football stadium. If somewhere along the line a vigorous disturbance is caused, this is transmitted quite fast along the line from person to person by jerky motions. The individuals are however hardly displaced from their mean positions.

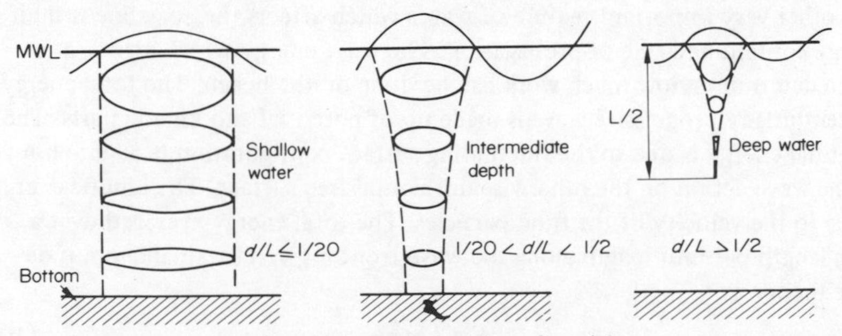

Fig. 10.6 Orbital motion of particles

Small amplitude wave theory has predicted two different orbital motions for particles subjected to deep water progressive waves and those subjected to shallow water progressive waves. In both cases the path for an individual particle is closed and the particle maintains a constant mean position. The situation is different with respect to finite amplitude waves as will be discussed in Section 10.4.

In deep water an individual particle oscillates along a circular path whose centre is the mean position of the particle. The diameter of the circle at the surface equals the wave height but decays to zero at a depth given by $d/L = 1/2$.

This is illustrated in Fig. 10.6. At a finite depth (shallow water $d/L \leqslant 1/20$) the circular path is distorted into an elliptical one. The major axis of the ellipse in the plane of wave motion hardly changes with depth (its value being $2\,a/[kd]$) but the minor axis (maximum vertical displacement) varies from $2a$ at the surface to zero at the bottom. Thus the bottom particle oscillates to and fro along the bed. At intermediate depth the wave motion presents a more complex picture. Both the major and minor axes of the orbital path vary with depth. The general equations are:

$$\text{major axis: } A = 2a\,\frac{\cosh k(d-z)}{\sinh (kd)} \tag{10.8a}$$

$$\text{minor axis: } B = 2a\,\frac{\sinh k(d-z)}{\sinh (kd)} \tag{10.8b}$$

where z is the distance below mean surface level.

The to-and-fro motion of water particles in shallow and intermediate depth water pushes and pulls rolling sedimentary particles, sweeps off the bed and carries to and fro the saltating particles and oscillates orbitally the suspended finer particles. Due to the decay with depth the energy content of fluid particles in intermediate depth motion near bed level is not as significant as that of particles in shallow water. It is important to remember these points when considering material movement at coast lines.

10.3.4 Wave Energy

One other very important feature of waves which affects the coastline is their energy content and rate of propagation. The wave energy which arrives at a beach determines how much work can be done on the beach. The total energy content in any progressive wave is made up of potential and kinetic parts. The potential energy is due to the fluctuating surface configuration brought about by the wave action on the otherwise horizontal free surface. The kinetic energy is due to the velocity of the fluid particles. The total energy averaged over a wave length per unit length along the wave front is given by small amplitude wave theory as

$$E = \tfrac{1}{2}\gamma a^2 \tag{10.9}$$

where E is energy per unit area, γ is specific weight of the liquid medium and a is the wave amplitude. The potential and kinetic energies have the same magnitude, $\frac{1}{4}\gamma a^2$.

One point that must be emphasized is that the wave energy is not propogated in the direction of motion with the velocity of an individual wave C in the case of a series of successive waves known as a wave train. A wave train may result from a group of indentifiable individual waves in an infinite series of disturbances such as occurs in the interaction of waves of different length or a finite number of consecutive waves propagating in an otherwise undisturbed fluid. In either case the group wave (or beat envelope) which represents the net effect of the individual waves (see Fig. 10.7) moves at a group velocity C_G usually much smaller than the individual wave velocity C. Thus the individual waves appear to originate at the rear node of the group and apparently disappear at the front node.

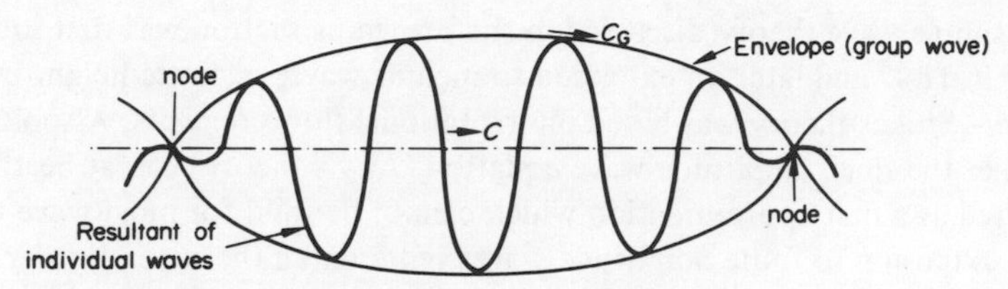

Fig. 10.7 Wave train (wave beat)

The group velocity C_G is given by

$$C_G = \frac{1}{2} C \left(1 + \frac{2kd}{\sinh(2kd)}\right) \tag{10.10}$$

From equation (10.10) we may deduce that $C_G = C/2$ in deep water but $C_G = C$ in shallow water. The energy is transmitted in the direction of the wave train at the group velocity. The rate of propagation of energy $\dot{E}$ per unit length of wave front, which is also equal to the rate at which work is done on an undisturbed fluid medium through which a finite number of progressive waves is being transmitted, is given by

$$\dot{E} = \frac{1}{2} \gamma a^2 C_G \tag{10.11}$$

Equation (10.11) enables the changes in wave height as the wave moves into shallow water from deep water to be calculated (see Section 10.5).

According to Parseval's theorem, from which the energy equation (10.9) is derived, superposition is allowed for certain conditions of composite waves. If the wave is composed of a number of waves, all *moving in the same direction* but having *different frequencies*, the total energy of the composite wave averaged over the beat period is the sum of the energies per unit surface area of each of the individual component waves. If the waves move in *opposite directions* the total energy is also obtained from the sum of the energies of the individual components.

i.e. $$E = \frac{1}{2} \rho g \left(a_1^2 + a_2^2 + \ldots + a_n^2\right)$$

If, on the other hand, the waves move in the *same direction* and also have the *same frequencies* the total average energy is that corresponding to the resultant amplitude i.e.

$$E = \frac{1}{2} \rho g (a_b)^2$$

where a_b is the amplitude of the beat envelope.

10.4 Finite Amplitude Waves

Waves which are met in engineering practice are hardly ever of small amplitude or sinusoidal in form. The errors introduced by using small amplitude wave theory for finite (large) amplitude waves must be appreciated and are accordingly summarized in this section.

The simple wave theory discussed in the preceding section was first introduced by Airy in 1842 and later on extended to include waves of finite height by Stokes. The Airy–Stokes theory was based on irrotational flow concepts. As pointed out earlier the small amplitude wave equation (Airy's theory) can at best be considered as a first approximation which cannot be valid for most wave forms. Stokes' extension to finite amplitude waves represented the second order approximation and was in series form which was proved in 1925 by Levi-Civita to be convergent for deep water. Stokes' theory predicts the amount of mass transport due to a progressive wave to be $H^2\sqrt{(g\pi/32L)}$ per unit width of wave crest (H is wave height).

Earlier, in 1802, Gerstner had proposed a theory based on the assumption that the wave form is trochoidal. In contrast to Stokes' infinite series results, Gerstner's results were in a closed form and represented an exact solution for the particular trochoidal wave form. In Gerstner's theory the flow is rotational. This theory does not predict the observable phenomenon of mass transport and predicts wave breaking at a peak angle of zero as opposed to the experimentally observed angle of $2\pi/3$ radians which was predicted by Stokes.

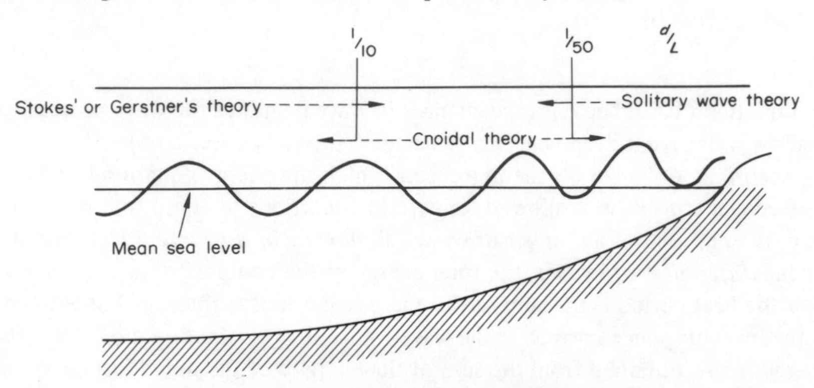

Fig. 10.8 Classification of finite amplitude waves

More recent developments are the cnoidal wave theory by Keulegan and Patterson and the solitary wave theory. The former applies over a range $1/50 < d/L < 1/10$ but most waves just before they break appear to attain a solitary form. A solitary wave lies almost wholly above the still water level and consists of a single 'mound' which apparently propagates at a constant speed and with unaltered form.

Ocean waves undergo a series of regular and irregular changes between their area of generation and the shoreline on which they break. They may have small amplitude characteristics far out at sea but on the continental shelf they change to finite amplitude waves of different forms (trochoidal, cnoidal and solitary). Dean and Eagleson have sought to summarize the situation and possible applicability of the various theories in a sketch similar to the one in Fig. 10.8.

Thus we see that as yet no entirely satisfactory theory exists to describe the conditions of waves approaching a sea shore. The problem is much more complicated by the fact that many shores have irregular submarine features beyond the existing scope of rigorous mathematical analysis. The coastal engineer can therefore at best use the existing theories as a guide. The following relationships for crest amplitude a_c and trough amplitude a_t based on Stokes' second order theory have found an extensive application.

$$\frac{a_c}{H} = \frac{1}{2} + \frac{\pi}{8}\frac{H}{L}\frac{\cosh kd\,(2 + \cosh 2kd)}{(\sinh kd)^3} \tag{10.12a}$$

$$\frac{a_t}{H} = \frac{1}{2} - \frac{\pi}{8}\frac{H}{L}\frac{\cosh kd\,(2 + \cosh 2kd)}{(\sinh kd)^3} \tag{10.12b}$$

These reduce for *deep water* to:

$$\frac{a_c}{H} = \frac{1}{2}\left(1 + 1{\cdot}57\frac{H}{L}\right) \tag{10.13a}$$

$$\frac{a_t}{H} = \frac{1}{2}\left(1 - 1{\cdot}57\frac{H}{L}\right) \tag{10.13b}$$

and for *shallow water* to:

$$\frac{a_c}{H} = \frac{1}{2}\left(1 + 12{\cdot}1\frac{H}{L}\right) \tag{10.14a}$$

$$\frac{a_t}{H} = \frac{1}{2}\left(1 - 12{\cdot}1\frac{H}{L}\right) \tag{10.14b}$$

Equations (10.12), (10.13) and (10.14) have shown that the height of the crest is higher than the depth of the trough for a finite amplitude wave. The wave speed is still given by equation (10.3) but the average energy per unit area is given by

$$E = \frac{\gamma H^2}{8}\left(1 - 4{\cdot}93\frac{H^2}{L^2}\right) \tag{10.15}$$

All the equations reduce to small amplitude relationships as H/L approaches zero.

10.5 Changes in Shallow Water

Both small and finite amplitude wave theories have shown that changes in depth affect many principal characteristics of a wave. The only principal wave characteristic which does not change with depth is the wave period. This statement must not be confused with an earlier statement that when waves leave their generating area their wave period changes. The change in that case results from the fact that waves of very many widely different periods interact to give a new wave form. A stable synthetic form is eventually achieved which then progresses at a constant frequency. As this wave enters shallow waters its amplitude, length and speed change. The theoretical prediction of variations of wave height, wave amplitude and C_G/C with depth relative deep water characteristics is shown in Fig. 10.9.

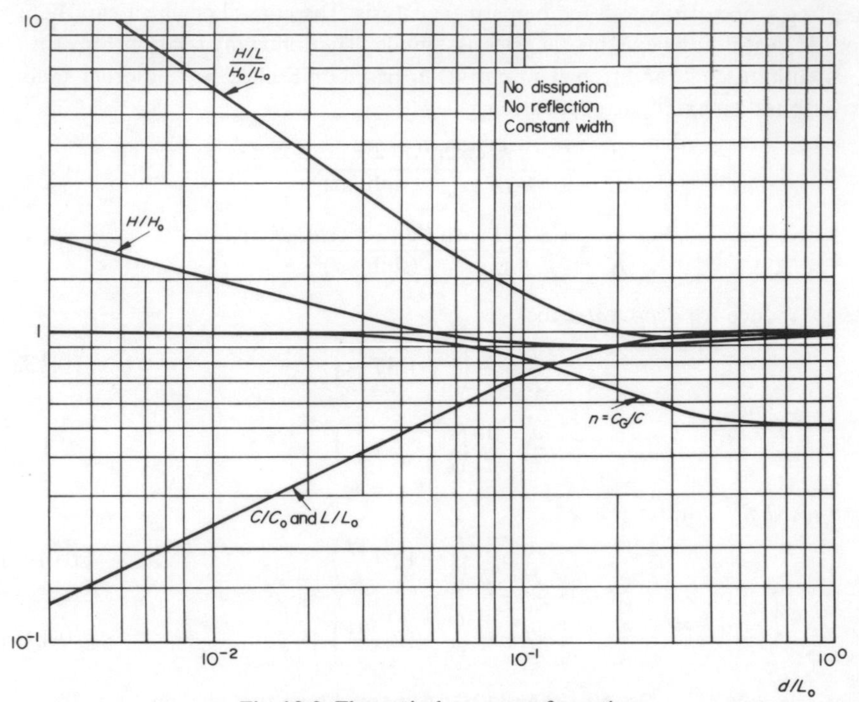

Fig. 10.9 Theoretical wave transformation

One very important feature of waves which determines their action on a beach is wave breaking. It is evident from Stokes' equations (and others) that a_c/H is much greater in shallow waters than in deep waters. A wave thus undergoes a transformation which steepens it as it approaches the shore. Stokes' criteria for wave breaking is that the water particle velocity at the crest is just equal to the velocity of the wave propagation. He determined theoretically that this condition existed when the wave crest angle attained 120°. Experimental observations seem to support Stokes' theory. Michell used Stokes' criteria to estimate that wave breaking would occur in *deep water* if

$$H_b/T^2 = 0{\cdot}267 \tag{10.16}$$

where H_b is wave height at point of breaking and T is wave period. McCowan used the same criteria and solitary wave theory to estimate the breaking condition in *shallow water* as

$$H_b/d = 0{\cdot}78 \tag{10.17}$$

Both equations match well with observations as reported by Reid and Bretschneider and reproduced in Fig. 10.10.

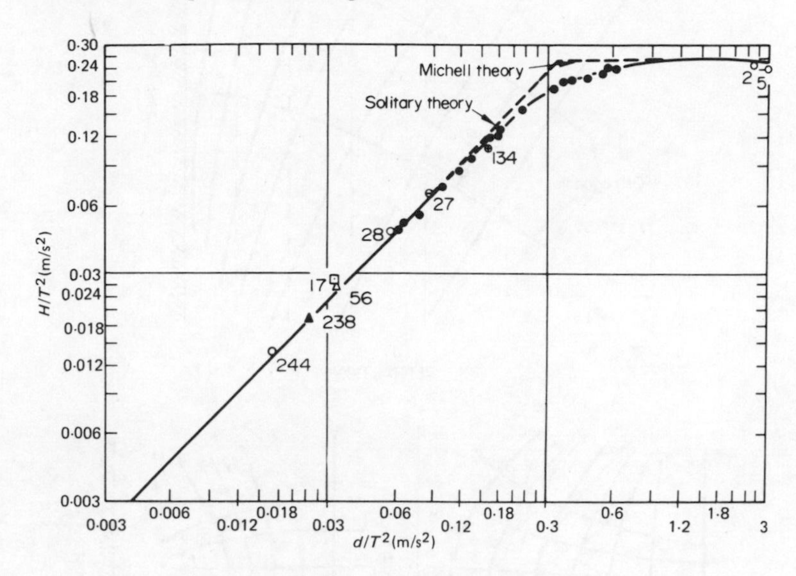

Fig. 10.10 Breaking index curve (After Reid and Bretschneider)
● Danel France ◻ WHOI
⊙ Estero Bay ⊖ Bed Tank
▲ SIO ⊕ Lake Superior
○ Berkely Tank (Beach slope 1·30, 1·50)
Numbers beside plotted points designate number of waves averaged

Energy changes as a wave enters a shallow water can be estimated using equation (10.11). Assuming no dissipation of energy through friction, seepage and other means

$$\frac{H_0^2}{8} C_{G0} B_0 = \frac{1}{8} H^2 C_G B$$

where the subscript '0' refers to deep water and B is the width of the wave front. Thus

$$H = K_r H_0 \sqrt{(C_{G0}/C_G)} \tag{10.18}$$

where $K_r = \sqrt{(B_0/B)}$ is known as the refraction coefficient.

As a wave front approaches a coast line obliquely (Fig. 10.11(a)) one end approaches the continental shelf before the other. According to equation (10.3) the part of the front which experiences depth changes is slowed down. The overall effect is to produce a curved front. Two orthogonals to the wave front (over

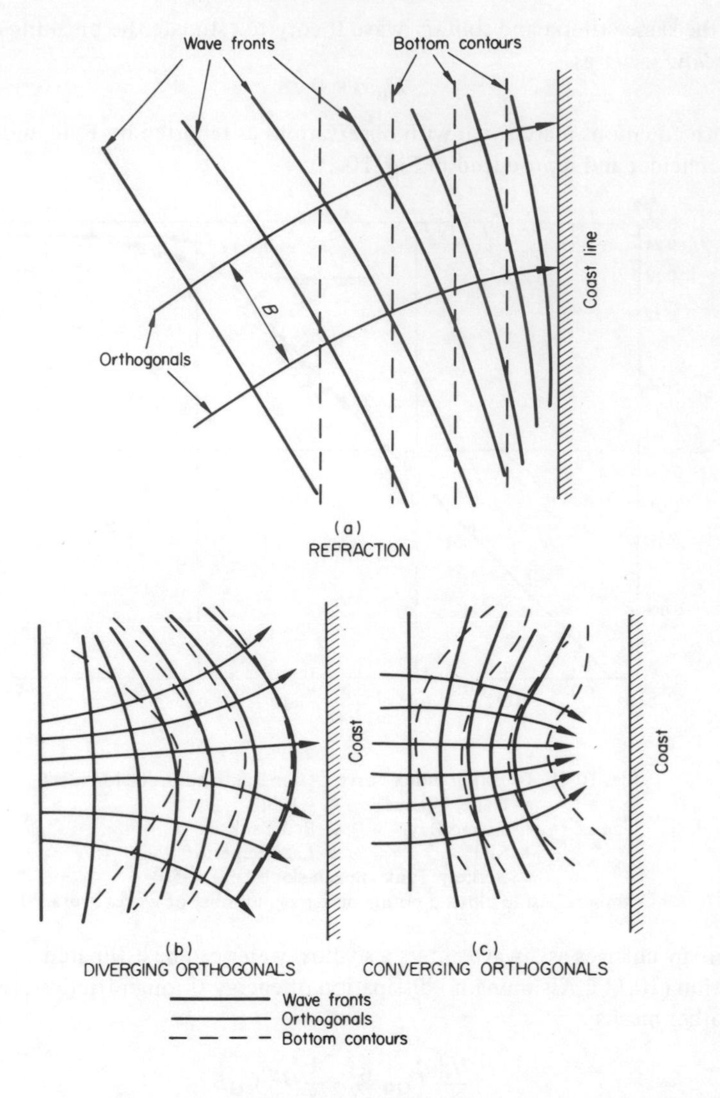

Fig. 10.11 Refraction of waves

which the same amount of energy is propagated) diverge. This phenomenon is known as *refraction* and the *refraction coefficient* is given by the degree of divergence of the orthogonals to the wave front. The submarine topography also produces refraction. While a submarine valley tends to produce dispersion of energy arriving at the coastline (Fig. 10.11(b) and (c)) a submarine ridge tends to produce a concentration of energy and therefore a more potentially effective coastal attack.

10.6 Wave Reflection and Diffraction

Objects standing in the path of a progressive wave constitute new sources of a disturbance which in turn send out other wave forms. A wall or screen reflects a progressive wave with which it interacts. The direction of propagation of the reflected wave depends on the angle of incidence of the original wave. If the incident angle is normal to the screen the reflected ray moves in a direction exactly opposite to the original wave. The reflection from the obstacle may be complete or partial. The degree of reflection is measured by a coefficient of reflection defined as the ratio of the reflected wave amplitude to the incident wave amplitude. The coefficient of reflection depends on the permeability of the wall, its height as well as on its energy absorption and transmission capacity. These considerations are important in the design of breakwaters, sea walls and model studies in which reflection of waves from the walls of the wave basin and from other objects must be controlled.

In the case of perfect normal reflection, the coefficient of reflection is unity and the reflected wave has the same amplitude (a) as the incident wave. The small amplitude wave theory which uses potential flow concepts allows us to superpose the amplitudes of the incident and reflected ways. The reflected and incident waves have the same frequency and since they propagate over the same depth of liquid their speeds will be the same magnitude and therefore their wave lengths will also be equal.

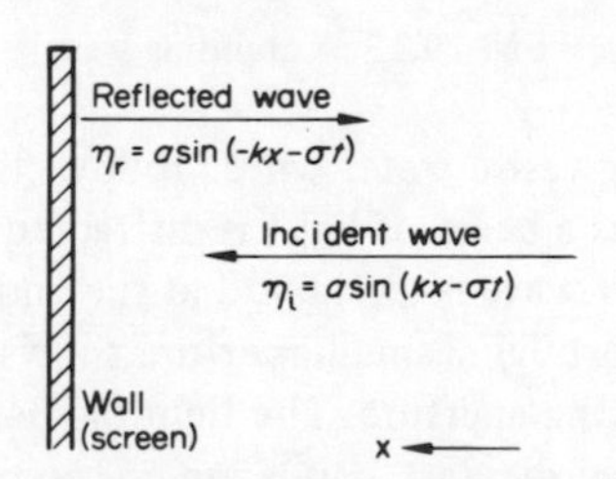

Fig. 10.12 Wave reflection

With reference to Fig. 10.12 and equation (10.1) the surface disturbance due to the incoming wave is given by

$$\eta_i = a \sin(kx - \sigma t)$$

and that due to the reflected wave is given by

$$\eta_r = a \sin(-kx - \sigma t)$$

the negative sign ($-kx$) is due to the fact that the reflected wave is moving in the negative x-direction (i.e. opposite to the incident wave). This can also be regarded as a change in phase angle of π. The total disturbance is then given by

$$\eta_t = a \sin (kx - \sigma t) + a \sin (-kx - \sigma t)$$

or
$$\eta_t = a \sin (kx - \sigma t) - a \sin (kx + \sigma t) \tag{10.19}$$

Equation (10.19) may also be written as

$$\eta_t = -2\, a \sin \sigma t \cos kx \tag{10.20}$$

which is a product of two terms, one independent of x and the other independent of t. The wave form represented by the equation is known as a *standing wave* in that the nodes are fixed in space. Standing waves may be observed behind many breakwaters. The nodes $\eta_t = 0$ are given by $\cos kx = 0$

or
$$x_n = \pm \frac{(2n + 1)\pi}{2k}; \text{ with } n = 0, 1, 2, \ldots$$

The graphical representation of a standing wave is shown in Fig. 10.13.

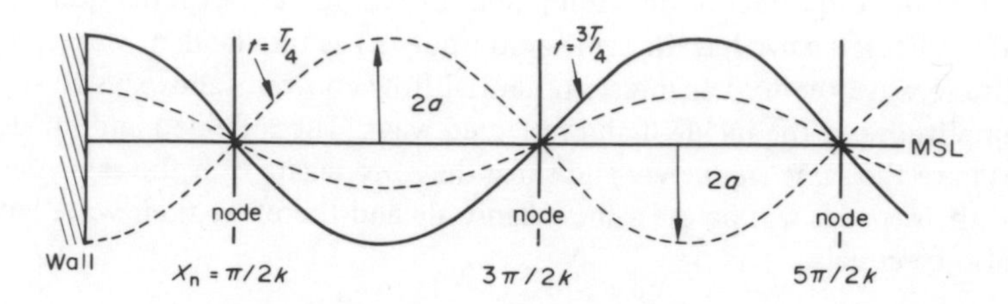

Fig. 10.13 A standing wave

The *diffraction* of progressive water waves follows the same principle as the diffraction of light. Just as a beam of light is diffracted around a corner of a screen, a progressive water wave is diffracted at the end of an obstructing wall. A beam of light passing through a small aperture shows alternate light and dark bands on a screen behind the aperture. The light bands are due to reinforcement of the diffracted rays while the dark bands are due to cancellation of the rays. The corresponding effect is a series of standing waves in a harbour due to progressive water wave diffraction by the harbour entrance gap. If the period of the standing wave system approaches the natural period of any moored system (ships and docks) within the harbour, resonance occurs. This is an important consideration related to the planning and operation of a harbour requiring a proper assessment of the dynamic mooring forces on ships and docks.

EXAMPLE 10.3

The diagram shows a section of a wide canal whose mean depth changes abruptly from 7·6 m to 4·9 m. The mean water level remains the same in both canals. The wave system before and after the rise has an amplitude of 50 cm. Assuming the waves are long and that no energy is dissipated, derive expressions for the forms of the component waves before the change in bed level.

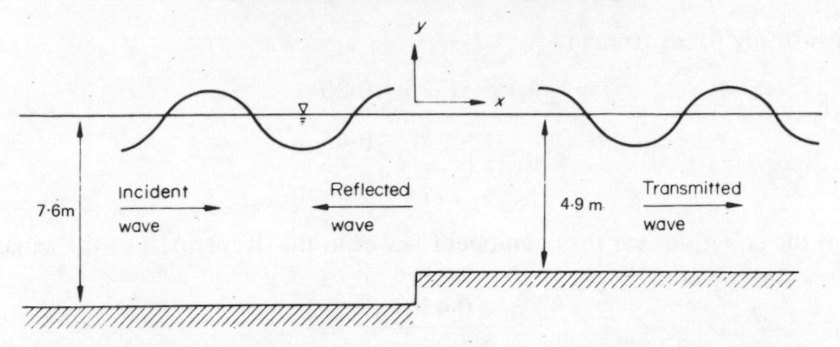

SOLUTION

There will be a partial reflection at the abrupt rise. Taking the coordinate axes as shown ($x = 0$ at the rise) the waves may be represented as

$$\eta_i = a_1 \sin(k_1 x - \sigma t) \text{ incident wave}$$

$$\eta_r = -a_2 \sin(k_2 x + \sigma t) \text{ reflected wave}$$

$$\eta_t = a_3 \sin(k_3 x - \sigma t) \text{ transmitted wave.}$$

The three waves originate from the same initial wave therefore they will wave the same period. Also $k_1 = k_2$ because the incident and reflected waves move in the same depth of water (see equation (10.4)).

The wave form in the deeper part of the canal is given by the superposition of the incident and reflected waves.

i.e. $\eta_1 = \eta_i + \eta_r = a_1 \sin(k_1 x - \sigma t) - a_2 \sin(k_1 x + \sigma t)$

$$= (a_1 - a_2) \sin(k_1 x) \cos(\sigma t) - (a_1 + a_2) \cos(k_1 x) \sin(\sigma t)$$

At $x = 0, \eta_1 = -(a_1 + a_2) \sin(\sigma t)$

But

$$a_1 + a_2 = 0{\cdot}50 \text{ m} = a_3 \tag{1}$$

Since energy loss is to be neglected,

$$\tfrac{1}{2}\rho g\, a_1^2\, n_1\, C_1 = \tfrac{1}{2}\rho g\, a_2^2\, n_1\, C_1 + \tfrac{1}{2}\rho g\, a_3^2\, n_3\, C_3$$

where $n_1\, C_1$ and $n_3\, C_3$ respectively are group velocities before and after the rise. However $n_1 = n_3 = 1{\cdot}0$.

Thus

$$a_1^2 - a_2^2 = \frac{C_3}{C_1} a_3^2 = \sqrt{\left(\frac{d_3}{d_1}\right)}\, a_3^2$$

$$a_1^2 - a_2^2 = 0{\cdot}8\,(0{\cdot}5)^2 = 0{\cdot}20 \tag{2}$$

Substituting for a_2 from (1)

$$a_1 - 0{\cdot}25 = 0{\cdot}20$$

$$\therefore \qquad a_1 = 0{\cdot}45 \text{ m}$$

and

$$a_2 = 0{\cdot}05 \text{ m}$$

Thus the equations for the component waves in the deeper part of the canal just before the rise are:

$$\underline{\eta_i = 0{\cdot}45\,(kx - \sigma t)}$$

$$\underline{\eta_r = -0{\cdot}05\,(kx + \sigma t)}$$

The coefficient of reflection

$$a_2/a_1 = \underline{0{\cdot}111}$$

The coefficient of transmission

$$a_3/a_1 = \underline{1{\cdot}11}$$

10.7 Coastal Processes

10.7.1 Currents

The physical changes in profile of any coastline result from erosion and deposition of materials. The most important factor which the coastal engineer deals with is perhaps the movement of sand. Rivers and streams and to a smaller extent winds constitute the major suppliers of sand from the land mass to the

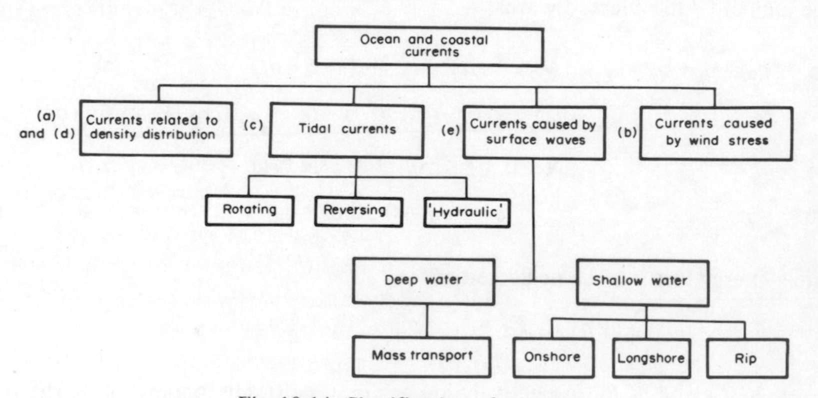

Fig. 10.14 Classification of currents

sea. The movement of these materials is closely tied up with the prevailing currents. A current may be defined as the drift of fluid particles in any direction. A clear understanding of coastal processes must start with adequate appreciation of the nature of ocean and coastal currents.

Johnson and Eagleson have summarized these currents (Fig. 10.14) into (a) those which are related to the changes in the density of the seas, (b) those caused directly by wind stresses, (c) tidal currents and others associated with internal waves, (d) local currents induced by river fresh water entering the sea and (e) currents and transport induced by surface gravity waves.

Currents associated with sea density changes are invariably large global currents such as the equatorial currents, the Gulf stream, the Benguela current, etc. Their effect on coastal sediment problems are generally limited to cases in which they are forced near the shore by islands and submarine land features.

Currents associated with offshore and onshore winds are generally too weak to contribute significantly to coastal sediment transport except on a localized scale.

Tides are produced by astronomical forces between the moon and the sun. They are swells of extremely long length and periods of 12 hours (semidiurnal) or 24 hours (diurnal). Tide induced currents are of three forms: the rotary form generally observed in the open sea and along extensive coastlines, the reversing type observed in estuaries and the 'hydraulic' type in straits connecting two independent tidal bodies of water such as the Panama canal. All three types are oscillatory in nature; following the pattern of high and low tides. Unless interfered with the net transport of water and sediment due to tides is usually zero.

Rivers with steep slopes and with constricted mouths discharge fresh water into the oceans at relatively high velocities. These create density currents in the nearshore region of the river mouth which sometimes produce significant local effects on the general current pattern and therefore on the movement of sedimentary material.

By far the most significant factor in coastal processes is the breaking of progressive water waves and the associated currents. The phenomenon of wave breaking and its induced currents is discussed in the following subsection.

10.7.2 Wave Breaking

The concept of wave breaking was referred to in the discussions on changes in shallow water (Section 10.5). The orbital paths described by fluid particles in finite amplitude waves are open, leading to a net drift of mass in the direction of wave propagation. Entrained sediment particles are carried along with the fluid mass and the bed load particles drift slowly toward the shore. As the wave approaches the shore its height increases (the length and speed decrease) until it becomes unstable and breaks. Just at breaking the instantaneous fluid particle velocity, the net drift velocity and the phase velocity C are all believed to approach each other in magnitude and direction. The momentum of the water particles at the instant of breaking is sufficient to make the mass of fluid rush into the surf zone as a surge. This is followed by a retreating backrush. The uprush is short and rapid but the retreat is much slower and therefore has

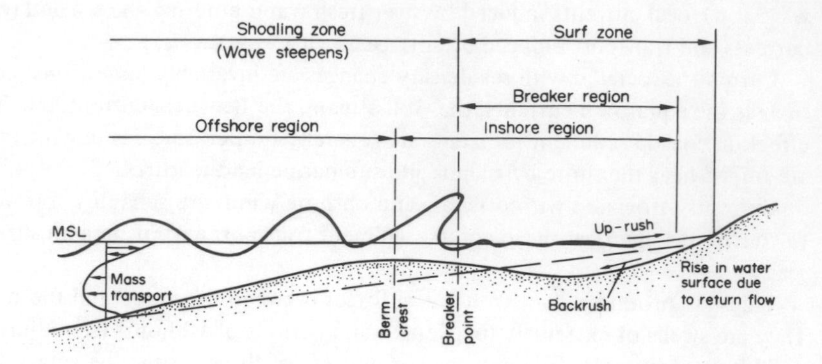

Fig. 10.15 Sketch of beach zones

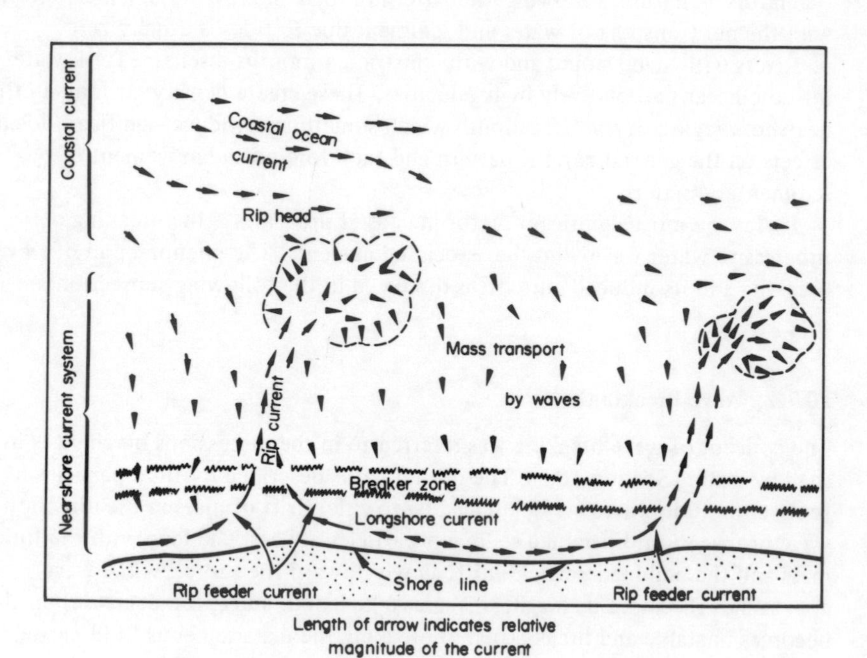

Fig. 10.16 Nearshore circulation pattern (After Shephard and Inman)

greater effect on material transportation. If the shore material is porous, such as stones and rubble, the slow retreating water sinks and the transportation effect is very much reduced. In Fig. 10.15 the terms used to describe the various zones associated with wave breaking are illustrated. If the time at which the wave breaks is quite different from the time of the arrival of the backrush of the preceding wave at the breaker point, substantial fluid exchange takes place across the breaker line and therefore substantial amount of material is transported one way or the other.

Waves which approach the shore line obliquely present a more complex current pattern than uprush followed by backrush. Indeed even a wave striking normally may produce, as a result of submarine topography and subsequent concentration of wave energy, discrete locations along which the retreat flow concentrates. These concentrated high velocity zones known as rip currents are very dangerous to swimmers and divers. An oblique wave has a velocity component along the shore and on breaking produces longshore currents primarily in the surf zone. Fig. 10.16 depicts a three-dimensional nearshore circulation pattern.

EXAMPLE 10.4

Waves of period 10 s were observed to break near a shore where the depth was 4 m. Estimate height of the waves when they break.

SOLUTION

The Reid-Bretschneider wave breaking index curve of Fig. 10.10 is used in the solution.

$$d/T^2 = 4/100 = 0{\cdot}04$$

Corresponding to this,

$$H_b/T^2 = 0{\cdot}024$$

$$H_b = 2{\cdot}4 \text{ m}$$

The waves break when their height is 2·4 m.

10.7.3 Movement of Beach Material

The movement of beach material is closely related to the current system as described above. Experiments have shown that sand movement outside the breaker zone is predominantly in the landward direction, apparently because of the orbital movement on the bottom where the maximum velocity occurs under the wave crest in a landward direction. The amount of sand movement in a fixed depth is dependent on the wave height and steepness, the wave period, the beach gradient and the nature of the bed. The quantity of material transported increases with increasing wave height and period. More material is moved on a gentler slope. Many beaches outside the breaker zone become rippled depending on the wave length and wave steepness and the beach normally becomes flatter as the bottom ripples. Since a beach is normally subjected to different types of waves in the year it goes through cycles of flatness and ripple and dune formation. The ripples and dunes are similar to those in erodible channels through which different discharges pass (see Chapter 7).

The movement of sand inside the breaker zone is very irregular and the direction of motion is determined by many factors. The direction of motion may be resolved in the on-offshore and longshore directions. As already pointed out the uprush of fluid after the wave breaks is short and brisk but the retreat is much slower and therefore may drag more sand along. The normal tendency therefore would be for more sand to be removed from the beach and transported seaward. The porosity of the beach tends to reduce this effect. The timing of the next wave breaking also controls the sand movement. If the next wave breaks before the backrush reaches the breaker line the new uprush slows down the oncoming retreat considerably and therefore reduces its effective action on the beach material.

The longshore current produces what is known as littoral drift. It tends to transport sand along the shore line. Thus sand which is eroded in one place may be transported many miles until the current becomes so weak that it cannot carry the full load and therefore deposits some of it. Net erosion occurs on a beach if there is more sand being removed than being deposited and net accretion occurs when more sand arrives than is removed. Most natural beaches are in a perpetual state of change due to the varying nature of the waves which work on them. The so-called stable beach is one which shows no net change over a long period of time (a year or a decade). Such a beach loses as much sand through erosion over the period of observation as it gains through deposition. Artificial marine structures such as harbours, groynes and sea walls tend to upset the equilibrium of coastlines in their vicinity.

One other dynamic feature of beaches is the sorting of materials. This process may also have components normal and parallel to the coastline. Fine particles which are carried in suspension are transported in a direction normal to the shore and deposited behind the breaker zone. However they can also be carried parallel to the shore by the littoral drift. The sand behind the breaker zone tends to become finer and finer in the seaward direction. Coarser materials in the form of shingle or pebbles and bigger sands tend to be concentrated in the foreshore and surf zones and they generally show a diverse gradation. This may be explained by the fact that pebbles can be carried in suspension only at the breaker point of intense agitation and can generally progress only by rolling on the bottom under the influence of greater acceleration in the direction of wave motion. It is generally observed that coarser sand is deposited on beaches when waves have high wave heights (energy) and therefore can move them. Finer particles are deposited in times of low waves when they can least easily be transported into the deep seas. The gradation varies both with time and space. Generally, at one time, finer sand is found in the deeper offshore regions and coarser sand in the shallower regions and on the berm crest. Along many shore lines bands of sediment are observed aligned parallel to the shore both on the beach and in the offshore regions. Each band shows distinctive characteristics.

Lateral sediment sorting is also sometimes observed especially on sheltered shore lines. Strong longshore currents may carry heavy loads of sediment.

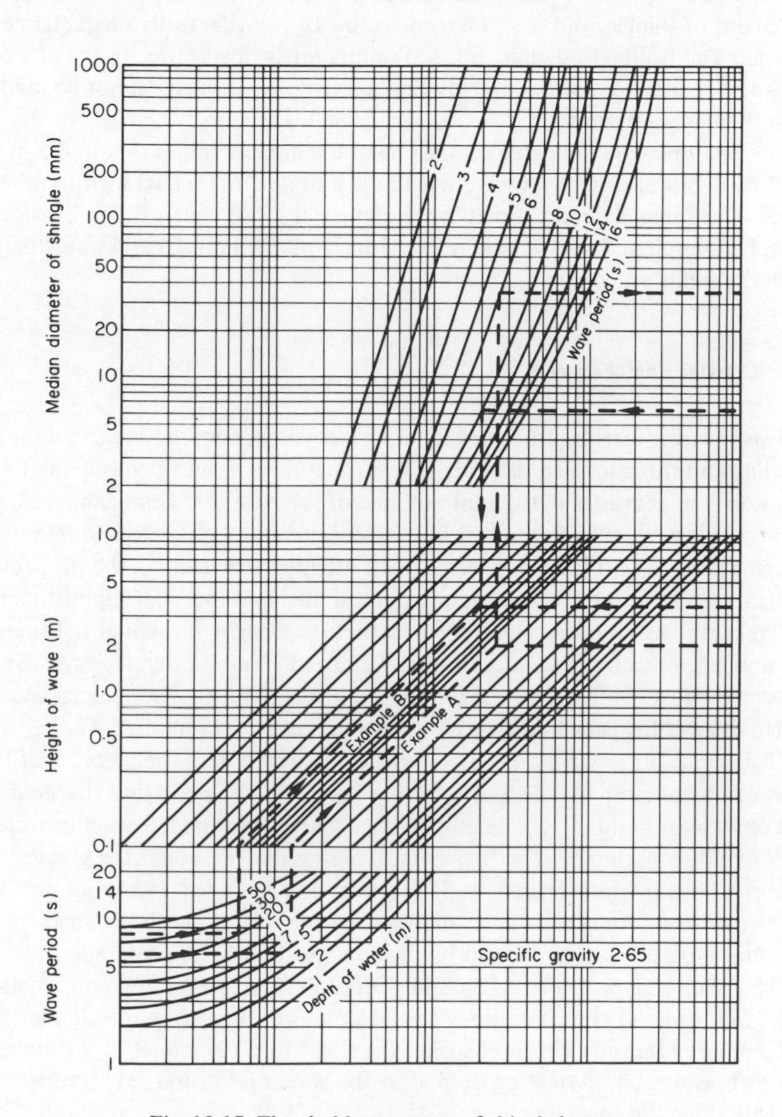

Fig. 10.17 Threshold movement of shingle in waves

However as they approach weaker zones they progressively deposit the heavier materials. This process is greatly enhanced by wind action which tends to augument current activity in the unsheltered end of the coast.

Recent experimental studies at the Hydraulics Research Station, Wallingford (England) have yielded the working diagram reproduced in Fig. 10.17 for the prediction of the wave conditions necessary to move beach pebbles (shingle) of a particular size. The same diagram can also be used to determine the size of material which will not be moved by a known set of wave conditions. The Wallingford results also confirm that grading is relatively unimportant in the movement of shingle, and that the median size D_{50} satisfactorily characterizes the material. The Wallingford chart is based on materials of specific gravity of 2·65 and is the counterpart of the Shields diagram (Fig. 7.6) which cannot be used when there is wave motion.

Two examples are indicated on the chart. Example A shows that in 4·5 m (13·7 ft) depth of water a 6s wave would need to be 2 m (6·1 ft) high to move shingle of 35 mm diameter and above. Example B shows that with an 8s wave, 3·5 m high in 20 m depth of water, fine shingle of medium size 6·4 mm would be on the point of moving.

10.8 Coastal Engineering

Civil engineering marine structures fall into two categories: those located on the coastline and those located offshore. The former have received considerable attention over centuries but the importance of the latter has been enhanced quite recently by attempts to find oil within the continental shelf areas and in deeper waters. The major problem connected with offshore structures is the proper evaluation of drag and lift forces on structural members (vertical support members and cross-bracing) and the avoidance of structural designs whose natural frequencies would approach the frequency of the induced vibration due to waves. Such resonance effect could produce structural failures of enormous financial and human costs as happened in the case of the Texas tower in the U.S.A.

Civil engineering works on coasts may be connected with the creation of 'still' water conditions in parts of the sea for the protection of structures (harbours are the commonest in this class), defence against wave attack and erosion or reclamation of land from the sea. The first two categories include by far the greatest number of major works on coasts. Sea reclamation is limited to heavily populated islands like Hong Kong and Japan and reclamation of salt marsh silt areas for agricultural purposes after the salt has been washed out is also limited.

The underlying principles of coastal structures may involve any combination of the following: (1) interference with and therefore changes in the characteristics of the progressive wave before it strikes the shore line, (2) reduction of the effect of the uprushing and retreating fluid after the wave breaks and (3) interception of the drift along the coast line.

The breakwater is the main structure used to interfere with waves before they strike the coastline. The object is to reduce the energy content of the waves behind the breakwater. The breakwater is a wall which brings about total or partial reflection of the oncoming waves. Most harbours have permeable breakwaters of blocks of stone or concrete.

Partial reflection of the oncoming waves and viscous damping of the transmitted waves are sufficient to allow waves of negligible amplitude in the harbour. Some wave breaking occurs on breakwaters and it is necessary that they should be high enough to avoid over-topping by the uprush. A breakwater may be completely submerged if it is not required to produce complete destruction of the oncoming waves. Breakwaters are particularly useful in encouraging beach formation where it is known that beach materials are coming to the shore from somewhere but it is necessary to stop the local waves from carrying them away. Military operations have resulted in the development of mobile breakwaters. This type of breakwater is composed of floatable prefabricated sections. These are towed to site, arranged and moored, used for a limited time only and removed to be used elsewhere. The advantages of utility, economy and time with mobile breakwaters over conventional breakwaters are obvious as is the major disadvantage of reduced efficiency.

Wave refraction diagrams are very important in the design of breakwaters. It is desirable that the breakwater should be aligned as far as possible with the front of the principal waves as they approach the shore line. Siltation is a major problem in harbours. The still sea water has a smaller capacity for carrying already entrained soil particles and therefore deposits them. *Regular* dredging is therefore required in most harbours. The major source of circulation of fluid mass in a harbour is through diffraction as discussed in the latter part of Section 10.6.

The destructive and constructive actions of waves as they break on the shore have already been discussed. All coastlines are undergoing a continuous change but the effect of change on a stable beach over a long period of time (a year or even a decade) is nil. It would appear that most permanent changes on a beach come about as a result of man's interference with the natural cycle of events. An artificial harbour may introduce significant changes in the wave and current patterns in its neighbourhood and therefore cause permanent erosion or deposition at other parts of the coast line. If this process is not controlled by man the coast line may attain a new equilibrium level after some years. Large volumes of sand are removed from beaches every year for building purposes. The sand is replaced fully or partially from somewhere else along the coast line only for it to be removed at the next low tide. The overall effect is a pronounced beach erosion somewhere along the coast initiated by the sand digging process.

The problem often confronting the civil engineer is how to stop the erosion caused by a new structural feature on the coast line (harbours, topographical changes due to an earth tremor, etc.) or by sand quarrying. The second cause appears easier to solve. Legislation against digging sand or feeding sand back onto the beach would remedy the situation in some cases.

Sea walls have been known in many cases to reduce erosion considerably. A sea wall is a special type of breakwater located such that the wave breaks on or near it and it also provides a solid surface over which the uprush and backrush occur. Sea walls with flat slopes are observed to be better than those with steep slopes. The flat slope minimizes wave reflection (reflected waves increase orbital velocities of water particles in proportion to the amplitude of the reflected waves). Many sea walls are provided with berms and sometimes steps with a view to reducing the height of the uprush and reflection. A gently sloping sea wall also allows the wave to break on the wall itself thus reducing possibilities of sand entrainment due to turbulent agitation at the foot of the wall. Slopes not steeper than 1 in 2 are required for the wave to break on the wall itself. The major structural requirement for a sea wall is its stability. The foundation must be firm, preferably on rock or piles and adequate protection must be provided for the foot. The main disadvantage of sea walls is that they are expensive to construct and maintain. Overtopping of sea walls is a major problem and an adequate drainage system must be provided behind a sea wall to minimize the danger of slope failure and undermining.

Aprons of rock, concrete blocks or gabions (steel wire bags filled with stones and other heavy material) are another common method of coastal defence. These form rough surfaces over which the wave breaks and rushes and therefore offer resistance to the moving sea water. A section of a rock fill protection at Sekondi (Ghana) is shown in Fig. 10.18. The important features to note are the sizes of the rocks, the gradation especially at the filter bed, the slopes and the rocks at the foot and their location. The block must be of sufficient weight, size and strength to withstand the hydrodynamic lifting and breaking-up forces due to the waves. Keulegan has employed semi-analytical methods to propose a formula for determing the weight of breakwater blocks required to prevent dislodging:

$$\frac{(\rho_s g)\,H^3}{(\rho_s/\rho - 1)^3\,W} = \kappa \cot\alpha \tag{10.21}$$

where ρ_s is the density of block material, ρ the density of sea water, H the height of the oncoming wave, α the slope of the apron (breakwater) and W is the weight of a block. The constant coefficient κ is reported to be 3·2 for quarry stones and 9·5 for tetrapods (special four-legged concrete blocks which interlock through wave agitation and are therefore less likely to dislodge).

The heavier blocks at the apron foot provide anchor. They must be located below the low water mark to avoid undermining. The filter bed is necessary for draining rain water from the land and for preventing the backwash from seeping through the sand foundation and undermining the apron. Apron blocks tend to sink into the sand especially where there is inadequate drainage, and may cause stability problems. If they sink deep and become covered with new layers of sand they lose their effectiveness. Rock aprons are generally easier to construct and maintain than sea walls.

Sheet piles are also sometimes used for coast protection. They are known to

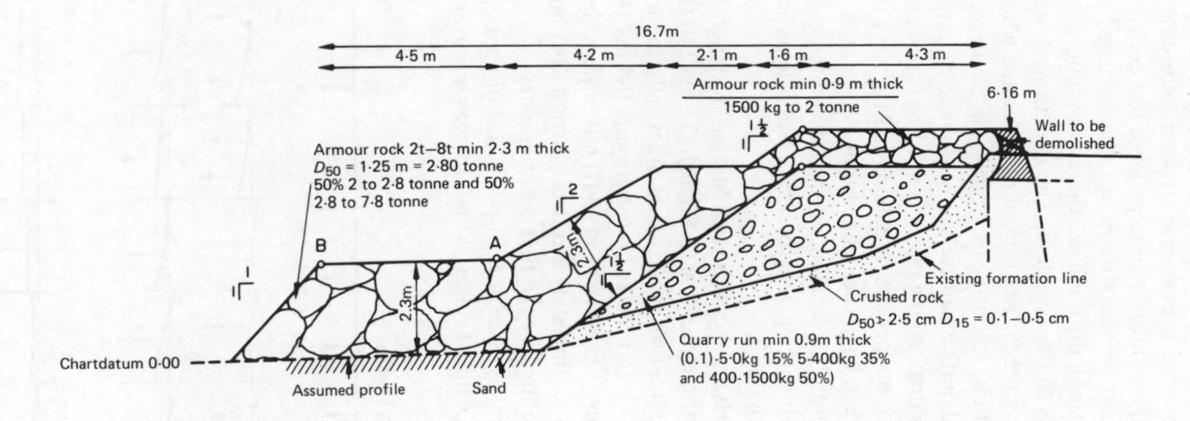

Fig. 10.18 Typical section of rock protection work at Sekondi (Ghana P.W.D.)

encourage beach formation behind them in some cases (for instance at Keta (Ghana) in the initial stages). Experience from Keta has however suggested that steel piles must be used with caution. Heavy corrosion and possible undermining caused the Keta piles to fail in many places in less than five years. Iron sheet piles are less likely to be subjected to corrosive attack if fully submerged in water. Similarly, fully submerged timber piles experience less decay.

The last group of coast defense structures to be discussed are groynes. Groynes are used to intercept littoral drift and are therefore useful only when there is a drift of material parallel to the coastline. The direction of littoral drift may be determined by observing the distribution of rare material whose source of supply is known. Tracers (artificial colouring of sand grains, flourescent materials and radio-isotopes) are used quite extensively nowadays to determine the direction and quantity of littoral drift. Along many beach fronts however, drifting materials show random movement in opposite directions and the net littoral drift may be a small difference between big movements in opposite directions.

Model studies are helpful in deciding on the suitability of groynes at a receding beach, as well as on their spacing, length, height and orientation. They are usually built perpendicular to the shore.

Experiments at the Hydraulics Research Station, Wallingford, U.K., were reported to indicate that groynes 1 m (3 ft) high, 75 m (240 ft) long to sea and spaced at 55 m (180 ft) intervals could arrest practically the whole drift moving along inshore of the low water mark. Lower groynes 0·5 m (1½ ft) high, similarly spaced or groynes of same height i.e. 1 m (3 ft) but more widely spaced at 110 m (360 ft) intervals did not completely arrest the littoral drift but showed no sand washing seaward as exhibited by the first set of groynes.

Sand accumulation takes place at the updrift side of a groyne and erosion at the downdrift side at the initial stages. At later stages however all compartments

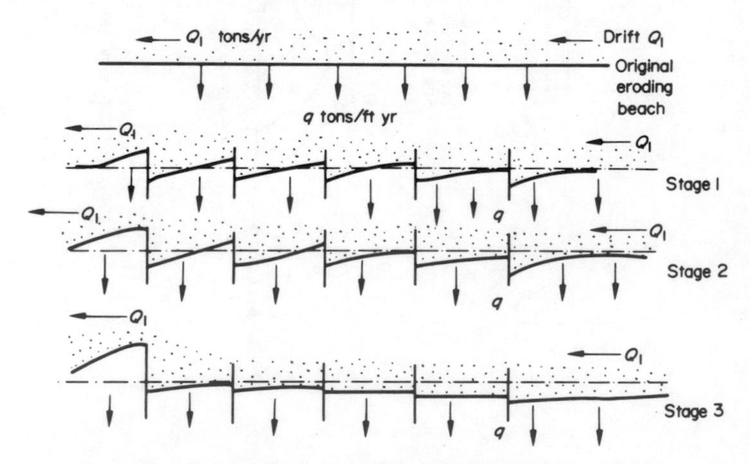

Fig. 10.19 The charging plan of an originally eroding beach when groynes are erected on it (After Russell)

normally show a net build up as illustrated in Fig. 10.19. The downdrift side of the last groyne experiences significant erosion. The existence of locally increased velocities and eddies is generally believed to be responsible for sand removal at the downstream side of structures. Russell has suggested that in order to maintain the drift with the groynes in place at the same level as before sand must be removed from the back of the last groyne. According to this hypothesis the size of the hole at the downdrift side of the last groyne increases at a rate nql where n is the number of groyne compartments of length l and q is the net rate of loss of material per unit length of beach to sea.

Groynes may be constructed of timber, steel, stone, concrete, or combinations of them. Impermeable groynes are often solid structures which prevent littoral drift passing through the structure but permeable groynes have sufficient openings to allow the passage of appreciable littoral drift. Recent experiments tend to suggest that permeable groynes are effective in reducing small drifts (usually at low water) but their effectiveness is limited when the drift is great (along the upper beach). They show very little scour at the downdrift side of the last groyne, apparently because of the smallness of the arrest.

FURTHER READING

Hydraulic Research Station (Wallingford) HRS Notes No. 15, December, 1969.

Ippen, A. T. (ed.), *Estuary and Coastal Hydrodynamics*, McGraw-Hill, New York.

Keulegan, G. H., 'Wave Motion' in *Engineering Hydraulics*, ed. H. Rouse, Wiley, New York.

King, C. A. M., *Beaches and Coasts*, Edward Arnold, London.

Milne-Thomson, L. M., *Theoretical Hydrodynamics*; Macmillan, New York.

Russell, R. C. H., 'Coast Erosion and Defence' *Hydraulics Research Paper No. 3*, Her Majesty's Stationery Office, London.

Stoker, J. J., *Water Waves* (*The Mathematical Theory and Applications*), Interscience Publishers, New York, London.

Wiegel, R. L., *Oceanographical Engineering*, Prentice Hall, New Jersey.

11 Fundamental Economics of Water Resources Development

11.1 Introduction

The primary aim of any eingineering activity is to convert resources of given forms and locations into resources of other forms and locations. In the performance of this task the engineer is often limited by the prevailing economic, political and social circumstances of society. He is or must be conscious of the basic economic requirements of his productive activity. These demand that the end products and their locations must provide more useful opportunities for further production or consumption.

For example, the resources of the human brain, natural river flow, earth, rocks, land, machinery, etc. have been combined to produce the Volta lake. The reservoir is then used to produce hydro-electricity at Akosombo. The power is conveyed 100 km to Tema where it is needed by the Volta Aluminium Company (Valco) for the extraction of aluminium, which is deemed (by the producers) to be still more useful for further production. It is important that the power used at Valco must be in the required form (correct voltage) and it must be supplied in adequate quantity and at the time the company requires it.

The principal resources involved in water resource planning may be listed as:

(1) naturally available water in various forms
(2) land to be drowned, excavated or developed
(3) construction material
(4) manpower
(5) equipment etc.

Out of these resources are to be developed (produced):

(1) water at location and time required for irrigation
(2) domestic and industrial water supply
(3) hydroelectric power
(4) waterways for navigation and general water transportation

(5) low water regulation
(6) flood control
(7) water facilities for recreational purposes
(8) water quality control, etc.

It is apparent from the above list that monetary capital considerations are overwhelming. This capital is in short supply and the situation is particularly crucial in the so-called developing world. The need for some basic appreciation of the optimum use of resources by the modern engineering graduate cannot therefore be overemphasized. The purpose of this brief chapter is to introduce the undergraduate to some of these considerations without any attempt at thorough coverage of the subject. The problems of economic, political and social constraints in water resources planning are too complex to be adequately tackled at the undergraduate level. The advent of the computer and the latest advances of linear and dynamic programming methods have led to sophistication in the approach to multi-purpose water resources utilization well beyond formal undergraduate engineering education. The young graduate in the developing world should nevertheless try to understand the basis of economic thinking as he will soon be called upon to participate in decision making in planning national development programmes.

Water resources planning, like other engineering design, involves choosing from two or more technically feasible possibilities. While basically concerned with technical efficiency in the design, the engineer must nevertheless be concerned with the design's economic efficiency. Technical efficiency, according to Libsey, measures the use of inputs (resources) in physical terms while economic efficiency measures the use in terms of costs. They are not incompatible, and are, in fact, generally complementary. Economic efficiency basically involves choosing from among the already established technically efficient combinations, the one that involves the least sacrifice from the owner of the project. Each technically efficient possibility should be considered on economic grounds if a rational choice is to be made. This is particularly important because of the many political and social implications of water resource projects. Ironically, many of the political and social attitudes run contrary to purely technical and economic considerations. The modern engineer must therefore be sufficiently aware of the social and political consequences which he would otherwise consider unimportant in his design.

There are only a relatively few suitable dam sites on any river and once they are appropriated the possibilities of further economic development of the same basin become limited. For example, the Akosombo dam creates over a 300 km long lake on the Volta River. Should it be decided in the near future that a higher dam at a location upstream of the existing dam could have resulted in a more economic multi-purpose development of the Volta basin, it is obvious that reconstruction of the system would involve unreasonable financial investment. The Volta project would have been unsoundly planned in the light of the new realization but very little could be done about it. In 50 years time, however, it might be considered economically feasible to abandon the existing dam and build a new complex. This shows that water resources planning is really a multi-dimensional problem with time as one of the important dimensions.

Political and social factors vary enormously from locality to locality and are very susceptible to time. Basic economic principles and objectives related to engineering planning are however universal and, notwithstanding the increasingly powerful methods of realization, remain geared to maximum output from any input.

Economic engineering planning must start with the identification and definition of each promising solution and as far as possible its cost. The money estimates must take into account the relevant interest rates and where necessary they must all be reduced to an annual cost through the capital recovery factor. The capital recovery factor, CRF is the annual payment which is sufficient to repay a unit of the capital plus its compound interest at the rate I (minimum attractive rate of return) in N years. This is given as follows

$$\mathrm{CRF} = \frac{I(1+I)^{\mathrm{N}}}{(1+I)^{\mathrm{N}}-1} \tag{11.1}$$

Table 11.1 gives the CRF for different interest rates over N years (up to 100). For example, the CRF for a capital of 80 000 (units) at 6% interest for 25 years is 0·07823 (units) per unit per year. Thus a yearly payment of 0·07823 × 80 000 = 6258·40 (units) is required to pay the capital plus all interest in 25 years. The total payment is 156 460 (units).

Table 11.1
Capital recovery factors
(from Linsley and Franzini)

Years	*Interest rate (%)*						
	1·5	*2*	*2·5*	*3*	*4*	*6*	*8*
5	·20909	·21216	·21525	·21835	·22463	·23740	·25046
10	·10843	·11133	·11426	·11723	·12329	·13587	·14903
15	·07494	·07783	·08077	·08377	·08994	·10296	·11683
20	·05825	·06116	·06415	·06722	·07358	·08718	·10185
25	·04826	·05122	·05428	·05743	·06401	·07823	·09368
30	·04164	·04465	·04778	·05102	·05783	·07265	·08883
35	·03963	·04000	·04321	·04654	·05358	·06897	·08580
40	·03343	·03656	·03984	·04326	·05052	·06646	·08386
45	·03072	·03391	·03727	·04079	·04826	·06470	·08259
50	·02857	·03182	·03526	·03887	·04655	·06344	·08174
60	·02539	·02877	·03235	·03613	·04420	·06188	·08080
70	·02317	·02667	·03040	·03434	·04275	·06103	·08037
80	·02155	·02516	·02903	·03311	·04181	·06057	·08017
90	·02032	·02405	·02804	·03226	·04121	·06032	·08008
100	·01937	·02320	·02731	·03165	·04081	·06018	·08004

The final step in engineering design and planning is to recommend a choice based on the actual monetary considerations and political, social and other factors which cannot necessarily be reduced to monetary units. In the next paragraphs we will discuss the basic economic and technological concepts which assist in such decision making.

11.1.1 Salvage Values

Money recovered as a salvage value at the end of the life of a hydraulic machine or structure reduces the annual cost of capital recovery. The portion of the first cost of machinery or structure equal to the prospective salvage value is considered to earn interest but no principal payments are required. The difference between the first cost and the prospective salvage value on the other hand has to be recovered with interest. Thus if P = the first cost of the machinery or structure whose life of operation is N years, S = the prospective net salvage value at the end of N years and I = the interest rate, the annual cost of capital recovery (CR) is expressed as:

$$\mathrm{CR} = (P - S)\,\frac{I(1+I)^N}{(1+I)^N - 1} + SI$$

or

$$\mathrm{CR} = (P - S)\,(\mathrm{CRF}) + SI \tag{11.2}$$

Equation (11.2) is applicable to a zero salvage value as well as to a negative salvage value. A negative salvage value means that instead of being able to recover money at the end of the project through, for example, sale of some recoverable machinery or other items, extra money would be required in order to demolish the structure in excess of that which could be recovered by selling some constituent parts.

11.2 Basic Economic and Technological Concepts (Decision Theory)

11.2.1 The Production Function

In strictly technological terms it must be recognized that given a certain set of resources x only a certain range of benefits y can be produced in a given system. For example, given a certain river basin and human resources only a specific range of various combinations of hydro-power, water supply, irrigation water and the others can be produced. In other words the given resources x and the technologically feasible benefits y form a technologically feasible pair. If x is changed through a modification in one or all of the resource elements a new set of benefits y in the system is feasible. All such technologically feasible pairs can be regarded as a technologically feasible region as illustrated in Fig. 11.1. A point outside the defined region is technologically impossible.

Ideas of technological feasibility and technical (technological) efficiency may also be illustrated by the example of a reinforced concrete bridge. The design must take into account physical knowledge about the effects of stress on reinforced concrete structures. A design in which too little steel is used in relation to cement and other materials is ruled out because it produces a bridge which is likely to collapse. Thus with a fixed amount of money certain combinations of steel and concrete which would not produce a sufficiently strong bridge are not feasible. However certain other combinations of steel, cement, aggregates and

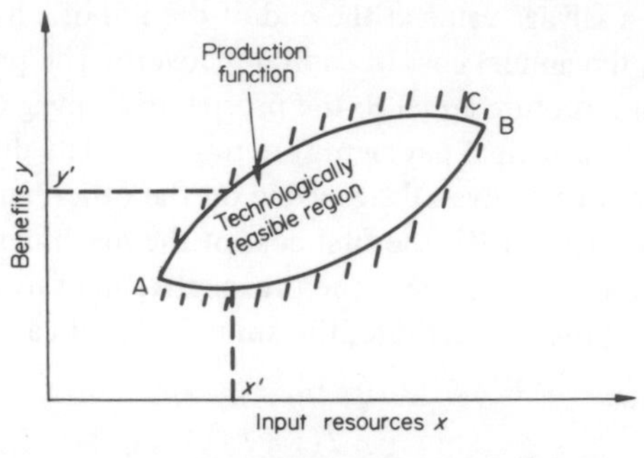

Fig. 11.1

other materials can produce a strong enough bridge, but not necessarily the strongest possible one. The combination which gives the strongest possible bridge for the available money is technically the best and therefore represents the most technically efficient use of the available resources.

In other words, for a given set of resources x' only one of the many possible benefits y' is technically efficient. Any other benefit y'' from the input x' is either technically unsound or not technologically feasible. Thus (x', y') is more efficient than (x', y'') if the latter is at all technologically feasible. By definition, therefore, (x', y') is most technically efficient if in the technologically feasible region there is no other pair (x, y) such that either $x < x'$ for $y = y'$ or $y \geqslant y'$ for $x = x'$. The lower curve AB represents the lower limit of technological feasibility below which no useful result can be obtained. The upper curve ACB passes through the points representing the best technical results which can be obtained from certain available inputs. The choice of the point of operation on the curve ACB is determined by economic considerations.

The locus of the technically efficient points ACB (Fig. 11.1) is known as the *production function.* It may be represented by the equation

$$f(x', y') = 0 \tag{11.3}$$

Since the input x' is itself made up of different elements $x_1', x_2' \ldots x_n'$ and y' may also be made up of $y_1', y_2' \ldots y_m'$ (multi-purpose utilization), equation (11.3) may be assumed to be equivalent to

$$f(x_1', x_2' \ldots x_n'; y_1', y_2' \ldots y_m') = 0 \tag{11.4}$$

Since only efficient points will be considered in the discussions which follow thc dashes can be dropped from the equation for the production function without any danger of confusion. Thus the production function is given by

$$f(x_1 \ldots x_n, y_1 \ldots y_n) = 0 \qquad (11.5)$$

It must be emphasized here that the efficient points may not be practicable or desirable from the point of view of political and social considerations. It must also be noted that on a long term basis the probabilistic nature of the inputs (water input varies with time; and maintenance, operation and repairs play a major role in water resources) demands that the expected output benefits must be considered. In other words, the relationship between x and y is stochastic. Detailed consideration of stochastic behaviour of the production function is however beyond the scope of this chapter and equation (11.5) will be treated as a short-run production function in which the values do not vary significantly with time.

11.2.2 **The Optimum Net Benefit** (The Economically Efficient Point)

Referring to Fig. 11.1 it is observed that only one point C on the production function yields maximum benefits. One may be tempted to conclude that it is the most economically efficient point of operation. Since, however, point C may not yield the maximum net benefit per unit input it does not necessarily correspond to the most economic point. The problem needs further examination to determine this point. If the unit cost of the *i*th element of the resources (river water or technical personnel in monetary value) is p'_i and the unit value of the *j*th benefit (irrigation water or recreational benefits) is p_j the net benefit u of the system is

$$u(x,y) = \sum_{j=1}^{m} p_j y_j - \sum_{i=1}^{n} p'_i x_i \qquad (11.6)$$

Equation (11.6) is the *net benefit function* in which the unit values p' and p are generally functions of the input and output quantities (x, y). In particular the unit price of the product p follows what economists call the law of diminishing returns, i.e. the more of a particular type of goods is produced in a locality the cheaper they become if all other factors such as population, tastes, etc., are held constant.*

Further appreciation of the production and benefit functions may be gained by reference to Fig. 11.2 on which the vertical and horizontal scales are equal. It must be observed that although the gross benefits continue to increase with increasing costs, net benefits exist only between A and B. The maximum net benefits occurs at C. One may be led to conclude that the project corresponding to C therefore is the most economic since it offers the maximum net physical

* The law of diminishing returns states that each additional unit of input yields a smaller increment of output than the previous one until a point is reached when the marginal cost is equal to the marginal revenue.

benefits for an investment. The evidence so far is however not conclusive and a further examination is required. This is provided by the ratio of benefits to costs. The highest ratio occurs at D which means that there one gets the maximum rate of return on a unit investment. Thus D offers the optimum net benefit and limits the investment into the project if it is desired to obtain a maximum rate of return

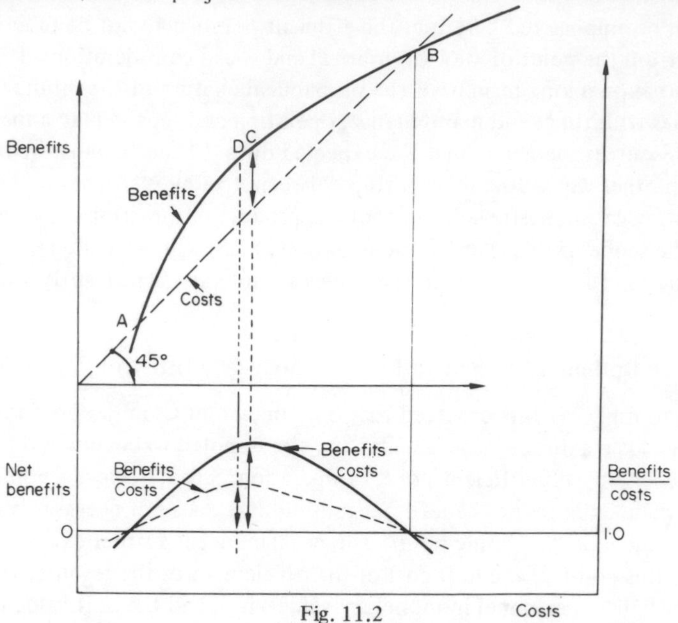

Fig. 11.2

on the investment. It is also necessary to examine the ratio of incremental benefits to incremental cost. Further investment is considered justifiable as long as this ratio exceeds unity.

EXAMPLE 11.1

Five proposals have been submitted for the control of floods on an agricultural field cultivated with maize (Table 11.2). If a bushel of maize costs 0·50 units of currency and the annual interest rate is 4%, which project would you recommend? There is no prospective salvage value.

Table 11.2

Proposal	*Capital (Money units)*	*Life expectancy (Years)*	*OMR (Money units)*	*Average maize harvest (bushels)*
1. Channel improvement alone	10 000	5	4 000	24 000
2. Dam A alone	100 000	15	8 000	82 000
3. Dam B alone	150 000	20	8 000	86 200
4. Dam A plus channel improvement	110 000	20	12 000	88 000
5. Dam B plus channel improvement	160 000	20	12 000	88 500

OMR = Operation, maintenance and repairs

The first step toward solution is to reduce all quantities to a comparable financial basis. Using the capital recovery factor (CRF) of Table 11.1, the capital investment and the interest can be reduced to an appropriate annual cost. This is listed in the second column of Table 11.3. The total annual cost made up of the annual capital recovery and the OMR costs is given in column 4. The gross revenue from maize harvest is listed in column 5 as annual benefits. All five projects show substantial net annual profit and column 6 shows that from net benefits point of view alone, there is very little to choose between 2, 3 and 4; project 3 being slightly better. The benefits/costs ratios of column 7 however clearly demonstrates the superiority of project 2 over the others. The ratios of incremental benefits to incremental costs are shown in the last column. Marginal results are obtained in moving from the second project to the third. Project 2 would be preferred although project 3 is also economically efficient.

Table 11.3

Project	*Annual capital recovery*	*OMR*	*Total annual costs*	*Annual benefits*	*Net annual benefits*	*Benefits / Costs*	*Δ Benefits / Δ Costs*
1	2 246	4 000	6 246	12 000	5 754	1·98	–
2	8 994	8 000	16 994	41 000	24 006	2·41	2·70
3	11 037	8 000	19 037	43 100	24 063	2·26	1·03
4	8 094	12 000	20 094	44 000	23 906	2·20	0·94
5	11 773	12 000	23 773	44 250	20 477	1·86	0·07

11.2.3 Alternative Substitution Possibilities

In the above example we lumped together all input parameters as x and all output parameters as y. It is quite apparent from the earlier discussions that both the inputs and the outputs may have different elements. Normally the production function tells us that there are numerous input factor combinations which produce the same output and the benefit function tells us that there are numerous output factor combinations which can be produced from the same input. For example, for a desired irrigation water production there are various combinations of water resources, dam sites, distribution pipe network and personnel which will achieve the same purpose. Similarly for a fixed investment in a water resource development project various combinations of irrigation scheme, hydro-power generation, domestic water supply, etc. are possible. It is desirable to examine in detail the economic possibilities of the various combinations. This is done by examining the *marginal rates of substitution* and *marginal rates of transformation.*

The net benefits function is defined in equation (11.6) as

$$u = \Sigma p_j y_j - \Sigma p_i' x_i \qquad (11.6)$$

Let us continue to substitute the ith input (x_i) for the hth input (x_h) holding all other factors constant until the maximum benefit is achieved. Mathematically, this is represented by

$$\frac{\partial u}{\partial x_i}\,\delta x_i + \frac{\partial u}{\partial x_h}\,\delta x_h = 0$$

or

$$\frac{\partial u}{\partial x_i}\Big/\frac{\partial u}{\partial x_h} = -\frac{\mathrm{d}x_h}{\mathrm{d}x_i} \tag{11.7}$$

Similarly the substitution of the jth output (y_j) for the kth output (y_k) keeping other factors constant until the maximum benefit is achieved gives

$$\frac{\partial u}{\partial y_j}\,\delta y_j + \frac{\partial u}{\partial y_k}\,\delta y_k = 0$$

or

$$\frac{\partial u}{\partial y_j}\Big/\frac{\partial u}{\partial y_k} = -\frac{\partial y_k}{\partial y_j} \tag{11.8}$$

It is also necessary to examine the rate at which an input x_i should be changed to produce a particular output y_i if all other factors are held constant. Thus

$$\frac{\partial u}{\partial x_i}\,\delta x_i + \frac{\partial u}{\partial y_j}\,\delta y_j = 0$$

or

$$\frac{\partial u}{\partial x_i}\Big/\frac{\partial u}{\partial y_j} = -\frac{\partial y_j}{\partial x_i} \tag{11.9}$$

$\partial u/\partial x_i$, which is the rate of change of the net benefit function with respect to x_i is defined in economics as the *marginal cost* of x_i(MC_i). For a constant unit price of the ith input, MC_i equals $-p_i'$ (from equation (11.6)). Similarly $\mathrm{MC}_h = -p_h'$.

$\partial u/\partial y_j$, the rate of change of the net benefit function with respect to y_j is defined as the *marginal benefit* of y_j, and $\mathrm{MB}_j = p_j$ for a constant unit price of y_j.

($-\mathrm{d}x_h/\mathrm{d}x_i$) is the *marginal rate of substitution* of x_i for x_h, MRS_{hi} and represents the rate at which the hth input can be reduced for a unit increase in the ith input or the rate at which the hth input must be increased for a unit decrease in the ith input, all other inputs and all outputs remaining fixed. Equation (11.6) stipulates that the substitution of x_i for x_h must continue as long as the ratio of the specific net benefits exceeds the marginal rate of substitution

$$\frac{\mathrm{MC}_i}{\mathrm{MC}_h} = \mathrm{MRS}_{hi}$$

($-\mathrm{d}y_k/\mathrm{d}y_j$) is the *marginal rate of transformation* between y_j and y_k, MRT_{kj}. It represents the rate at which the kth output must be reduced per unit increase in the jth output, all inputs and other outputs being held fixed and vice versa.

$$\frac{MB_j}{MB_k} = MRT_{kj}$$

$(-dy_j/dx_i)$ is the rate at which the jth output can be increased or must be decreased for each unit increase or decrease in the ith input. It is called the *marginal productivity* of x_i with respect to y_j, MP_{ij}.

$$MC_i/MB_j = MP_{ij} = -p_i'/p_j \text{ for constant unit prices}$$

Further appreciation of the MRS concept may be gained through an illustration using a dam/channel improvement flood control (dual input, single purpose) system. The curves in Fig. 11.3(a) are equal-product (equal-output) curves. They give the various combinations of annual channel improvement costs x_2 and annual dam construction, operation and maintenance costs x_1 required to produce a fixed set of benefits (flood control), $y = q$. Fig. 11.3(b) shows equal cost (h) lines for various combinations of the costs of x_1 and x_2. The lines are straight because constant unit (factor) prices are assumed and are thus represented by the linear equation

$$h = p_1'x_1 + p_2'x_2$$

In Fig. 11.3(c) the desired equal-output curve is superimposed on the family of equal-cost lines. A project designer will keep moving along the concave equal-output curve from right to left (that is substituting more channel improvement x_2 for dam construction x_1) as long as it crosses over to lower cost lines. The minimum possible cost line makes a tangent at C after which further substitution of x_2 for x_1 becomes uneconomic since costs will go up. At C the slope of the equal output curve equals the slope of the equal-cost lines $(-p_2'/p_1')$. The locus of the tangent point for different outputs gives an expansion path along which the dam/channel improvement design for different required benefits should move.

A similar illustration (Fig. 11.4) can be given for MRT using a single reservoir, dual purpose (irrigation and hydropower) water resources development. The full curves are equal-cost curves representing various combinations of water utilization from a fixed input h. The broken lines, $q = p_1y_1 + p_2y_2$, represent equal-benefit lines using constant unit prices. To find the optimum water utilization combination a designer would move along an equal-cost line (right to left) transforming more of y_1, into y_2 as long as higher equal benefit lines are crossed. The tangent touches at C beyond which it is economically unwise to transform more y_1 into y_2 since the benefits fall. The locus of C gives the expansion path.

It is possible in some cases that the equal-product curves of Fig. 11.3 will be convex instead of concave or the equal-cost curves of Fig. 11.4 will be concave instead of convex. These give extremum conditions (Fig. 11.5) in which only one input or only one output may provide an economically optimum condition.

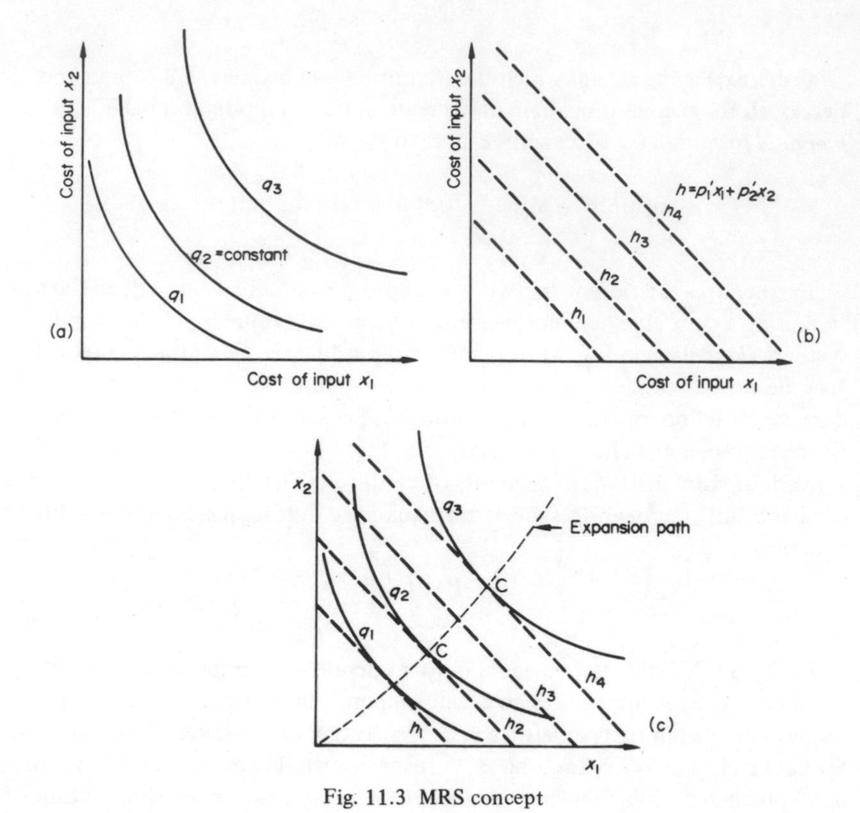

Fig. 11.3 MRS concept

Fig. 11.4 MRT concept

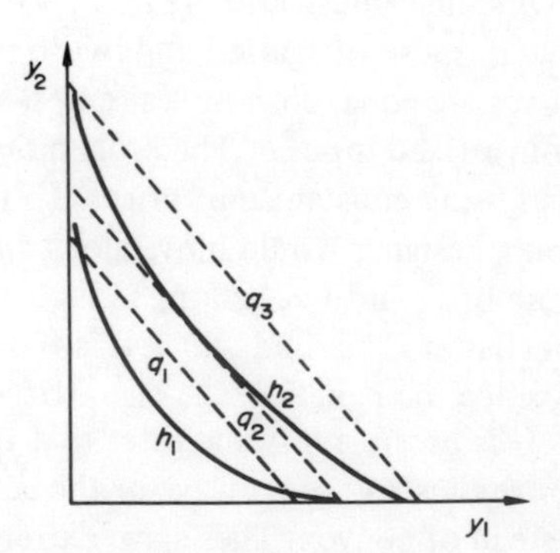

Fig. 11.5 Extremum conditions

FURTHER READING

Carter, C. F., *The Science of Wealth*, Edward Arnold, London.

Grant, E. L., *Principles of Engineering Economy*, The Ronald Press Company, New York.

Libsey, R. G., *An Introduction to Positive Economics*, Weidenfeld and Nicolson, London.

Linsley, R. K., and Franzini, J. B., *Water-Resources Engineering*, McGraw-Hill, New York, London.

Mass, A. *et al.*, *Design of Water-Resource Systems*, Harvard University Press, Massachusetts.

Samuelson, P. A., *Economics–An Introductory Analysis*, McGraw-Hill, New York, London.

Problems

Note

Although most of these problems use metric units some have been set in Imperial units in order to familiarize the user with both systems.

Chapter 1

1.1. Two parallel plates are 6·3 mm apart. The lower plate moves at 1·5 m/s and the upper one at 6 m/s. If a force of 3·57 N/m^2 is needed to maintain the upper plate in motion, find the dynamic and kinematic viscosity of the oil whose density is 880 kg/m^3 contained between the plates. State units clearly. (5×10^{-3} Ns/m^2, $5{\cdot}8 \times 10^{-6}$ m^2/s)

1.2. Air flows over a very porous wet flat plate and the concentration of moisture in the air at a section x is given by

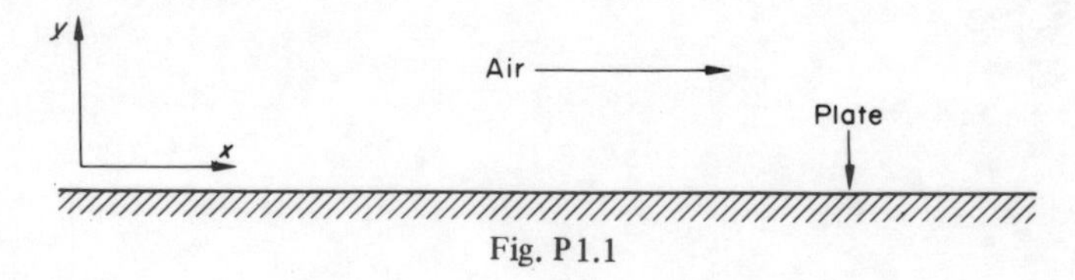

Fig. P1.1

$$C_w \text{ (kg/m}^3) = \rho(5 - kx)e^{-2y}$$

(see Fig. P1.1) where y is measured perpendicular to the plate. If the mass diffusivity α_m = 0·0083 m^2/h and the constant k = 0·15 m^{-1}, calculate the rate of evaporation from such a plate 3 m long and 0·3 m wide if the air blows along the length of the plate. State units clearly. Assume that the appropriate molecular transfer law holds in this range. U.S.T. Part I. (71·4 kg/h)

1.3. Use the principle of diffusion to write down the rate of flow of electricity. Explain the significance of all physical quantities and give their units. With a simple circuit diagram show the analogy between momentum transfer in fluid flow and electricity transport.

1.4. Establish that the average velocity v of viscous flow through two parallel-plates $2a$ apart is given in terms of the pressure gradient dp/ds by

$$\overline{v} = -\frac{2}{3}a^2 \frac{1}{2\mu}\frac{dp}{ds}$$

(a) The chamber shown in Fig. P1.2 contains crude oil (μ = 1 poise) at 138 kN/m^2 gauge pressure. If the annulus is 0·05 mm, and the length of the spindle is 150 mm and its diameter 50 mm, calculate the rate of leakage of oil from the chamber.

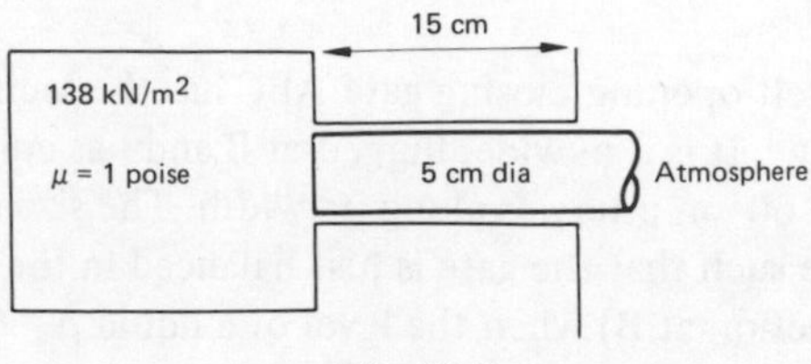

Fig. P1.2

(b) A 50 mm diameter shaft runs in a bearing of diameter 50·25 mm and length 25 cm. The clearance space is filled with oil of specific gravity 0·9 and kinematic viscosity 2·0 stokes. Calculate the power lost in the bearing when the shaft rotates at 40 rev/min. (15 x 10^{-9} m^3/s, 0·31 W)

1.5. Water at 60°F flows over a flat plate. The velocity profile at a point is given by v(ft/s) = $1 + 3y - y^3$, where y(ft) is measured perpendicular to the plate at the point of interest. Find the shear stress at that point. Comment on the form of the velocity profile.

If the temperature profile at the point is given by T(°F) = 4 sin ([π/2] y) + 60 for $0 \leqslant y \leqslant 2$ (ft) find the heat flux (quantity and direction) through the wall at the point. The properties of water at 60°F are

$$\rho = 62{\cdot}4 \text{ lbm/ft}^3$$
$$C_p = 1 \text{ Btu/lbm °F}$$
$$K = 0{\cdot}34 \text{ Btu/h ft °F}$$
$$\mu = 2{\cdot}71 \text{ lbm/h ft}$$

(70·5 x 10^{-6} lbf/ft^2, 2·13 Btu/h ft^2)

1.6. A special type of U-tube manometer is shown in Fig. P 1.3. Each limb has a reservoir at the top, the cross sectional area of the reservoir being 50 times that of the tube. The manometer contains oil having a specific gravity of 0·88 and water, the surface of separation being in the limb occupied by the oil. A small

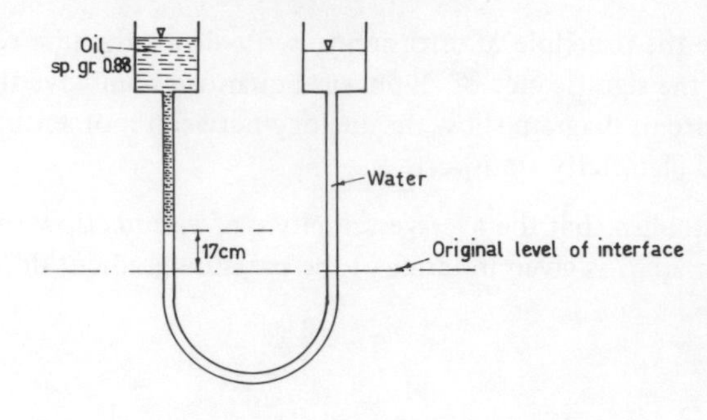

Fig. P1.3

pressure difference is applied between the two reserviors and the surface of separation rises by 17 cm. Calculate the applied pressure in N/m². U.N.Z.A, 1974. (264 N/m²)

1.7. A weightless self-opening/closing gate ABC has the section (Fig. P1.4) described by $x = 0{\cdot}82y^2$. It is 3 m wide, hinged at B and has cubic granite counterpoises W hanging at 60 cm intervals along its width. The size of each weight and its line of action are such that the gate is just balanced in the position shown (neglect vertical reaction at B) when the level of a liquid $\rho = 960$ kg/m³ is at B. Determine the size and position of W. If the liquid level rises 30 cm above B, calculate the minimum tension in a horizontal rope tied at C, 61 cm above B, if the gate should be kept closed. Assume the specific gravity of granite as 2·70. U.S.T. Part I. (0·545 m, 0·366 m, 45 000 N)

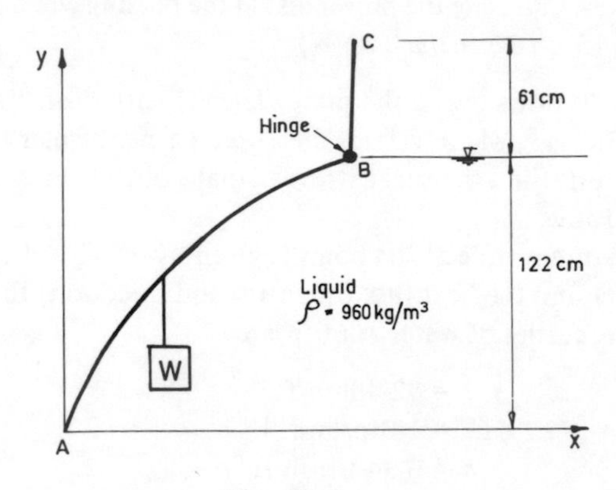

Fig. P1.4

1.8. A cylindrical timber log 30 cm in diameter, 1·22 m long and of specific gravity 0·55 rests with its axis parallel to two opposite sides of a 1·22 m square tank (see Fig. P1.5). Oil of specific gravity 0·80 is poured to a depth of 23 cm on one side of the cylinder. Determine the depth h of a liquid of specific gravity 1·2 required on the other side and the vertical force necessary to prevent the

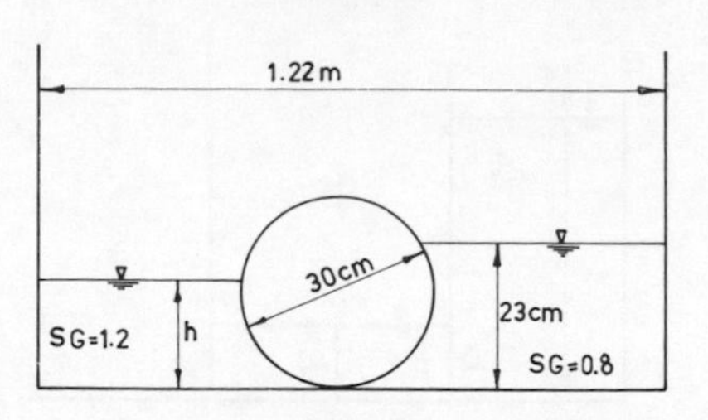

Fig. P1.5

cylinder from moving and to keep it in contact with the bottom of the tank. Neglect friction between the cylinder and the walls of the tank but assume that there are no leaks between log and tank. U.S.T. Part 1. (144 N)

1.9. A 2 m long hollow cylinder is placed in an upright position in water. The Outside and inside diameters of the cylinder are 1·22 m and 0·61 m respectively. Determine the metacentric height when floating 1 m and 30 cm deep. State in each case with appropriate reasons whether or not the cylinder is stable.

1.10. Discuss the general conditions which govern the stability of an object floating in a body of liquid and relate these particularly to the stability of a floating vessel.

A buoy has the upper portion cylindrical, 1·9 m diameter and 1·2 m deep (Fig. P1.6). The lower portion displaces a volume of 0·378 m^3 and its centre of buoyancy is situated 1·275 m below the top of the cylinder. The centre of gravity of the whole buoy and the load it holds up is situated 0·9 m below the top of the cylinder and the total displacement weight is 25 kN. Find the metacentric height when in sea water of density 1025 kg/m^3. U.N.Z.A, 1974.

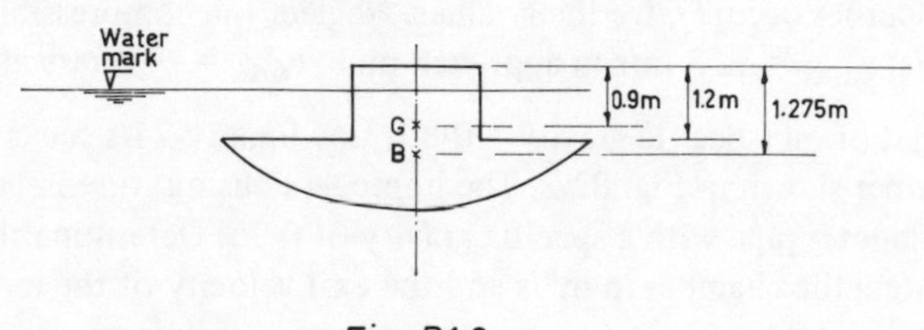

Fig. P1.6

Chapter 2

2.1. In Fig. P2.1 the cross sectional area of the tank on the left is A_1 and of that on the right is A_2. If the orifice O is left open but valve V is closed write appropriate differential equations describing the filling of the tanks when:

(*a*) $h_1 > d$ and $h_2 < d$ and
(*b*) $h_1 > d$ and $h_2 > d$

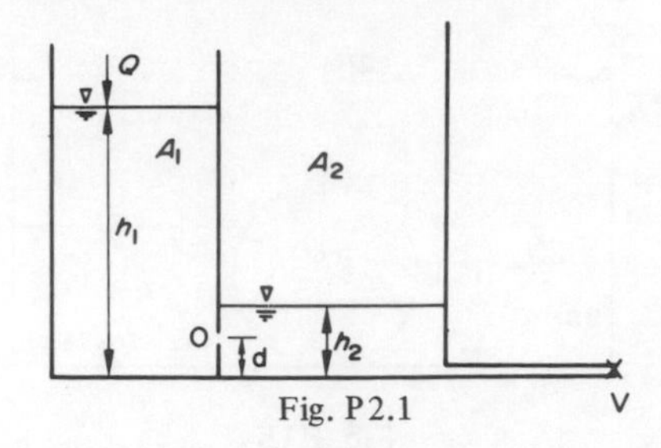

Fig. P2.1

2.2. If initially $h_1 = h_2 = d$ (Fig. P2.1) plot a graph showing the difference in levels in the two tanks against time with the valve closed. Q = 14·1 l/s, the orifice diameter is 2·5 cm and C_d = 0·60. The tank on the left is 0·91 m square and that on the right is 1·2 m square.*

$$t = \frac{2}{b^2}\left\{bh^{\frac{1}{2}} - a\log_e\left(1 + \frac{b}{a}h^{\frac{1}{2}}\right)\right\} \text{ where } a = Q/A_1 \text{ and } b = a_0 C_d \sqrt{(2g)}\left(\frac{1}{A_1} + \frac{1}{A_2}\right)$$

2.3. With Q = 14·1 l/s, A_1 and A_2 and the orifice as described in question 2.2, and V open, determine how long it would take for the net inflow into the system to be just zero, if the rate of change of difference in levels in the tanks is maintained constant at 0·3 cm/s. Initially $h_2 = d = 0$. The diameter of the valve is 3·8 cm and its C_d = 0·60. (1·4 x 10^3 s)

2.4. Air flows through a 15 cm diameter pipe at 0·57 m^3/s. The pipe has two rows of uniformly spaced holes along 0·3 m of its length. If air is injected through one row of holes at a rate given by $4x^2$ (m^3/s m) and is withdrawn from the other at $1{\cdot}5(0{\cdot}3-x)$ m^3/s m over the length $x = 0$ and $x = 0{\cdot}3$ m, determine the velocity of air in the pipe beyond the holes. Where do the maximum and minimum velocities occur? Give their values. Neglect the compressibility of air. (30·5 m/s; v_{max} = 32·3 m/s in approach pipe; v_{min} = 29·4 m/s after holes.)

2.5. A jet of oil (specific gravity = 0·80) flowing at 57 l/s sucks in water into the chamber shown in Fig. P2.2. The homogeneous mixture issues out through a 0·3 m diameter pipe with a specific gravity of 0·95. Determine the rate at which water enters the chamber in m^3/s and the exit velocity of the mixture. (0·171 m^3/s; 3·24 m/s)

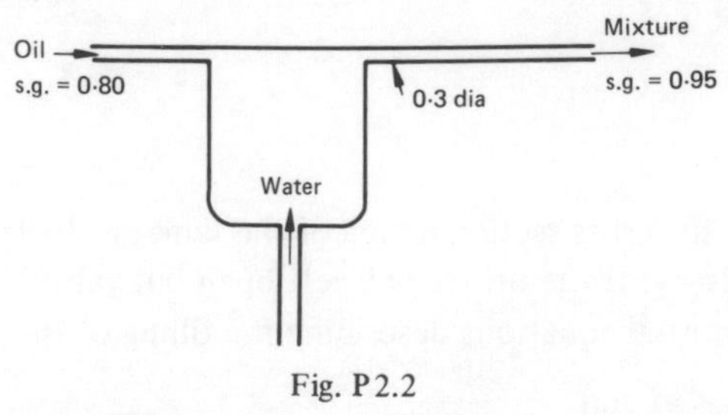

Fig. P2.2

* Orifice discharge = $C_d a_0 \sqrt{(2gh)}$.

2.6 The orifice at the bottom of the cistern in Fig. P2.3 opens automatically when the water level in the cistern is 1·2 m or higher. If the hose discharges at a constant rate of 5·6 l/s, sketch a graph showing the water depth against time when initially the cistern is empty. The cistern is 0·61 m in diameter and the orifice is 3·9 cm in diameter. The coefficient of discharge = 0·60. (The orifice opens at t = 62·7 s when h = 1·2 m; equilibrium is attained when h = 3·14 m)

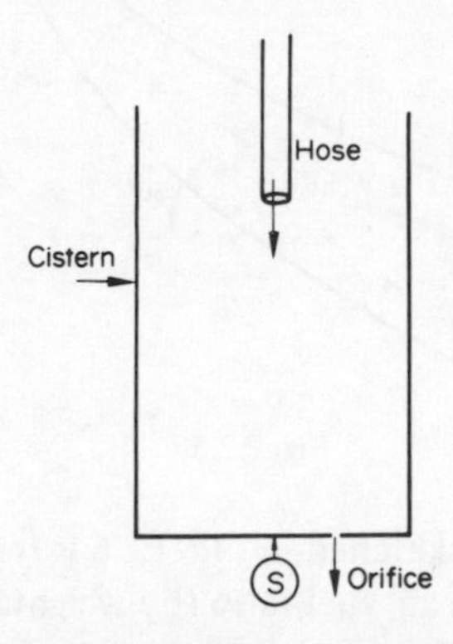

Fig. P2.3

2.7. If in question 2.6 the scale S initially reads zero, what will be the corresponding readings when the depths of water are 1 m and 1·6 m if the water hose is 2·5 cm in diameter? (2·91 kN; 4·77 kN)

2.8. Two tanks X and Y have plan areas of 300 m^2 and 400 m^2 respectively. They are connected together by means of three horizontal pipes each 150 m long and of diameter D, laid in parallel (see Fig. P2.4). The initial difference in

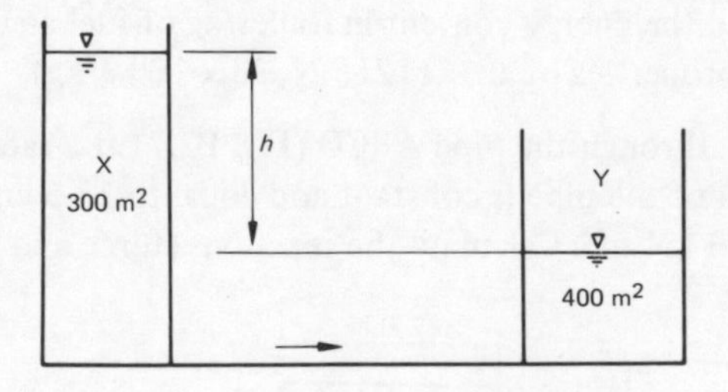

Fig. P2.4

the level of water surfaces between tanks X and Y is 3 m. If 0·34 m^3/s of water is now admitted into tank X, what should be the diameter of the connecting pipes if the difference in the water surface levels in X and Y is to remain unchanged at 3 m during filling? The flow through each pipe is given by h = 0·01 LQ^2/D^5. What is the ratio of the flow through the pipes to the water admitted into the system? (0·296 m; 0·572)

2.9. Figure P2.5 shows the plan view of a combined bend and reducing piece in a pipeline carrying 21 l/s of water. The gauge pressure at entry to the bend is 240 kN/m^2. Neglecting degradations of energy, calculate the resultant force on the bend. (1500 N inclined at 35·5° to the x-axis)

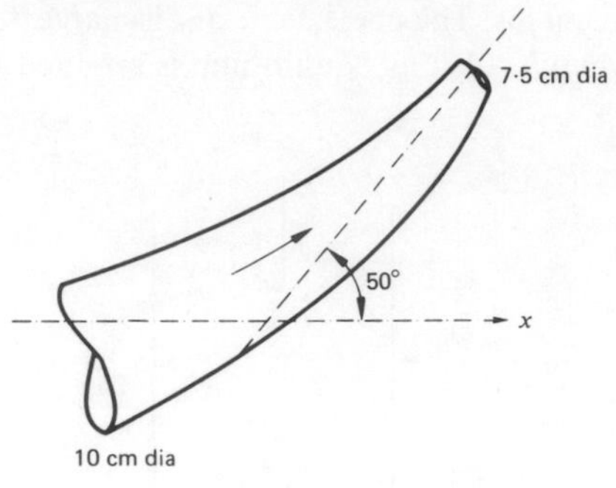

Fig. P2.5

2.10. The propulsion unit sketched in Fig. P2.6 is travelling at 61 m/s at sea level (temp. = 15°C) with an air–fuel ratio (by weight) of 30:1. The density of the exhaust is 0·44 kg/m^3. The density of the air is 1·22 kg/m^3 and the pressure

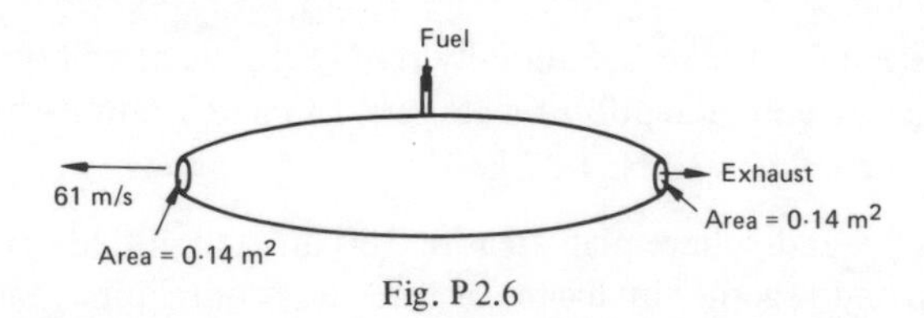

Fig. P2.6

over both inlet and exit is 101 kN/m^2. Estimate the thrust developed by the propulsion unit and the energy content in joules/kg of fuel assuming the mixture (exhaust) has the properties of air. (1215 N; 23 x 10^6 J/kg)

2.11. Water flows through the pipe ABCD (Fig. P2.7) at a rate of 14 l/s. The cross sectional area of the pipe is constant and equal to 18·5 cm^2. The water pressure at A is 103 kN/m^2. Calculate the reactions (force and moment) which

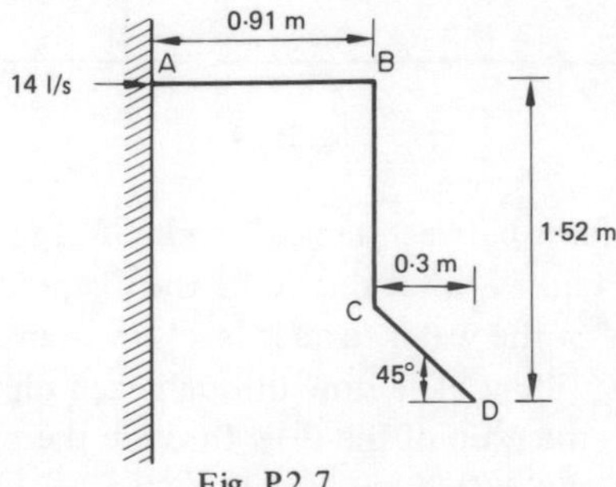

Fig. P2.7

must be provided at A to keep the pipe in equilibrium. The pressure given at A is gauge pressure and atmospheric pressure at D is 101 kN/m^2. The pipe alone weighs 58 N/m. (250 N at $\tan^{-1}$ 0·53 to the horizontal; 175 Nm)

2.12. Steam flowing through a main pipe is to be trapped into a side container through a small tapping at the side of the main pipe. If flow conditions denoted by the subscript 'p' in the main pipe are not affected by the trapping process and can therefore be assumed constant, derive an expression for the heat transfer into the container during the process (neglecting friction and potential energy). Use the subsripts 1 and 2 to denote initial and final conditions respectively, U is the internal energy, h_p is the enthalpy, v is the velocity and m is the mass of steam inside the container at any time.

2.13. Solve problem 2.11 with the flow direction reversed; the pressure at A being − 103 kN/m^2 gauge and the water level in the space to right of the wall 1·22m above D. (see Fig. P2.8). (198 N at 24·5° to the horizontal; 157 Nm)

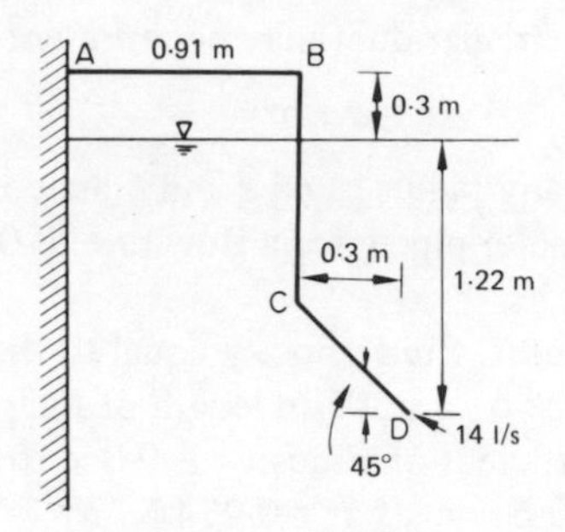

Fig. P2.8

2.14. A 2 ft long V-shaped vessel has its sides sloping at 45° such that all horizontal planes cutting through it show different sizes of rectangular sections each 2 ft long. Water flows into it at the rate $e^{-\alpha t}$ ft^3/s where $\alpha = 1\ h^{-1}$. The vertex of the vessel which is downward has an $\frac{1}{8}$ in wide slot along the entire length with a coefficient of discharge of 0·70. At time t = 35 min the depth of water in the vessel is 16·65 ft. Estimate the maximum depth and the time it occurs taking time increments of 2 min which may be assured small. Sketch the variation of depth with time. U.S.T. Part I. (16·98 ft; 44 min)

Chapter 3

3.1. Water is discharged from a tank through a mouthpiece in its side as shown in Fig. P3.1. The exit is rounded and entry losses may be neglected. The throat of the mouthpiece is 5·1 cm in diameter. If the water head above the centre line of the mouthpiece is 1·83 m, what must be the minimum diameter at the mouthpiece exit for the absolute pressure at the throat to be 2·4 m of water?

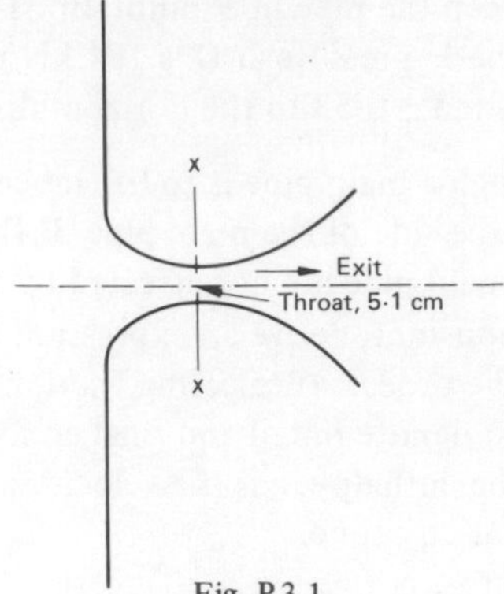

Fig. P 3.1

What is the rate of discharge in this case and what would it be if the piece to the right of XX were removed? (7·76 cm; 0·028 m³/s; 0·012 m³/s)

3.2. The flow through a circular duct may be represented by

$$v = a + br^2$$

where v is the velocity at any radius r and a and b are constants.

A test on a 30 cm diameter pipe gives a flow rate of 0·085 m³/s. Determine a and b and specify their units.

At what radius is the actual fluid velocity equal to the mean fluid velocity? What is the wall shear force over a 305 m length of the pipe?

Take the effective coefficient of viscosity as $9{\cdot}0 \times 10^{-3}$ Ns/m². (a = 2·42 m/s; $b = -107{\cdot}5$ m^{-1} s^{-1}; r = 10·6 cm; shear = 0·274 N/m)

3.3. Fluid enters the viscosity pump shown in Fig. P3.2 at A, flows through the annulus and leaves at B. Assuming that the annular space h is very small compared with the drum radius R, derive an expression for the discharge per unit length of

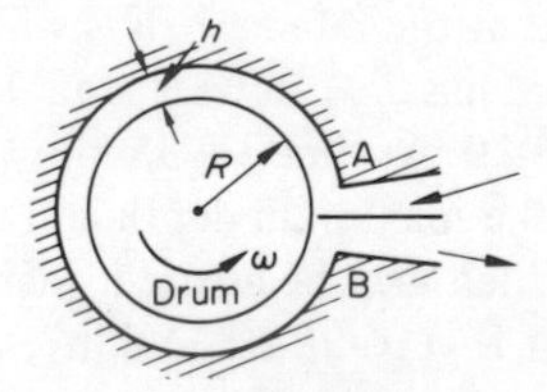

Fig. P 3.2

the pump in terms of pressure rise (between A and B), R, h, angular speed ω and viscosity μ.

With R = 1·5 in, ω = 120 rev/min, h = 0·005 in and a pressure difference of 10 lbf/in² calculate the h.p. (per unit length of the pump) of a motor driving it at 80% efficiency. The oil being pumped is castor oil for which μ = 0·0206 lbfs/ft².

$$(Q/\text{ft} = \frac{\omega R h}{2} - \frac{h^3}{24\pi R\mu}\,\Delta p,\ 0{\cdot}22\ \text{h.p./ft})$$

3.4. Air flowing at 30·5 m/s in a 76 mm dia. pipe suddenly enters an 204 mm dia. pipe. Calculate (a) the pressure change in mm of water and (b) loss of energy in watts. Take the density of air as 1·45 kg/m^3.

If after a short length of the 204 mm dia. pipe the air again enters a 76 mm dia. pipe find the loss of energy for a coefficient of contraction C_c = 0·64. Comment on the relative magnitude of the two energy losses. (16 mm; 69 W; 29·5 W)

3.5. A 0·61 m dia. water main is 7·6 km long and has a coefficient of friction f = 0·008. The head loss is 36·6 m at a certain discharge capacity and it is required to increase the capacity by 20% through a booster pump. Find the power of the motor required to drive the pump at 65% efficiency. What is the head developed by the pump? (124 kW; 16·2 m)

3.6. In the pipe system shown in Fig. P3.3, L_1 = 910 m, D_1 = 30·5 cm, f_1 = 0·005; L_2 = 610 m, D_2 = 40·8 cm, f_2 = 0·0045; L_3 = 1220 m, D_3 = 20·4 cm and f_3 = 0·0043. The gauge pressure at A is 550 kN/m^2, its elevation is 30·5 m

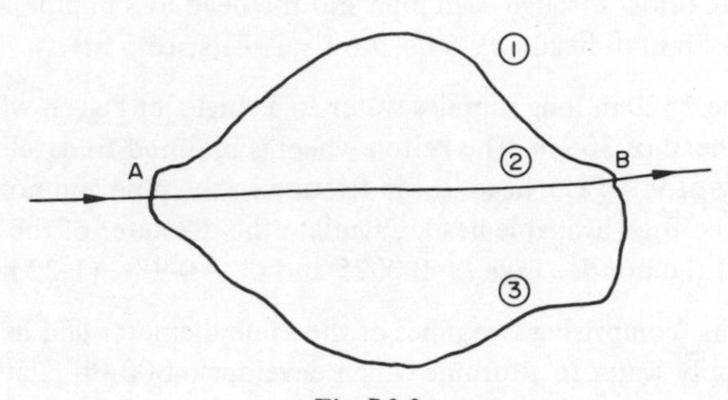

Fig. P3.3

and that of B is 24·5 m. For a total flow of 0·34 m^3/s, determine the flow through each pipe and the pressure at B. (0·09 m^3/s; 0·22 m^3/s; 0·03 m^3/s; 570 kN/m^2)

3.7. In Fig. P3.4 A and B are connected to a reservoir and C and D to another lower reservoir. If the discharge velocity through pipe AJ is 4 ft/s calculate the

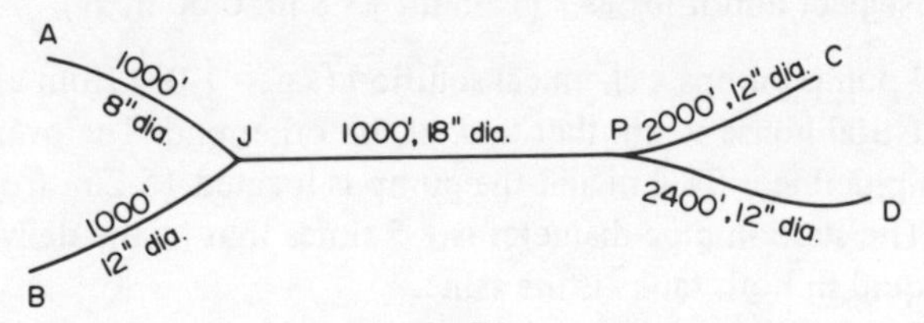

Fig. P3.4

flow through the system and the head H between the reservoirs. Take f = 0·006 for all pipes. Neglect minor losses. (1·39 ft^3/s; 3·84 ft^3/s; 5·23 ft^3/s; 2·50 ft^3/s; 2·73 ft^3/s; 20 ft)

3.8. A pump P delivers 22·6 l/s of water from reservoir R into other reservoirs A, B and C. The water levels in A and B are 30 m and in C 39·7 m above the water level in R. The pipes leading to A, B and C have a common joint J. Pipes JA, JB and JC are each 152 m long and 76 mm in diameter. The friction factor for all pipes is 0·006. (See Fig. P3.5.)

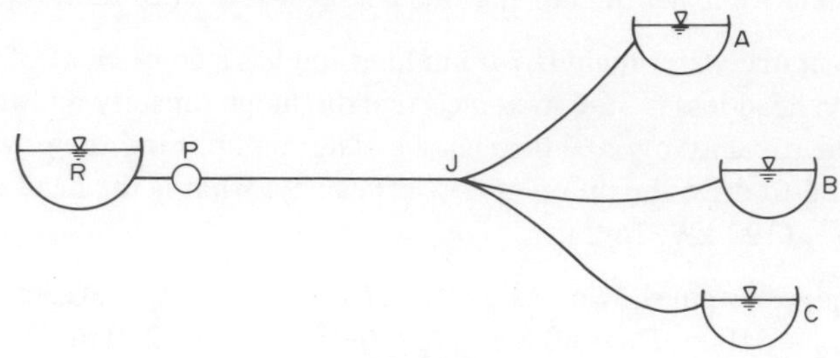

Fig. P3.5

Determine the flow through each pipe and the head loss in pipe RPJ if the pump develops 58 m of head. (9·6 l/s; 9·6 l/s; 3·4 l/s; 16·5 m)

3.9. A pipeline 1520 m long supplies water to a single jet Pelton wheel under a total available head of 305 m. The Pelton wheel is required to develop 7·46 MW at an efficiency of 90%. The head lost in friction in the pipe and nozzle is not to exceed 8% of the total available head. Calculate the diameter of the pipeline and the diameter of the nozzle. Take $f = 0{\cdot}0075$ and $C_v = 0{\cdot}98$. (1·22 m; 22·9 cm)

3.10. A pipeline comprising two pipes of the same diameter laid in parallel is required to supply water to a turbine which develops 6 000 Bh.p. at 85% efficiency under a gross head of 550 ft. Each pipe is 12 000 ft long with $f = 0{\cdot}006$. Determine the diameter of the pipes if 95% of the gross head is to be available at the machine. (3·91 ft)

3.11. A pipeline 76 cm in diameter and 6·44 km long delivers 0·56 m^3/s; $f = 0{\cdot}006$; Find the loss of pressure due to friction. If the last half of the pipe is replaced by two 61 cm dia. pipes in parallel, what will be the total loss of head for the same total delivery? What will be the delivery if the original friction loss is permitted? Neglect minor losses. (15·8 m; 13·8 m; 0·60 m^3/s)

3.12. A small pump pumps a chemical solution (s.g. = 1·05) from a tank at one end of an industrial house to another tank at the other end. The overall length of the horizontal pipeline is 91·4 m and the pump is located 15·2 m from the suction tank. The suction pipe diameter is 1·5 times that of the delivery pipe. The level of liquid in both tanks is the same.

Make a well-proportioned diagram showing clearly the energy and hydraulic grade lines. Write down appropriate energy and continuity equations and solve them for $f = 0{\cdot}006$ and a 76 mm suction pipe. The pump develops 6·1 m of head. What is the power of the motor driving the pump at an efficiency of 80%? (Q = 3·6 l/s; 280 W)

3.13. Two reservoirs A and B are connected by a pipe 1·22 m in diameter and 3050 m long. Initially the level of water in reservoir A is 30·5 m above that in B. If both reservoirs have a constant cross sectional area of $0{\cdot}9 \times 10^6$ m^2 for A and $0{\cdot}45 \times 10^6$ m^2 for B find the time required for the water in reservoir A to fall 3·5 m.

Take f as 0·0075 and neglect all losses except pipe friction. (12 days)

3.14. The pipeline from a reservoir to an impulse turbine consists of a 1220 m long, 30·5 cm dia. pipe followed by a 610 m long, 22·8 cm dia. pipe at the end of which is a 44·4 mm nozzle. C_v for the nozzle is 0·95, $f = 0{\cdot}006$ for the pipes and K for all sharp edge contractions may be taken as 0·5. If the nozzle is 670 m below the reservoir water surface, calculate the discharge in m^3/s, the jet power and the overall efficiency of the transmission system. (0·158 m^3/s; 870 kW; 80%)

3.15. Water is supplied to reservoir C from A and B (Fig. P3.6). the water levels in both A and B are 30·5 m above that in C. The pipes have the following characteristics.

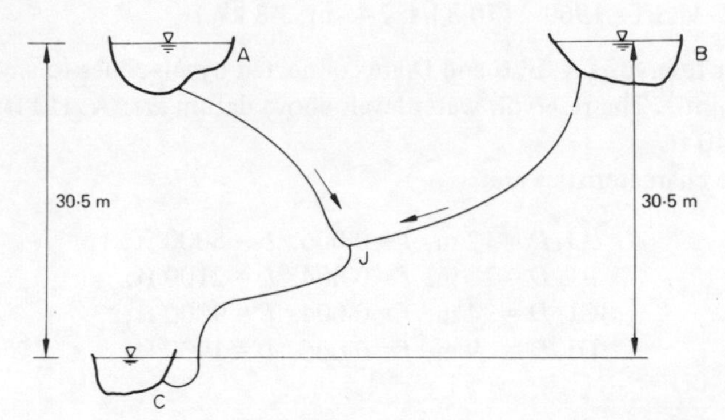

Fig. P3.6

Pipe	*Length*	*Diameter*	*f*
AJ	610 m	30·5 cm	0·006
BJ	1525 m	45·7 cm	0·006
CJ	305 m	61 cm	0·006

Neglecting all minor losses, determine the rate of flow into C. To what level will water rise in a piezometer inserted at the junction J? (0·67 m^3/s; 3·2 m above level in C)

3.16. A water pipe system ABCD (see Fig. P3.7) has pressures maintained as follows: at A 33·6 m; at B 30·5 m at C 24·4 m, at D 12·2 m above datum. The constants K in the pipe friction equation: friction head (m) = 10^3 K flow^2 (m^3/s)^2 are; AX 18·2; BX 4·2; XY 1·9; YC 3·0; YD 18·2. Determine the flow in XY. (0·041 m^3/s; H_x = 28·1 m; H_y = 25·0 m).

3.17. If in Fig. P3.7 maintained heads are as follows: at A 36·6 m; at B 33·6 m; at C 9·0 m and at D 0·0 m above datum and a small turbine in XY develops 1·35 kW. at a constant 80% efficiency, what will be the flow through XY? What head will be consumed by the turbine? If a throttle valve in XY restricts the

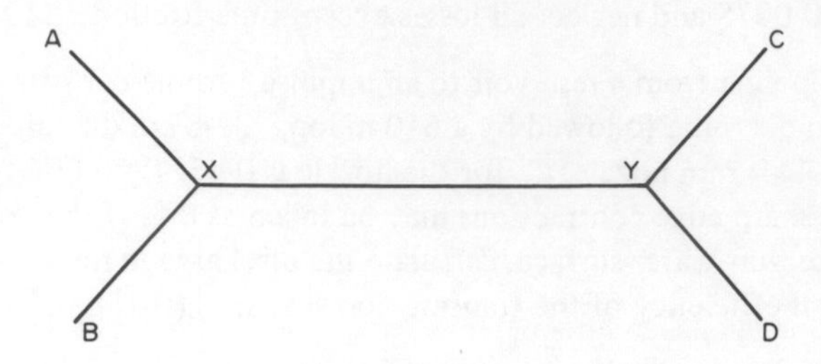

Fig. P3.7

flow through it to 28·3 l/s, calculate the power lost across the valve with no flow through YC. U.S.T., 1969. (70·8 l/s; 2·43 m; 3·8 kW)

3.18. Four reservoirs A, B, C and D are connected by pipelines to a common junction point J. The reservoir water levels above datum are: A, 110 ft; B 100 ft; C 80 ft; D 40 ft.

The pipe characteristics are:

AJ, D = 12 in, f = 0·006, L = 8000 ft;
BJ, D = 18 in, f = 0·004, L = 2100 ft;
CJ, D = 24 in, f = 0·004, L = 9000 ft;
DJ, D = 9 in, f = 0·006, L = 1900 ft;

(a) Determine the flow in each pipe.
(b) A fifth pipe JE, 8 inches in diameter, f = 0·008, L = 4000 ft, is later attached to J and discharges to a reservoir at level 70·0 ft above datum. A flow of 1 ft^3/s is to pass through it. At which outlet should a throttling valve be placed, and what will be the loss of energy and power there? U.L. (2·1; 10·07; 8·97; 3·19 ft^3/s; JC; 6·29 h.p.)

3.19. A pipeline network ABCDEF has inflow and outflow as shown in Fig. P3.8. The values of K in the head loss equation $h = 10^3 KQ^2$ are as follows:

Pipe	AB	BC	CD	DE	EF	FA	BE
K	1·9	3·8	7·6	3·8	1·5	0·8	0·8

Estimate the flow in each pipe using the Hardy Cross method. If the pressure at A is 689·5 kN/m^2, the elevation of A is 63·7 m above datum and the elevation of C is 110 m above datum, determine the water pressure at C. What would be the effect of closing a valve in: (a) CD; (b) BE?

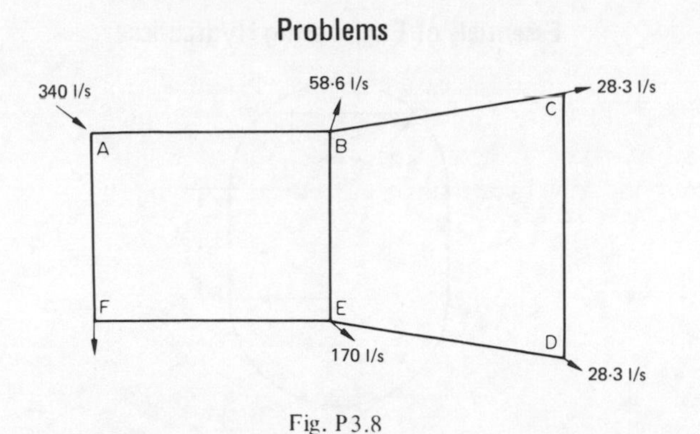

Fig. P3.8

3.20. The pipe length and diameter of each pipe of the Kumasi high-tower water supply are given in square blocks in Fig. P3.9. Approximately 700 gal/min is supplied from the reservoir. Supplies to the various communities are indicated on

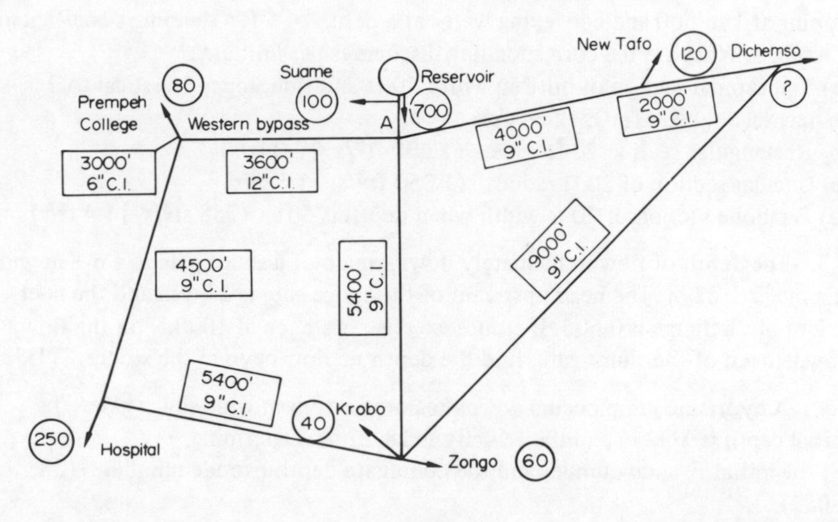

Fig. P3.9

the diagram in circles in gal/min. Calculate the flow in each pipe. What will be the pressure at Prempeh College which is 10 ft above the pipe level at the main junction A where a pressure gauge indicates 31 lbf/in^2? Take $h_f = 0{\cdot}002Q^2L/D^5$ where Q is in gal/min, L is in ft and D is in inches. U.S.T., 1968, Part III.

Chapter 4

4.1. A culvert has vertical sides and semi-circular top and bottom as shown in Fig. P4.1. Assuming that C in the Chezy formula is the same for all depths of

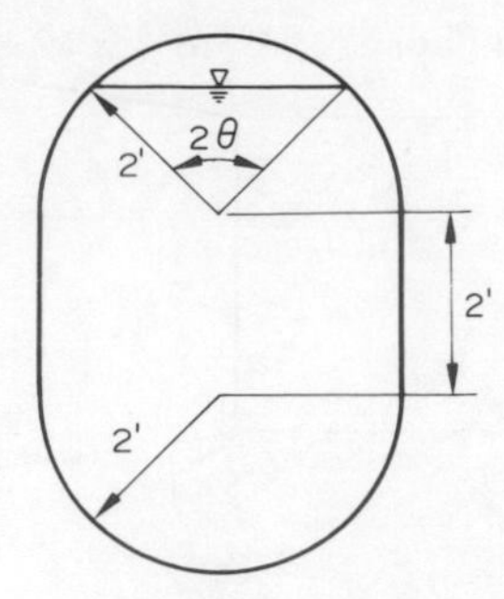

Fig. P4.1

flow, show that the velocity is a maximum when the depth of water in the culvert is approximately 5·05 ft. (Hint: plot m vs. θ)

4.2. What is the normal discharge in each of the following sections of channel sloping at 1 in 500 and conveying water at a depth of 5 ft? Manning's coefficient $n = 0{\cdot}012$. What are the corresponding discharges per unit area?
(a) Trapezoidal section of bottom width 20 ft and side slopes 1 vertical to 2 horizontal? (1900 ft^3/s; 12·6 ft/s)
(b) Rectangular section 20 ft wide. (1 200 ft^3/s; 12·0 ft/s)
(c) Circular section of 20 ft radius. (1 050 ft^3/s; 11·5 ft/s)
(d) Parabolic section of 20 ft width when depth is 5 ft. (758 ft^3/s; 11·4 ft/s)

4.3. The depth of flow immediately downstream of a sluice gate in a 6·1 m wide channel is 1·22 m. The head upstream of the sluice gate is 14·6 m and the coefficient of discharge is 0·605. If stones exert a resistance of 100 kN on the flow downstream of the sluice gate, find the depth of flow beyond the stones. (1·5 m)

4.4. A hydraulic jump occurs downstream of a 15·5 m wide sluice gate. The initial depth is 1·22 m and the velocity is 18·2 m/s. Determine
(a) the initial Froude number and the conjugate-depth Froude number. (28; 0·082)
(b) the energy dissipated in the hydraulic jump. (31 MW)
(c) illustrate the situation on a specific energy sketch.

4.5. A 15·5 m wide rectangular channel slopes at 1 in 6000. The discharge is 28·3 m^3/s. The Chezy coefficient is 55 m$^{\frac{1}{2}}$/s. If the depth at a certain point is 2·74 m, calculate the slope of the water surface. Is the flow subcritical or supercritical? (0·000 093; $F = 0{\cdot}016$)

4.6. A 1·55 m wide rectangular channel conveys 1·13 m^3/s of water at a normal depth of 61 cm. A hump of height Z m is installed on the channel bed. Draw a curve to show the variation of Z/H with y/H where y is the depth of flow over the hump and H is the total head immediately upstream of the hump. Determine the value of Z which would produce a critical depth on the hump. (11 cm)

4.7. A V-shaped channel has a vertex angle of 60°. For a discharge of 0·91 m^3/s estimate the normal and critical depths. Manning's $n = 0{\cdot}012$ and the bed slope $S_0 = 0{\cdot}0001$. (1·8 m; 0·87 m)

4.8. A trapezoidal channel of bottom width 20 ft and side slopes 1 vertically to 2 horizontally conveys 400 cusec of water. The bed slopes at 1 in 250 and Chezy's $C = 100$ $ft^{\frac{1}{2}}/s$. Calculate the normal and critical depths. (2·03 ft; 2·14 ft)

4.9. A horizontal rectangular stream is 6·1 m wide. It is desired to measure the flow using a Venturi flume ($C_d = 0{\cdot}90$) which produces a critical depth at the throat. If the upstream depth must not exceed 2·1 m when the flow is 14·1 m^3/s, determine the required width at the Venturi throat. What would happen if a narrower throat were used? (2·9 m)

4.10. An irrigation canal is to be of trapezoidal cross section with natural lining of channel banks and bottom. To prevent scour, average velocities are to remain below 0·61 m/s. Stability of channel sides requires slopes of more than 2 horizontal to 1 vertical.

If 56·5 m^3/s are to be conveyed over a distance of 64 km with a total elevation difference of 21·5 m, determine the depth and the bottom width of a suitable trapezoidal channel section. Assume Manning's roughness coefficient as $n = 0{\cdot}020$. What is the minimum right of way to be secured? Investigate the feasibility of periodic drop structures. (0·554 m; 167·7 m)

4.11. As an alternative to the natural lining of question 4.10 a trapezoidal section with concrete lining is to be investigated for which the hydraulically most favourable b/y_0 may be assumed as well as a reduced roughness coefficient of $n = 0{\cdot}014$ for the given average slope and a side slope of 1:1. What is the reduction in excavation per metre of canal? (64·8 m^3/m)

4.12. A circular conduit of 2·43 m diameter when flowing full is hydraulically more efficient for a given resistance coefficient and slope than a square section of equal area. With a free surface, the conduit will carry a still higher discharge when the depth is about 95% of the diameter. What discharge by comparison will the square section carry at 95% of full depth? What will be its discharge at 98% depth? Assume $C = \sqrt{(2g/f)} = 66$ and $S_0 = 1{:}500$. (0·98, 11·5 m^3/s)

4.13. A trapezoidal channel with a side slope 1:1 is concrete lined, so that the Manning coefficient n may be assumed as 0·013. The slope of the channel is determined by the topography as 3 ft in 10 000 ft. What width of channel and what depth is required for the hydraulically most efficient section to carry 1 200 cusec? ($b = 8{\cdot}7$ ft; $y_0 = 10{\cdot}5$ ft)

4.14. A trapezoidal channel with a bottom width of $b = 15$ ft and sides sloping 2 horizontal to 1 vertical is laid to a slope of 1:4 000. The natural lining consists of sand and gravel, for which the absolute roughness (effective hydraulic roughness) has been evaluated as $k_e = 0{\cdot}03$ ft. What is the capacity of this channel for a depth of 5 ft and of 15 ft?

What is the change of the capacity for 15 ft depth if it is calculated for this depth on the basis of the f obtained for the 5 ft depth?

What are the values of Manning's n for either case? Assume:

$$1/\sqrt{f} = 4 \log_{10} m/k_e + 4{\cdot}68.$$

(374 ft^3/s; $n = 0{\cdot}0178$; 3 540 ft^3/s; $n = 0{\cdot}0183$; 3 150 ft^3/s; -11%)

4.15. The velocity in a wide, rectangular stream of 10 m depth was measured as given in the table below. Determine the average velocity $\bar{v}$ and the discharge per unit width. Calculate the total kinetic energy passing the section as well as the momentum and determine the correction factors α and β.

Depth, y (m)	0	2	4	6	8	10
Velocity, v (m/s)	0·00	4·00	5·50	6·00	6·00	5·50

Explain the reason for the reduction in velocity near the surface. (4·93 m/s; 1·25; 1·11)

4.16. Show that the ratio of the average velocity $\bar{v}$ to the maximum velocity v_{max} for a wide rectangular channel is given by

$$\bar{v}/v_{max} = 1 - 2{\cdot}5\sqrt{(gy_0S_0)}/v_{max}$$

Hence show that the velocity v at any depth is given by

$$\frac{v}{\bar{v}} = 1 + 2{\cdot}5\,\frac{\sqrt{(gy_0S_0)}}{\bar{v}}\left(1 - 2{\cdot}3 \log_{10}\frac{y_0}{y}\right)$$

4.17. In laboratory tests, the shear τ_0 exerted on the bottom of a wide rectangular channel was measured directly by means of an isolated bottom panel as 12·0 N/m^3. The average velocity was 1·08 m/s and depth of flow was 0·61 m.

Calculate the slope of this channel and determine the velocity at points 0·20 y_0 and 0·80 y_0 from the surface. Compare the average of the two velocities with the average velocity as given. (0·002; 1·29 m/s; 0·91 m/s; 1·02)

4.18. A channel of semi-circular cross section (radius r = 10 ft) is carrying water at critical velocity when flowing half full (y_0 = 5 ft). Determine

(a) what is the discharge for this depth?

(b) what is the slope of this channel, if Manning's roughness coefficient may be assumed as $n = 0{\cdot}012$?

(c) what is the discharge for this channel when flowing full, and what is the ratio of this discharge to the critical flow for the full semi-circular channel? (654 ft^3/s; 0·001 74; 2390 ft^3/s; 0·956)

4.19. The discharge carried by a trapezoidal channel is 102 m^3/s. The cross section is given by the surface width B = 24·4 m, the bottom width b = 12·2 m and the depth y_0 = 3·05 m. For this flow

(a) determine the critical velocity v_c for this section

(b) determine the depth of flow for the alternate stage
(c) calculate the critical depth for the above rate of flow. (4·72 m/s; 1·1 m; 1·74 m)

4.20. A spillway bucket is constructed with a radius of 10 m. The stream arriving at the foot of the chute has a depth normal to the surface of 1·5 m and a velocity of 30 m/s. What is the maximum pressure exerted on the bottom assuming no submergence? (158·7 kN/m^2)

4.21. A high velocity stream characterized by a Froude number $F_1 = v_1^2/(gy_{01}) = 16$ enters a stilling pool over an abrupt drop of height. The ratio of the depth in the pool y_{02} to the depth of the stream y_{01} was fixed by design as $y_{02}/y_{01} = 5{\cdot}40$. Determine the magnitude of the drop Δz in terms of y_{01}. (0·22 y_{01}; for 0·79 y_{01} the foot will be at drop)

4.22. Water is discharged at a rate of 5·9 m^3/s per unit width into a wide, horizontal concrete channel from under a sluice gate. Due to friction the depth increases with distance from the sluice gate and changes practically linearly from 0·46 m at 9·2 m to 0·76 m at 39·6 m. A change of flow conditions further downstream causes a jump to be formed with $y_{02} = 3$ m. Where will this jump become stationary? Give a sketch of the changes in water surface profile. (29·46 m)

4.23. At one point in a very long, wide rectangular section channel, the downhill bottom slope suddenly decreases from 1:100 to 1:900. The water flow is 1·85 m^3/s per metre width. In the steeper part of the channel the Chezy C is 55 m$^{\frac{1}{2}}$/s, and in the other part 69 m$^{\frac{1}{2}}$/s. Sketch possible free surface curves for this case, and decide which of these in fact occurs. Estimate the positions, relative to the break in the bottom slope, where a normal depth occurs. U.L. (Approximately 30·8m downstream including length of jump = 6 $(y_{02} - y_{01})$; also on steep slope).

4.24. At one point in a very long, wide, rectangular channel the downhill bottom slope suddenly decreases from 1:256 to 1:2500. The water flow is 1·85 m^3/s per metre width. In the steeper part the Chezy C is 55 and in the other part 69. Sketch possible free surface curves for this case, and decide which of these in fact occurs. Estimate the positions, relative to the break in the bottom slope where normal depth flow occurs. U.L. (115 m upstream and in the channel downstream of the break).

4.25. Water is flowing at the rate of $Q = 56{\cdot}6$ m^3/s in a channel of surface width $B = 12{\cdot}2$ m and of maximum depth $y_0 = 3{\cdot}2$ m. The cross sectional area is given by the expression: $A = Ky_0^{\frac{3}{2}}$. What is the ratio of the discharge Q to the critical discharge Q_c corresponding to the same depth? What would be the ratio of the slope to the critical slope for the same depth of flow? Assume friction to be the same. (0·476; 0·227)

4.26. A very long flume has two relatively rough sections interconnected by a long smooth section. The ratio of the corresponding Chezy coefficients is

$1{:}2\sqrt{2}$. If the normal depth on a rough section is 1·22 m, sketch the water surface profile showing the three sections of the flume. Assume a constant bed slope and q = 2·8 m^3/s per metre width.

4.27. An open rectangular canal 6·1 m wide is supplying water at a velocity of 0·91 m/s to a reaction turbine developing 2·23 MW under 12·2 m net head with a turbine efficiency of 85 percent at full load. If the turbine gates are suddenly partially closed, corresponding to a sudden reduction of flow to one half of full load flow, estimate the initial depth and velocity with which the resulting surge will travel along the supply canal. State clearly all assumptions and work from first principles. U.L. (4·3 m; 5·65 m/s)

Chapter 5

5.1. Use the normalization process to derive three dimensionless numbers for dynamic similarity from the Navier–Stokes equations. Assume μ and ρ as constants and adopt a constant reference pressure (p_0). What is the physical significance of each dimensionless number?

5.2. Derive two dimensionless numbers from the small-amplitude gravity wave equation

$$C^2 = \left(\frac{\sigma}{\rho}\frac{2\pi}{L} + \frac{gL}{2\pi}\right) \tanh\frac{2\pi d}{L}$$

where C is wave speed (celerity), L is wave length, d is the depth of undisturbed liquid, ρ is the density of liquid and σ is the surface tension of liquid.

A liquid whose surface tension is one quarter that of water and whose specific gravity is 1·02 is used to simulate small amplitude wave motion in water. What should be the length ratio for dynamic similarity to be attained? (0·496)

5.3. The two-dimensional incompressible, steady state, viscous flow equation concerned with heat transfer is given by

$$\rho c\left(u\frac{\partial T}{\partial x} + v\frac{\partial T}{\partial y}\right) = K\left(\frac{\partial^2 T}{\partial x^2} + \frac{\partial^2 T}{\partial y^2}\right) + \mu\left(\frac{\partial u}{\partial y} + \frac{\partial v}{\partial x}\right)^2$$

Show that the temperature distribution in the space between coaxial cylinders is given by

$$T' = y' + \frac{\mu(R\Omega)^2}{2K\Delta T}\, y'\left(1 - y'\right) + \frac{T_i}{\Delta T}$$

where the surfaces of the inner and outer cylinders are maintained at temperatures T_i and T_0 respectively ($\Delta T = T_0 - T_i$). The annulus width a is small compared with the radius R of the outer cylinder. The outer cylinder rotates with angular velocity Ω and the inner cylinder is stationary.

5.4. For each of the following cases, list the dimensionless quantities that must have the same value in a hydrodynamic model and its prototype:

(a) incompressible, steady flow when (i) there is free surface with minor friction effects and (ii) the flow is enclosed;
(b) incompressible, unsteady flow when (i) there is free surface and (ii) the flow is enclosed;
(c) compressible, steady flow;
(d) compressible, unsteady flow.

5.5. A model study is to be made of a problem involving the effects of friction on waterhammer. What laws would you use? Investigate the feasibility of using a model with length ratio $L_r = 1/5$. What fluid would you recommend for the model?

5.6. A plane travels at 805 km/h through air at 0°F and 69 kN/m^2 absolute. A model with a scale ratio 1 to 30 is to be tested in an insulated pressurized wind tunnel. If the tunnel operates at an air temperature of 70°F and a pressure of 340 kN/m^2 absolute, what should the wind velocity in the tunnel be? (6450 km/h neglecting compressibility)

5.7. The velocity of fall under gravity of droplets in a fluid is found to be inversely proportional to the viscosity of the fluid. Show by dimensional analysis that the velocity must also be proportional to the square of the radius of the droplet.

5.8. A test is performed in a laboratory during which balls of specific gravity S_1 and diameter d are allowed to roll down straight tubes of internal diameter D inclined at angle θ to the vertical and filled with a liquid of specific gravity S and viscosity μ. The results of series of such experiments give a unique plot of $\gamma d^2 \cos\theta\,(S_1 - S)/\mu V$ against d/D, where ρ is the density of water and V is the terminal velocity of the ball. Justify the groupings of the plot and state under what conditions one would expect the relationship to be unique.

5.9. A torpedo model is to be tested in a wind tunnel to determine the drag characteristics. The scale ratio is 1:5 and the prototype speed is to be 6·1 m/s 15·5°C water. If the wind tunnel is maintained at 15·5°C and is capable of a maximum speed of 61 m/s find the density of air in the wind tunnel and comment on the feasibility of the test. (8 kg/m^3)

5.10.
(a) Why is it generally impossible to fulfil the requirements of Froude and Reynolds modelling simultaneously?
(b) Why is it often difficult or impossible to achieve exact Reynolds modelling?
(c) Why are wind tunnels often pressurized?

5.11. Eddies are produced at a frequency f behind a circular cylinder placed with its axis perpendicular to a uniform stream. What variables would affect f? Arrange them in a suitable dimensionless form.

Tests in a wind tunnel show that eddies are shed at a frequency of 120/s behind a 5·1 cm diameter cylinder when the air speed is 30·5 m/s. At what frequency would they be shed behind a bridge pier of 30·5 cm diameter standing in a river? What is the corresponding river speed? How could results of tests in air be used to predict the frequencies at other river speeds. (0·238/s; 0·36 m/s)

5.12. Explain the terms 'skin friction' and 'form drag' and indicate the circumstances under which one may be more important than the other in relation to an object standing in a following fluid.

A vertical strut is elliptical in cross section with 1:3 ratio of minor to major axes. Experiments on a model of the strut with the major axis aligned parallel to the flow gave the following results:

R (based on minor axis)	10^3	10^4	10^5	10^6	10^7
C_D	1·0	0·5	0·12	0·12	0·14

Comment on the results.

Determine the drag on the strut whose major axis is 1·83 m when in 6·1 of water ($\upsilon = 1{\cdot}1 \times 10^{-6}$ m²/s) flowing at 1·82 m/s. Would you expect any difference in the flow pattern around the strut if it had been aligned with the minor instead of the major axis parallel to the flow? U.S.T. Part I. (740 N)

5.13. A dam is modelled to a linear scale of 1/60. The prototype is designed to carry a flood of 3 200 m³/s. What is the required model discharge for the design flood? What time in the model represents 1 day in the field? (0·115 m³/s; 186 min)

5.14. A stretch of the Wiwi stream on the U.S.T. campus resembles in all physical respects a stretch of the lower Volta river. It has therefore been decided to use observations in the Wiwi to estimate the drag on a pier of the lower Volta bridge at Tefle. All physical lengths of the Wiwi studies are 1/120th of the corresponding dimensions in the Volta. The drag on a geometrically similar pier in Wiwi is measured to be 1·78 N at a flow speed at 0·37 m/s. Estimate the corresponding speed of the lower Volta and the drag on a pier of the Volta bridge. U.S.T. Part I. (4·05 m/s; 3070 kN)

5.15. A dam is to be furnished with 18 equal spillway openings of 9·2 m width. The discharge coefficient for the spillway is to be determined by model tests. A maximum flow of 113 l/s is available in the laboratory and the flood flow of the prototype is 2830 m³/s. What kind of model would you propose and what would be the scale ratio? ($L_r = 1/18$)

5.16. A harbour breakwater is to be tested for destructive wave action by means of a model. It is proposed to use 100 kN blocks, approximately cubical, having a specific gravity of 2·60 in the prototype. The model blocks which weigh 8·9 N and have a specific gravity of 2·50 are damaged if the wave height exceeds 30·5 cm. Estimate the wave height which would cause similar damage to the prototype. (6·65 m)

5.17. After the release of flood waters over the spillway of a high dam, cavitation damage was noted on baffle piers in the stilling basin. The velocity of the stream entering the stilling pool was estimated at 100 ft/s. A model of the stilling pool was built to a scale ratio of 1:25 to investigate this condition in a vacuum tank. What absolute pressure must be maintained in the tank to simulate cavitation conditions of the prototype? The atmospheric pressure at the prototype location is +32·2 ft (abs.), the vapour pressure of water for the model and prototype is +0·55 ft (abs.). (1·816 ft of water)

5.18. Discuss methods of measuring fluid velocities in hydraulic models of rivers and estuaries.

A spherical bead on a fine thread is suspended in a stream to measure water velocities by the angular deflection θ of the thread from the vertical. Determine the form of the calibration equation connecting θ with velocity; plot the curve, and comment on its usefulness and likely errors. U.L.

5.19. Discuss and compare the conditions of dynamic similarity applicable to models involving changes of water surface level, with those involving sand movement on the bed of a stream.

A model river is built and operated to scales of 1:25 vertically and 1:100 horizontally. The average sand on the bottom of the full size river is 1 mm diameter and of specific gravity 2·65, and has a settling velocity in still water of 10 cm/s. What must be the settling velocity of the model sand so that movements in the model are accurately foretold? What diameter and specific gravity must the sand grains have for similarity of flow around them while so settling? Comment on the validity of your method of solution in actual practice. U.L. (8 cm/s; 1·25 mm; 1·845) (see Chapter 7, especially Section 7.4).

5.20. Reynolds and Froude numbers represent ratios of forces influencing fluid motions. Define these force ratios and hence determine the conventional expression for each ratio. Illustrate the distinctive applications of these dimensionless numbers to engineering practice.

A ski-jump sillway is to discharge a maximum flow of 2500 m^3/s and tests are to be made on a geometrically similar model with a linear scale of 1 to 60. Determine the corresponding scales of velocity, discharge, pressure, force and time and calculate the necessary rate of water flow for the model. In what respects, if any, may the flow pattern of the full-size spillway be expected to differ from that of the model?

River models are often distorted and have their surfaces roughened artificially. Account for these practices. U.L. ($v_r = 1/7{\cdot}75$; $Q_r = 1/27\,800$; $p_r = 1/60$; $P_r = 1/216\,000$; $t_r = 1/7{\cdot}75$; $Q_m = 0{\cdot}09$ m^3/s)

5.21. A distorted rigid boundary model of a reach of a river has the linear scale ratios $X_r = 1/720$ and $Y_r = 1/72$. The average value of the hydraulic mean depth ratio is $m_r = 1/80$. In the prototype Manning's n is 0·020. Find the corresponding n for the model. What are the discharge and time ratios? (0·029; $2{\cdot}3 \times 10^{-6}$; 0·0118)

5.22. A study is to be made of backwater effects of bridge piers in a trapezoidal flood channel. The water depth is 9·15 m, the bottom channel width 36·6 m, and the slopes are 1 vertical on 2 horizontal. The mean velocity is 1·22 m/s and $n_p = 0{\cdot}014$.
(a) Consider an undistorted model with $X_r = 1/60$.
(b) Consider a distorted model with $Y_r = 1/60$ and $X_r = 1/240$.
Which of the two would you recommend? Give model dimensions, material to be used, and quantity of water for model operation. (Undistorted: $n_m = 0{\cdot}0071$; $v_m = 0{\cdot}158$ m/s, $Q_m = 21{\cdot}9$ l/s; Distorted: $n_m = 0{\cdot}0106$; $v_m = 0{\cdot}158$ m/s; $Q_m = 5{\cdot}5$ l/s)

5.23. A study is to be made of the passage of flood waves in a reach of a river. The cross section of the river may be represented by a parabola (vertex at the middle of the channel). At flood stage the maximum depth is 7·33 m and the surface width is 122 m. The roughness of the channel may be represented by $n_p = 0{\cdot}025$.
(a) Calculate the discharge ratios and the model roughnesses for the following models assuming similarity according to both gravity and resistance forces:
(i) $X_r = 1/100$, $Y_r = 1/25$; (ii) $X_r = 1/400$, $Y_r = 1/40$.
(b) After passage of the flood wave the maximum channel depth is 3·66 m. Calculate the error in the discharge ratios computed in (a) for each model.
((a) (i) 0·000 08; 0·027; (ii) 0·000 01; 0·032; (b) Q_r(full)/Q_r(half) = 1·1)

Chapter 6

6.1. A small impulse water turbine is to drive a generator for a 50 cycles per second power supply. The generator has four pairs of poles. The available water head is 61 m and the discharge is 57 l/s. The table below gives the relationship between specific speed N_s and the speed factor, $\phi = u/\sqrt{(2gH)}$ for the most efficient operation of the turbine. u is the bucket speed.

N_s	7·6	11·4	15·2	19·0	22·8
ϕ	0·47	0·46	0·45	0·44	0·433

Determine the ratio of the mean diameter of the wheel to the jet diameter. Assume $C_v = 0{\cdot}98$ for the nozzle and that the overall efficiency is 82%. What is the head loss in the nozzle? U.S.T. Dip. Mech. (8·25; 2·44 m)

6.2. A centrifugal water pump has an impeller with $r_2 = 30{\cdot}6$ cm, $r_1 = 10{\cdot}2$ cm, $\beta_1 = 20°$, $\beta_2 = 10°$. The impeller is 5·1 cm wide at $r = r_1$ and 1·9 cm wide at $r = r_2$. For 1500 rev/min neglecting losses and assuming the vane thickness to cover 8% of peripheral passage ways, determine
(a) the discharge for shockless entrance $\alpha_1 = 90°$;
(b) α_2 and the theoretical head H;
(c) the power required at 90% hydraulic efficiency;

(d) the pressure rise through the impeller and
(e) the specific speed for an efficiency of 80%.
(175 l/s; 15·7°; 92 m; 175 kW; 73 m; 25 rev min^{-1} s$^{-\frac{1}{2}}$ m$^{+\frac{3}{4}}$)

6.3. A Kaplan axial flow water turbine with 915 cm blade tip diameter and 305 cm boss diameter is to run at 60 rev/min developing 22·4 MW. under 6·1 m head. Assume a hydraulic efficiency of 90% and neglect mechanical losses. What must be the angles between the plane of rotation and the tangents to the blades at the entry and exit to the rotor, both at the tips and at the boss? State all the assumptions made.

If half the power is required at the same head and speed, by how much must the blades be twisted by the governor so that the efficiency is preserved? U.L. (At blade tip: β_1 = 14°50′; β_2 = 13°56′; at boss: β_1 = 60°56′; β_2 = 36°40′; angle of twist = 11° Using mean of values at tip and boss.)

6.4. An axial flow pump, having 1 ft tip and 3 in hub diameters, delivers 2000 gall/min of water when running at 1450 rev/min, the specific speed being 11 550 rev gal$^{\frac{1}{2}}$ min$^{-\frac{3}{2}}$ ft$^{-\frac{3}{4}}$.

The hydraulic efficiency of the pump is 91%. The water enters the impeller without whirl and the velocity of flow can be assumed constant throughout the impeller. The whirl component of the absolute velocity at the exit from the impeller forms a free vortex. Calculate the impeller blade angles at inlet and exit for the hub and tip diameters. Assume that the relative velocity of the water is tangential to the blade. (Entry: 5·5° (tip); 20·9° (hub); exit: 5·8° (tip); 86·8° (hub)).

6.5. In a Francis-type turbine the inlet angle of the moving vanes is 70° and the outlet angle 20°. Both fixed and moving vanes reduce the area of flow by 15%.

The runner is 61 cm outside dia. and 40·6 cm inside dia. and the widths at entrance and exit are 5·1 cm and 7·6 cm respectively. The pressure at entry to the guides is + 26·5 m head and the kinetic energy there can be neglected. The pressure at discharge is −1·83 m head.

If the losses in the guide and moving vanes can be neglected, determine
(a) the speed of the runner in rev/min for tangential flow on to the runner vanes;
(b) the power given to the runner by the water. (493 rev/min; 84 kW)

6.6. A centrifugal pump is required to deliver 6300 gal/min against a head of 20 ft at a speed of 600 rev/min. Assuming that all the velocity head is lost, and that the actual head is 75% of the theoretical head, find the diameter and breadth of the impeller at outlet. The velocity of flow, taken as constant is 10 ft/s and the blades are curved back 30° to the tangent at outlet. Also determine the inlet blade angles, if the inlet diameter is made half the outlet diameter. (d = 1·25 ft; b = 0·43 ft; β = 27°)

6.7. A pipeline joining a lake and a reservoir whose water levels differ by 91·4 m has a diameter of 45·7 cm and a length of 6100 m and the friction factor f is

0·008. The gravity flow is inadequate and it is necessary to install a pump half-way along the pipeline to act as a booster. The pump equation is $H = 183 - 760Q^2 + 32{\cdot}5Q$ where H is the head developed by the pump in metres and Q the flow in m^3/s. Estimate:

(a) the percentage increase in the flow due to the pump;

(b) the daily operating cost of the pump, if the pump overall efficiency is 70 per cent, and the cost of power is 2 pesewas per kilowatt-hour (kWh). U.S.T. Part II. (26%; ¢ 188/day)

6.8. Tests on a single centrifugal pump running at 600 rev/min gave the results indicated below. This pump is to be used to deliver 2·72 m^3 of water per minute continuously against a head (including friction) of 15·2 m.

At what speed should it be driven, and what input power will be required? If it is required later to run the pump at maximum efficiency under a head of 18·3 m, what input power will be required and at what speed should it be run?

Q (m^3/min)	0	0·45	0·91	1·36	1·59	1·82	2·04	2·28
Head H(m)	15·2	16·3	16·3	14·6	12·8	10·4	7·6	4·6
η (%)	0	30	61	81	85	80	67	47

U.S.T. Part II. (820 rev/min; 9·7 kW; 6·8 kW; 720 rev/min)

6.9. A model centrifugal pump ¼ full size is tested under a head of 7·6 m and 500 rev/min. It was found that 7460 W. was required to drive the model.

Calculate the speed and power required by the prototype when the head is 44 m. What is the ratio of the quantities pumped by the large pump and by the model under these conditions. (300 rev/min; 1650 kW; 38:1)

6.10. At a particular dam site the flow available is 20 000 cusec under a head of 30 ft. Find the least number of turbines required, all of the same size, if their synchronous speed is 100 rev/min, and their overall efficiency is 80% in order to develop fully the potential hydro-power. For the turbines N_s = 100 rev/min^{-1} h.p.$^{\frac{1}{2}}$ ft$^{-\frac{5}{4}}$. (11)

6.11. A model is to be built for studying the performance of a turbine having a runner diameter of 3 ft and maximum output 3000 h.p. under a head of 164 ft. What is the appropriate diameter of the runner of the model if the corresponding power is 12 h.p. and the available head is 25 ft? Show that the corresponding model speeds will be about 50% greater than those of the prototype. (0·78 ft)

6.12. A water turbine to develop 37·3 MW. under 24·4 m head is to run at 75 rev/min and 90% efficiency. It is desired to make a 1/20 scale model to run on compressed air at 5 atmospheres, the pressure difference across the model being 34·5 kN/m^2. What will the model speed be? If the model is estimated to have an efficiency of 85% what power will it generate, and what will be the air flow? What are the advantages and disadvantages of using compressed air as a testing fluid. (7 600 rev/min; 60 kW; 2·04 m^3/s)

6.13. Write a note (about 300 words) on the nature and effect of cavitation.

Give sketches showing the form now commonly adopted for the discharge pipe of a turbine and explain why this reduces the risk of cavitation as compared with the older forms of draft tube.

Explain the use of the Thoma cavitation parameter for a pump, $\sigma = (h_a - h_v - h_s')/H$.

	Atmospheric pressure	*Vapour pressure corresponding to water temperature*	*Suction lift including friction in suction pipe*	*Overall head*
Large pump	14·7 lbf/in^2	0·5 lbf/in^2	10 ft	220 ft
Model	14·5 lbf/in^2	0·48 lbf/in^2	x ft	160 ft

Calculate the static height x, including friction, of the model above the suction tank level so that the suction conditions of the two pumps correspond. (15·8 ft)

6.14. A propeller pump on test at its designed speed of 1450 rev/min gave the following results:

H (ft)	29	20	13·5	17·0	12·5	8·0
Q (gal/min of water)	0	250	500	750	1 000	1 250
η (%)	0	20	35	60	69	64

Plot on squared paper an estimated head-quantity curve for the pump running at 1000 rev/min. Explain the significance of the kink in the curve.

The pump is to run continuously at its design speed and is to deliver water through 40 ft of pipe against a static lift of 5 ft. The only pipes available are 6 in, 8 in and 10 in diameter all with $f = 0{\cdot}007$. Which will be the most suitable pipe for this duty? Estimate the horse power required to drive the pump and calculate the specific speed of the pump at maximum efficiency. U.L. (6 in pipe; 5·5 h.p.; 6 890 rev min$^{-\frac{3}{2}}$ ft$^{-\frac{3}{4}}$ gal$^{\frac{1}{2}}$)

6.15. What is the specific speed of a rotodynamic pump which is rated at 18·2 m^3/min under a head of 27·4 m at 1760 rev/min? Determine the head and capacity of this pump if operated at 1480 rev/min at rated efficiency. Specify in each case the maximum tolerable suction lift for critical cavitation number of 0·40. Take vapour pressure of water as 3447 N/m^2 absolute. (81 rev/min^{-1} s$^{-\frac{1}{2}}$ m$^{-\frac{3}{4}}$; 19·4 m 15·3 m^3/min; −1·01 m; 2·19 m)

6.16. What should be the specific speed of the pump required to lift 20 ft^3/s of water through 10 000 ft of 24 in diameter pipe with $f = 0{\cdot}005$? The speed of rotation is 1750 rev/min.

What would be the specific speeds if two identical pumps were used in series and in parallel? (6900; 11 500; 4870)

6.17. Calculate the power output in kilowatts from a turbine (η = 90%) supplied with 7·08 m^3/s of water. The level of water in the supply reservoir is 29 m above the tailrace and water is supplied through a 1·52 m diameter pipe 305 m long and having an average roughness of 0·9 mm. If the inlet to the draft tube which is vertical and changes from 1·83 to 3·05 m diameter in a length of 5·5 m is 4·59 m above tailrace, calculate the absolute pressure head at the inlet to the draft tube. Would you expect cavitation in the turbine under these conditions if σ_c = 0·03? (1640 kW; 5·5 m; $\sigma = 0{\cdot}0077 < \sigma_c$)

6.18. A pump delivers water to reservoir B from reservoir A. The pump is 6 ft above the level of water in both reservoirs. At a speed of 1200 rev/min the rate of water discharged to B is 90 gal/min and the losses (including inlet and exit losses) in the suction and delivery pipes are 8 ft and 30 ft respectively. How much can the speed and discharge be increased before cavitation occurs for σ_c = 0·045? Assume that the pump works at a constant maximum efficiency and take p_v = 0·84 ft of water. What type of a pump would you recommend for the above job and why? Head loss = constant x Q^2. U.S.T. Part III. (Q = 151 gal/min; N = 2200 rev/min; n_s = 738)

6.19. The tables below give the characteristics for two pumps A and B. Construct a head–discharge–efficiency graph for the pumps when connected in parallel and in series.

Pump A

H (ft)	70	60	55	50	45	40	35	30	25	20
Q (ft³/s)	0	2·0	2·56	3·03	3·45	3·82	4·11	4·38	4·59	4·73
η (%)	0	59	70	76	78	76·3	72	65	56·5	42

Pump B

H (ft)	80	70	60	50	40	30	20
Q (ft³/s)	0	2·60	3·49	4·96	5·70	6·14	6·24
η (%)	0	54	70	80	73	60	40

Chapter 7

7.1. Find the median size D_{50}, the geometric standard deviation σ_g and the uniformity coefficient for the soil sample A the results of whose sieve analysis are given below.

Sieve opening (mm)	*Grammes retained*	
	A	*B*
0·495	0·85	–
0·417	1·56	0·16
0·351	3·88	0·74
0·295	3·82	2·71
0·246	5·35	8·39
0·208	5·69	9·61
0·175	4·31	21·92
0·147	5·06	17·32
0·124	2·37	22·41
0·104	1·16	9·98
0·088	0·21	6·32
0·074	0·12	2·22
0·061		0·66
Pan	0·04	–

(0·23 mm; 1·50; 1·65)

7.2. Composite sediment is made consisting of two parts of sediment A and one part of B. From the frequency distribution of the size diameters of the composite sediment, find the median diameter, the geometric standard deviation and the uniformity coefficient. (0·18 mm; 1·56; 1·76)

7.3. Find the ratio of the fall velocities in water at 24°C of two sand particles with a sieve diameter of 0·8 mm, one having a shape factor of 0·7 and the other being spherical. (0·85)

7.4. Plot the frequency distribution of the fall velocities of sediment A assuming a shape factor of 0·7 for all particles and 20°C water temperature. Determine the median fall velocity w_{50} and the standard deviation of fall velocities σ_w. How does σ_w compare with σ_g?

Repeat the exercise for sediment B and the composite sediment of question 7·2. Do the results suggest any definite pattern of relationship between σ_w and σ_g? (Sample A: w_{50} = 2·7 cm/s, σ_w = 1·84, σ_w/σ_g = 1·23)

7.5. The depth of flow of a certain river ranges from 1·22 to 3·66 m and the discharge per unit width from 0·74 to 3·25 m^3/sm. The width is 122 m. Taking the average sand sediment size to be approximately equal to the median size D_{50} = 0·246 mm, prepare a stage-discharge curve assuming that the Strickler formula holds. Also determine the bed load transport rate using the DuBoys–Straub formula and then the Shields' formula and plot the results together.

7.6. The critical shear stress for sand movement was determined in a laboratory experiment using a wide rectangular channel with a fixed slope and 1·0 mm mean sediment size to be 0·033 lbf/ft^2 at a 3 inch flow depth. Assuming the value of Manning's n to be 0·015, estimate the sediment transport rate when the discharge in the flume is doubled. You may assume that the DuBoys formula is valid. (q_b/q = 0·042% by weight)

7.7 Two wide canals are to be dug through an alluvial deposit. One is to have a slope of 0·000 2 and the other 0·000 3. If the expected depth of flow is 3·05 m in the first canal what should it be in the second in order to limit the sediment transport rate to not more than 10% over that in the first canal? The mean sediment size is 0·244 mm. Use Shield's transport equation. (2·16 m)

7.8. The table below gives the velocity distribution and the suspended sand distribution in a wide river. On the day of measurement the depth was 7·8 ft and the flow was approximately uniform. The mean particle size was 0·105 mm and the water temperature was 20°C. Plot the velocity profile on a semi-logarithmic graph paper and concentration profile on log-log paper (c vs. $(y_0 - y)/y$) and determine shear velocity, slope of bed, mean velocity, Darcy friction factor, Z (the exponent in the suspended load equation) and β (in $\epsilon_s = \beta\epsilon$).

Determine the rate of transport of this particular size of sand in lbf/ft s by graphical integration of the product vc.

y (from bottom) (ft)	v (ft/s)	c (g/l)
0·7	4·30	0·411
0·9	4·50	0·380
1·2	4·64	0·305
1·4	4·77	0·299
1·7	4·83	0·277
2·2	5·12	0·238
2·7	5·30	0·217
2·9	5·40	–
3·2	5·42	0·196
3·4	5·42	–
3·7	5·50	0·184
4·2	5·60	–
4·8	5·60	0·148
5·8	5·70	0·130
6·8	5·95	–

(0·286 ft/s; 0·00033; 5·31 ft/s; 0·0063; 0·358; 0·64; 0·48 lbf/ft s)

7.9. You plan to use a distorted model to investigate bed erosion in a reach of the Prah river. Laboratory conditions limit your scales to $X_r = 1/150$ (horizontal) and $Y_r = 1/50$ (vertical). The following prototype conditions are to be simulated:

S_0 = bed slope = 0·0004
v = velocity of stream = 2·14 ms
y_0 = depth of river = 3·64 m
B = average width of river = 15·2 m
d = average diameter of bed sand = 0·32 mm

What diameter of the same type of sand is to be used in your model to achieve similarity of sediment transport? How much water will be circulated for your model tests?

State the main assumptions you may make or imply in your solution. U.S.T., 1968, Part III. (0·09 mm; 2·25 l/s)

7.10. A model river is to be built and operated to scales 1 : 100 horizontally and 1 : 25 vertically. The average sand on the bottom of the full size river whose bed slope is 0·000 3 and average depth is 3·05 m when flowing at 1·84 m/s is 1·0 mm and specific gravity 2·65. What must be the size of the model sand (same specific gravity) to achieve similarity of sediment movement? Is the model feasible? (0·1 mm; feasible; $w/v' = 2{\cdot}17$)

Chapter 8

8.1. Explain briefly what is meant by the hydrological cycle. Sketch a typical hydrograph of storm run-off and indicate how the recession curve can vary with the nature of the catchment area.

A reservoir has a constant plan area of $5{\cdot}56 \times 10^6$ m². Discharge from the reservoir takes place over a spillway the characteristic of which is $Q = 135 H^{\frac{3}{2}}$ m³/s. Storm run-off entering the reservoir is estimated at the following figures:

Time (h)	0	6	8	10	12	14	16	18	20
Run-off (m³/s)	0	113	170	283	226	170	113	56·5	0

It is noted that at 12 h the head H over the spillway is 0·813 m and the reservoir surface level is still rising. Estimate how much further the level will rise. Sketch the general shape of the outflow hydrograph and explain what is meant by the lag effect of the reservoir. U.L. (0·154 m)

8.2. Describe very briefly how to obtain a reliable estimate of the necessary reservoir storage of a hydroelectric scheme, allowing for both long-term and annual storage.

The flow (in m³/s) of a tropical river over 17 months (each of 30 days) was:

April	M	J	J	A	S	O	N	D	J	F	March
8·7	18·8	79·1	104·4	240	249	218	65	28·3	22·6	18·8	8·5
41·6	70·8	147	218	164							

A hydroelectric plant using a net head of 26 m, and developing 15·6 MW. at 138·5 rev/min and 86% efficiency is to take water from a reservoir on this river. An exceptionally wet winter preceding the first April would have left the reservoir full. What reservoir capacity would be necessary to supply 15·6 MW. continuously? What volume of water will escape over the spillway in the above period? How many turbines having a specific speed of 298 (SI units) will be required? U.L. ($0{\cdot}68 \times 10^9$ m³; $1{\cdot}23 \times 10^9$ m³; one)

8.3. A flood entering a reservoir of area 0·13 km^2 has the following hydrograph:

Time (h)	0	2	4	6	10	14	20	30
Flow (m^3/s)	0	14·1	42·5	85·0	70·8	42·5	14·1	0

The dam of the reservoir has a spillway 61 m long (coefficient 2·21 m$^{\frac{1}{2}}$/s), and a culvert (ground sluice) with its centre line 30·5 m below crest level. At time zero the water level in the reservoir was just at crest level.

Estimate

(a) the maximum discharge through the culvert, if the water is never to rise more than 30·5 cm above the crest;
(b) the size of the culvert, assuming it acts as a simple orifice, C_d = 0·64
(c) the period during which the culvert is flowing. U.L. (62·4 m^3/s; 2·25 m; 15·5 h)

8.4. A hydroelectric station is fed from a reservoir whose surface area of 81 ha may be assumed constant within the normal range of variation of top water level. The supply from the catchment is increased by diverting 14·1 m^3/s from streams in adjacent catchments. The turbines require a constant flow of 28·3 m^3/s and any excess inflow to the reservoir is discharged over a spillway 61 m long.

It is anticipated that a catastrophic flood could linearly increase the inflow from the catchment from a base flow of 14·1 m^3/s to 141 m^3/s within 4 h. This peak would be maintained for two hours, and the recession curve would fall at a constant rate to 28·3 m^3/s after a further 4 h. Assuming the reservoir is full to spillway crest level when the storm commences, determine the height of the dam to provide a minimum free board of 61 cm. The weir constant may be taken as 1·85 m$^{\frac{1}{2}}$/s. (1·63 m above spillway crest)

8.5. Sketch a typical hydrograph of storm run-off from a catchment area and discuss how the shape varies with characteristics of the catchment area.

Given below are the observed flows from a storm of 3 hours duration at a gauging station where the drainage area is 310 km^2.

Time (h)	0	3	6	9	12	15	18	21	24	27	30
Flow (m^3/s)	28·3	56·5	158·3	248·6	222·3	182·5	141·1	101·7	62·1	39·6	28·3

Derive the 3 h unit hydrograph assuming a constant base flow of 28·3 m^3/s. Discuss how you would use your result to estimate the hypothetical design flood from the catchment caused by a major 3 h storm.

8.6. The inflow to a reservoir, of constant surface of 200 acres, is increased by diverting a continuous flow of 200 cusec from a neighbouring catchment. The spillway is 180 ft long with a coefficient of 3·34 ft$^{\frac{1}{2}}$/s; and a constant flow of 800 cusec is taken from the reservoir for power production. Assuming that the worst flood expected from the catchment takes the form shown below, and the reservoir is full to spillway crest level when the storm commences, estimate the difference in elevation between the top of the dam and the spillway crest to give a free board of 2 ft.

Discuss other methods of discharging flood water from a reservoir, with their advantages and disadvantages.

Time (h)	0	1	2	3	4	5
Inflow (cusec)	600	2200	3800	3800	2600	1400

(4·5 ft approx.)

8.7. The monthly flow rate for a proposed hydroelectric project is given below. If each month has 30 days, determine the power that can be produced continuously if the available head is 30·5 m and the overall efficiency of the plant is 80%. Find the total quantity of water wasted during the year. The average area of the reservoir is 970 ha, its useful capacity is 24·6 x 10^6 m^3 and it is assumed to be full at the beginning of the first month.

Month	J	F	M	A	M	J	J	A	S	O	N	D
Mean flow (m^3/s)	7·6	8·8	6·9	3·5	2·8	2·3	1·7	2·0	4·8	8·8	9·7	7·9

Assume an evaporation loss of 15 cm/month and a mean monthly seepage loss of 0·28 m^3/s.

If the cost of installation is 360 000 cedis and the total annual charge (fixed operating costs) is 10%, estimate the cost of power in pesewas/kWh. (810 kW; 65 x 10^6 m^3; 0·50 p/kWh)

8.8. A drainage basin is related to a hydrometric station, where from long term observations the mean annual discharge is obtained as 2806 ft^3/s.

From an isohyetal map of the region the following areas closed by isohyets and the divide of the drainage basin were determined

Isohyets (in)	*Area* (ml^2)	*Isohyets*	*Area* (ml^2)
36–38	108	44–46	608
38–40	186	46–48	254
40–42	274	48–50	156
42–44	412	50–52	26

Compute:

(a) mean annual precipitation depth over the basin;
(b) mean annual run-off depth;
(c) run-off coefficient (run-off/precipitation);
(d) mean annual depth of evaporation. (43·4 in; 18·9 in; 44%; 24·4 in)

8.9. In the 30-year period 1930–59 the number of peak river discharges causing flooding exceeded the number of years of observation. The 30 highest from all registered peak discharges are given (in ft^3/s) as:

45 300;	38 200;	37 800;	37 000;	36 000;	30 800;
29 200;	29 000;	28 400;	24 600;	22 200;	21 400;
20 000;	20 000;	19 000;	17 800;	17 000;	16 800;
16 600;	16 000;	15 800;	15 000;	14 600;	14 200;
14 100;	14 000;	13 900;	13 500;	13 100;	13 000;

Determine the floods of the return periods 1, 5, 10, 20 and 25 years.

8.10. The table below refers to discharges through a river due to a 24-h precipitation on the basin above the point of measurement. The drainage area is 3810 sq. km. What is the direct run-off volume in mm over the drainage area?

Derive the unit hydrograph. Estimate the peak flow through the river due to four successive 24-h rainfalls whose run-off values are 10·2, 17·9, 5·1 and 30·5 mm respectively.

Date	*Hour*	*Flow* (m^3/s)	*Date*	*Hour*	*Flow* (thousand ft^3/s)
18 July	2400	42·5	22 July	1200	198
19 July	0600	42·5		2400	161
	1200	45·6	23 July	1200	133
	1800	56·8		2400	111
	2400	87·7	24 July	1200	90·5
20 July	0600	147		2400	79·3
	1000	209	25 July	1200	68·0
	1200	272		2400	62·3
	1800	354	26 July	1200	56·6
	2400	342		2400	53·7
21 July	0600	314	27 July	2400	48·0
	1200	289	28 July	2400	42·5
	1800	263	29 July	2400	39·7
	2400	240			

(18·2 mm; 760 m^3/s)

8.11. Derive the Muskingum equation for flood routeing in a river basin. Explain clearly the significance of the factors x and K and indicate how they can be estimated.

The inflow hydrograph for a reach of a river is given below. Find the peak of the outflow hydrograph and its time of occurrence if $K = 1{\cdot}4$ days and $x = 0{\cdot}20$ for the reach. Plot the complete inflow and outflow hydrographs for the given flood. Assume that the outflow from the reach is equal to the inflow at the start of the flood.

Date	Hour	Inflow (m^3/s)
8 February	1200	32·4
9 February	0600	32·4
	1200	35·3
	1800	46·1
	2400	80·3
10 February	0600	141
	1200	236
	1800	342
	2400	323
11 February	0600	301
	1200	279
	1800	256
	2400	229
12 February	0600	204
	1200	189
	1800	178
	2400	156
13 February	1200	122
	2400	101
14 February	1200	59·3
	2400	53·1

8.12. The table below gives the inflow and outflow hydrographs for a reach of a river. Determine the Muskingum factor K for the reach if x has been estimated fairly accurately to be 0·20. How does your value compare with the time lag between inflow and outflow peaks?

Date	Hour	Inflow (m^3/s)	Outflow (m^3/s)
9 February	0600	42·4	42·4
	1200	45·3	41·9
	1800	56·6	41·2
	2400	87·8	40·3
10/2	0600	147	42·6
	1200	272	48·0
	1800	352	82·9
	2400	340	138
11/2	0600	314	181
	1200	289	208
	1800	265	227
	2400	241	237
12/2	0600	212	241
	1200	198	237
	1800	182	230
	2400	161	222
13/2	1200	133	200

Chapter 9

9.1. The concrete dam shown in Fig. P9.1 rests on a homogeneous, isotropic soil of infinite depth, and $k = 10^{-3}$ cm/s

(a) Sketch the flownet and discuss how the quantity of seepage per unit width of dam may be estimated.

(b) Determine the pressure distribution on the base of the dam.
(c) With h_1 = 8 m and h_2 = 3 m plot a graph showing the variation of the seepage gradient behind a 15 m wide dam.

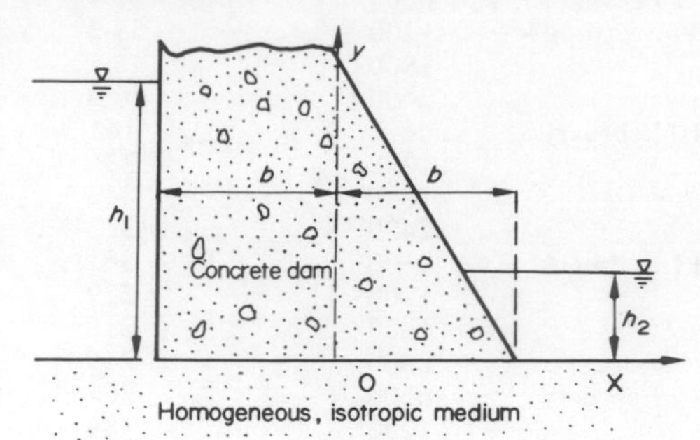

Fig. P9.1

9.2. Draw the flownet for seepage through the body of pervious soil shown in Fig. P9.2 and estimate the seepage rate. Use four flow channels and a scale of 15 ft = 1 in. $k = 8 \times 10^{-3}$ cm/s.

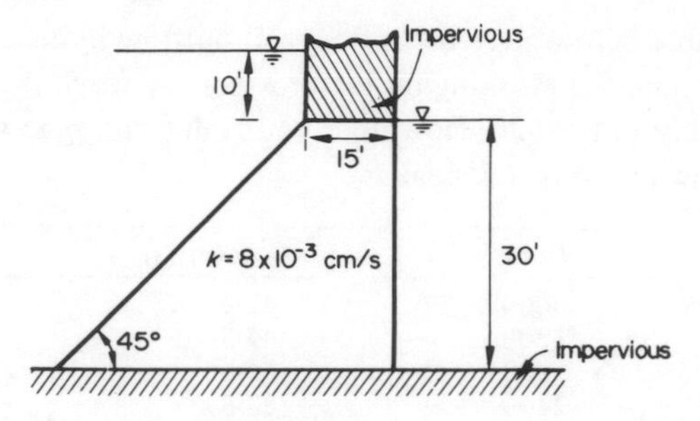

Fig. P9.2

9.3. Steady state, two-dimensional flow is occurring into the double row of sheet piles shown in Fig. P9.3. Draw the flownet and compute the rate of flow per metre of wall length. Determine the maximum exit seepage gradient and the corresponding factor of safety against piping. Plot to scale the water pressure along both sides of the sheet pile wall.

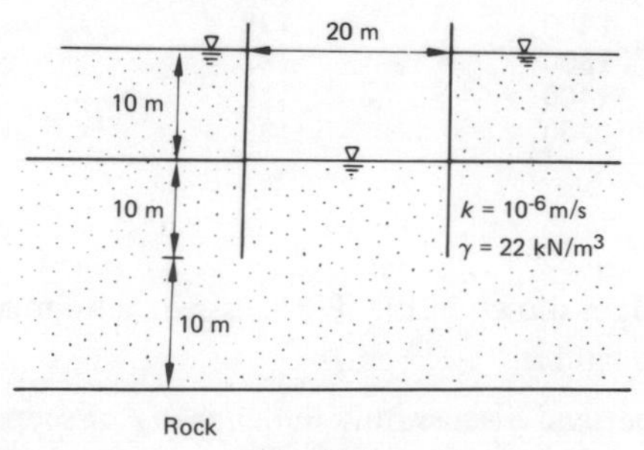

Fig. P9.3

9.4. If in Fig. P9.3 one sheet pile wall is removed and the water levels on both sides of the remaining pile are maintained as in the figure, draw the new flownet and determine the rate of seepage. Determine the new maximum exit seepage gradient and where it occurs. What is the new factor of safety against piping?

9.5. The ground levels on the two sides of a vertical retaining wall are 64 m and 60 m respectively and the foot of the wall which is 60 cm wide and semi-cylindrical is at 59 m. The soil rests on an impervious clay bed at level 57 m. Sketch the flownet approximately to scale when the soil is completely saturated. Use your sketch to estimate whether there is a risk of the ground lifting on the lower side. What is the leakage under the wall if the permeability of sand is $0{\cdot}30 \times 10^{-3}$ m/s?

9.6. The Laplace equation in cylindrical coordinates is given by

$$\frac{1}{r}\frac{\partial}{\partial r}\left(r\frac{\partial h}{\partial r}\right) + \frac{1}{r^2}\frac{\partial^2 h}{\partial \theta^2} = 0$$

Assuming purely radial flow, solve the above equation for h. Using the Darcy equation for the radial velocity find the total rate of flow to a well in a confined aquifer. Discuss the boundary information needed to obtain a numerical solution.

9.7. In the soil profile shown in Fig. P9.4 steady vertical seepage is occurring. A piezometer whose sensing element is at elevation −30 m has water standing to elevation −12 m.

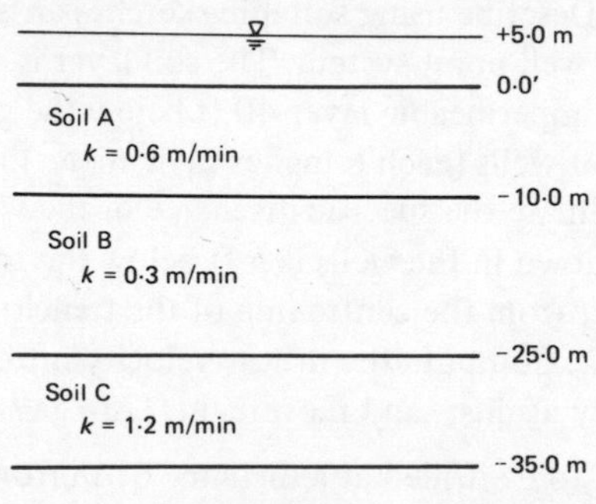

Fig. P9.4

Taking the datum elevation as zero, make a scaled plot of elevation versus total head, pressure head, and elevation head.

9.8. Show that by adopting a coordinate $z' = (k_x/k_z)^{\frac{1}{2}}\, z$ a two-dimensional flow in an anisotropic medium can be converted into an equivalent two-dimensional flow in an isotropic medium. Calculate the equivalent average coefficient of permeability $\bar{k}_e = (\bar{k}_x \bar{k}_z)^{\frac{1}{2}}$ for the soil profile shown in Fig. P9.4 and the corresponding scaling factor which converts it into an equivalent homogeneous isotropic soil medium taking the z axis as vertical. The permeability in the horizontal x direction is twice that in the vertical direction for all the soils. (0·786 m/min; 1·66)

9.9. Figure P9.5 shows the drawdown curve for an extensive confined aquifer from which water is pumped at a steady rate. Derive an expression for determining the radial distance r_0 from the well at which the drawdown may be assumed as zero.

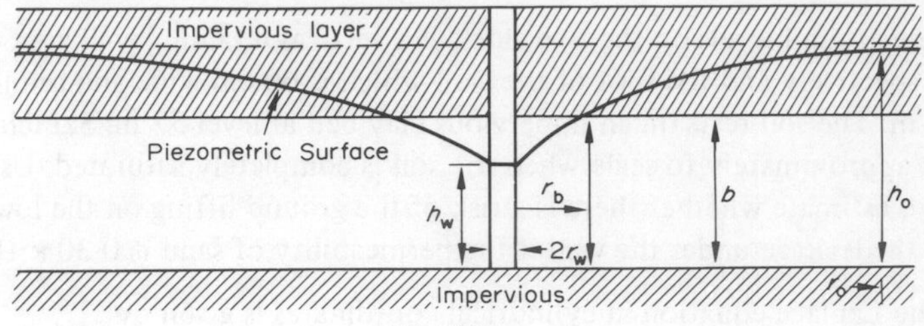

Fig. P9.5

Water is pumped from such a well at a steady rate of 137 m^3/day. If h_0 = 7·6 m at r_0 = 152 m, calculate the point where the piezometric surface intersects the upper boundary of the aquifer assuming b = 6·1 m and k = 7·3 m/day. What is the corresponding level of water in the well whose diameter is 15·2 cm? (6·98 m; 3·78 m)

9.10. A trench 100 ft long and 3 ft wide at the bottom is to be excavated to a depth of 8 ft in an area where the water table is 3 ft below the ground surface. It is required that the water table be temporarily lowered by 1½ ft below the bottom of the trench. Describe using suitable sketches an arrangement that can be employed using the well point system. The soil layer is sandy with k = 0·008 ft/s and is underlain by an impermeable layer 40 ft below the ground surface. Suppose a line of five equidistant wells (each 6 inches in diameter) is used on one side of the excavation and estimate the specific discharge of the wells in gal/min. Assume that the drawdown in the wells is 5 ft below the trench bottom and that the line of wells is 10 ft from the centre line of the trench. You may assume $r_0 = 1650\,(h_0 - h_w)\sqrt{k}$. Compute the inflow velocity into the wells and evaluate the coefficient of safety against sand movement. (234 gal/min; 0·0148 ft/s; 0·54)

9.11. A 15 cm well is to be drilled at a distance 61 m from the shore of a deep lake, in a zone where the lake penetrates a permeable layer (k = 1·52 cm/min) as in Fig. P9.6. The maximum drawdown in the well is to be 6·1 m.

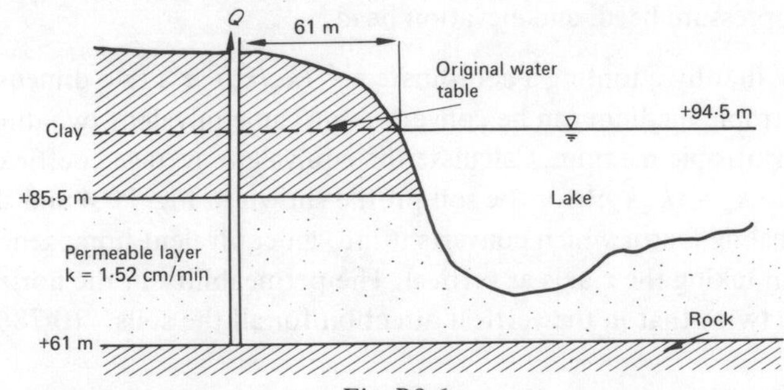

Fig. P9.6

(a) Estimate the discharge of the well in m^3/h.
(b) Compute the drawdown at a point midway between the well and lake and 30·5 m from the line joining them.
(c) Sketch the plan view flownet and indicate clearly the method of obtaining it.
(115; 0·41 m)

9.12. An earth embankment resting on an impervious rock foundation slopes at 1·0 vertical to 1·5 horizontal at its front and back. The base is 370 ft wide. Water is impounded to 100 ft above foundation level on one side with zero tailrace on the other. It is estimated that the potential slip surface has a radius of 125 ft and its centre is in a vertical plane 50 ft from the back heel of the embankment. Using an appropriately drawn flownet show the variation of the pore pressure along the slip circle and determine its maximum value and point of occurrence.
(55 ft approx. about 100 ft from heel)

Chapter 10

10.1. The average wind speed in the central regions of Ghana at certain times of the year may be taken as 8 mph blowing in a direction along the length of the Volta lake which is approximately 200 miles long. Estimate the significant height and period of the waves generated in the Volta lake at those times of the year. Specify the wind duration and average depth of the lake which would make your estimates valid. What other conditions could make your estimates erroneous?
($H_{1/3}$ = 1·1 ft; $T_{1/3}$ = 3·7 s; $t > 35$ h; $d > 35$ ft)

10.2. The velocity potential ϕ for small amplitude wave motion in the positive x direction is given by

$$\phi = \frac{ag \cosh [k(d+z)]}{\sigma \cosh (kd)} \cos (kx - \sigma t)$$

where z is measured vertically upward from the mean water level. If pressure is related to ϕ by

$$p/\rho = \frac{\partial \phi}{\partial t} - gz$$

show that the pressure distribution due to the propagation of a small amplitude progressive wave is

$$p = \rho g \left[\frac{\eta \cosh [k(d+z)]}{\cosh (kd)} - z \right]$$

Sketch the variation of pressure (relative to hydrostatic conditions) with depth under the crest and trough of a small amplitude wave.

10.3. A lake is 15·2 m deep. A small amplitude progressive wave of 61 cm height and 24·4 m length passes through it. Calculate the range of pressure fluctuations indicated by pressure sensing instruments located at 3·05 m and 6·1 m depths. (2·84 – 3·12 N/cm²; 5·87 – 6·05 N/cm²)

10.4. A two-component deep water wave system is given by

$$\eta_t = a_1 \sin (k_1 x - \sigma_1 t) + a_2 \sin (k_2 x - \sigma_2 t)$$

A boat travels with the speed of the first component wave. Show that the wave pattern relative to the boat is given by

$$\eta_{tb} = 3 \sin (0{\cdot}436\, \sigma_1 - \sigma_2)\, t$$

when $a_1 = 0{\cdot}6$ m, $L_1 = 24$ m, $a_2 = 1{\cdot}0$ m and $L_2 = 55$ m.
Simplify further the equation assuming the waves to be short.

10.5. What is the rate of total energy transport per metre length in deep sea water (s.g. = 1·025) of the wave system of problem 10.4? Use $L = \frac{1}{2}(L_1 + L_2)$. (27000 W/m)

10.6. The wave amplitude and length of a wave measured at a point where the mean water depth is 122 m are 1·0 m and 55 m respectively. Calculate, neglecting energy dissipation, the difference in energy per square metre of surface area when the wave moves to a point where the mean depth is 2 m. Assume that the wave does not break and the specific gravity of the fluid is 1·025. (252 J/m²)

10.7. Show that the composite wave derived from two small amplitude waves with the same amplitude a but with different lengths and periods and moving in the same direction is of the form

$$\eta_t = 2a \sin (k'x - \sigma' t) \cos (k''x - \sigma'' t)$$

Deduce that $k' = (k_1 + k_2)/2$ is equivalent to the envelope wave number while $k'' = (k_1 - k_2)/2$ is the resultant wave number.

10.8. Using superposition concepts and the expressions for the velocity potential and pressure given in problem 10.2, show that the pressure fluctuation at a wall ($x = 0$) reflecting a small amplitude wave completely is given by

$$p = -\rho g \left[2a \sin \sigma t \, \frac{\cosh [k(d + z)]}{\cosh (kd)} + z \right]$$

An iron sheet pile wall brings about such a reflection in 15·2 m depth of water. If the incident wave is of height 30·5 cm and period 15 s, what will the pressure variation be at 9·1 m depth below the mean water level at the wall? (Hint: Obtain L from equation (10.4)). (5800 N/m²).

10.9. Waves of 1·83 m height and 10 s period were observed off the shore of Sekondi on a particular day. The mean depth at the measuring point was 6·1 m and tidal effects could be ignored in the analysis of the results. Calculate the wavelength at the measuring station and far at sea. What would be the wave height far at sea? How much power would be transmitted to the shore line by the breaking waves? (77·2 m; 156 m; 1·8 m; 31600 W/m)

10.10. If the waves of problem 10.9 break when the depth is 10 ft, what will the height of the waves be at breaking? (7·8 ft)

10.11. A progressive wave system, once generated, suffers decay of wave height after leaving the generating area. What factors contribute to the wave decay? Discuss their relative importance.

10.12. List the factors which are responsible for the transportation and selective distribution of granular sediments of different grain sizes on a beach. What effects govern the evolution of a stable beach profile?

10.13. How can an engineer recognize a beach under direct wave attack or under the influence of a near shore current system? Discuss concisely with sketches possible methods of protection in either case. U.S.T., 1969.

10.14. The standing waves formed by partial reflection by a section of the breakwater of Tema harbour has the appearance shown in the figure. The node of the envelope has an amplitude of 1·83 m and the antinode 3·05 m. The incoming wave is observed to have a period of 10 s and the water is 9·1 m deep at the

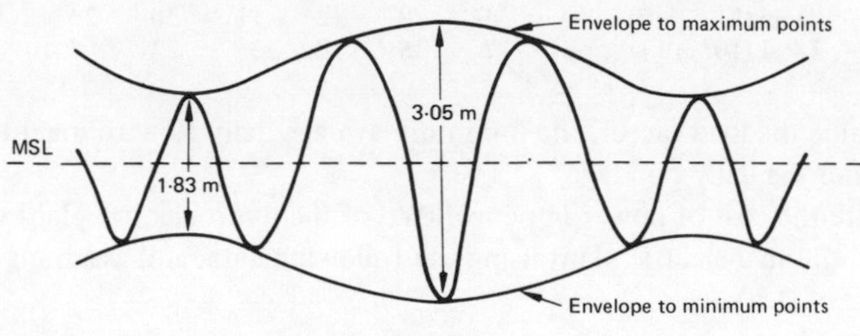

Fig. P10.1

relevant section of the breakwater. What is the rate at which energy is absorbed per unit length of the breakwater? Discuss briefly how you think this absorbed energy is dissipated (s.g. = 1·025) (54600 N/m; frictional damping; transmission to the harbour; breaking).

Chapter 11

11.1 Solve Example 11.1 (p. 352) using an interest rate of 6% but assuming negligible salvage values.

11.2. Solve Example 11.1 using an interest rate of 6% and the following salvage values

Project	*Units of currency*
1	–100
2	5 000
3	10 000
4	4 800
5	9 800

11.3. Find the cost per kWh for a hydroelectric scheme with a plant installed capacity of 74·6 MW. and a plant load factor of 50% using the following information:

(a) *Construction (dam, land, houses, etc.)*

Cost: ₵20 m
Interest: 6%
Life: 50 years

(b) *Installations (plants, etc.)*

Cost: ₵270 per kW of installed capacity
Interest: 6%
Life: 20 years

(c) OMR 2% of total construction and installation costs.

11.4. The maximum estimated daily load to be met by a run of a river hydroelectric plant with overall efficiency of 88% and effective head of 148 ft is as follows:

Time (h)	0	4	8	12	16	20	24
Load (10^3 kW)	8	12	28	12	38	22	8

Determine the load factor, the minimum average daily flow to meet this demand, and the daily pondage.

Compare the cost of power in pence/kWh of the hydroelectric plant with an alternative thermal-electric plant using the following data, and assuming an annual load factor of 0·50.

Thermal plant		*Hydro plant*	
Unit cost on installation	= £60/kW	Unit cost of installation	= £120/kW
Annual fixed charges	= 13·5%	Annual fixed charges	= 8%
Annual operation and maintenance: fixed	= £1·0/kW	Annual operation and maintenance	= £0·5/kW
plus variable fuel;			
calorific value	= 12000 Btu/lb		
cost	= £2/ton		
performance	= 15000 Btu/kWh		

11.5. Which of the following three motors would you select for a pump requiring 80 h.p.? Give your reasons including possible irreducibles. The interest rate on investment is 8%.

(a) A 95 h.p. diesel motor costs 25·50 cedis per horse power. Its life expectancy is 10 years and its salvage value is 500 cedis. OMR costs = 5000 cedis per year.
(b) A 100 h.p. electric motor costs 35·00 cedis per horse power. OMR costs = 4200 cedis per year. Its life expectancy is 20 years and salvage value is 550 cedis.
(c) A 95 h.p. petrol motor costing 2500 cedis. OMR costs = 4500 cedis per year. Its life expectancy is 10 years and salvage value is 450 cedis.

11.6. 22 500 m^3 per day of water are to be delivered against a static head of 83 m through 900 m length of a pipe. If the pump operates for 18 h/day, determine the most economical pipe size for the job. The pump efficiency is 80%, the friction factor f is 0·007, and the interest rate on investment is 8%. All pipes have the same life expectancy of 20 years. The table below gives the costs for purchase and laying of different sizes of pipes. Electric power costs 2 pesewas per kWh.

Pipe size (cm)	45	60	90	120
Cost (cedis/m)	18·35	20·0	22·94	24·95
Excavations and laying (% pipe cost)	8	10	13	17

11.7. A canal is to carry 500 cusec of water for 300 days during the year with 1 ft free board. The canal will have the most efficient trapezoidal section with bottom width–depth ratio $b/y_0 = 2\sqrt{(1 + k^2)} - 2k$. The sude slopes are $k = 2$ horizontal to 1 vertical. The canal can be excavated at a price of 40 pesewas per cubic yard. A concrete lining, if provided, will cost 1·20 cedis per square yard.

The annual maintenance cost of the unlined canal is estimated at 20 pesewas per yard. Maintenance charges of 2·50 cedis per yard every 12 years are estimated for repair of the lined canal. The estimated seepage loss from the unlined canal is 0·5 ft^3 per square yard on a horizontal plane per day, and the price of water is 2·50 cedis/acre foot. If interest rates are unimportant and the life of either alternative is 60 years, would you recommend lining the canal?

Take the Chezy constant for the lined canal is 100 $ft^{\frac{1}{2}}$/s and for the unlined canal as 80 $ft^{\frac{1}{2}}$/s. The longitudinal slope $S_0 = 0{\cdot}0025$. U.S.T., 1967, Part III. (Costs: lined – ₵28·15/yd; unlined – ₵20·78/yd)

11.8. The flood which occurs in the northern sector of Accra every rainy season has been estimated to cost 400 000 new cedis of damage to property and public utilities annually. The following five schemes have been proposed for the flood control.

(a) Dredging Korle lagoon and diverting streams into it.
(b) Diverting the streams to a pumping station from where the water will be pumped into the sea.
(c) Scheme (a) plus pumping from Korle lagoon into the sea.
(d) Diverting the streams through a tunnel into the sea.
(e) Constructing an earth dam in a valley north of Accra to impound waters diverted from streams and subsequently using the water for irrigation purposes.

Improved sanitation and recreational possibilities of (a) and (c) and irrigation benefits of (e) provide secondary benefits. The table below lists the annual costs (capital, interest, operation and maintenance) and the expected reduced flood damages for the various schemes. Draw the appropriate benefit/costs graphs and determine the most economically satisfactory of the five schemes and its expected benefits.

Scheme	Annual costs (₵)	Average annual flood damage (₵)	Secondary benefits (₵)
No control	0 0	400 000	0
a	128 720	305 000	65 000
b	154 950	190 000	–
c	206 600	195 500	70 500
d	283 670	100 000	–
e	335 320	100 000	40 000

What effect do you think doubling the interest rate on capital cost will have on your decision? U.S.T., 1968, Part III.

11.9. It has been proposed that the recently failed Weija dam should be replaced by a bigger dam in order to meet Accra–Tema water requirements of 40 million gallons per day. The new dam and ancilliary facilities are estimated to cost 7 million new cedis. The life expectancy of the Weija project is 30 years.

An alternative arrangement is to expand facilities in connection with Kpong waterworks to meet the entire demands in the Accra–Tema area. The life expectancy of the Kpong project is 20 years. If water supply from Kpong costs 2·46 pesewas per 1000 gallons more than that from Weija, what maximum capital investment will make the Kpong scheme competitive with the Weija scheme?

Assume an interest rate (minimum attractive rate of return) of 6% in both cases. What irreducibles can you think of in connection with each scheme. U.S.T., 1969, Part III. (₵1·72 m)

11.10. A hydroelectric scheme has an installed turbine capacity of 100 000 kW and a plant load factor of 0·5. Power can be sold at 0·90 p per kWh. Using the information below, determine the maximum expenditure x on the turbine draft tube in order to make the project justifiable. By how much would the value be altered if the life expectancy of plants was 25 instead of 20 years?

Construction

Cost: ₵$(18{\cdot}5 + x)$ m
Interest: 6%
Life: 50 years

Plants and Installations

Cost: ₵200 per kW of installed capacity
Interest: 6%
Life: 20 years (or 25 years)

Annual OMR

2% of gross construction and plant installation costs.

U.S.T., 1970, Part III. (₵3·08 m; + ₵2·14 m)

(U.S.T. = University of Science and Technology, Kumasi; UNZA = University of Zambia; U.L. = University of London.)

Appendix
Notes on Flow Measurement

As a result of many different demands, a great variety of methods has been developed for measuring flow of fluids. Some of these require elaborate apparatus and are complicated and expensive. Others are simple, inexpensive but usually not very accurate. The best method to use in a given situation depends on the volume of flow, conditions of measurement and the degree of accuracy required.

A high degree of accuracy in measurement is, for instance, essential in determining empirical coefficients. Also the testing of turbines and other machines require the use of the most accurate and refined methods of flow and other measurements, because these measurements form the basis of determination of plant efficiencies. On the other hand, continuous discharge records of streams require periodic measurements of flow for which extreme accuracy in individual measurements is not essential, provided errors compensate and are not cumulative.

In general, methods of measuring water flow rate may be classified in one of two groups: velocity-area methods and direct discharge methods.

Velocity-area methods may require the determination of discharge by measurement of the mean velocity of flow. The discharge is then the product of the mean velocity and the conduit area. Alternatively the discharge may be obtained graphically or by integrating an expression of the form $Q = \int v \mathrm{d}A$ where v is the measured flow velocity across an area δA.

Direct discharge methods do not involve velocity measurement. In some cases the determination of velocity is a step in the proceeding but no actual measurement of velocity is made.

A.1 Velocity-Area Methods

A.1.1 The Pitot-static Tube

The Pitot tube is a simple device for measuring total head accurately in a flowing fluid. With reference to Fig. A.1(a), let the approach velocity of flow to the nose

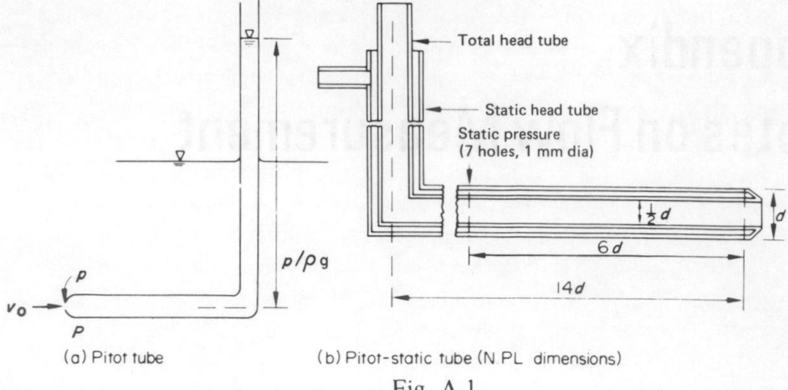

Fig. A.1.

P of a simple Pitot tube be v_0. At equilibrium, there is no flow into the tube and therefore the fluid velocity at the nose is zero. Applying Bernoulli's equation,

$$p_s/\rho g + v_0^2/2g = p/\rho g \qquad \text{(A.1)}$$

where p_s is the static pressure in the approach flow and p is the total pressure at the nose of the Pitot tube. For hydrostatic balance at *P*, fluid will rise up in the vertical tube to a height $p/\rho g$, which gives the total head for the fluid. The flow velocity is given by

$$v_0 = C_v\sqrt{[2g(p - p_s)/\rho g]} \qquad \text{(A.2)}$$

where C_v is a coefficient of velocity which accounts for errors introduced in equation (A.1) by neglecting friction, shock and other effects on the flow.

The Pitot-static tube, a typical standard design of which is shown in Fig. A.1(b), combines devices for measuring both the total head ($p/\rho g$) and the static head ($p_s/\rho g$) or their difference from which the velocity can be calculated. A number of holes connect the surrounding fluid medium into an outside chamber surrounding the total head tube thereby subjecting the chamber to the fluid's static pressure. By connecting the total head tube and the surrounding static head tube to two limbs of a differential manometer, the velocity head, $h = (p - p_s)/\rho g$ can be measured and the velocity calculated using equation (A.2).

The Pitot-static tube is used mainly in laboratory channels and in pressure conduits of small dimensions. It requires prior knowledge of the direction of flow to minimize errors due to improper alignment. It is suitable only for high velocity measurements (greater than 5 ft/s (about 1·5 m/s)) since low velocities give a head difference too small to measure accurately.

A.1.2 Current Meters

The essential features of a current meter are a set of cups or a propellor which rotates when immersed in flowing water and a device for determining the number of revolutions per unit time. The rotation is usually converted into electrical impulses which are counted using earphones (for picking up clicking sounds) or flashes of a lamp which the electric pulse switches on. Electronic recorders are now generally used. The velocity of flow is found from a calibration chart or equation supplied by the manufacturers.

Each current meter is usually calibrated for each possible type of suspension with which it may be used. Common types of suspension are; from a bridge, from a cableway, from a boat or supported by a wading rod. Since the accuracy of the current meter is greatly affected by dirt or injury it should be rated at least once a year and more often if inaccuracy is suspected.

The current meter is used extensively in river gauging (see subsection 4.1.3). The cross section of the river is divided in certain vertical sections usually between 1 and 3 metres apart. The distance should be smaller near the banks than in the middle of the river. Normally measurements are taken at between 5 and 10 points in each vertical. The points should be closer near the stream bottom than near the surface. Knowing the velocities in the grid the discharge can be found either by drawing lines of equal velocity (similar to Fig. 4.4) and adding the discharges between them or by finding the mean value of velocity in each vertical and multiplying the area of each vertical strip between the measurement verticals with the average velocity.

The above procedures have to be followed for accurate measurements. However, experience has shown that the mean velocity in a vertical (within a maximum error of 1%) is given by the mean of the velocities at 0·2 and 0·8 of depth respectively. Alternatively the mean velocity in a vertical (within a maximum error of 3%) may be assumed to occur at 0·6 of the depth measured from the surface.

The principal sources of error in a current meter measurement of discharge result from rating of the meter, observation of soundings, observation of time, placing of the meter in position, use of insufficient measurement points and general conditions of measurement. These errors should not exceed 5% generally and can be kept below 2% with good care.

A.1.3 Floats

If no other method is feasible, the velocities in a channel can be determined approximately by means of floats. This involves timing a floating object over a known distance and estimating its velocity of drift. Surface floats travel with the flow velocity near the water surface. Therefore, the mean velocity in the vertical is obtained by applying a coefficient of 0·8–0·95 (average being 0·85) to the velocity recorded from the float. Two types of technique are commonly used.

One method relies on a surface float connected by a rope to a larger submerged float at about 0·6 of the mean depth along the path followed. The submerged float is comparatively large so that the effect of the smaller surface float can be neglected.

Rod floats (or chain floats) are made from wooden poles or hollow metal cylinders weighted at one end so as to float in an approximately upright position with the unweighted end slightly above the water surface. Chain floats are pieces of wood connected by a chain. They also float approximately upright. Rod or chain floats should reach as close as practicable to the bottom of the channel without touching it at any point on their path. The velocity of the floats are generally multiplied by 0·92–0·94 to get the mean velocity in the vertical of the path.

A.1.4 Salt-velocity Method

A concentrated salt solution (or brine) is introduced into the flow. This increases the electrical conductivity of the water. Electrodes connected to an ammeter or any other electrical recorder are installed at one or two points of observation below the place where the salt solution is introduced. An increase in the electric current is indicated when the prism of water containing salt passes the electrodes. By registering the time of travel of the brine over a fixed distance the flow velocity can be calculated.

In a pipe or conduit of uniform cross sectional area the length of the reach divided by the time of travel gives the mean velocity and the discharge is the product of this mean velocity and the cross sectional area. If the cross section of the conduit is not uniform, the volume of the reach must be determined. This volume divided by the time of travel gives the discharge.

The main difficulties in the salt method are correction for dispersion or diffusion of the salt in solution and the volumetric effect of the solution on the normal flow rate.

A.2 Direct Discharge Methods

A.2.1 Venturi Meter, Orifice Plate and Dall Tube

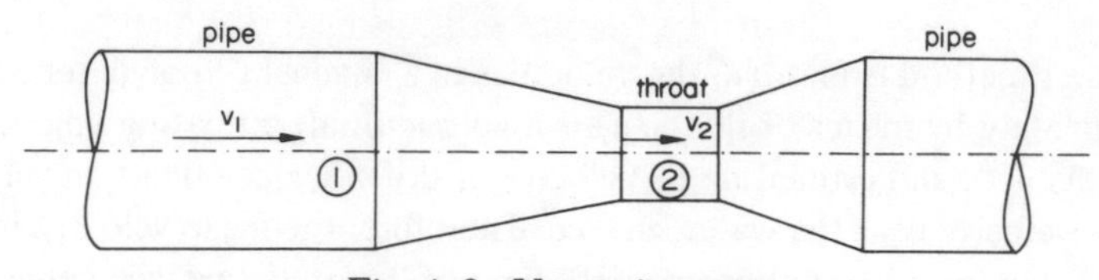

Fig. A.2. Venturi meter

The Venturi meter is appropriate for measuring discharges in pipes. It is principally a short length of a pipe which tapers from a known diameter to a throat from where it again expands to a fixed diameter.

With reference to Fig. A.2, let the pipe cross-sectional area be a_1 and the venturi throat area be a_2. Also let the pressure head at inlet to the venturi tube be h_1 and at the throat be h_2. The corresponding velocities are v_1 and v_2 respectively. Applying the energy equation,

$$v_1^2/2g + h_1 = v_2^2/2g + h_2 \qquad \text{(A.3)}$$

Combining equation (A.3) with the continuity equation yields

$$Q = a_2 v_2 = C_d a_1 a_2 \sqrt{[2g(h_1 - h_2)/(a_1^2 - a_2^2)]} \qquad \text{(A.4)}$$

where the coefficient of discharge C_d (less than unity) corrects for losses due to resistance to flow, flow contraction in the meter and other minor effects. By measuring $(h_1 - h_2)$ on a differential manometer and knowing C_d from a previous calibration, the discharge through the pipe can be calculated from equation (A.4).

Other instruments for measuring flow through pipes which are based on the same principle of flow contraction through a section are the orifice plate and the Dall tube. Discharge through them is also given by equation (A.4). In general their coefficient of discharge depends on the shape of the instrument, the form of orifice edge, the ratio of orifice diameter to the pipe dimensions and the roughness of the pipe walls.

On the whole the orifice plate is simple and cheap and occupies a small length but produces a large degradation of energy. The venturi meter is relatively complex and expensive but produces a much smaller degradation of energy. The Dall tube is of moderate cost, gives a greater head difference and provides less energy degradation than either the orifice plate or the venturi tube. It is accordingly much more efficient. All three instruments are suitable for measurement of discharge of a liquid or a gas in a closed conduit. However, they are unsuitable for measurement in the laminar flow range since their coefficients of discharge are extremely variable in this range.

A.2.2 The Venturi Flume

The venturi flume is based on the venturi principle discussed in subsection 4.3.4 and is used for flow measurement in open channels. The value of its coefficient of discharge, which accounts for friction and contraction effects, ordinarily lies between 0·95 and 1·00. It can be kept at 0·98 or above if care is taken to have smooth surfaces and to round off all corners so as to lead the flow to the throat of the flume without contractions or unnecessary turbulence.

The venturi flume is suitable for measuring flow in small rivers in which the discharge does not exceed 20 m^3/s (710 ft^3/s). However, the largest known was reported to have been suitable for measuring 50 m^3/s (1800 ft^3/s). One type of a standard design which combines the principle of the venturi with that of the broad-crested weir is the Parshall flume in which geometrical ratios and other conditions of installation are specified.

A.2.3 Weirs

Weirs, especially sharp-crested ones, are commonly used to measure small quantities of water flow, generally less than about 0·25 m^3/s (9 ft^3/s). However, although sharp-crested weirs are primarily thought of in connection with water flow measurement, they appear in many structures as channel control and also serve as the basis of spillway design.

The nappe, immediately after leaving the sharp crest, suffers a contraction which corresponds to the contraction of a jet issuing from an edged orifice. According to test results the shape of the nappe can be determined as a function of the overflow head. Generally given data in this regard and for flow measurement are valid only if the upper and lower nappe are subject to full atmospheric pressure. Insufficient aeration causes a reduction of pressure beneath the nappe due to removal of air by the over-falling jet and a change in the form of the jet.

Many formulae for the overflow coefficient for sharp-crested weirs with a horizontal crest have been developed. However few are accurate. One of the most accurate seems to be that given by Rehbock. This gives discharge Q(m^3/s) as:

$$Q = \left(1{\cdot}782 + 0{\cdot}24\frac{h_e}{h_0}\right)Bh_e^{3/2} \tag{A.5}$$

where

h_0 = height of the vertical weir plate (m)
$h_e = h + 0{\cdot}0011$ (m)
h = overflow head above weir crest (m)
B = width of the weir (m)

Equation (A.5) is valid only within the limits

$$0{\cdot}15 < h_0 < 1{\cdot}22 \text{ (m) and } h < 4h_0$$

and under the condition that the lower as well as the upper nappe is fully exposed to atmospheric pressure.

Triangular (or *V*-notch) weirs permit a more accurate measurement of smaller discharges because the discharge of a *V*-notched weir increases more rapidly with head in comparison with a horizontal crest weir. The equation for discharge over a *V*-notched weir is given by theory as:

$$Q = \frac{8}{15}\sqrt{(2g)}\tan\theta/2 \,.\, h^{5/2} \tag{A.6}$$

where θ is the included angle and h is the water head above the vertex.

Equation (A.6) has to be corrected by a factor expressing the influence of the deviations from ideal conditions assumed by theory. Thus

$$Q = \mu \,.\, \frac{8}{15}\sqrt{(2g)}\tan\theta/2 \,.\, h^{5/2} \tag{A.7}$$

where

$$\mu = 0{\cdot}565 + 0{\cdot}0087\, h^{-1/2} \tag{A.8}$$

when metric units are used.

The coefficient for weirs with a crest made of a circular arc can be computed from Rehbock's empirical formula:

$$\mu = 0{\cdot}312 + \sqrt{[0{\cdot}3 - 0{\cdot}01\,(5 - h/r)^2]} + 0{\cdot}09\, h/h_0 \tag{A.9}$$

where r is the radius of curvature of the rounded top. Equation (A.9) is valid within the limits

$$0{\cdot}02 < r < h_0 \text{ (m)}$$

and

$$h < r\,[6 - 20r/(h_0 + 3r)]$$

If the crest of a weir is sufficiently wide to prevent the nappe from springing free, it is classified as broad crested. Flow over such a weir must pass through a critical depth somewhere near the downstream corner (see subsection 4.3.4). The location of this point varies appreciably with the head and weir proportion. The discharge is given by equation (4.46). The value of C_d generally depends on the overflow head, the shape of the weir and the approach velocity. Broad-crested weirs are seldom used nowadays for measurement purposes but serve mainly for river improvement and transition purposes.

A.2.4 Gravimetric and Volumetric Measurements

These methods require determination of the weight or volume of flow discharged in a certain time. They are used primarily in experimental work in laboratories and for calibration purposes. Furthermore, they can be used only for measuring comparatively small flows.

A 2.5 California Pipe Method

The discharge is determined at the open end of a partially filled horizontal pipe discharging freely into air. It is used primarily to measure comparatively small flow in pipes. However, it can also be used to measure flow in small open channels if the water is diverted into a pipe which it does not completely fill and which discharges without any submergence at the outlet.

Figure A.3 illustrates the California pipe method. Other designs may be possible. With such an arrangement the only measurements necessary are the inside diameter of the pipe and the vertical distance from the upper inside surface of the pipe to

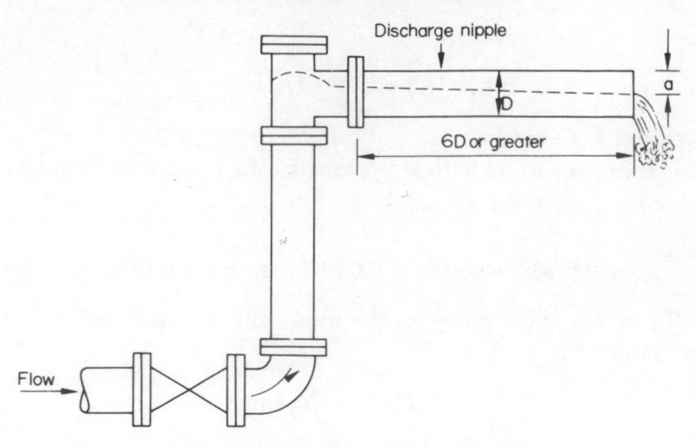

Fig. A.3. The California method

the surface of the flowing water at the outlet of the pipe. With this information the discharge is computed from:

$$Q = 8{\cdot}69\,(1 - a/D)^{1{\cdot}88}\,D^{2{\cdot}48} \tag{A.10}$$

where Q = discharge (ft^3/s)

$(D - a)$ = depth of flow at outlet of the pipe (ft)

D = internal diameter of the pipe (ft)

Equation (A.10) was developed from experiments with pipes 3 to 10 inches in diameter and therefore gives accurate values of discharge for that range of size.

A.2.6 Chemical Gauging

Chemical gauging involves determining a discharge by introducing a chemical at a known rate into flowing water and determining the quantity of chemical in the stream at a section far enough downstream to ensure thorough mixing of the chemical with the water. Common salt (NaCl) is usually employed because of its cheapness and harmlessness. For convenience the salt is dissolved in water before its introduction into the stream. The method is usually applied at very irregular river channels where no other method can be effectively and cheaply used.

Let

$Q(L^3/t)$ = discharge of the river (to be determined)

$q(L^3/t)$ = rate of introduction of brine (salt solution)

$c_0(M/L^3)$ = concentration of brine

$c_1(M/L^3)$ = natural salt concentration in river (determined by sample analysis)

$c_2(M/L^3)$ = concentration of salt in river after introduction of brine (determined by sample analysis)

By the principle of conservation of salt mass,

$$q \,.\, c_0 + Q \,.\, c_1 = (q + Q) \,.\, c_2$$

Thus

$$Q = q \,.\, (c_0 - c_2)/(c_2 - c_1) \tag{A.11}$$

Example: In a typical case the following results were obtained

$$q = 1{\cdot}637 \text{ (l/s)}, \; c_1 = 31{\cdot}8 \text{ (mg/l)}$$
$$c_0 = 218300 \text{ (mg/l)}, \; c_2 = 37{\cdot}1 \text{ (mg/l)}$$
$$Q = 1{\cdot}637 \times (218300 - 37{\cdot}1)/(37{\cdot}1 - 31{\cdot}8) \text{ (l/s)}$$
$$= 67500 \text{ (l/s)} = 67{\cdot}5 \text{ m}^3\text{/s}$$

A comparative measurement by current meter yielded 65·33 m^3/s.

A.2.7 Radioactive Isotope Methods

The use of radioactive isotopes in flow measurement and other areas of hydrologic and hydraulic investigations is becoming more and more common and cheap. The main advantages of the use of radioactive isotopes in flow measurement are that:

(1) the tracer can be selected to show the same behaviour as the fluid
(2) the amount of tracer is so small that its addition does not interfere with the nature of flow
(3) the measurement can often be carried out without disturbing the water
(4) the accuracy is sometimes better than in conventional methods.

The main disadvantages are:

(1) most measurements are not suitable for continuous readings
(2) some of the methods are applicable only in flows with high turbulence
(3) health and safety considerations. Even if investigations can be carried out without exceeding the maximum permissible concentration of radioactive tracers the application will always be limited by the necessary provisions against possible negligent handling of the radioactive substances
(4) difficulty and cost of obtaining the appropriate radioactive substances.

Methods for determining flow rates with radioactive isotopes are:

(1) peak to peak method. This is identical with the salt-velocity method of Section A.1.4 if instead of salt brine a radioactive substance is used;
(2) dilution method. This is identical with chemical gauging method of Section A.2.6 if instead of a chemical a radioactive substance is used and its radiation measured;

(3) total count method. Consider a stream of uniform cross section and velocity, say in a pipe (Fig. A.4). If the tracer is introduced in such a way that it is mixed thoroughly across the pipe, a radioactive atom passing a counter station downstream has a certain chance p of emitting a gamma ray that will excite the counter. This chance is not affected by the presence or absence of neighbouring radioactive atoms. Hence the number of counts recorded must be independent of the distribution of the tracer *along* the stream. If A atoms are in the tracer pulse, the probable number of counts to be observed is pA. But if the flow rate Q increases, the time an atom is likely to be in the vicinity of the counter decreases and p accordingly decreases, and the number of counts decreases in proportion. The total count is inversely proportional to the rate of flow:

$$N = FA/Q$$

or

$$Q = (A/N)F \tag{A.12}$$

where Q = discharge,

A = amount of tracer (radioactivity)

N = number of counts

and F = constant of proportionality.

Thus if F is known, Q can be calculated from equation (A.12). F is generally determined as follows.

A piece of pipe similar to the one used in the flow measurement and slightly longer than the counter tube is filled with a solution of tracer at a known concentration (see Fig. A.4). The counter is fixed to the tube in the same geometrical shape as in the field test and the counting rate R is measured. This rate is proportional to the concentration.

$$R \text{ (counts/min)} = FC$$

with C measured in millicuries per cubic metre (mc/m³).

Thus the counting factor F is given by

$$F = R/C \left[\left(\frac{\text{counts}}{\text{min}}\right) \cdot \left(\frac{\text{m}^3}{\text{mc}}\right)\right]$$

It has been shown elsewhere that the factor F is identically the same as that used in equation (A.12).

Thus

$$Q = FA/N \quad (\text{m}^3/\text{min})$$

When the tracer stream is divided into separate streams with the flow rate (rQ) in a branch, the amount of tracer carried through the branch will be (rA). Hence the number of counts observed will be

$$N = F(rA)/(rQ) = FA/Q$$

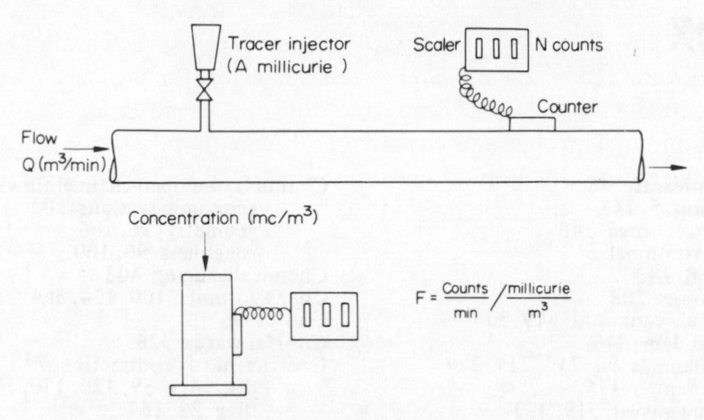

Fig. A.4. Flow rate by total count.

This remarkable result shows that the number of counts observed on a branch of a stream is independent of the branching ratio. The same number of counts that would be recorded on a branch stream is exactly the same as would be recorded in the whole stream. Thus the fraction r need not be known. It is only necessary that it remains constant during the test. However the constant factor F has to be determined for the appropriate branching stream.

Acknowledgment: Parts of the notes in this Appendix were adapted from original notes compiled by Professor Gunther Garbrecht and his kindness to permit their use is gratefully acknowledged.

FURTHER READING

B.S. 1042, *Flow Measurement*, 1943.
Linford, A., *Flow Measurement and Meters*, Spon Ltd.
Pao, *Fluid Mechanics*, Wiley, New York.
Streeter, V. L., *Fluid Mechanics*, McGraw-Hill, New York.

Index